WATER

A COMPREHENSIVE TREATISE

Volume 2

Water in Crystalline Hydrates

Aqueous Solutions of Simple Nonelectrolytes

WATER
A COMPREHENSIVE TREATISE

Edited by Felix Franks

WATER
A COMPREHENSIVE TREATISE

Edited by Felix Franks

Unilever Research Laboratory
Sharnbrook, Bedford, England

Volume 2
Water in Crystalline Hydrates
Aqueous Solutions
of Simple Nonelectrolytes

ℚ PLENUM PRESS • NEW YORK–LONDON • 1973

Library of Congress Catalog Card Number 78-165694
ISBN 0-306-37182-0

© 1973 Plenum Press, New York
A Division of Plenum Publishing Corporation
227 West 17th Street, New York, N.Y. 10011

United Kingdom edition published by Plenum Press, London
A Division of Plenum Publishing Company, Ltd.
Davis House (4th Floor), 8 Scrubs Lane, Harlesden, London, NW10 6SE, England

"No one has yet proposed a quantitative theory of aqueous solutions of non-electrolytes, and such solutions will probably be the last to be understood fully."

J. S. ROWLINSON

in *Liquids and Liquid Mixtures*, First Edition, 1959, and Second Edition, 1969.

Preface

When the concept of "Water—A Comprehensive Treatise" was first discussed it was suggested that three volumes be produced of which the second would cover aqueous solutions of simple electrolytes and nonelectrolytes. Even at the planning stage this still seemed a reasonable undertaking, but during the preparation of the manuscripts it soon became clear that this course of action would lead to a very heavy volume indeed. Reluctantly it was decided to divide the volume into two, one dealing with crystalline hydrates and solutions of nonelectrolytes, and the other with solutions of electrolytes. This division may appear somewhat arbitrary but in order to do the subject justice, it was felt to be the right course of action.

The present volume provides what is hoped to be a balanced account of the physical properties of crystalline hydrates and aqueous solutions of simple nonelectrolytes. One is struck by the apparent wealth of experimental information; however, on closer study it becomes clear that very few systematic efforts have as yet been made to characterize fully a number of related aqueous systems. Probably the aqueous solutions of simple alcohols are the most thoroughly investigated systems, but there are still considerable gaps in our knowledge which must be filled before a successful attempt at a molecular interpretation of the properties of such solutions can be made.

As our understanding of the peculiarities of water as a solvent gradually improves, it becomes more and more obvious that past efforts to force aqueous mixtures into the straitjacket of existing solution theories have not proved too successful and there is still no theory capable of explaining the well-established physical properties of such mixtures. It seems likely, however, that the next few years will see significant advances in the elucidation of structures and interactions in aqueous solutions and it is hoped that the information collected and discussed in this volume may help toward the achievement of such an objective.

I should like to express my debt of gratitude to the authors who have contributed to this volume. Editing a work of this nature can strain long-established personal relationships and I thank my various colleagues for bearing with me and responding (sooner or later) to one or several letters or telephone calls. My special thanks once again go to Mrs. Joyce Johnson, who bore the main brunt of this seemingly endless correspondence and without whose help the editorial and referencing work would have taken several years.

F. FRANKS

Biophysics Division
Unilever Research Laboratory Colworth/Welwyn
Colworth House, Sharnbrook, Bedford
January, 1973

Contents

Chapter 1

The Solvent Properties of Water

F. Franks

Chapter 2

Water in Stoichiometric Hydrates

M. Falk and O. Knop

Chapter 3

Clathrate Hydrates

D. W. Davidson

Chapter 4

Infrared Studies of Hydrogen Bonding in Pure Liquids and Solutions

W. A. P. Luck

Chapter 5

Thermodynamic Properties

F. Franks and D. S. Reid

Chapter 6

Phase Behavior of Aqueous Solutions at High Pressures

G. M. Schneider

Chapter 10

NMR Spectroscopic Studies

M. D. Zeidler

Chapter 11

Molecular Theories and Models of Water and of Dilute Aqueous Solutions

A. Ben-Naim

Contents of Volume 1: The Physics and Physical Chemistry of Water

Contents of Volume 3: Aqueous Solutions of Simple Electrolytes

Contents of Volume 4: Aqueous Solutions of Macromolecules; Water in Disperse Systems

CHAPTER 1

The Solvent Properties of Water

F. Franks

Unilever Research Laboratory Colworth/Welwyn
Colworth House
Sharnbrook, Bedford, England

1. WATER, THE UNIVERSAL SOLVENT—THE STUDY OF AQUEOUS SOLUTIONS

To point out that water existed on this planet long before life evolved is, of course, stating the obvious, but the uniqueness of water as the universal solvent is still not generally appreciated. Until not so many years ago experimental studies of the physical and inorganic chemistry of solutions were restricted almost completely to aqueous systems and solution theories were largely based on results so obtained. It was mainly Hildebrand and Scatchard who, together with their many colleagues, drew the attention of the scientific and technological world to the fact that solutions need not necessarily be aqueous, and that the study of other solvent media would be more likely to help in the elucidation of solvent–solute interactions in condensed phases. Ironically, though, the development of solution theories based on experiments performed with "simple" solvents has often led to generalized claims regarding the role of the solvent, irrespective of the fact that water does not conform to some or all of the predictions of such theories. Even now there is a widely held view that water need not in any way be treated as a unique substance, but that its somewhat nonconformist behavior as a solvent can be allowed for by the introduction of various specific parameters into the well-established equations which adequately represent the behavior of simple nonaqueous mixtures.

This sort of approach will often provide a fit to experimental results

1

but is unlikely to advance our understanding of the true nature of the aqueous medium. Furthermore, it may just be able to produce acceptable explanations for the processes involving simple systems, but it is sure to founder when trying to account for one of those processes where water behaves differently from any other solvent. For instance, none of the accepted theories of polymer solutions is likely to explain how a number of apparently random polypeptide chains will, in aqueous solution only, adopt a complex but highly specific conformation known as a native protein.

While it should not be claimed that a better understanding of the solvent participation will provide a complete insight into the mechanisms and energetics of physicochemical processes, it is probably safe to predict that a total neglect of the solvent will not prove helpful in providing such understanding. Recent years have seen a gradual realization that valuable information can be obtained from systematic studies of solute–water, solute–solute and water–water interactions in dilute solutions of electrolytes and nonelectrolytes. Looking back over published work since the beginning of this century, one is struck by the enormous efforts made by chemists to document the properties of aqueous solutions. Unfortunately, one is also struck by the fact that few of the available experimental data originating from before 1955 can be utilized in basic studies of solution processes on a molecular scale. The reasons are twofold: Much of the experimental work was performed on fairly concentrated solutions and extrapolations to "infinite dilution" frequently gave rise to incorrect values of limiting thermodynamic quantities and limiting concentration dependences.[636] The other difficulty encountered in some earlier work relates to a lack of experimental precision, e.g., few of the calorimetric results reported before 1950 have withstood the test of time. This is, of course, easily understandable and should not be taken as an implied criticism of the experimental prowess of many of the earlier workers in the field of solution chemistry. Rather, it reflects the state of development of instrumental accessories which, at the present time, are very much taken for granted.

An ideal approach to an understanding of the nature of aqueous solutions would consist in coordinated experiments aimed at characterizing all aspects of the systems under study. This would include the bulk and microscopic properties of such systems. Basically we require answers to the following questions involving pair potentials $U(r)$ and pair correlations $g(r)$: What is the nature of $U(r_{12})$ and $g(r_{12})$, where subscripts 1 and 2 refer to solvent and solute, respectively? Does the introduction of a solute molecule effect $U(r_{11})$, and is $U(r_{22})$ affected by an increase in the solute concentration? It may be said that such fundamental information can hardly be obtained while little is yet known about $U(r_{11})$ or $g(r_{11})$ in pure liquid water. One of the aims of Volume 1 of this Treatise has been to show that some progress has been and is being made in providing acceptable answers to these questions. In Chapter 14 of Volume 1 Frank tried to differentiate, on the basis of the available experimental evidence, between what water *must* be like and what water *might* be like. Until further progress is made in the theories of liquids such a semiempirical approach to water will have to serve. Although not as rigorously and quantitatively described as some of the earlier models, it nevertheless seems to be nearer the truth, since it tries to reconcile *all* the available experimental information.

Therefore, as regards realistic progress which can be made and indeed has been made, one requires first of all some thermodynamic data on dilute aqueous solutions. These should include activity coefficients, partial molar heats, heat capacities (preferably of a calorimetric origin), and entropies. Further information is obtained from a complete characterization of *PVT* relationships in such mixtures: This involves determinations of partial molar volumes, expansibilities, compressibilities (preferably isothermal), and the thermal pressure coefficients. All thermodynamic measurements should extend down to very low concentrations such that extrapolations to infinite dilution can be performed with reasonable confidence. Such requirements often make stringent demands on the necessary experimental precision and it is therefore not surprising that no aqueous system has yet been studied in such a thorough manner. The most completely documented systems include aqueous alkali halides, tetraalkylammonium halides, and the lower monohydric alcohols.

Having established the required thermodynamic properties, other useful bulk properties, indicating long-range order, and hence providing information about $g(r)$, are obtainable from X-ray-, neutron-, and light-scattering experiments. Little has been done in these areas, which would appear promising fields for further study. The close-range environment of water molecules and its perturbation by solute molecules or ions is currently receiving a good deal of attention with the application of NMR and dielectric relaxation techniques, and of Raman and, to a lesser extent, infrared spectroscopy.

To complete the picture, the dynamic behavior of solutes and solvent in aqueous solutions and the effects of aqueous solvent on the kinetics of reactions need to be investigated. For experimental reasons most effort has been concentrated on solutions of electrolytes.[738] As regards non-electrolytes, much attention has been given to viscosity measurements (notoriously difficult to interpret in molecular terms), but high precision self-diffusion results are now gradually finding their way into the literature.

Systematic studies of solvent effects on reaction mechanisms have been and are being pursued in depth by Robertson, Arnett, Hyne, Kohnstam, Caldin, and others.

Apart from the choice of a particular physical technique, careful thought must be given to the system to be studied and the amount of meaningful information which can hopefully be extracted from raw experimental results. The simplest systems amenable to physicochemical studies are aqueous solutions of rare gases, but at the present time experimental difficulties still set obvious limits to such investigations, which have therefore been confined almost exclusively to solubility determinations. The same is true for the slightly more complex solutions of diatomic gases and hydrocarbons, although from theoretical considerations systems in which the solute molecules have no hydrogen bonding capabilities are the most attractive ones. It is to be hoped that eventually the sensitivity of spectroscopic techniques will be improved so that the effect of nonpolar moieties on water–water interactions, which undoubtedly exists, can be studied more directly.

For practical reasons the classes of solutes which have been studied most extensively include soluble electrolytes and nonelectrolytes capable of interacting with water by hydrogen bonding, but the picture is far from complete. The next stage in complexity includes three-component systems in which the third component is often used simply as a probe to monitor changes in the aqueous solvent as its composition is changed. Since our knowledge of alcohol–water mixtures seems to be on a comparatively firm footing, the effects of varying the composition and the type of alcohol on the chosen probe form favorite subjects for experimental studies, some of which are discussed in this volume. Unfortunately, in many cases the results are not extensive enough for the most to be made of them, and many interesting effects still remain obscure. Another way of employing three-component systems is in the study of solute–solute interactions. This method has been applied with success to mixed electrolyte solutions (see Volume 3, Chapter 2). In order to be able to extract the most useful information from such experiments, the total solute concentration is kept constant for any one experiment while the solute ratio is changed.

Unfortunately, the scientific status of the subject of the intermolecular nature of water and its perturbation by solutes has not been enhanced by the many published reports of experimental investigations dealing with multicomponent systems of some complexity, where the authors have nevertheless been audacious enough to claim solvent structural changes as being responsible for certain observed effects. The dangers inherent in ascribing curious experimental results to changes in "water structure" cannot be overemphasized. This practice has been harshly criticised by Holtzer and Emerson[678] in what has become a well-known polemic against the "water structure diviners."* Finally, for the process of ionic and molecular hydration to be properly characterized, more extensive data on the behavior of nonaqueous solutions are required. It is only thus that the unique nature of aqueous solutions becomes apparent. During the last ten years much valuable work on ionic solvation phenomena in a variety of solvents has been reported and has led to a reexamination of earlier work on aqueous solutions. Another solvent suitable for a comparison with water is, of course, D_2O, and solvent isotope effects have been and are being investigated in an attempt to elucidate the function of hydrogen bond energies in equilibrium and transport processes. The first requirement for such comparative studies is the isolation of types of behavior common to aqueous and nonaqueous solutions, or common to solutions in H_2O and D_2O. Starting from such a baseline which might be provided by one of the acceptable models for liquid mixtures (regular solution, lattice solution, scaled particle theory, free volume model, etc.), specific effects can then be characterized in terms of dipolar, hydrogen bonding, cooperative, electrostatic, or hydrophobic contributions.

The remainder of this chapter consists of a brief summary of recent progress in our understanding of the peculiar nature of water as a solvent. Many important aspects will be only briefly mentioned because they are dealt with in depth in succeeding chapters.

2. AQUEOUS SOLUTIONS OF NONELECTROLYTES

From an experimental point of view solute–solvent interactions in nonelectrolyte solutions are more amenable to investigations than are ionic solutions in which long-range Coulombic interactions tend to mask any short-range hydration effects. The concentration range in which limiting

---

* It is, however, the author's personal opinion that, on balance, Holtzer and Emerson have been less than fair in the selection of their case histories, namely, the comparison of H_2O with D_2O (mainly in terms of dielectric constant and viscosity), and the thermodynamic properties and viscosity of water–urea mixtures. While making the point that water structural theories cannot be tested except by "definitive" X-ray or neutron diffraction measurements, they do not mention instances where the applications of these very methods have indeed indicated structuring effects, nor do they make any reference to the extensive spectroscopic information which supports many of the earlier speculations based on thermodynamics only.

behavior might be expected should therefore be more easily accessible in the case of simple nonelectrolytes. On the other hand, there does not exist an adequate solution theory by means of which the limiting concentration dependence of thermodynamic quantities can be established. As a first approximation, limiting activity coefficients are often obtained from virial equations based on the regular solution theory:

$$\log \gamma_1 = Bx_2{}^2 + Cx_2{}^3 + \cdots \tag{1}$$

but the weakness of this approach lies in the fact that the observed solution behavior is expressed solely in terms of $U(r_{22})$ and cannot therefore further our understanding of the role of water (if any) in determining solution properties. There are also many known instances where at low concentrations marked deviations occur from the behavior predicted by eqn. (1), and careless extrapolation can lead to large errors in computed limiting quantities. Thus, in the absence of equations suitable for purposes of extrapolation it is probably safest to extend the experimental concentration ranges as far as the precision will allow.

2.1. Apolar Solutes

As has already been mentioned, nonpolar solutes seem more likely than other substances to provide us with experimental data capable of interpretation. The first indications that the situation is rather complex are provided by the solubilities of a series of simple apolar substances. Figure 1 shows the mole fraction (x_2) solubility as a function of temperature in water and cyclohexane, respectively.[977] It is seen that the aqueous solubilities are lower by at least an order of magnitude and that the partial molal entropy, given by $\bar{S}_2(x_2) - S_2{}^{\ominus}(\text{gas}) = R \, \partial(\ln x_2)/\partial(\ln T)$, in aqueous solution is negative, irrespective of the solute, but is also a function of the temperature. In other words, the introduction of apolar solutes into water is associated with marked heat capacity effects. The partial molal entropies of solution reduced to a common (limiting?) mole fraction of 10^{-5} are given, together with some limiting partial molal heat capacities at 25°C, in Table I. Such negative entropies and large heat capacities of solution are peculiar to aqueous mixtures but not necessarily to nonpolar solutes. In the light of modern relaxation studies the original explanations by Eley[446] and Frank[504] probably turn out to be quite near to the truth. These are based on the simple recognition that the four-coordinated nature of water makes for a comparatively large free volume and interstitial rather than

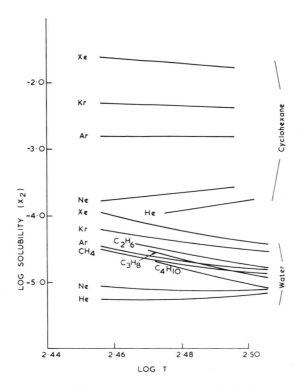

Fig. 1. Mole fraction (x_2) solubilities of simple gases in cyclohexane and water, as function of temperature; redrawn from Ref. 977.

TABLE I. Mole Fraction (x_2) Solubility, Partial Molar Entropy, and Heat Capacity of Solution of Gases in Water at 25°C[(977)]

	$10^5 x_2$	$\bar{S}_2(x_2) - \bar{S}_2^{\ominus}(\text{gas})$, eu	$\bar{S}_2(x_2=10^{-5}) - S_2^{\ominus}(\text{gas})$, eu	$\Delta \bar{C}_{P_2}$, cal mol^{-1} deg^{-1}
He	0.68	−1.35	−2.0	30
Ne	0.82	−3.5	−3.9	—
Ar	2.54	−9.1	−7.3	—
Kr	4.32	−10.8	−7.9	—
Xe	7.71	−14.5	−10.4	65
CH$_4$	2.48	−10.5	−8.7	55
C$_2$H$_6$	3.10	−13.4	−12.0	66
n-C$_3$H$_8$	2.73	−17.5	−15.3	—
n-C$_4$H$_{10}$	2.17	−20.3	−18.8	—

substitutional solution. Frank and Evans[508] went a stage further and accounted for negative solution entropies in terms of a solvent ordering process; hence arose the concept of the now famous "icebergs," a term which has in turn been used, misused, criticized, reinterpreted, and misinterpreted.*

The fact remains that no *credible* alternative explanation of the observed thermodynamic properties of apolar solutes in water has been forthcoming but that a wealth of supporting evidence now exists for what has become known as apolar or hydrophobic hydration. This concept implies that the solvent medium is inhomogeneous, because water molecules affected by a solute molecule are "different" from those in the unperturbed state. The nature and magnitude of this "difference" have been the subject of intensive studies over the last ten years. It must, however, be emphasized that only models based on such inhomogeneities in the solvent medium can at present account for experimental findings such as the mutual "salting-in" of hydrogen and nitrogen[608] or the specific effects of nonelectrolytes on the aqueous solution thermodynamics of argon[144] and hydrocarbons.[1398]

Following on from the general suggestions regarding the origin of negative limiting solution entropies of apolar solutes, various attempts have been made to provide a quantitative basis for the experimental observations. These have adopted the common starting point of water as a mixture of distinguishable molecular species in equilibrium. One of the species is extensively hydrogen bonded, with predominantly fourfold coordination (as, for instance, in ice) and therefore possesses a low density ("bulky water"), whereas the important feature of the other species is the existence of nonhydrogen-bonded OH groups, such that O—O distances shorter than the normally accepted hydrogen bond length are permitted ("dense water"). The equilibrium existing in the liquid and in aqueous solutions can then be written as

$$(H_2O)_b \rightleftharpoons (H_2O)_d, \qquad \Delta H_1^\ominus > 0, \quad \Delta V_1^\ominus < 0 \qquad (2)$$

where b and d refer to the "bulky" and "dense" states, respectively. The magnitudes of $\Delta H_1^\ominus$ and $\Delta V_1^\ominus$, the standard enthalpy and volume changes, respectively, associated with equilibrium (2), are severely limited by the boundary process of the fusion and sublimation of ice. The various models for liquid water which have been advanced on the basis of equilibrium (2)

---

* It.is a strange fact that this study, which laid the groundwork for most of the subsequent experimental and theoretical developments, went unnoticed for almost 15 years. Its popularity dates from a brief mention in the second edition of Robinson and Stokes's famous work, "Electrolyte Solutions."[1151]

have been critically discussed in Volume 1, Chapter 14. The present interest lies in the effect of solutes on this equilibrium and at its simplest it can be stated that apolar solutes appear to shift the equilibrium from right to left.

Any model which sets out to describe quantitatively the solution behavior of, say, hydrocarbon/water mixtures needs some further specification of $(H_2O)_b$ and $(H_2O)_d$. It is here that thermodynamics can only speculate and suggest various alternatives which could be reconciled with experimentally observed solution behavior. For example, the properties of $(H_2O)_b$ have in turn been identified with those of ice[1018,1026,1174] and those of a clathrate lattice.[509,510,973] The feature possessed in common by both these crystal structures is the tetrahedral coordination of nearest neighbor oxygen atoms, and this element of order has also been used by Ben-Naim in his recent model for aqueous solutions (see Chapter 11). The clathrate hydrates have been found to be so ubiquitous and to exist in many, structurally very similar varieties[705] that it is tempting to speculate that even in aqueous solution water molecules in the immediate neighborhood of apolar solute molecules are (over a time average) disposed in a clathratelike arrangement.*

The specification of $(H_2O)_d$ is somewhat more uncertain, mainly because its experimental characterization is difficult. However, for any statistical thermodynamical model to be developed, the entropy of mixing of the species must be accessible to calculation and this requires some assumptions to be made regarding the spatial disposition of, and interactions between, water molecules in the dense configuration. Nemethy and Scheraga[1025] have simply regarded water as a mixture of icelike clusters and a nonhydrogen-bonded dense fluid. Namiot,[1018] using the ice lattice for $(H_2O)_b$, allows for single water molecules to occupy interstitial sites. Frank and Quist, adopting the well-known type I clathrate structure (see Chapter 3) for $(H_2O)_b$, assign molecules of the $(H_2O)_d$ species to the cavities normally occupied by the guest molecules. All these models can then be developed in a more or less rigorous manner. Extending these models to allow for the introduction of apolar solutes requires additional information (or assumptions) as to the manner in which the solute molecules mix with the water species and how equilibrium (2) is affected. Nemethy and Scheraga[1026]

---

* Support for the validity of such speculation comes from recent molecular dynamics simulation experiments on water[1126] which have been extended to study the changes in the $g(r)$ pair correlations resulting from the addition of a neon atom to water. The resulting radial distribution function indicates 12 nearest neighbors (compared with about five in water), and there is a marked tendency for first-layer water molecules to form hydrogen-bonded pentagons, i.e., incipient clathrate cage formation (F. H. Stillinger, private communication).

postulate that the presence of a hydrocarbon molecule would favor energetically the four-coordinated hydrogen-bonded arrangement of water, i.e., a shift of the equilibrium to the left. In turn, the concentrations of water molecules having respectively three, two, one, or no hydrogen bonds, and making up $(H_2O)_d$, decrease. The model specifically excludes the possibility of interstitial solution and characteristically it has been found to describe the properties of aqueous solutions of butane and benzene reasonably well but not those of methane or ethane. The Nemethy–Scheraga formalism gives rise to certain conceptual problems, e.g., four-bonded molecules can only occur in a cluster and therefore a four-bonded molecule cannot exist as nearest neighbor of a solute molecule or, indeed, of an unbonded water molecule. These and other limitations are not allowed for in the mathematical development. Theirs was, however, the first attempt to express in molecular terms the interactions between water and solutes and drew the attention of many scientists to the unique nature of aqueous solutions. Concurrently Frank and Quist[510] extended Pauling's suggestion of a "water-hydrate" model[1079] to encompass aqueous solutions of methane. The solute molecule was allowed to exist inside the clathrate cavities and the known structure and stoichiometry of methane hydrate, $CH_4 \cdot 5.76H_2O$, were applied to the aqueous solution.

For a system of N_1 water molecules a fraction f of which exist as the bulky modification, providing interstitial sites for the single molecules of the dense variety, the total number of distributions of $(H_2O)_d$ among the sites is

$$W = (N_1 f/v)! \, [N_1(1-f)]! \, [(Nf/v) - N(1-f)]! \qquad (3)$$

where v is the number of molecules of $(H_2O)_b$ required for the provision of a site. The Gibbs free energy of this system is given by

$$G = G_d{}^\ominus + \alpha + f(\Delta G^\ominus - \alpha) - RT \frac{f}{v} \ln \frac{f/v}{(f/v) - (1-f)}$$
$$-(1-f) \ln \frac{1-f}{(f/v) - (1-f)} \qquad (4)$$

where $\Delta G^\ominus = G_b{}^\ominus - G_d{}^\ominus$ and α is a parameter to account for the fact that $G_b{}^\ominus$ refers to the "empty" lattice and must be modified as cavities become occupied. Numerical application of the simple model has shown that the volumetric, expansibility, and compressibility properties of liquid water can be accounted for quite adequately. This is, however, not the case with the specific heat.

If now N_2 solute molecules are included in the system and they are allowed to compete on an equal basis with the $(H_2O)_d$ molecules for occupation of the interstitial sites, then eqn. (3) becomes

$$W = (N_1 f/v)! \, [N_1(1-f)]! \, N_2! \, [N_1(f/v) - N_1(1-f) - N_2]! \quad (5)$$

From eqn. (5) the various thermodynamic properties of the solution can be calculated in terms of f, v, and the concentration. In particular, the solute excess partial molal entropy at infinite dilution is given by

$$\bar{S}_2^{\circ E} = f(1-f)(v+1)\frac{\Delta H_1^{\ominus} - \alpha_1}{T} + R \ln\left[\frac{f}{v} - (1-f)\right] \quad (6)$$

where $\Delta H_1^{\ominus}$ refers to equilibrium (2). Adopting the known stoichiometry of methane hydrate, the calculated $\bar{S}_2^{\circ E}$ is certainly of the right order of magnitude. Equation (6) indicates that the low $\bar{S}_2^{\circ E}$ is due to two different effects; the first term is a measure of the shift in the water equilibrium (2), while the second term arises from the barring of some water from fulfilling its solvent function. The model cannot explain the corresponding heat capacity $\bar{C}_{P_2}^{\circ E}$, which is hardly surprising in view of its inability to reproduce C_P for pure water.

Perhaps the main weakness of the model lies in its limited range of application. Thus, with rising temperature the clathrate lattice is progressively destroyed, leading to an increased degree of occupancy of the cavities. The limits are set by the conditions $f \to 1$ and $f \to v/(v+1)$, corresponding to the empty lattice and complete occupancy of cavities, respectively. Frank and Quist therefore provide for a further modification of water as "state III," the proportion of which increases with rising temperature. This species is not defined in detail and cannot therefore be used in the calculation of ΔS^M and other derived properties.

Although this clathrate model has been partially successful in accounting for entropies of solution and compressibilities, it also has certain shortcomings and has left unexplained a number of experimental results, notably the large $\Delta \bar{C}_{P_2}$. Also, it does not really make use of the mixture model concept, as previously proposed by Frank. The question of hydrocarbon solubility and the effect of urea on the aqueous solution thermodynamics of hydrocarbons was reexamined by Frank and Franks[509] in terms of a model based on equilibrium (2), with the following specifications:

1. $(H_2O)_b$ is not identical to ice but resembles a clathrate framework.
2. $(H_2O)_d$ is a quasilattice liquid which mixes ideally with $(H_2O)_b$.
3. Hydrocarbons are partitioned between the water species, forming interstitial solutions in $(H_2O)_b$ and regular solutions with $(H_2O)_d$. The

total number of configurations in such a system is written as

$$W = \frac{N_1}{(N_1 f)!\,[N_1(1-f)]!} \frac{(N_1 f/v)!}{(N_2 g)!\,[(N_1 f/v) - N_2 g]!}$$
$$\times \frac{[N_1(1-f) + N_2(1-g)]!}{[N_1(1-f)]!\,[N_2(1-g)]!} \qquad (7)$$

where g is the fraction of solute molecules existing in an interstitial environment.* Via the evaluation of ΔS^M, $\mu_2{}^\circ$ (the solute limiting chemical potential) and $\bar{S}_2{}^\circ$ can be obtained as

$$\mu_2{}^\circ - RT \ln x_2 = \mu_r{}^* + RT \ln[(1-g)/(1-f)] \qquad (8)$$

and

$$\bar{S}_2{}^\circ + R \ln x_2 = \bar{S}_r{}^* - R \ln \frac{1-g}{1-f} + (g-f)\frac{\Delta H_1{}^\ominus}{T} + g\,\frac{\Delta H_2{}^*}{T} \qquad (9)$$

where $\mu_r{}^*$ and $S_r{}^*$ are, respectively, the standard potential and partial molar entropy of the solute in the regular (quasilattice) solution, i.e., in "dense" water. Similarly, $\Delta H_2{}^*$ is the standard molar enthalpy change accompanying the transfer of hydrocarbon from dense to bulky water.

Once again it is instructive to consider the shift said to be produced in the water equilibrium by the nonpolar solute. In eqn. (9) this effect is accounted for by the third term on the right-hand side, which has, however, no counterpart in eqn. (8). In terms of this model the solubility of a hydrocarbon does not therefore show up this structural shift. On the other hand, the exclusion of water molecules from their solvent function, already referred to in connection with eqn. (6), appears in both eqns. (8) and (9) and therefore is reflected in the solubility. The final term in eqn. (9) relates to the entropy loss suffered by the solute molecule upon transfer from the regular solution in $(H_2O)_d$ to an interstitial site in $(H_2O)_b$.

The numerical test of the above formalism against the known thermodynamic properties of water–hydrocarbon mixtures requires the assignment of four parameters the magnitudes of which are adjustable within certain limits. As for the Frank and Quist model, $\Delta H_1{}^\ominus$, f, and v need to be fixed, but in addition, g, relating to the partition of hydrocarbon between the two solution environments, and $\Delta H_2{}^*$ must be considered. The former is as-

---

* One difference between the formalisms expressed by eqns. (5) and (7) is that in the latter the possibility of interstitial water molecules within the lattice framework is excluded.

TABLE II. Thermodynamic Properties of Alkanes in Infinitely Dilute Aqueous Solution According to the Model of Frank and Franks[509]

Property[a]	CH_4	C_2H_6	C_3H_8	C_4H_{10}
g	0.781	0.642	0.474	0.309
$-\Delta H_2{}^*$, cal mol^{-1}	8780	8300	9350	14200
ν	5.76	5.76	17	17
$\Delta G_2{}^* - RT \ln \nu$, cal mol^{-1}	-105	305	715	1130
$-[\Delta S_2{}^* + R \ln \nu]$, cal mol^{-1}	29.0	28.9	33.8	51.5
$\Delta H_{2,\mathrm{vap}}^{\circ}$(exptl.)	3100	4300	5800	6100
$\Delta G_{2,\mathrm{vap}}^{\circ}$(exptl.)	6290	6120	6230	6370
$\Delta S_{2,\mathrm{vap}}^{\circ}$(exptl.)	31.5	35.0	40.4	41.8
$H_2{}^{\ominus}(g) - H_r{}^*$	-3850	-470	2580	2890
$-[G_2{}^{\ominus}(g) - G_r{}^*]$	6370	5920	5790	5770
$S_2{}^{\ominus}(g) - S_r{}^*$	8.5	18.3	28.1	29.1
$H_2{}^{\ominus}(g) - H_c{}^*$	4930	8130	11930	17090
$S_2{}^{\ominus}(g) - S_c^{**}$	37.5	47.2	61.9	80.6

[a] Experimental and computed thermodynamic properties of hydrocarbons in clathrate (c) and regular solution (r) environments. Values for ν are those of the type I and II clathrate hydrate structures $R_n C_{2n+1} \cdot \nu H_2O$, and $S_c^{**} = S_c{}^* - R \ln \nu$.

sumed to bear a simple relationship to the number of carbon atoms in the solute molecule, and the latter has been derived from the observed standard enthalpy of transfer, $\Delta H_{\mathrm{tr}}^{\ominus}$ of hydrocarbon from water to 7 M aqueous urea.[1398]

Table II summarizes the calculated properties of the lower hydrocarbons in their hypothetical clathrate and regular solution environments. The various contributions making up the observed $\bar{S}_2^{\circ}$ of propane are displayed in Fig. 2. According to the formalism employed and with the numbers assigned to the various parameters, Frank and Franks conclude:

"It becomes evident that propane does not make, but breaks structure when it dissolves, and that the effect of the structure shift, so produced, is not large, but small, and is not to lower, but to raise $\bar{S}_2^{\circ}$. The effect which is here responsible for the large negative partial molar entropy is, instead of the structure promotion, postulated by Frank and Evans, the entry of the fraction g of the solute into the quasiclathrate environment. That is, the need which those authors recognized to find some physical effect to offset

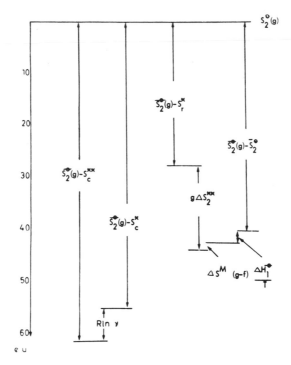

Fig. 2. Contributions to the observed partial limiting entropy $\bar{S}_2$ of propane in aqueous solution, according to Ref. 509; see also eqn. (9) and Table II.

the entropy-raising process characteristic of the solutions of nonpolar gases in "normal" liquids is here met by allowing a fraction of the solute molecules to go into an entirely different kind of dissolved state, where the neighboring solvent molecules can no longer relax away from them, because they are held by the solvent–solvent bonds which the quasiclathrate contains."

The observed effects of urea on the hydrocarbon/water systems[1398] can also be accounted for on the basis of the preceding model and reference will be made to urea–water systems in Section 2.2.2. The formalism of Frank and Franks is again of limited value because it applies only in the limit of infinite dilution. On the other hand, to describe the concentration dependence of thermodynamic variables would require the introduction of additional adjustable parameters which, for aqueous solutions of apolar solutes, could not be tested experimentally.

This model, as that of Frank and Quist, has been tested only at room temperature. It provides a sensible picture of two possible solution environments which can be tested against known reference states, i.e., the crystalline clathrate and the regular solution. Although the concentration dependence of the various properties described can of course be predicted, the low solubility of the substances to which this model can be applied makes such predictions impossible to verify. Although the postulate of a clathrate-type solution might also be applicable to other types of solutes (Chapter 3), this is unlikely for the quasiregular solution, where additional parameters describing specific interactions might have to be included.

The problems of concentration dependence and "state III" have been further developed by Mikhailov,[972,973] who, however, takes as his starting point the model for water described by Samoilov[1174] according to which $(H_2O)_b$ is equated to the ice lattice in which $(H_2O)_d$ molecules may occupy interstitial sites.* The Mikhailov model has the following features:

1. Interstitial solution occurs in the bulky, icelike component.

2. A large solute molecule which cannot be accommodated in a single cavity may occupy several adjacent sites.

3. State III, i.e., nonavailability of interstitial voids, is allowed for by the definition of a third water species which is structureless and assumed to mix regularly with apolar solutes.

4. Additional equilibria of the type (solute)$_i$ $\rightleftharpoons$ (solute)$_r$ and (water)$_i$ $\rightleftharpoons$ (water)$_r$ (subscript i refers to interstitial solution) are thus introduced.

The approach adopted by Mikhailov, originally based on the Frank and Quist model, is thus a combination of the two formalisms described by eqns. (5) and (7), with the additional feature of allowing for the accommodation of large molecules in solution. As is usual with models of increasing complexity, more adjustable parameters have to be introduced, and in the end the student of such models must decide for himself whether the good fit to experimental data has been obtained at the cost of an unreasonable hypothesis or too large a number of adjustable parameters. In the case of the Mikhailov model, the above criticisms do not seem to apply,† and it is therefore regrettable that many workers with an interest in aqueous solution chemistry and biochemistry are apparently unaware of this rather useful approach.

---

* The realism of this type of model has sometimes been questioned because of the notorious reluctance of ice to form solid solutions, while, on the other hand, a whole host of clathrate hydrates are known to exist.

† Apart from the credibility of the ice lattice as a likely host to solute molecules. However, the quantitative aspects of the model could equally well be applied to the clathrate lattice.

According to the Mikhailov model, G of a binary system, where the solute concentration is low, is given by

$$G = nG_b^{\ominus} + n_2 G_2^{\ominus} + n_2 RT(\ln y) + (r-1)n_2 RT \ln\left[\frac{f}{v} - (r-1)y\right]$$

$$-rn_2 RT \ln\left[\frac{f}{v} - (1-f) - ry\right] + n_1 RT \ln\left[\frac{1-f}{(f/v) - (1-f) - ry}\right]$$

$$(10)$$

where $y = N_2/N_1$ and r is the length of channels produced by the fusion of adjacent cavities. The computed values of $f(y)$ for various r given in Fig. 3 show that $\partial f/\partial y$ is positive, indicating a structure-promoting process. It also follows that large molecules are more effective structure promoters than are smaller molecules.

As y increases and eventually all interstitial sites become occupied, provision is made in the model for $N_1\alpha$ and $N_2\beta$ molecules of water and

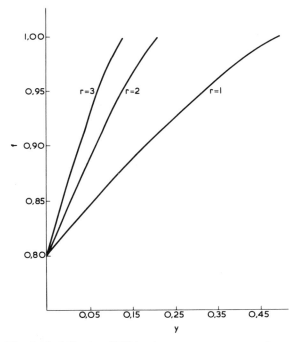

Fig. 3. Stabilization $(\partial f/\partial y)$ of water structure at 25° by apolar solutes of varying size (reflected by r), according to Mikhailov.[973] For $r = 1$ this model reduces to that of Frank and Quist.[510]

solute, respectively, to be located outside the lattice and to form a regular solution. Since this structureless region is spatially separated from the lattice, the expression derived for G predicts an eventual phase separation. It can be shown that the concept of such a microheterogeneity does not seriously affect the validity of the model, provided that v is large.

By minimizing G with respect to f, α, and β, expressions can be derived for the equilibrium constants governing the respective equilibria

$$H_2O(\text{lattice}) \rightleftharpoons H_2O(\text{interstitial}) \qquad K$$

$$H_2O(\text{interstitial}) \rightleftharpoons H_2O(\text{regular solution}) \qquad K_\alpha$$

$$\text{solute}(\text{interstitial}) \rightleftharpoons \text{solute}(\text{regular solution}) \qquad K_\beta$$

By using these expressions, the free energy of the system can be written as

$$G = n_1 G_b{}^{\ominus} + n_2 G_2{}^{\ominus} + n_2 RT \ln[y(1 - \beta)]$$

$$+ (r - 1)n_2 RT \ln\left[\frac{f}{v} - (r - 1)y(1 - \beta)\right]$$

$$- rn_2 RT \ln\left[\frac{f}{v} - (1 - f - \alpha) - ry(1 - \beta)\right]$$

$$+ n_1 RT \ln\left[\frac{1 - f - \alpha}{(f/v) - (1 - f - \alpha) - ry(1 - \beta)}\right]. \qquad (11)$$

By subtracting a term relating to the regular solution ($f = 0$, $\alpha = \beta = 1$), the excess Gibbs free energy G^E can be evaluated.

The model has been successfully applied to the system water–ethanol over the whole range of composition and it will be further explored in Section 2.2.1.

The range of application for apolar solutes is for very low values of y. Provided that K_β is not too large, small additions of solute stabilize or promote the $(H_2O)_b$ structure and the water molecules displaced from the cavities by the solute are incorporated into the $(H_2O)_b$ lattice in a manner suggested by Frank and Quist. The structure-promoting capacity of a given solute depends on the size of its molecules, on $U(r_{12})$, and hence on K_β.

At higher concentrations or temperatures the model predicts the expulsion of water and solute from the interstitial solution, giving rise to a different solution environment, as suggested also by Frank and Franks. The interesting point made by Mikhailov is that this process is accompanied by an onset of microheterogeneity which may in certain cases lead to a complete phase separation. Although this cannot be tested with hydrocarbons, lower critical demixing is known to occur in a number of systems forming solid interstitial solutions (clathrates) whose behavior can well be described by this model.

However, the existence of solute–solute interactions, even in very dilute solutions of hydrocarbons, has been discussed for many years, and the concept of the hydrophobic bond has its origin in this type of thinking.[735] The reasoning follows the line that since a transfer of solute from an apolar to an aqueous environment, although energetically favored, $\Delta H < 0$, is nevertheless accompanied by an unfavorable entropy change, thus making the transfer process nonspontaneous ($\Delta G > 0$), then the reverse process, i.e., the creation of a solute–solute contact, must be spontaneous. There is ample evidence from aqueous solutions of large and small molecules that $u(r_{22})$ is indeed affected by solute concentration but more detailed studies of this phenomenon suggest that solute–solute interactions are not just a reversal of the solution process but that the situation is rather more complex.[523]

Finally, it must again be emphasized that the experimental basis of the solution behavior of apolar solutes is extremely weak. Complete reliance has in the past been placed on solubility determinations, and ΔH and ΔS obtained indirectly. In view of the magnitude of experimental errors involved, all such data must be treated with extreme caution, as must also predictions about the nature of these solutions obtained by extrapolation from data on rather more complex systems. As an example can be cited the widely quoted "result," much used in discussions of hydrophobic bonding, that the dissolution of hydrocarbons in water is an exothermic process. This is based on the van't Hoff treatment of solubility which will be more closely examined in Chapter 5. Whatever the intrinsic merits or shortcomings of the van't Hoff enthalpies, solubility determinations on sparingly soluble gases are subject to such large errors, systematic as well as random, that the differentiation of such results is largely meaningless. Indeed, the first calorimetric determination of heats of solution of alkanes[1141] has shown that the process is in fact *endothermic at 25°.* As it happens, ΔH_{sol}, although positive, is quite small, so that this experimental finding does not materially affect the arguments regarding hydrophobic bonds, except in so far as at 25° this type of interaction, in the case of *real* apolar solutes, is endothermic rather than exothermic, as is widely quoted in the literature.

---

* In view of the very large heat capacity changes involved (see Table I), it is quite probable that at 10° ΔH is large and negative. The adoption of 25° as universal standard state is quite arbitrary and may not be too good a basis for comparing different solutes (see Chapter 5).

It is to be hoped that as experimental techniques become more sensitive and sophisticated, solutions of hydrocarbons will receive increasing attention so that their actual properties can be established directly rather than by doubtful comparisons with other, usually highly polar molecules.

2.2. Polar Solutes

From an experimental point of view the study of polar molecules in aqueous solutions is more rewarding because their higher solubility and the presence of functional groups make analysis rather easier. It has generally been stated that the solubility is due to hydration (hydrogen-bonding) interactions between solute and water and that these interactions are stronger than those which operate between solute molecules in the pure liquid state or in the crystal lattice. Thus it follows that on ascending a homologous series, the solubility decreases with carbon number and that the introduction of more polar groups into a molecule enhances its solubility. However, the position is not quite so simple: For instance, closely similar compounds possessing several OH groups and differing only in minor stereochemical detail have very different solubilities. Examples are the pairs mucic and saccharic acid, and scyllo- and myo-inositol. The first compound in each pair is insoluble, while the second is quite soluble; yet the only difference between the two members of the pairs is in the conformation of one —OH group (i.e., axial versus equatorial).

```
           COOH                              COOH
            |                                 |
    H——|—OH                          H——|—OH
   HO——|—H                          HO——|—H
   HO——|—H                           H——|—OH
    H——|—OH                          H——|—OH
            |                                 |
           COOH                              COOH

       mucic acid                        saccharic acid
```

scyllo-inositol

myo-inositol

TABLE III. Proton Donor and Acceptor Potentials of Functional Groups in Organic Molecules

Functional group	Exchangeable protons	Proton acceptor sites
—O—	0	2
—OH	1	2
$>C=O$	0	2
—NH$_2$	2	1
$>NH$	1	1
$>N$	0	1
—CONH$_2$	2	3
—COOH	1	4

The common functional groups which are capable of interacting with water by virtue of their proton-donating or -accepting properties are summarized in Table III. It can be seen that secondary amines, R_1R_2NH, occupy a unique position because, like water, they are capable of forming an equal number of hydrogen bonds by proton donating and accepting.

It is becoming apparent that aqueous solution behavior is determined by several features of the solute molecule[514]: (1) the number of hydrogen-bonding sites in relation to the size of the nonpolar residue; (2) the number of internal degrees of freedom of the solute molecules; (3) the degree of unsaturation or aromaticity of the hydrocarbon residue; and (4) the relative positions of polar groups, and their freedom of rotation and conformation. For the sake of this discussion solutes can be broadly classified into two groups: The solutions of monofunctional solutes exhibit most of the features discussed in Section 2.1, although not to the same extent. Thus the solution thermodynamics (e.g., deviations from ideal behavior) are dominated by large negative entropies rather than by enthalpies of mixing, there are pronounced heat capacity effects, and, in addition, the volumetric behavior is rather peculiar. Another common feature of these solutions is the pronounced concentration dependence of all physical properties in the very dilute solution range. These solutions also show evidence (large excess ultrasound absorption) of interactions giving rise to long-range order which diminishes with rise in temperature. It may also be relevant that many of

TABLE IV. Solute Properties which Determine Aqueous Solution Behavior

"Aqueous" behavior (apolar group predominates)	"Nonaqueous" behavior (hydrogen-bonding sites predominate)
Number of carbon atoms	Proton donor/acceptor ratio
Internal degrees of freedom	Molecular conformation
Unsaturation, aromaticity	Spacing between proton acceptor sites
Freedom of rotation of polar sites	Freedom of rotation of polar groups

the compounds of this type are known to form clathrate hydrates[705] (see also Chapter 3). As the ratio of polar to nonpolar groups is increased the solution behavior changes to a pattern which is typical for most non-aqueous solutions, i.e., deviations from ideal behavior are governed by enthalpic factors, heat capacities of mixing are small, specific interactions can be observed between polar groups and water molecules, the limiting concentration dependence of physical quantities is almost zero, and there is no evidence of temperature-sensitive structural relaxations. Table IV shows the different characteristics observed for the two types of aqueous mixtures described above, and Fig. 4 is a classification of common solutes according to their solution behavior.

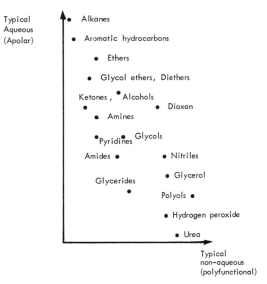

Fig. 4. Classification of solutes according to their dilute aqueous solution behavior.

2.2.1. *Typically Aqueous Solutions—Monofunctional Solutes*

It is seen from Table IV and Fig. 4 that monofunctional solutes give rise to solutions which resemble closely those of hydrocarbons, and this type of behavior is peculiar to water as solvent and can therefore be termed "typically aqueous." Although it used to be believed that such solutes affect water largely via the polar group, the observation that the main factor determining the solution behavior is the size of the alkyl group rather than the hydrogen-bonding groups, led to the suggestion that direct hydrogen-bonding effects might only provide a minor contribution to the observed behavior, which was dominated by the alkyl group–water interactions. If this is the case, then there should be evidence of an ordering process, as predicted by equilibrium (2), and it should also be possible to apply the various model treatments discussed in Section 2.1 to such solutions and compare the predicted and observed concentration dependences of thermodynamic quantities.

Unfortunately, not many X-ray diffraction and scattering studies have yet been undertaken on aqueous solutions, so that direct evidence for "structures" is limited. However, it has been shown that the observed diffraction pattern of water–ethanol mixtures differs considerably from that expected for a simple mixture in which the ethanol dilutes the water and destroys its long-range order.[276] It is clearly shown in Fig. 5 that for all

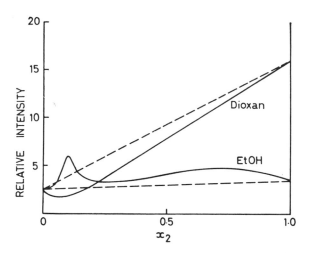

Fig. 5. Intensities of X-ray diffraction peaks observed for water–ethanol and water–dioxan mixtures. Broken lines indicate expected intensities for ideal mixtures. From Ref. 276.

compositions the degree of order (as indicated by the intensity of the diffraction peak) is higher than would be expected, with a most pronounced structuring occurring near 10 mole %. The opposite is true for mixtures of water and dioxan with the additional feature that, as dioxan is added to water the position of the observed diffraction peak shifts such that for a 20 mole % mixture it has reached the position which it occupies for pure dioxan. In other words, in such mixtures the 80 mole % of water contributes nothing to the order, as monitored by X rays, and cannot therefore be distinguished from water vapor.

A related technique, low-angle X-ray scattering, has also been applied to the study of aqueous mixtures,[78] although again to a very limited extent. Results on mixtures of water with *tert*-butanol show very clearly that pronounced scattering intensity peaks occur at a concentration of 10–12 mole %, and that the scattering intensity increases with rising temperature. Since the excess scattering intensity arises from local concentration fluctuations (Volume 1, Chapter 8), the observed pattern is clear evidence for very definite clustering of solute and water in fairly dilute solution, disappearing as the concentration is increased. A summary of the results is shown in Fig. 6. The increase in scattering intensity with temperature provides a clear indication of lower critical demixing in such solutions and this is in agreement with predictions based on the relative magnitudes of the thermodynamic mixing functions of the compounds described in Fig. 4 as "typically aqueous." The occurrence of lower critical demixing has, of course, long been established in some aqueous systems, notably solutions of amines and glycol ethers, but the X-ray scattering results suggest that solutions of alcohols and ethers* also exhibit this phenomenon and that the "seeds" of critical demixing are observable at temperatures far below the critical temperature, which may not always be easily accessible, e.g., in the system of water/tetrahydrofuran it occurs above the boiling point of the solute.[954] The main point of interest, however, is that a pronounced microheterogeneity is observed in dilute solutions of quite simple substances, disappearing at higher concentrations, and this lends support to the Mikhailov solution model (see Section 2.1).

Another technique which has been applied to the study of aqueous solutions is that of ultrasound attenuation. A considerable amount of data has been accumulated and is discussed in Chapter 9. Like the X-ray scattering experiments, ultrasound absorption results clearly show marked qualitative differences in the behavior of dilute and concentrated aqueous

* D. Atkinson, unpublished work at this Laboratory.

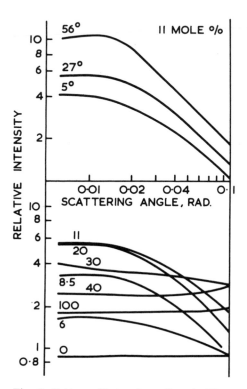

Fig. 6. Bottom: Plots of small-angle X-ray scattering intensities versus scattering angle observed for water (0), t-BuOH (100), and various mixtures of indicated mole percent t-BuOH. Top: Scattering intensity versus scattering angle isotherms for 11 mole % aqueous t-BuOH solutions.

solutions of monofunctional solutes, such as alcohols, ethers, and amines. Thus, in dilute solutions pronounced excess absorptions are observed which undergo maxima at given concentrations, characteristic of the solute species. The intensity of the excess absorption decreases with rising temperature and this suggests that the effect observed is not directly related to the lower critical demixing behavior reflected in the small-angle X-ray scattering results. In fact there is as yet no general agreement about the exact nature of the molecular relaxation process which gives rise to the observed experimental results.[203]

    Evidence for unique types of interactions in aqueous solutions is also provided by the PVT behavior.[516] The quantities of interest are the partial molar volumes $\bar{V}_2(x_2)$ and their limiting values $\bar{V}_2^\circ$, the partial molar expansibilities $\bar{\alpha}_2 = \partial \bar{V}_2(x_2)/\partial T$, adiabatic compressibilities $\bar{\beta}_2 = \partial \bar{V}_2(x_2)/\partial P$, and the thermal pressure coefficient $(1/P)\,\partial P/\partial T$. These properties are discussed in detail in Chapter 5, so that for present purposes only some general observations need be made. For the compounds discussed in this section the excess limiting partial molar volume, $\bar{V}_2^{\circ E} = (\bar{V}_2^\circ - V_{\text{liq}})$, where V_{liq} is the molar volume of the pure liquid solute, is always negative, i.e., the molar contribution of the solute to the total volume of the system is less in aqueous solutions than in the pure liquid solute. Some values are given in Table V, showing $\bar{V}_2^{\circ E}$ to depend on the size of the solute molecule and its configuration, the volume anomaly becoming larger with increasing

TABLE V. Limiting Excess Partial Molar Volumetric Properties of a Range of Simple Solutes in Aqueous Solution[a]

	t, °C	$\bar{V}_2^{\circ E}$, cm³ mol⁻¹	$10^2\bar{\alpha}_2^{\circ E}$, cm³ mol⁻¹ deg⁻¹	$10^4\bar{\beta}_2^\circ$, cm³ mol⁻¹ bar⁻¹
MeOH	20	$-2.4^{(532)}$	$-3.9^{(532)}$	—
EtOH	20	$-3.4^{(532)}$	$-4.4^{(532)}$	—
PrOH	20	$-4.5^{(532)}$	$-5.2^{(532)}$	—
i-PrOH	20	$-4.8^{(532)}$	$-4.9^{(532)}$	—
BuOH	25	$-5.0^{(524)}$	$-3.8^{(524)}$	—
i-BuOH	25	$-6.1^{(524)}$	$-3.8^{(524)}$	—
s-BuOH	25	$-5.6^{(524)}$	$-6.0^{(524)}$	—
t-BuOH	25	$-7.0^{(524)}$	$-11.8^{(524)}$	$-20.5^{(520)}$ (5°)
Tetrahydrofuran	25	$-4.7^{(518)}$	$-3.0^{(518)}$	$+7.7^{(520)}$
Tetrahydropyran	25	$-6.5^{(518)}$	$-6.0^{(518)}$	$+6.4^{(520)}$
1,4-Dioxan	25	$-4.7^{(518)}$	$+2.1^{(518)}$	$+9.3^{(520)}$
Tetrahydrofurfuryl alcohol	25	$-3.6^{(518)}$	$-1.0^{(518)}$	$+4.4^{(520)}$
Tetrahydropyran-2-carbinol	25	$-5.4^{(518)}$	$-1.7^{(518)}$	$-2.0^{(520)}$
Me$_2$NH	0	$-7.7^{(1336)}$	—	—
Me$_3$N	0	$-19.3^{(1336)}$	—	—
Pyridine	25	$-3.4^{(333)}$	—	$+2.2^{(333)}$
2-Me-pyridine	25	$-4.9^{(333)}$	—	$+5.3^{(333)}$
2,6-diMe-pyridine	25	$-6.0^{(333)}$	—	$+7.3^{(333)}$

[a] The pure liquid solute has been adopted as the reference state.

carbon number. It has also been established that for dilute solutions $\bar{\alpha}_2^{\circ E}$ is negative (see Table V), i.e., that the volume anomalies increase with rising temperature.[524] With rising concentration $\bar{\alpha}_2^E$ increases and eventually becomes positive at a concentration characteristic for a given solute. The negative $\bar{\alpha}_2^E$ are related to the well-known maximum density effect, and the property possessed by a number of simple solutes of raising the temperature of maximum density of water has been studied in detail.[45,526,1363,1364] A rather interesting distinction between structure-promoting and -breaking effects in terms of trends of $\partial\bar{\alpha}_2/\partial T$ has been made by Neal and Goring,[1019] although only limited reliance can be placed on second derivatives of medium-precision density measurements. Unfortunately, for a comparison involving a range of solutes $\bar{V}_2$ and $\bar{\alpha}_2$ must be related to reference states because they contain sizeable contributions due to the intrinsic volumes of the solute molecules. It is for this reason that excess functions are employed* but this problem does not arise with $\bar{\beta}_2$, which is due solely to changes in the interactions within the system. Thus the fact that the solutes discussed in this section exhibit negative $\bar{\beta}_2$ values at low temperatures and concentrations indicates that the solutions are less compressible than pure water at the same temperature. This need not necessarily arise from any changes in the long-range order or the extent or the nature of the hydrogen bonding in the solvent, but the results can be interpreted in such a manner, especially in view of the observation that $\partial\bar{\beta}_2/\partial T > 0$ and that $\bar{\beta}_2^\circ$ is very sensitive to the size of the alkyl residue (see Table V).

The other interesting aspect of the partial molar properties of aqueous solutions concerns the concentration dependence of $\bar{V}_2^E$ and its derivatives. At low temperatures $\partial\bar{V}_2^E/\partial x_2$ is negative and passes through a minimum at a concentration which is characteristic for a given solute. This means that the contribution per mole solute to the total volume of the system decreases with concentration, indicating the importance of solute–solute effects, even in very dilute solutions. As the size of the alkyl group increases, the minimum moves to a lower x_2 and becomes more sharply defined. There are also more subtle effects, such as the limiting value of $\partial\bar{V}_2^E/\partial x_2$, the nature of the second derivative, and the effect of temperature on $\bar{V}_2(x_2)$, all of which again accentuate the unique behavior of aqueous solutions.[524] Once the concentration is increased beyond the value corresponding to the extrema in the volumetric properties, the solutions assume the type of behavior normally associated with nonaqueous systems.

---

* It is very questionable where a comparison of different solutes in terms of the properties of the pure liquids at a common temperature has any meaning.

None of the techniques so far described is able to discriminate between the different molecular species in the mixture, and the results discussed cannot therefore directly provide evidence that the observed effects in the presence of solutes are due to changes in the nature of the long-range order of the solvent. More definitive information can be derived from NMR experiments by means of which solvent and solute species can be examined independently. Most published data refer to ^{1}H chemical shift measurements and an upfield shift in hydrogen-bonded systems is generally equated with a net decrease in the degree of hydrogen bonding. Until recently the results for aqueous systems had indicated that upfield shifts of the water proton resonance always occurred in the presence of solutes, and this was taken to signify that the effect of solutes was to diminish the hydrogen bonding in the solvent. However, by means of very careful experiments, Glew and his colleagues were able to show that dilute solutions of monofunctional solutes do, in fact, give rise to downfield shifts[571,576] and these results have since been confirmed.[1388] Glew interprets the observed concentration dependence of chemical shifts and indeed of several other measured quantities in terms of the solute in a clathratelike environment which is stabilized in a cooperative manner by occupancy of neighboring cavities. He points out that the observed extrema in the measured quantities occur at concentrations which correspond to the clathrate stoichiometry. Although it has been argued that the clathrate analogy can be pushed too far, a mounting weight of evidence suggests that the solute molecule does, indeed, influence the geometry of or interactions between the surrounding water molecules. This is particularly well demonstrated by NMR relaxation studies involving different nuclei. Thus it has been demonstrated most convincingly[558,582] (see Chapter 10) that in dilute aqueous solutions of, e.g., tetrahydrofuran or acetone an increase in the solute concentration is accompanied by a general slowing down of the diffusional motions (rotational and translational) of water molecules. The rotation of the solute molecules, on the other hand, is hardly affected and resembles that which is characteristically found in the vapor state. These observations have been considerably extended and various rules governing the motional properties of molecules in aqueous solutions have been established by Goldammer and Hertz.[581] Studies of molecular dynamics using electron paramagnetic resonance techniques[716] have led to rather similar conclusions. The results are consistent with an interstitial type of solution where the solute molecule has almost unrestricted freedom of rotation while the rotational diffusion of the solvent molecules is inhibited. Of course the methods employed, although indicating that the observed results are caused by changes in the *water* molecule motions, cannot supply direct evidence for the promotion of long-range order or the type of such order. A somewhat more complete picture can be obtained when the results are seen in the context of other experiments which measure lattice vibrations and diffusional motion in liquids. The method of neutron inelastic scattering as applied to liquid water, which has already been discussed in Volume 1 (Chapter 9), can provide information (although at present only in a very qualitative manner) about the molecular motions of water in aqueous solutions as compared to those in ice or in pure water. It is of interest that the inelastic scattering spectra of ice and liquid water at 2° are almost identical,[832] suggesting common intermolecular modes (hydrogen bond stretching, libration, etc.). The addition of *tert*-butanol produces a marked change in the spectrum, with the discrete structure disappearing and being replaced by a very broad, intense scattering maximum.[519] At the same time the self-diffusion coefficient of the water decreases and the diffusional period increases. These results are hardly consistent with the concept of quasicrystalline structure promotion as developed in Section 2.1. Rather, it appears as though the solute acts by raising the microviscosity of water which takes on some of the properties of a supercooled liquid. The picture of a "glassberg," rather than an "iceberg," is quite consistent with the observed thermodynamic and relaxation behavior, although there seems to be a conflict with the older X-ray diffraction data.

Before examining the applicability of a number of solution theories to aqueous mixtures of simple monofunctional compounds it is helpful to summarize their salient features:

1. In dilute solutions they resemble hydrocarbons rather than polyfunctional polar compounds.

2. The magnitude of the negative excess mixing entropy determines their thermodynamic behavior, with corresponding positive deviations from Raoult's law.

3. The solute molecules have a pronounced effect on the surrounding water molecules, lowering their rate of reorientation and diffusion. These effects reach a well-defined maximum at a given solute concentration.

4. There is evidence for microheterogeneity and lower critical demixing.

5. The phenomena are confined to a narrow concentration range which depends mainly on the size of the alkyl group. More concentrated solutions behave normally, i.e., like nonaqueous mixtures of polar molecules.

In view of these observations it is of interest to examine to what
extent the available solution theories can account for such apparently
bizarre behavior. The Nemethy–Scheraga model for aqueous hydrocarbon
solutions[1026] has been extended to solutions of alcohols[822] mainly by
the incorporation of an extra assumption, i.e., the close similarity between
the alcohol OH group and the water molecules in the aqueous environment.
Thus alcohols are allowed to be incorporated in the water clusters by
forming three, two, or one hydrogen bond(s). By the setting up of a partition function, the limiting thermodynamic properties of the lower alcohols
in aqueous solution have been calculated and agree reasonably well with
the corresponding experimental values. The main criticism of this model,
as of the one on which it is based, concerns the credibility of the assumptions
and some discrepancy between the description of the model and its mathematical formulation. It would be interesting to see whether this model can
correctly account for the concentration dependence of activity coefficients
and other partial molal properties.

The Mikhailov formalism outlined in Section 2.1 has been successfully
applied to solutions of ethanol on the assumption that alcohols form
interstitial solutions which can be compared to those of inert gases and
hydrocarbons, but that the OH group might replace an $(H_2O)_b$ molecule
with a corresponding distortion of the lattice. The various parameters in
eqn. (11) were obtained from the experimental $\Delta G^E(x_2)$ data and Fig. 7
shows the dependence of f, α, and β on the concentration. It is shown that
maximum stabilization of the long-range order occurs at $0.06 < y < 0.07$.
From eqn. (11) and experimental data, standard enthalpies, entropies, and

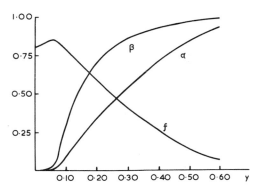

Fig. 7. Changes in the degree of "structure"
induced in water by the addition of ethanol; f, α,
and β are defined by eqn. (11) and in the text;
$v = 50$. After Mikhailov.[973]

TABLE VI. Calculated Thermodynamic Functions at 25° (Mikhailov Model[(972,973)])
for Water–Ethanol Mixtures

Water:	Bulky $\underset{}{\overset{K}{\rightleftharpoons}}$ Interstitial $\underset{}{\overset{K_\alpha}{\rightleftharpoons}}$ Regular solution
	$(H_2O)_b$ $\qquad$ $(H_2O)_d$ $\qquad\qquad$ $(H_2O)_r$
Solute:	Interstitial $\underset{}{\overset{K_\beta}{\rightleftharpoons}}$ Regular solution
	$(EtOH)_d$ $\qquad\qquad$ $(EtOH)_r$

$v = 50,\ K = 0.7071,\ K_\alpha = 0.8414,\ K_\beta = 0.2961$

$H_{1d}^{\ominus} = H_{1b}^{\ominus} = -5000\ \text{cal mol}^{-1}$

$H_{1\alpha}^{\ominus} - H_{1b}^{\ominus} = -4330\ \text{cal mol}^{-1}$ (depends on solute)

$S_{1\alpha}^{\ominus} - S_{1b}^{\ominus} = -14.9\ \text{eu}$

$H_1(\text{pure}) - H_{1\alpha}^{\ominus} = -670\ \text{cal mol}^{-1}$

$S_1(\text{pure}) - S_{1\alpha}^{\ominus} = -2.55\ \text{eu}$

$V_{1\alpha}^{\ominus} = 16.55\ \text{cm}^3\ \text{mol}^{-1}$

$H_2(\text{pure}) - H_{2d}^{\ominus} = -2120\ \text{cal mol}^{-1}$ (for $0.013 < x_2 < 0.37$)

$S_2(\text{pure}) - S_{2d}^{\ominus} = -9.55\ \text{eu}$ (for $0.013 < x_2 < 0.37$)

volume changes for the equilibria involving the various water species can be computed and these are summarized in Table VI. Figure 8 shows the application of the model to the evaluation of excess partial molal enthalpies and volumes of ethanol in aqueous solutions and Fig. 9 shows the calculated

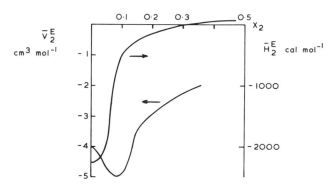

Fig. 8. Excess partial molar volume and enthalpy of ethanol in aqueous solutions, calculated by Mikhailov[(972)] from pressure and temperature derivatives of eqn. (11). Values for f, α, and β are those shown in Fig. 7.

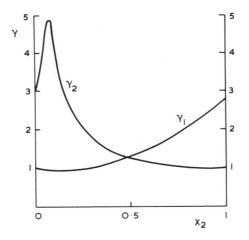

Fig. 9. Activity coefficients of water (1) and ethanol (2) in binary mixtures, calculated according
to the Mikhailov solution model.[972,793] The
maximum in $\gamma_2(x_2)$ has not yet been confirmed experimentally, although the minimum in
$\bar{V}_2(x_2)$ (see Fig. 8) is well established.

activity coefficients of the components in the mixtures.* The agreement with
experimental results is remarkable and suggests that the model incorporates
the right assumptions and the relevant features of aqueous solutions.

Another useful approach to an interpretation of the behavior of
aqueous solutions has been made by Kozak et al.,[786] who, from an analysis
of the virial coefficients in the activity coefficient equation [see Eq. (1)]

$$\log \gamma_1 = Bx_2{}^2 + Cx_2{}^3 + \cdots$$

have been able to account for the limiting concentration dependence of
activity coefficients of a large number of aqueous solutions as functions of
the alkyl chain length and the number of polar groups in the solute molecule. They examined the lattice solution models of Flory–Huggins and
Guggenheim–McGlashan (for relevant references see Ref. 822) and the
theory of McMillan and Mayer.[924] This theory relates the osmotic virial
coefficients in the equation

$$\pi/kT = \varrho + B^*\varrho^2 + C^*\varrho^3 \cdots \tag{12}$$

where ϱ is the number density, to solute pair, triplet, etc. potential functions.

* The extrema in the partial molal properties are *not* discontinuous, since γ_2 and $\bar{V}_2$
can be differentiated over the whole composition range.

The coefficients B^* and C^* are related to B and C in eqn. (1) by means of $\bar{V}_1^\circ$, $\bar{V}_2^\circ$, $\bar{V}_1(x_2)$, and $\bar{V}_2(x_2)$. One weakness of the theory is that reliable analytical values for $U(r)$ are not yet available and that for all the calculations the molecules are treated as hard spheres. However, from the expressions for B^* and C^* in terms of integrals of the radial distribution function, the attractive and repulsive components of $U(r)$ can be isolated and their minimum values computed. The relative importance of pair and triplet interactions can be assessed by comparing the three-particle distribution function with the three pair distribution functions of which it is composed.

As regards the experimental origin of the various quantities required for the computation of $U(r)$, B and C in eqn. (1) are obtained from polynomials or graphical fitting of freezing point or isopiestic data. In addition, enthalpies and heat capacities of mixing must be known as functions of composition. As has already been mentioned, the conversion of B and C into B^* and C^* requires very accurate information about the concentration dependence of $\bar{V}_1$ and $\bar{V}_2$. Table VII summarizes some of the data used and computed by Kozak *et al.*, who demonstrate quite convincingly that

TABLE VII. Virial Coefficients in Activity Coefficient and Osmotic Pressure Equations and Computed Minimum Attractive Contribution to Pair and Triplet Potential Functions[a]

Solute	t, °C	B	C	NB^*	N^2C^*	$A_{2,\text{min}}$, cal mol^{-1}	$A_{3,\text{min}}$, cal mol^{-1}
Methanol	0	-1.0	$+5.5$	$+47$	-600	110	15,000
Ethanol	0	$+1.0$	-24	$+28$	$+7,800$	192	22,000
1-Propanol	0	$+2.3$	-43	$+20$	$+12,000$	261	38,000
2-Propanol	0	$+1.4$	-57	$+37$	$+18,000$	250	34,000
2-Me-1-propanol	0	$+5.2$	-150	-17	—	360	—
2-Me-2-propanol	0	$+0.3$	-90	$+70$	—	265	—
1-Butanol	0	$+4.3$	-100	$+1$	—	347	—
2-Butanol	0	$+1.6$	-80	$+48$	—	295	—
1,4-Dioxan	0	-1.5	—	$+99$	—	221	—
Urea	25	$+1.8$	-13	$+2.2$	$+3,800$	175	16,000
Glucose	0	-4.3	—	$+180$	$+5,300$	198	45,000
Fructose	0	-7.6	—	$+240$	—	200	—
Glycerol	25	-1.3	$+3$	$+86$	—	198	—
Mannitol	25	-0.7	-14	$+139$	—	401	—
Sucrose	25	-5.2	-42	$+286$	—	558	—
Raffinose	25	-7.9	-16	$+495$	—	950	—

[a] According to Kozak *et al.*[786] See eqns. (1) and (12).

the predictions of lattice theories cannot be reconciled with the experimentally observed trends in B and C. An examination of $A_{2,min}$, the minimum attractive contribution to the pair potential function, shows that for homologous series of alcohols, ketones, and amines $A_{2,min}$ becomes more negative with increase in the aliphatic chain length, and this is interpreted in terms of pairwise interactions (hydrophobic bonding?). The contribution of the functional group is either equal to zero or the same for OH, NH_2, NH, $C=O$, etc. It is of special interest to note that $\partial A_{2,min}/\partial T$ is negative, indicating (as do the small-angle X-ray scattering studies) that microheterogeneity becomes more pronounced at higher temperatures. The minimum attractive contribution $A_{3,min}$ to the triplet potential function derived from C^* is subject to large experimental uncertainties, but it appears that this coefficient also becomes more negative with increasing alkyl group size.

2.2.2. *Typically Nonaqueous Solutions—Polyfunctional Hydrophilic Solutes*

As has already been pointed out, the interactions which predominate in solutions of polyfunctional solutes are of a rather different nature from those which give rise to the typical hydrophobic behavior discussed in Section 2.2.1. We are now concerned with specific interactions between polar groups and water molecules, and one of the main problems is the reconciliation of such interactions with any long-range three-dimensional order in liquid water. Unfortunately, experimental studies have been rather less systematic and no comprehensive picture has as yet emerged from such investigations. Much of the available data derive from solutions of amides,[54] simple carbohydrates,[520,1275] dimethylsulfoxide[1061,1171] and urea.[485,1189,1252]

Whereas hydration effects in solutions of predominantly apolar solutes have been shown to be largely nonspecific and characterized by pronounced microheterogeneities (reflected in the marked concentration dependences of physical properties), the opposite is the case for highly polar solutes. Thus limiting thermodynamic properties are sensitive to minor stereochemical differences in the solute molecules, but observed concentration dependences are minimal. There is no evidence of solute clustering but instead marked short-range solute–water interactions are apparent.[837] These effects are elaborated in several chapters of this volume and, for purposes of this introductory account, only a summary of the current views and developments is required.

As regards the interpretation of thermodynamic properties, several alternative formalisms have been employed to explain the small deviations from ideal behavior normally exhibited by solutions of polyfunctional compounds. Thus Stokes and Robinson used the semiideal solution concept to account for the solution thermodynamics of simple sugars.[1253] This model is based on stepwise solvation equilibria of the type

$$S_{i-1} + H_2O \rightleftharpoons S_i \ (i = 1, 2, \ldots, n) \tag{13}$$

from which mean hydration numbers can be computed which account reasonably well for the concentration dependence of activity coefficients. Another formalism is based on solute–solute association equilibria of the type

$$U_{n-1} + U \rightleftharpoons U_n \tag{14}$$

and this particular treatment has been employed to account for the solution thermodynamics of urea up to quite high concentrations.[1189,1252] Its credibility has been questioned[485] on the grounds that equilibrium constants evaluated by this method do not take into account the fact that water, too, is highly associated and that it may be unrealistic to neglect this feature, even if the calculated equilibrium constants are of a reasonable magnitude and provide a good fit to experimental activity coefficients. A third alternative formalism was developed[509] to explain the nature of aqueous urea solutions and the effect of urea on hydrocarbon solubility and takes as its basis the simple water structural equilibrium (2). It is based on the assumption that urea, by virtue of its geometry, is barred from the predominantly tetrahedrally coordinated $(H_2O)_b$ but can mix with $(H_2O)_d$ with which it forms a regular solution. Urea therefore lowers the chemical potential of $(H_2O)_d$ and thus produces a shift from left to right in the structural equilibrium (2). The role of urea is therefore envisaged as that of a statistical structure breaker and this model provides a good fit to the experimental activity coefficient data up to high urea concentrations.

Finally, the Kozak et al.[786] treatment, based on the McMillan–Mayer solution theory, has also been applied to some solutions of polar compounds for which the necessary experimental results are available. The analysis of the virial coefficients suggests (and this is of course inherent in the theory) that deviations from ideal behavior are caused by pairwise solute–solute interactions which become more pronounced as the number of polar substituents increase. Thus urea, glycolamide, lactamide, and glycerol, each with three functional groups, are found to have similar $A_{2,\min}$ values (~ -200 cal mol^{-1}), but as the size of the molecules and the number of functional groups increase, e.g., mannitol (six), sucrose (eight), and raffinose (eleven), $A_{2,\min}$ becomes markedly more negative.

Not much is available in the way of spectroscopic information by which the validity of the various thermodynamic treatments might be tested. Walrafen studied the effect of sucrose and urea on the hydrogen bond stretching vibration in water, as observed at 175 cm^{-1} in the Raman spectrum.[1376] He found that neither solute produced a frequency shift, nor was the band shape altered. The main change was in the integrated intensity of the band, indicating that the solute does not affect the nature of the water–water hydrogen bonds giving rise to this band, but that the degree of hydrogen bonding is enhanced by sucrose and diminished by urea. More recently the hydration of monosaccharides has been examined by ^{17}O NMR and time-domain dielectric relaxation methods.[1275] This latter technique enables a reliable analysis of the dielectric spectrum to be performed (see Chapter 7) and in aqueous solutions of glucose three relaxation processes have been identified with characteristic relaxation times of 1.9, 6.9, and 25 psec at 5° and having different activation energies. On the basis of their temperature and concentration dependences they have been assigned to rotations involving the unperturbed water, solute, and the hydrate water. Both the NMR and the dielectric results are fully compatible with an extended form of a recent water structure model.[1061] Thus for a solution in which water can interact specifically with given OH sites,

$$(H_2O)_{hydr} \underset{k_2}{\overset{k_1}{\rightleftharpoons}} (H_2O)_b \underset{k_4}{\overset{k_3}{\rightleftharpoons}} (H_2O)_d \qquad (15)$$

$$\downarrow k_5$$

$$\text{reoriented } (H_2O)_d$$

where the rate constants are related by

$$k_3, k_4 \gg k_5 \gg k_1, k_2$$

Physically, this scheme is interesting in a number of ways. It suggests, for example, that in the final analysis the observed relaxations attributed to hydration water and unperturbed water arise through the reorientation of the same species. It also implies that glucose is surrounded essentially by $(H_2O)_b$ and in the sense that the concentration of $(H_2O)_d$ is decreased through the equilibria of scheme (15) glucose can be classified as a structure promoter. The relaxation time of the bulk water can also be used as a criterion to examine how glucose modified water structure; the observed small increase over that of pure water may not be very significant in the structure-making sense but is large enough to dispel any ideas about glucose acting as an electrostrictive-type structure breaker. As such, these conclusions are directly contrary to an earlier hypothesis[1128] which suggests that glucose and other hydrophilic solutes increase the local concentration of $(H_2O)_d$. Ben-Naim's observation[144] that glucose decreases the aqueous solubility of argon apparently does not indicate, as he suggests, that glucose destabilizes water structure, but only that the hydration of glucose and argon are incompatible, and in the light of the results from relaxation measurements this is hardly surprising.

It is also of interest to consider the differing hydration properties of the monosaccharides studied, i.e., glucose, mannose, galactose, and ribose. The interpretation of the relaxation data suggests that a glucose molecule in solution is surrounded by $(H_2O)_b$ and, without necessarily defining its precise nature, one can justify the use of a three-dimensional form of the model shown in Fig. 10 to approximate the local environment of the monosaccharide molecule. The apparently similar behavior of glucose, galactose and mannose, and the quite different behavior of ribose, indicates that compatibility with the "lattice" water is the major factor in determining the extent of hydration. Thus glucose, galactose, and mannose can be accommodated into such a model with a majority of their hydroxyl groups hydrogen-bonded into the water "lattice." On the other hand, the various forms of ribose cannot be so accommodated. It has been tentatively suggested that those conformers of ribose which do not have a majority of their hydroxyls capable of hydrogen bonding to the lattice (about 75%) interact preferentially with $(H_2O)_d$ to form a monohydrate (as suggested for all the monosaccharides at temperatures high enough for the lattice to have

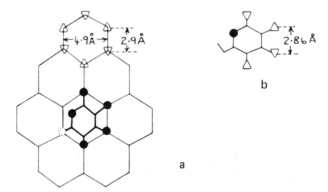

Fig. 10. The O—O spacings in (a) hypothetical ice lattice at 25°C and (b) β-glucose. Oxygen atoms denoted by ▽ lie in one plane; O atoms marked △ lie in another plane. After M. J. Tait *et al.*[1275]

largely disappeared in liquid water). Thus it would follow that hydrophilic solutes can inherently be water structure promoters or breakers, depending upon the detailed stereochemistry of the solute molecule. It is interesting to note[1253] that the thermodynamically derived hydration numbers are in good agreement with the hydration numbers calculated from relaxation studies for glucose but have been found to be incompatible with excess heat and volume data for sucrose solutions. It has been suggested[1204] that the simple hydration equilibrium should be extended to include the equilibrium between water and "hydration water," and this is of course the concept on which the interpretation of the relaxation data is based.

Apart from the sugars, the case of urea also merits a few comments. Urea is of great interest as an effective protein denaturant and a considerable literature deals with the mechanism of such denaturing processes. Until recently no molecular data were available against which the various thermodynamic theories could be tested, and on balance the Schellman picture of urea dimerization and polymerization[1189] (see also Chapter 5) was the generally accepted one. A recent NMR study of urea–water systems based on chemical shift and relaxation measurements on ^1H, ^{14}N, ^{15}N, and ^{17}O nuclei[485] has provided evidence that proton exchange between the species is very slow at pH 7, although the process is strongly acid- and base-catalyzed. Furthermore, the experiments have shown that the long-range order characteristic of $(H_2O)_b$ is destroyed and is not replaced by extended ordering involving urea. From the magnitude of the rotational reorientation time ($\sim$3 psec in 1 M urea) and its concentration dependence, it is clear that no long-lived urea aggregates exist in solution. On the other hand, there is evidence of extensive short-range (hydrogen bonding), short-lived interactions between water and urea. The results and conclusions are in agreement with those derived from dielectric relaxation measurements* on the same systems and it is becoming clear that the curious manner in which urea affects aqueous solutions of simple molecules, surfactants, and polymers is due to its power of destroying long-range order of the aqueous medium without at the same time interacting strongly with water.

Another solute of general interest is dimethylsulfoxide, and to a lesser degree dimethylsulfone. Although thermodynamic and viscosity measurements have been available for several years, it is only recently that X-ray and neutron inelastic scattering results have become available[1171] and made possible the interpretation of such bulk data as existed. In addi-

---

* R. Pottel and D. Adolph, personal communication, cited in Ref. 485.

tion the detailed NMR studies of Packer *et al.*[1061] (referred to in Chapter 10) have provided us with a fairly complete and convincing molecular picture of the interactions and dynamics in dimethylsulfoxide–water mixtures. As distinct from the urea–water systems, which are characterized by their lack of strong interactions, dimethylsulfoxide and water interact specifically and strongly through the S → O dipole. Dimethylsulfoxide must therefore be classed as a water structure breaker, since it disrupts the normal type of long-range order; this structure-breaking effect is nevertheless accompanied by a solute–water ordering, as evidenced by a sharpening of the intermolecular water frequency peaks in the neutron scattering spectrum and by an enhanced 8-Å peak in the X-ray radial distribution function.

Such apparent contradictions in many published reports could be avoided if the nature of the "structure" was more carefully specified, leaving less of the interpretation to the imagination of the reader. It is also becoming clear that discussions of "structures," interactions, binding, complexing, etc. based on thermodynamic data alone (and sometimes on equilibrium constants alone) and so common in the literature, have reached the point of diminishing returns, with so many molecular techniques now available for supplementing macroscopic information.

2.3. Ionic Solutes Containing Alkyl Residues—"Apolar Electrolytes"

As has been discussed in the foregoing sections, the solution behavior of substances containing substantial alkyl residues is largely dictated by the phenomenon of apolar or hydrophobic hydration. The main features are the apparent interference with the motional properties of water molecules in the hydration region and the microheterogeneity of such solutions due to relatively long-range solute–solute interactions within specific narrow concentration ranges.

The alkyl-substituted quaternary ammonium salts and the alkyl sulfates are two classes of compounds in which this type of behavior is particularly marked, although in the latter the effects described are usually obscured by the onset of micelle formation.* The various physical aspects of aqueous solutions of tetraalkylammonium halides are discussed in several of the chapters of this volume, but it is instructive to trace the development of

---

* The process of reversible aggregation, or micellization, itself arises as a consequence of the phenomena of apolar hydration and solute alkyl group interactions in aqueous solution.

current views and the light that studies of such systems has thrown on electrolyte solution theories.

It would appear to be a valid assumption that the tetraalkylammonium halides $R_4N^+X^-$ (where $R = Me, Et, Pr, Bu, \ldots$ and $X = F, Cl, Br, I$) should be "well-behaved" electrolytes, since the cations are singly charged and spherical, and that they should obey the Debye–Hückel and Onsager–Fuoss theories at least as well as do the alkali halides. Although some of their thermodynamic properties had already been studied many years ago by Lange,[445,826] the realization that these compounds did not conform to "normal" electrolyte behavior seems to date from 1957, when it was observed[512] that their partial heat capacities of solution $\Delta \bar{C}_{P_2}$ were large and positive and compared in magnitude with those of hydrocarbons or alcohols in aqueous solution. Reports of systematic studies of these compounds began to appear in 1964 and by 1968 the macroscopic picture was fairly clear. Initial thermodynamic measurements[851] had revealed a complex behavior, in that the osmotic coefficients of the chlorides increased with the size of the cation, whereas the bromides and iodides exhibited the reverse order. The magnitudes of ΔG^E were, however, small throughout the series. Subsequent calorimetric experiments[847] showed the heats and entropies to be very much larger and much more systematic (see Fig. 11),

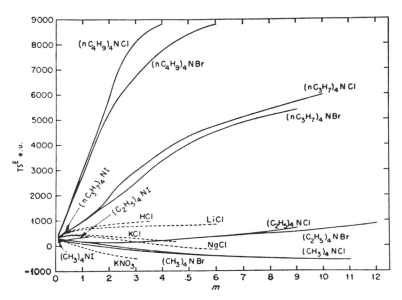

Fig. 11. Plots of TS^E at 25°C of a series of electrolytes in aqueous solutions. Reproduced from Ref. 847 with the author's permission.

thus once again emphasizing the dangers inherent in attempts to interpret solution behavior from free-energy data alone. The initial thermodynamic measurements of Lindenbaum have been supplemented by $\Delta \bar{C}_P$ determinations,[849,990,1180] by calorimetric studies on the more complex R_3N^+–$(CH_2)_nR_3N^+$ salts,[849] and by investigations of the transfer of alkylammonium halides from water to other polar solvents.[1438] Of particular value have been the investigations of mixtures of R_4NX with simple electrolytes and these are discussed in detail in Volume 3, Chapter 2.

The comparative volumetric behavior of aqueous solutions of R_4NX has also been studied in detail. The impetus was provided by Wen and Saito,[1392] who found a negative concentration dependence of $\bar{V}_2$, with a marked minimum at a composition corresponding to the stoichiometry of the clathrate hydrates which had been well characterized by Jeffrey and his colleagues.[170,705] These observations appeared to be in direct contradiction to the $\bar{V}_2(m^{1/2})$ behavior predicted by the Debye–Hückel theory and led to more refined measurements at lower concentrations.[523] It became apparent that the limiting $\bar{V}_2(m^{1/2})_{m \to 0}$ slope was indeed that predicted by theory but that none of the extensions of the Debye–Hückel equation could account for the observed negative deviations from the limiting behavior. The picture was completed by high-precision compressibility[336] and expansibility[983] measurements, which confirmed that, as regards thermodynamic properties at finite concentrations, these mixtures do not conform to the behavior expected for uni-univalent electrolytes. On the basis of the thermodynamic evidence, a general view became established* that R_4N^+ ions resembled hydrocarbons rather than typical ions and that cation association due to apolar hydration effects was probably at the root of the observed phenomena.[334,850,852]

This view has been considerably strengthened as a result of detailed molecular relaxation and bulk transport studies. The NMR evidence,[657,658,659,853] which strongly points to apolar hydration effects, is based on the independent examination of the three components R_4N^+, X^-, and H_2O by different types of relaxation and chemical shift measurements involving nuclei ^1H, ^2H, ^{14}N, ^{17}O, and ^{79}Br; a detailed account of these studies is given in Chapter 10. Confidence in the interpretation of the NMR spectra must be strengthened by the fact that they can be reconciled with the observed dielectric relaxation behavior as analyzed by Pottel[1113] and with observed temperature and concentration shifts of the near-infrared water band (0.97 μm) in solutions of R_4NBr salts.[252]

---

* There were, however, and still are, notable dissenters—see Refs. 1122 and 1419.

Although bulk transport studies (viscosity, conductance, diffusion) cannot directly provide information about molecular motion, nevertheless very comprehensive studies, particularly those of Kay and his colleagues,[738] have convincingly demonstrated the unconventional behavior of alkyl-ammonium ions in aqueous solution (see Volume 3, Chapter 4). Some of the summarized findings are shown in Table VIII, where it is seen that the R_4N^+ ions (provided that $R \geq Et$) have a mobility deficiency in aqueous as compared with nonaqueous solvents but that this deficiency decreases with rising temperature; the interpretation of this observation in terms of temperature-sensitive hydration structures can be fully reconciled with thermodynamics and relaxation data. The concentration and temperature dependence of viscosity[741] also show trends which are opposite to those of simple electrolytes, and the existence of hydration structures has also been inferred from ultrasound relaxation studies.[185] It is of interest to note that the introduction of polar groups, e.g., OH at the terminal CH_3 group, converts the substituted ammonium compounds into well-behaved electrolytes,[460] thus indicating that the observed eccentric behavior of R_4N^+ ions is not solely due to size or low surface charge density effects.

Apart from the studies on binary R_4NX–water mixtures which have provided us with a fair insight into the nature of hydrophobic interactions, various published reports deal with thermodynamic properties of ternary systems, such as Me_4NBr–urea–H_2O,[1387] R_4NCl–EtOH–H_2O,[836] R_4NX–Ar–H_2O,[145] and R_4NBr–alkane–H_2O;[1389] some of these are discussed in Chapter 5.

TABLE VIII. Classification of Electrolytes According to Their Effect on Long-Range Order in Water[738]a

	Charge/size	$(\lambda\eta)_{aq}/(\lambda\eta)_{nonaq}$	$\partial(\lambda\eta)_{aq}/\partial T$	Examples
Electrostrictive structure promoters	Large	—	~ 0	Li^+, La^{3+}, F'
Structure breakers	Small	>1	<0	Cs^+, Me_4N^+, I', ClO_4'
Apolar structure promoters (hydrophobic)	Very small	<1	>0	Bu_4N^+, Ph_4B'

a $\lambda\eta$ is the limiting Walden product.

3. AQUEOUS SOLUTIONS OF ELECTROLYTES

The previous section on aqueous solutions of nonelectrolytes will have given the impression, and rightly so, that our understanding of observed phenomena is somewhat sketchy. By comparison, the state of our knowledge as regards the behavior of ions in solution looks rather healthier, but here again most of the experimental and theoretical effort has been directed to the study of interionic effects and the realization that ions cause remarkable changes in the solvent has come only recently, and even now a large section of the electrolyte community prefers not to delve too deeply into the mysteries of ion–solvent effects.

There are several ways in which a systematic approach can be made to a discussion of electrolyte solutions, and that used by Blandamer[179] seems to us a logical one. This involves separate consideration of single ionic properties, of hydration effects—as reflected, for instance, in the properties of water in the presence of ions—and of ion–ion effects, as reflected in the concentration dependence of measured properties.

Classical electrolyte theory concerns itself mainly with electrostatic effects and takes cognizance of the solvent only as a bulk medium the dielectric permittivity of which modifies ionic interactions. This is particularly well illustrated by the Debye–Hückel equation, which predicts the concentration dependence of mean ionic activity coefficients $f_\pm$ from a consideration of ionic distribution and potential functions. The solvent enters into the calculations only by virtue of its bulk permittivity, which appears in the constants A and B of the expression

$$\log f_\pm = -A \, | \, z_+ z_- \, | \, I^{1/2}/(1 + BaI^{1/2}) \qquad (16)$$

where z_+ and z_- are the valences of the cation and anion and a is the distance of closest approach of the ions in solution. Since this latter quantity cannot be experimentally determined, various assumptions have been made as to its probable magnitude.[1151]

3.1. Single Ion Properties

Physical studies of electrolyte solutions are complicated by the fact that one is inevitably concerned with systems of at least three components, and frequently the solvent interactions of the anion and cation differ not only in magnitude but also in kind. For the sake of clarity, therefore, it is desirable to study independently the various ionic species in solution. There are a limited number of ways in which this can be done rigorously, e.g., by NMR methods, one can observe single ionic species such as ^{23}Na, ^{39}K, ^{35}Cl, or ^{79}Br. The disadvantage of the technique lies in its relatively low sensitivity which makes it necessary for fairly concentrated solutions to be used. This in turn leads to results in which ion–ion and ion–solvent effects are superimposed. Infrared methods do not suffer from this drawback but they can be applied only to polyatomic ions.

Absolute ionic quantities which can be obtained experimentally include the mobility (from conductance and transference number measurements), the partial molal volumes (from ultrasonic vibration potentials),[1454] and the self-diffusion coefficient (from isotopic tracer diffusion studies). In these cases measurements can be performed in very dilute solution, approximating the infinite dilution condition. There exist several methods for evaluating relative ionic quantities; thus standard partial molal entropies of ions can be referred to $\bar{S}_i(\text{H}^+\text{aq}) = 0$, or the assumption is sometimes made that K$^+$ and Cl$^-$, which are isoelectronic and approximately of the same size, contribute equally to measured physical properties of KCl(aq).[335,347,521]

Where acceptable models exist for ionic solution processes experimental data can be fitted to equations and single ion quantities derived. In this way the standard Gibbs free energy change of solvation of an ion can be evaluated in terms of the Born equation, in which the variables are the ionic radius r and the permittivity of the solvent medium, i.e.,

$$\Delta G^\ominus = (ze^2/2r)[1 - (1/\varepsilon)] \tag{17}$$

where e is the ionic charge. This method has been extensively applied in studies of the changes in thermodynamic quantities arising from the transfer of ions from some standard solvent medium to water or an aqueous mixture.[155,156,518,533,596,800,1437] Whereas the free energy of transfer $\Delta G_t^\ominus$ is generally found to be a well-behaved function of r, marked deviation from such simple relationships are found in $\Delta H_t^\ominus(r)$ and $\Delta S_t^\ominus(r)$ which apparently cancel out in the free-energy term (see Fig. 12).

3.2. Ion–Water Interactions

A wealth of experimental data exist concerning the thermodynamic and spectroscopic properties of ionic solutions but their exact interpretation has highlighted many problems which are fully discussed in Volume 3. Spectroscopic studies of the aqueous component are complicated by marked effects which ions exert on the type of long-range order existing in un-

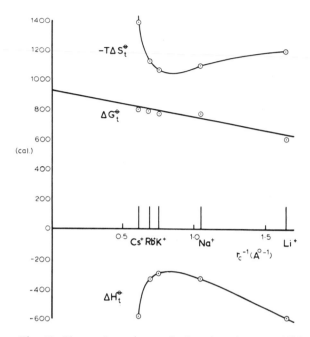

Fig. 12. Thermodynamic transfer functions (water → 20%
aqueous dioxan) of alkali metal ions. $\Delta G_t^{\ominus}$ apparently
obeys eqn. (17), but this is not the case for $\Delta H_t^{\ominus}$ or $\Delta S_t^{\ominus}$.
Reproduced from Ref. 155 with the author's permission.

perturbed water.[910] The measured spectral band shapes, frequencies, and
intensities, whether derived from infrared, Raman, NMR, or neutron
inelastic scattering experiments, are therefore made up of superimposed
contributions from unperturbed water and water affected by cations and
anions, respectively.

A simple dynamic picture of ion hydration was advanced by Samoilov[1176] based on exchange rates of water molecules between bulk water and
the ionic hydration envelope. If τ_0 is the diffusional period of a given water
molecule in bulk water and τ_i the residence time of the water molecule in
the neighborhood of an ion, then the term positive hydration was applied
to solutions for which $\tau_i/\tau_0 > 1$ (e.g., Li$^+$, Mg^{2+}, Al^{3+}), and the reverse
effect was termed negative hydration (e.g., Cs$^+$, I', ClO$_4'$). As a first approximation the magnitude of τ_i/τ_0 is proportional to the ionic surface
charge density and can be correlated with many observed effects exerted
by ions on solution, as well as with colloidal and biological processes.

Another ionic solution model based primarily on the partial molal
entropies of ions in infinitely dilute solution considers three distinct re-

gions[512]: In the neighborhood of the ion strong ion–dipole forces result in an oriented shell of water molecules; this is usually referred to as electrostrictive hydration. Far away from the ion the normal, predominantly tetrahedral order is largely intact, but there is an intermediate region in which the mutually incompatible effects of sp^3 tetrahedral hydrogen bonding and the radial Coulombic field interact to produce a random orientation of water molecules. The observed properties of solutions reflect the relative importance of electrostriction and randomization. The terms "structure making" and "structure breaking" have been applied to characterize the effect of ions on the observed properties of water. Care must be exercized in the use of such terms, for certainly the fairly long-lived electrostrictive hydration region in the vicinity of an Li^+ ion (normally classed as a structure promoter) is quite dissimilar from a type of hydrogen-bonded structure which exists in unperturbed water. The two models briefly referred to, although mutually compatible, still suffer the disadvantage of being only qualitative so that they cannot provide direct information about the effective radius of the "hydrated" ion, a quantity which is of such great importance in all phenomena involving ions. However, on a qualitative basis the models can account for the low mobility of Li^+ and for the low viscosity (lower than that of pure water) of K^+, Rb^+, and Cs^+ salts.

Problems arise when one tries to turn these or other models into quantitative molecular descriptions of ion–solvent effects. Volume 3 will illustrate that this subject is still in its early stages of development. Indeed, while there is still uncertainty about the correct assignment of observed spectral features, detailed interpretations of spectra can be at best speculative. Thus, although much more is known about electrostatic interactions between ions and dipoles than about intermolecular effects in nonelectrolyte mixtures, nevertheless only very limited success has been achieved in the inclusion of solvation effects in a general description of aqueous electrolytes.

The weaknesses of existing models and theories become very clear when one considers ionization phenomena

$$HA(aq) \rightarrow H^+(aq) + A'(aq)$$

There is an associated solvent reaction of the kind

$$xH_2O^{HA} + (y + z - x)H_2O \rightarrow yH_2O^{H^+} + zH_2O^{A'}$$

for any "simple" process and the problems posed in a separation of the reaction and hydration contributions have been analyzed by Ives and

Marsden.[401] For any process involving molecular as well as ionic hydration, complexities are likely to arise in the meaningful interpretation of emf or conductance data. Thus in a comparison of the ionization functions of acetic acid with those of a series of alkyl-substituted acids of the type $CH_3CH_2CHR_1CHR_2CR_3R_4COOH$, differences in $\Delta H^{\ominus}$ of up to 750 cal mol^{-1} have been found, depending on the position of the substituent group.[461] With the present state of knowledge there is as yet no reliable way of predicting the effect of alkyl substitution at a particular carbon atom on the thermodynamic functions of ionization.

The hydration of ions is also likely to affect their transport behavior in solution. The equations derived by Onsager, Fuoss, and many of their collaborators have provided a framework, of increasing sophistication and complexity as the years have passed,[538] which enables ionic mobilities to be accounted for in terms of a limiting term, reflecting ion–solvent interactions, and various concentration-dependent terms, reflecting interionic effects *only*. Basically, the Stokes concept of a sphere diffusing in a dielectric continuum has been employed. For a long period this approach has been very successful, but as physical techniques and concepts have become more refined the assumptions and approximations inherent in the Stokes–Onsager–Fuoss treatment and all its extensions have acted as a straitjacket inhibiting further progress. Spectroscopic studies (see Volume 3, Chapters 5–7) have shown beyond any doubt that both the intramolecular and intermolecular vibrations of the solvent molecules and their translational and rotational diffusional modes are severely perturbed in the presence of ions and these perturbations produce a feedback on the behavior of the ion in solution.

The weakness of the "structureless dielectric solvent" approximation becomes particularly serious when ionic transport in mixed aqueous solvents is considered. Thus an analysis of the conductance of KCl in water/dioxan mixtures suggests[846] that the a parameter in the conductance equation (i.e., the center-to-center distance for contact between rigid charged spheres in a continuum) increases from 2.84 to 5.61 Å as the dielectric constant of the solvent medium decreases from 78.54 to 12.74. The systematic studies of Kay and his colleagues have also brought to light some inconsistencies in the classical treatments (see Chapter 4). Most of these can be qualitatively resolved by taking into account order–disorder processes in the aqueous solvent media, and this would appear to be at least as realistic as is the manipulation of adjustable parameters. To cite Fuoss: "If one is determined to hold to the model at all costs, then one must conclude that the ions get bigger as ε decreases, say due to increased solvation (or some other rationalization). The alternative seems more reasonable; to conclude that the variable a's show that the sphere-in-continuum model is inadequate."[538]

3.3. Interionic Effects

Because of the long-range nature of Coulombic interactions the properties of ionic solutions at all concentrations, except those which are hardly accessible to experiment, will be influenced by interionic effects. In dilute solutions ($<0.01\ M$) these are adequately accounted for by the Debye–Hückel limiting law, so that we can infer that ions do not approach one another so closely that their hydration envelopes interfere with each other. Such interference would introduce an extra term into the activity coefficient expression which could not be described in terms of the normal electrostatic potential function. At higher concentrations the behavior of ionic solutions deviates from the simple concentration dependence ($c^{1/2}$) predicted by the limiting law. Various models have been advanced to account for the observed behavior, ranging from quasilattice distributions of ions (predicting a $c^{1/3}$ dependence of activity coefficients)[511] to the addition of "solvation number" terms to eqn. (16).[1185] Conceptually the most satisfying approach is that of Gurney,[609] which ascribes to each ion a hydration cosphere and accounts for deviations in the Debye–Hückel concentration dependence in terms of the overlap and interference of these cospheres. This model can account, although only qualitatively, for the observed trends in $\log f_{\pm}$ versus $c^{1/2}$ plots of different series of electrolytes and it also explains salting-in and salting-out effects. Friedman's recent reexamination of electrolyte solutions (see Volume 3, Chapter 1) in terms of several potential functions, including one for ion–solvent interactions, has transformed the Gurney concept into a semiquantitative tool, in that his $\log f_{\pm}$ versus c expressions contain only one empirical parameter, namely the coefficient of the "Gurney potential function.".In this way Friedman has been able to account for the experimental $\log f_{\pm}$ versus c curves over an extended concentration range without having to use values for the Gurney coefficients which are inconsistent with spectroscopic data on ion hydration.

A more stringent test for the validity of a model is its capacity to account for derivatives of $\log f_{\pm}$, e.g., partial ionic entropies and enthalpies. The free energy function, $\mu_i^{E} = RT \log f_{\pm}$ is quite insensitive to structural changes in the solvent medium, and those investigators who choose not to emphasize solvent effects need only restrict their measurements to free energies.[1150] This is illustrated in Fig. 13, which summarizes the effects of D_2O substitution on excess thermodynamic functions of alkali halide

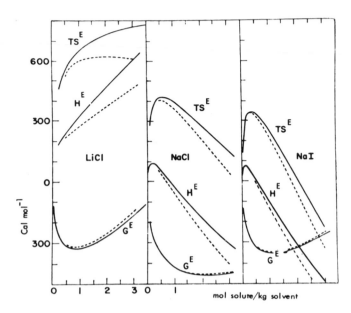

Fig. 13. Thermodynamic excess functions of alkali halides in H_2O (——) and D_2O (– –), demonstrating the insensitivity of ΔG^E to the nature of the solvent. Reproduced from Ref. 1437 with the authors' permission.

solutions.[1437] It is seen that G^E is quite nondiscriminating up to fairly high concentrations ($\sim 2\ M$), whereas H^E and TS^E are very sensitive to the nature of the solvent. The importance of hydration sphere overlap in determining ionic excess entropies is shown in Fig. 14; even the "improved" Debye–Hückel equation can adequately account for $\bar{S}_2{}^E$ only at very low concentrations ($<0.05\ M$).[506] Detailed molecular information about ion hydration envelopes in terms of the mobility of H_2O molecules has been provided by NMR proton relaxation rate measurements[454] (see Volume 3, Chapter 7) and this should provide a means of checking the predictions of recent developments in models of solvated ion interactions in solution.

4. COMPLEX AQUEOUS MIXTURES

It is now quite apparent that one way of subtly altering the aqueous solvent medium is by the addition of small amounts of cosolvents which can either promote or diminish the degree of long-range order in water or indeed promote a different kind of order from that which is characteristic of pure water. This approach is now often employed in comparative studies

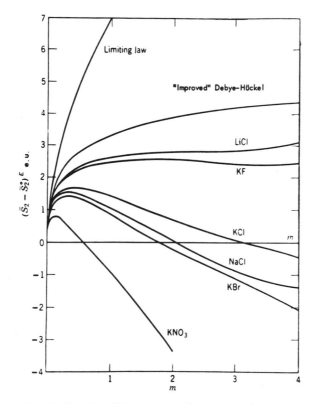

Fig. 14. Partial molal entropies of electrolytes in aqueous solution at 25°C as function of concentration. The curves predicted by the limiting and extended Debye–Hückel equations cannot account for the observed decreases of $(\bar{S}_2 - \bar{S}_2^\circ)^E$ with concentration. After Frank.[506]

of processes in solution and it highlights the inadequacies of existing theories in explaining the observed effects. For the purposes of this introductory discussion it will suffice to select a few typical examples in order to show the effect of solvent participation on a range of physicochemical and biochemical processes.

The role of solvent in affecting the kinetics and mechanisms of reactions in solution has been receiving much attention and several theories of solvent participation have been advanced. Generally it is proposed that observed effects arise from solvation changes of the transition state and for reactions where the mechanisms are thought to involve ionic transition states a useful solvent parameter has been the dielectric constant. The solvolysis of t-butyl chloride has received particular attention because its mechanism is well

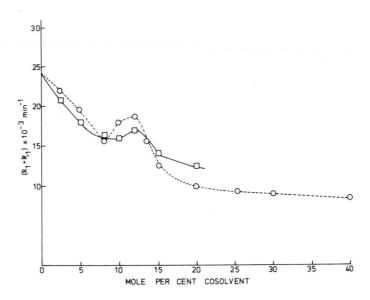

Fig. 15. Rate constants of mutarotation at 25°C of α-D-glucose in 0.1% aqueous solutions of *tert*-butanol (□) and tetrahydrofuran (○) as function of solvent composition (G. Livingstone, personal communication). The rate maxima correspond to the maxima in the X-ray scattering intensity (see Fig. 6) and the lower critical demixing composition.[78]

understood. Robertson first suggested that the large negative heat capacity of activation $\Delta C_P{}^{\pm}$ in aqueous solution should be attributable not to the transition state, but to changes in the structural integrity of the ground-state hydration shell with rising temperature.[948] The substitution of water by a mixed water/ethanol solvent results in complex changes in $\Delta C_P{}^{\pm}$, which reaches a sharp minimum of -166 cal mol^{-1} deg^{-1} in 10 mole % ethanol. As the ethanol concentration is increased to 15 mole %, $\Delta C_P{}^{\pm}$ increases to -49 cal mol^{-1} deg^{-1} and it reaches a limiting value of -34 cal mol^{-1} deg^{-1} in 25 mole % solution. Needless to say, this type of solvent composition dependence is not reflected in the dielectric constant.

Interesting solvent effects are also observed in the mutarotation of sugars, both in the kinetics and in the equilibrium relationships. Thus the rate of mutarotation does not decrease monotonically with increasing concentration of cosolvent.* In the presence of *t*-butanol or tetrahydrofuran the rate behavior observed is shown in Fig. 15. The maximum in the rate corresponds to the solvent composition at which small-angle X-ray scat-

* G. Livingstone, personal communication.

tering data indicate a maximum microheterogeneity (see Section 2.2). Like the scattering intensity, the maximum in the rate of mutarotation also becomes more pronounced with rise in temperature. Apart from affecting the rate, the solvent composition also affects mutarotation equilibria. Thus, for the case of glucose and fructose, the presence of alcohol in the solvent favors β-sugar formation, whereas the equilibrium is shifted toward the α-anomer in the case of galactose and maltose.[593] The equilibrium shifts are proportional to the solvent composition but are independent of the concentration of sugar.

Acid–base equilibria are also very sensitive to solvent composition in the dilute aqueous range, and it has been shown that for a series of isodielectric aqueous solvents the protonation of amines of triphenyl-carbinol depends on the nature of the cosolvent, and that for any one given solvent mixture pK is not a linear function of ε^{-1}.[1143] Experiments on the dissociation of neutral and cationic acids in a series of high dielectric solvents (water/N-methylacetamide) have also shown that ε is not the fundamental variable in determining the solvent effect on ionic dissocia-tion.[1144] On a more fundamental basis the changes accompanying the transfer of an ion from water to a mixed solvent have been investigated in terms of the behavior predicted by eqn. (17). As a result of systematic emf and calorimetric experiments Feakins concluded that, apart from the electrostatic contribution, a structural component must be considered.[155,473] Table IX summarizes the standard enthalpies of transfer of alkali halides from water to a 20% aqueous dioxan solution and compares the experi-mental values with those calculated by means of eqn. (17) (i.e., by allowing for electrostatic effects only). The disagreement between the theoretical and experimental values increases as the ionic size decreases.

The deviations from classical electrolyte theory which are encountered with mixed solvents have at times been interpreted in terms of specific solvation effects. If this were indeed the case, then ions such as Li^+ and I^- would be expected to exhibit rather different types of behavior. By a com-bination of the $\Delta G_t^{\ominus}$ measurements of Feakins et al.[474] for the transfer of alkali halides from water to water/MeOH mixtures with older calorimetric $\Delta H_t^{\ominus}$ data,[1227] the partial ionic entropies of solvation can be estimated,[521] as shown in Fig. 16. It is seen that although $\bar{S}_i(H_2O)$ differs for the various ionic species, the effect produced by the addition of MeOH is the same for all the ions. Further, $\bar{S}_i(H_2O/MeOH)$ does not follow the dielectric constant of the solvent mixture, but rather its ΔS^M, a property which is considered to be an index of the degree of long-range order (see also Fig. 5). If $\bar{S}_i$ provides an indication of the net effect of the ion on the structural integrity

TABLE IX. Experimental and Calculated (Born Model) Limiting Enthalpies of Transfer of Alkali Halides from Water (w) to 20% aqueous Dioxan (DI)[474]

$$\Delta H_t^{\ominus} = \Delta H_{DI}^{\circ} - \Delta H_w^{\circ}$$

$$\Delta H_t^{\ominus}(\text{calc}) = \frac{Ne^2}{2}\left(\frac{1}{r_+} + \frac{1}{r_-}\right)\left[\varepsilon_{DI}^{-1} - \varepsilon_w^{-1} + T\left(\frac{1}{\varepsilon_{DI}^2}\frac{\partial \varepsilon_{DI}}{\partial T} - \frac{1}{\varepsilon_w^2}\frac{\partial \varepsilon_w}{\partial T}\right)\right]$$

	$-\Delta H_t^{\ominus}$, cal mol^{-1}	
Salt	Experimental	Calculated
LiCl	585	1540
NaCl	330	1120
KCl	295	910
RbCl	335	860
CsCl	585	810
NaBr	745	1090
NaI	1165	1050
KBr	730	880
KI	1135	850

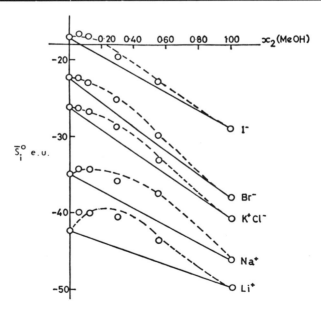

Fig. 16. Limiting partial ionic entropies of alkali metal and halide ions in H$_2$O/MeOH mixtures as function of solvent composition.[521] Qualitatively the same behavior is observed for strongly (Li$^+$) and weakly (I$^-$) hydrated ions.

(degree of association, hydrogen bonding, etc.) of the solvent, as discussed by Criss et al.[347] then the results in Fig. 5 suggest that mixed solvents behave exactly like pure solvents, in that $\bar{S}_i$ reflects only the internal order of the solvent and the ion cannot "see" two different molecular species. It is not surprising, therefore, that the observed thermodynamic (or molecular relaxation) properties do not follow the bulk dielectric constant.*

The theories of hydrophobic colloid stability and of electrolytes in solution are closely related and it is therefore likely that anomalies might be found in mixed aqueous–organic dispersions of hydrophobic sols. By application of the classical DLVO theory of colloid stability, it is possible to calculate in terms of the properties of the electrical double layer surrounding the disperse particles the concentration of univalent electrolyte required to flocculate a simple hydrophobic sol.[1220] The experimental results more or less bear out the predictions, although specific ion effects have been reported and rationalized in terms of the tendencies of various ions to approach more or less closely to the charged particle surface. However, flocculation experiments with polystyrene latices in water/ethanol mixtures have shown that initial additions of ethanol stabilize the sol against flocculation by NaCl and that a maximum stability is achieved in 8 mole % EtOH, although the dielectric constant of this solvent is 10% lower than that of water.[1348]

Some of the most striking effects of aqueous–organic solvent mixtures are encountered in the physical chemistry of processes involving biopolymers, such as enzyme–substrate interactions, protein denaturation, gelling phenomena, etc. A good example is the thermal denaturation of ribonuclease, which has been studied as a function of solvent composition in water/ethanol mixtures. The process is pH-dependent and initial experiments performed at 45–63° suggested that ΔH of denaturation was independent of the alcohol concentration,[1205] from which it was concluded that stoichiometric binding of alcohols to the protein was responsible for the decreased protein stability.† A more thorough investigation[215] covering a temperature range of 10–50° has shown that at low temperatures the thermodynamic functions of denaturation exhibit extrema at 15% ethanol (8 mole %). In particular, ΔH changes by 60 kcal mol^{-1} between 0 and 5 mole % ethanol. Similar results have also been reported for glutamate

---

* Even the *dielectric* relaxation time of water in dilute aqueous solutions of alcohols does not show the same concentration dependence as the static dielectric constant.
† It is still the widely held view that "preferential binding" is the main mechanism by which biopolymers can be made to undergo conformational changes.[250,1293]

decarboxylases: at $0°$ this enzyme is stabilized by low concentrations of ethanol which at $25°$ have a destabilizing effect. In this connection it is also of interest to note that the renaturation of trypsin at low temperatures is accelerated by low concentrations of ethanol.[1107] The influence of urea on ribonuclease is totally different;[215] at all concentrations the tendency is toward destabilization and ΔG is a linear function of urea concentration, whereas ΔC_P is constant over the range of urea concentration $0-4\ M$, indicating that thermally labile structures are not involved. This probably also applies to the stabilizing effect on ribonuclease and chymotrypsinogen A by sorbitol, erythritol, and glycerol (in that order).[549,550]

Finally, solvent medium perturbation experiments[1384] in an even more complex process, i.e., the binding of inhibitors to, and isomerization of substrates by, a steroid isomerase enzyme have also indicated that enzyme–substrate binding is critically affected by the solvent structural integrity.

The whole field of solvent effects in biopolymer conformational behavior is certain to receive more attention in the future as more subtle industrial processing methods for proteins, carbohydrates, lipids, and their complexes become increasingly important.

CHAPTER 2

Water in Stoichiometric Hydrates*

Michael Falk

Atlantic Regional Laboratory
National Research Council of Canada
Halifax, Nova Scotia, Canada

and

Osvald Knop

Department of Chemistry
Dalhousie University
Halifax, Nova Scotia, Canada

1. INTRODUCTION

This chapter is concerned with information obtainable from crystallographic and spectroscopic studies of the water molecule in stoichiometric hydrates.[†] Water molecules in a hydrate crystal interact with their surroundings through hydrogen bonding and cation-water coordination. We shall deal with these interactions and their consequences.

Stoichiometric hydrates may be either *true hydrates* or *pseudohydrates*. In true hydrates water is present as recognizable H_2O molecules ("water of crystallization"), while pseudohydrates contain water as hydroxyl or hydroxonium ions or as —OH and —H groups ("water of constitution"). Water of crystallization can be further classified as "water of coordination," which forms part of the coordination sphere of a cation, and "lattice water," which does not.[239] Suitable structural formulas may be written to indicate how many water molecules of each type are present. For example, the heptahydrate and the hexahydrate of magnesium sulfate, which both contain six molecules of coordination water per Mg^{2+}, can be represented

---

* NRCC No. 12257.

[†] Zeolitic hydrates are not included; clathrate hydrates are treated in Chapter 3 of this volume.

as $[Mg(OH_2)_6][SO_4] \cdot H_2O^{(112)}$ and $[Mg(OH_2)_6][SO_4]$,[1453] respectively. Where one stoichiometry corresponds to isomers differing in the ratio of coordination water to lattice water, the use of structural formulas is essential to avoid ambiguity. Thus, three different hexahydrates of chromic chloride are represented by the empirical formula $CrCl_3 \cdot 6H_2O$: the hexaquo complex $[Cr(OH_2)_6]Cl_3$, the pentaquo complex $[Cr(OH_2)_5Cl]Cl_2 \cdot H_2O$, and the tetraquo complex $[Cr(OH_2)_4Cl_2]Cl \cdot 2H_2O$.[1386]

Crystallographic studies of hydrates have been the subject of earlier reviews.[114,159,288,291,303,626,627,942,1209,1302,1386] Although some of these reviews are fairly recent, the more plentiful and reliable results of neutron (and to some extent X-ray) diffraction studies now available make it possible to improve classification and allow the relevant information to be assessed on statistical criteria. As for vibrational spectroscopy, the application of the isotopic-dilution technique to the study of the water molecule in hydrates has not been reviewed elsewhere.

Wideband nuclear magnetic resonance is also an important method for such studies. However, it has been reviewed extensively by Reeves[1138] and will therefore not be dealt with here. While thermodynamic and other static and dynamic properties of hydrates are relevant to our understanding of the behavior of the H_2O molecule in hydrates, they are outside the scope of this review.*

Our survey of the crystallographic literature extends to December 1970 and is limited largely to inorganic hydrates. Hydrates of organic compounds are included only if their structures have been studied by neutron diffraction or if they derive from the simplest organic acids (formates, acetates, oxalates).† We have located 282 hydrates with structures determined by diffraction methods in sufficient detail to reveal clearly the site symmetry and coordination of the H_2O molecules. These hydrates contain 663 crystallographically distinct H_2O molecules.

Our survey of the spectroscopic literature extends to July 1971. Eighty-six hydrates have been found for which infrared (and in many cases also Raman) spectra of good quality are available. This aggregate contains 54 D_2O and 38 HDO spectra.

* Physical properties in general have been reviewed by Wooster,[1431] dynamics of hydrogen atoms in crystals by Hamilton,[625] thermodynamic and transport properties by Barrer,[100] and lattice energetics and thermodynamic stability by Ladd and Lee[819-821] and by Ladd.[818]

† Organic hydrates have been reviewed by Clark[303] and by Jeffrey.[705] The classes of compounds which have been studied most extensively are sugars and other polyhydroxy compounds, nitrogen bases, amino acids and small peptides, carboxylic acids and their salts, and coordination compounds of organic ligands with heavy metals.

2. SYMMETRY AND TYPES OF ENVIRONMENT OF THE H_2O MOLECULE IN CRYSTALS

2.1. Site Symmetry

The free water molecule has the symmetry C_{2v}.* In the absence of orientational disorder or rotation, a water molecule in the crystal can therefore occupy a site having the symmetry of the point group C_{2v} or one of its subgroups C_2, C_s (two possible orientations), C_1.

Only 11 out of the 663 molecules covered by our survey occupy sites of symmetry C_{2v}. There are 34 cases of C_2 symmetry. In 18 of the 46 cases of C_s symmetry the mirror plane contains the H_2O molecule, in 22 cases it is perpendicular to the HOH plane, and in six cases the orientation with respect to the mirror plane is uncertain. Apart from the two cases, discussed below, of H_2O site symmetry involving a fourfold axis, the remaining 570 molecules (86% of the total) occupy sites of no symmetry, C_1. Most water molecules are thus asymmetric in principle, but the actual extent of distortion, as measured, for example, by the difference of the two OH bond lengths, is quite small—of the order of 1% (see Sections 3.4.2 and 4.3.3).

The two hydrates in which water molecules have been reported to occur at sites of fourfold symmetry are $K_2MnCl_4 \cdot 2H_2O$ (C_{4v})[711] and $Mo_6Br_{12} \cdot 2H_2O$ (C_4).[606] The water oxygens in these structures are located on fourfold axes with the HOH planes distributed statistically between two mutually perpendicular orientations. The occurrence of proven statistical disorder in true hydrates is thus quite rare. Additional, less certain examples of H_2O molecules stated to occupy sites of symmetries higher than C_{2v} have been recorded: D_{4h} in $K_4Ru_2OCl_{10} \cdot H_2O$[952] and $K_4Re_2OCl_{10} \cdot H_2O$,[1003] D_{2d} in $SrPb_2I_6 \cdot 7H_2O$,[478] C_{4v} in $K_3TlCl_6 \cdot 2H_2O$[675] and $NiCl_2 \cdot 4H_2O$,[1260] and C_{4h} in $KSnF_3 \cdot \frac{1}{2}H_2O$.[158]

2.2. Consequences of the Low Symmetry of the H_2O Molecule

Since the actual symmetry of the water molecule cannot be higher than C_{2v} (barring disorder or rotation), this requirement restricts the crystal symmetry of the hydrate. The crystallographic equivalence, or its lack, of the water molecules contained in a unit cell imposes further restrictions. For example, in gypsum, $CaSO_4 \cdot 2H_2O$, there are two H_2O molecules per

---

* To save space, only the Schoenflies symbols will be used to denote molecular and site symmetries. Explanation of symbols and a conversion table (Schoenflies to international notation) will be found, for example, in Ref. 696.

Ca atom (from chemical formula) and four Ca atoms in the monoclinic unit cell (from density). Assuming that gypsum does not have a defect structure and that the eight H_2O molecules in the unit cell are crystallographically equivalent (cf. Section 4.2), the admissible monoclinic space groups must contain at least one equipoint of multiplicity eight and site symmetry C_1, C_2, or C_s. Examination shows that this requirement is met in only two of the 13 monoclinic space groups, C_{2h}^3 and C_{2h}^6. Each contains one such equipoint of site symmetry C_1, so that the H_2O molecule can only be asymmetric. The actual space group of gypsum is C_{2h}^6. Another example is the tetragonal $K_2CuCl_4 \cdot 2H_2O$, whose unit cell contains four H_2O molecules. If we choose to assume that all four molecules are crystallographically equivalent and of symmetry C_{2v}, this condition can be realized in only 15 of the 68 tetragonal space groups (C_{4v}^{9-11}, $D_{2d}^{9,11}$, $D_{4h}^{1,3,5,7,9,10,12,14-16}$), and in eight of these uniquely. Symmetry considerations of this kind, where they apply nontrivially, provide a check on the correctness of the space group and crystal structure determination of the hydrate.

While the low symmetry of the water molecule is not the only factor, the effect of the presence of H_2O in a true hydrate is in general reflected in the lowering of the symmetry of the anhydrous crystal on hydration. Our survey contains 53 hydrates for which the crystal structures of the corresponding anhydrous compounds are known. In only seven cases is the point group symmetry of the hydrate higher than that of the anhydrous compound. In 41 cases it is lower, and in five cases it remains unchanged.

There is yet another consequence of the particular shape and symmetry of the H_2O molecule. Water in true hydrates is often found as water of coordination in fully hydrated $M(OH_2)_n^{m+}$ aquo complexes, where M is the central cation and n has most commonly the value of six. The highest symmetries which these complexes may attain in crystals are lower than those ideally possible for MX_n configurations involving monatomic ligands. Specifically, an $M(OH_2)_n$ ($n = 4, 8, 10$) complex in a crystal cannot be situated at a site of strictly cubic symmetry. There is no way of arranging the four, eight, or ten internal H—H vectors of the H_2O molecules around M to yield coordination polyhedra of cubic symmetry when the hydrogen atoms are essentially localized, i.e., in "fixed" positions. If the orientations of these vectors are completely randomized over the crystal, $M(OH_2)_4$ and $M(OH_2)_{10}$ may attain an apparent (diffraction) symmetry as high as T_d and $M(OH_2)_8$ one as high as O_h. The use of a symmetry sensitive local probe at the M site, such as a Mössbauer- or NMR-active atom, may however reveal the true symmetry or distribution of symmetries.

In $BeSO_4 \cdot 4H_2O = [Be(OH_2)_4][SO_4]$ the symmetry of the tetraquo complex is S_4[125,358,1224] but the metric distortion from a regular tetrahedral arrangement is relatively small. The H_2O molecules in the complex are asymmetric. No $M(OH_2)_8$ complex containing the water molecules at the corners of a cube seems to be known. In $SrO_2 \cdot 8H_2O = [Sr(OH_2)_8][O_2]$ (space group D_{4h}^2—$P4/mcc$) the coordination figure is a square anti-prism[1330] However, the symmetry of the *ideal* square antiprism is non-crystallographic, D_{4d}—$\bar{8}2m$, and thus incompatible with any space group. The symmetry reported for the $Sr(OH_2)_8$ complex is D_4, the water molecules being asymmetric.

An $M(OH_2)_6$ complex with essentially localized hydrogen atoms can, in a true hydrate, attain at most one of the following symmetries: T_d, T_h, T. The six pairs of hydrogen atoms cannot define a polyhedron of symmetry O_h or O, although these could be the apparent (diffraction) symmetries if the directions of the H—H vectors were completely randomized over the crystal. To satisfy the particular symmetries T_d, T_h, or T the 12 hydrogen atoms would have to be arranged, respectively, at the vertices of a truncated tetrahedron (the dual of the tristetrahedron), an icosahedron of symmetry T_h [the dual of the crystallographic (cubic) pentagonal dodecahedron], and an icosahedron of symmetry T (the dual of the tetartoid). In the alums (space group T_h^6—$Pa3$) both the $M^I(OH_2)_6$ and the $M^{III}(OH_2)_6$ complexes are trigonally distorted octahedra (S_6) and the water molecules are asymmetric.[71,348,349]

In the arrangement of highest symmetry possible for $M(OH_2)_{12}$ the localized hydrogen atoms would be situated at the vertices of a truncated octahedron of symmetry O_h or O.

While arrangements of high symmetry are geometrically possible, they may be energetically unfavorable because of repulsions between H atoms or between the lone pairs of electrons, and of course they would have to be compatible with the overall lattice energy requirements for the hydrate crystal.

2.3. Types of Environment

Water molecules in hydrates may be classified according to the number of nearest H_2O neighbors. Such a classification has been proposed by Wells (Ref. 1386, p. 577). A slightly modified version is shown in Fig. 1 for a water molecule in tetrahedral environment consisting of two electron acceptors and two proton acceptors.

An "isolated" water molecule has no water molecules among its immediate neighbors. Water molecules with one water neighbor may be of

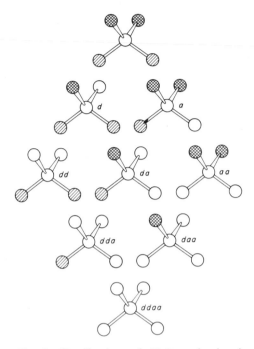

Fig. 1. Classification of H_2O molecules in tetrahedral surroundings according to their nearest neighbors. Line-shaded circles, electron-acceptor groups (OH, NH, or metal cation); cross-shaded circles, proton-acceptor groups other than water; unshaded circles, water oxygens (either proton donors or proton acceptors). The top configuration represents an "isolated" water molecule.

type *d* (proton donor) or *a* (proton acceptor). When all the water molecules in a crystal have one water neighbor each, H_2O pairs result. Each pair contains one H_2O of type *a* and one of type *d*. Three types of water molecules with two H_2O neighbors are possible, *dd*, *da*, and *aa*. When every water molecule in the structure has two H_2O neighbors an infinite unbranched chain or a ring of three or more H_2O molecules may form. For every molecule of type *dd* there must be a complementary molecule of type *aa*.

A water molecule with three H_2O neighbors can be either of type *dda* or *daa*. When all the water molecules belong to these two complementary types an infinite two-dimensional network results. Finally, when all the water molecules are of type *ddaa*, with four H_2O neighbors, an infinite three-dimensional network results. All of the types shown in Fig. 1 have been found in hydrates.

An analogous classification can be made for water molecules in trigonal and other, less common environments. Of the 12 possible types of trigonal environment three occur in the sample surveyed here. They satisfy the requirement that each water molecule have two proton-acceptor neighbors.

A classification of water environment in hydrates based on the number and type of lone-pair ligands of water molecules has been proposed by Chidambaram et al.[288] (cf. also Section 3.3).

2.4. Water Networks in Hydrates

When several types of water molecules occur in the same structure finite or infinite hydrogen-bonded H_2O networks may be formed. In surveying the networks which have been found to occur in hydrates included in this review (Figs. 2 and 3) it will be convenient to refer to pairs of H_2O

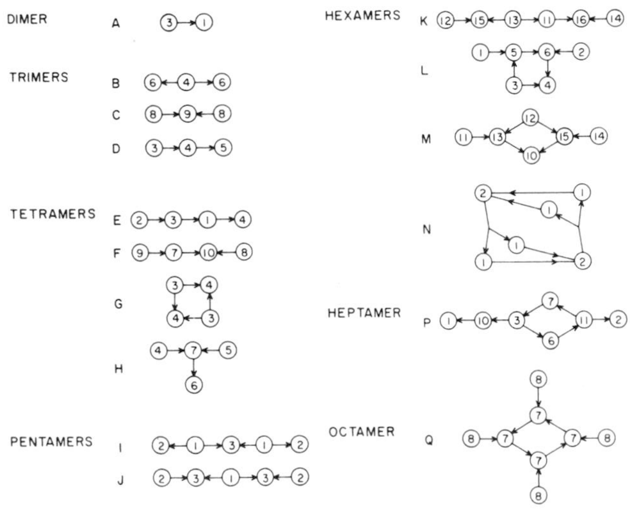

Fig. 2. Examples of finite networks of water molecules in hydrates. The numbering of the water molecules (circles) follows the original reference. Hydrogen bonds (arrows) point from donor to acceptor. (A) $MgS_2O_3 \cdot 6H_2O$.[77] (B) $Na_2Ca(CO_3)_2 \cdot 5H_2O$.[423] (C) $2Cu(NO_3)_2 \cdot 5H_2O$.[998] (D) $CaCO_3 \cdot 6H_2O$.[424] (E) Yugawaralite, $Ca_2Al_4Si_{12}O_{32} \cdot 8H_2O$.[752] (F) $3CdSO_4 \cdot 8H_2O$.[859] (G) $NaOH \cdot 4H_2O$.[171] (H) $FeSO_4 \cdot 7H_2O$.[113] (I) $KCr(C_2O_4)_2 \cdot 5H_2O$.[1331] (J) $Th(NO_3)_4 \cdot 5H_2O$.[1282] (K) Vauxite, $FeAl_2(PO_4)_2(OH)_2 \cdot 6H_2O$.[118] (L) $MgCl_2 \cdot 12H_2O$[1184] and $Zr(SO_4)_2 \cdot 7H_2O$.[121] (M) $Fe_2(SO_4)_2O \cdot 7H_2O$.[1269] (N) Cyclic hexamer held by bifurcated bonds, $NiCl_2 \cdot 6H_2O$.[765] (P) $Na_5P_3O_8 \cdot 14H_2O$.[993] (Q) Meta-torbernite, $Cu(UO_2PO_4)_2 \cdot 8H_2O$.[1159]

molecules hydrogen-bonded together as "dimers," to groups of three as "trimers," etc. However, it must be borne in mind that such groups of water molecules, while isolated from each other, are linked to the surrounding atoms or ions by forces at least as strong as the hydrogen bonds within the groups. Also, the $H_2O \cdots H_2O$ distances in question vary widely, some of them corresponding to very weak hydrogen bonds.

Dimers are very common, occurring in 33 out of the 251 hydrates of our sample; five of these 33 hydrates contain two dimers. Trimers are almost equally common. They are found in 23 hydrates, three of which contain two trimers. In a trimer the central H_2O molecule may donate two hydrogen bonds (one case), accept two hydrogen bonds (17 cases), or donate one and accept one hydrogen bond (eight cases).

Higher "oligomers" are far less common. Only two of the four distinguishable linear tetramers have been observed, two of the ten linear pentamers, two of the sixteen linear hexamers, and none of the thirty-six linear heptamers. All of the cyclic oligomers involve four-membered rings approximately square in shape.

Figure 3 shows examples of H_2O networks of infinite size. For most of these only one or two examples occur in unrelated structures among the hydrates examined. An exception is the simplest linear chain in which all water molecules are equivalent. It is found in six unrelated structures. Numerous three-dimensional H_2O networks occur in ices and clathrates (cf. Chapter 3 of this volume).

The variety of H_2O networks found in hydrates may be used as a basis of classification, in a manner similar to that known from the crystal chemistry of the silicates. Bernal[159] distinguishes hydrates containing three-dimensional water lattices (*tectohydrates*), two-dimensional water layers

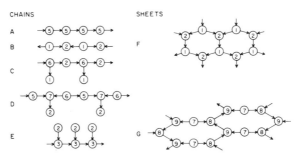

Fig. 3. Examples of infinite water networks in hydrates (cf. Fig. 2). (A) $Li_2SO_4 \cdot H_2O$.[1232] (B) $MoO_3 \cdot 2H_2O$.[787] (C) Euchroite, $Cu_2(AsO_4)(OH) \cdot 3H_2O$.[486] (D) $MgSO_4 \cdot 7H_2O$.[112] (E) Beraunite, $Fe^{II}Fe_5^{III}(OH)_5(PO_4)_4 \cdot 6H_2O$.[468] (F) $SnCl_2 \cdot 2H_2O$.[761] (G) $UO_2(NO_3)_2 \cdot 6H_2O$;[1281] a similar network occurs in paravauxite, $FeAl_2(PO_4)_2(OH)_2(H_2O)_6 \cdot 2H_2O$.[115]

(*phyllohydrates*), one-dimensional water chains (*inohydrates*), and isolated water molecules or small groups of molecules (*nesohydrates*). A weakness of Bernal's original classification is a lack of clear definition of what constitutes a link in the water lattice, layer, or chain. For example, $LiClO_4 \cdot 3H_2O$ and $CaSO_4 \cdot 2H_2O$ were classified respectively as an inohydrate and a phyllohydrate because in the first case the water molecules are linked into chains through Li^+ cations and in the second case into layers through hydrogen bonds to sulfate oxygens.[1386] If water molecules separated by ions are to be considered as connected, *every* hydrate may be classified as tectohydrate. Bernal's classification would be useful if only water molecules hydrogen-bonded to each other were considered linked.

3. INFORMATION FROM X-RAY AND NEUTRON DIFFRACTION

3.1. Quality of the Data

The quality of the structure determinations included in our survey varies widely. For X-ray diffraction results the estimated standard deviation σ on the distance between the water oxygen and its nonhydrogen neighbors varies from 0.001 to 0.1 Å, i.e., by a factor of 100. For neutron diffraction results the distances involving hydrogen atoms carry estimated standard deviations from 0.002 to 0.06 Å, i.e., varying by a factor of 30. This does not include the uncertainty due to the effect of thermal motion,[625] which may amount to an additional 0.03 Å or more.

It is convenient to divide the total of the 282 structures into three categories as shown in Table I, depending on the type and quality of the data. When σ is not stated and cannot be estimated the structure has been placed in the category of lower accuracy.

TABLE I

	Number of hydrate structures	Different H_2O molecules
Neutron data	41	69
High-quality X-ray data ($\sigma \leq 0.03$ Å)	151	412
Low-quality X-ray data ($\sigma > 0.03$ Å)	90	182
All data	282	663

TABLE II. Geometry of H_2O Molecules and Hydrogen Bonds in Hydrates—Neutron Diffraction Data[a]

Compound		Proton acceptor Y	Distances, Å			Angles, deg		
			O···Y	O—H	H···Y	O—H···Y	H—O—H	Y···O···Y
$AlCl_3 \cdot 6H_2O^{(242)}$		Cl	3.020(20)	1.040(40)	1.980(30)	180.0(30)	113.0(40)	105.2(2)
		Cl	3.030(10)	0.990(60)	2.050(60)	168.0(60)		
$BaCl_2 \cdot 2H_2O^{(1063)}$	$H_2O(1)$	Cl	3.182	0.967	2.237	165.4	105.8	111.2
		Cl	3.131	0.974	2.173	167.3		
	$H_2O(2)$	Cl	3.180	0.965	2.223	171.1	102.5	76.5
		Cl	3.238[b]	0.953	2.492	135.1		
$Ba(ClO_3)_2 \cdot H_2O^{(1225)}$	([e])	O	2.891(5)	0.958(11)[c]	1.991(10)	163.6(9)	110.7(14)	128.8(3)
$BeSO_4 \cdot 4H_2O^{(1224)}$		O	2.684(5)	0.971(8)	1.719(6)	172.3(6)	112.7(8)	114.6(2)
		O	2.617(5)	0.967(6)	1.656(8)	171.7(8)		
$C_4N_2O_4H_4 \cdot H_2O^{(345)}$ (Dialuric acid monohydrate)		O	2.830(7)	0.949[c]	1.913(10)	162.3(9)	—	87.9(2)
		O	2.834(7)	0.941[c]	1.914(7)	169.8(9)		
$C_4N_3O_4D_3 \cdot D_2O^{(346)}$ (Violuric acid monodeuterate)		O	2.756(8)	0.970(20)	1.820(20)	162.0(16)	106.1(15)	115.1(3)
		O	2.786(7)[b]	0.940(20)	2.070(20)	131.0(14)		
$C_5N_8H_{12} \cdot 2HCl \cdot H_2O^{(628)}$ (Methylglyoxal bisguanylhydrazone dihydrochloride monohydrate)		Cl	3.090(20)	1.020(30)	2.140(30)	153.0(21)	104.0(22)	126.5(5)
		Cl	3.150(20)	0.990(30)	2.190(30)	166.0(22)		
$\alpha\text{-}(COOH)_2 \cdot 2H_2O^{(1169)}$		O	2.864(5)	0.964(7)	1.917(8)	166.9(6)	105.9(7)	83.4(1)
		O	2.881(4)	0.956(9)	1.979(9)	156.6(7)		
$\alpha\text{-}(COOD)_2 \cdot 2D_2O^{(338)}$		O	2.879(2)	0.954(2)	1.939(2)	167.7(2)	105.8(2)	83.5(1)
		O	2.906(2)	0.954(2)	2.008(2)	156.0(2)		

Compound			O···X	O–H	H···X	∠	∠	∠
β-(COOD)₂ · 2D₂O(338)		O	2.855(2)	0.944(2)	1.960(2)	157.3(2)	108.8(2)	87.8(1)
		O	2.834(3)	0.947(2)	1.895(2)	170.3(1)		
CaSO₄ · 2H₂O(611) (Gypsum)		O	2.816(14)	1.002(30)	1.815(40)	177.7(31)	105.6(45)	104.6(19)
		O	2.824(15)	0.981(24)	1.843(121)	178.2(32)		
Cd(NO₃)₂ · 4D₂O(912)	D₂O(4)	O	2.967(17)	0.980(20)	1.990(18)	175.2(19)	103.2(25)	106.9(8)
		O	2.872(26)	0.936(26)	1.951(23)	167.5(17)		
	D₂O(5)	O	2.945(16)	0.965(15)	2.004(16)	164.6(14)	104.6(15)	103.2(5)
		O	2.865(17)	0.963(19)	1.917(17)	167.6(12)		
CoCl₂ · 2H₂O(341,996) (298°K)	(e)	Cl	3.222(7)	0.980(20)	2.250(20)	171.3(22)	103.2(10)	91.8(7)
CsAl(SO₄)₂ · 12H₂O(348) (CsAl alum)	H₂O(1)	O	2.822(6)	0.955(26)ᶜ	1.902(25)	165.0(20)	107.6(24)	97.1(2)
		O	2.766(7)	0.968(29)ᶜ	1.811(29)	171.0(20)		
	H₂O(2)	O	2.648(7)	0.974(26)ᶜ	1.692(27)	166.0(30)	107.2(20)	105.0(2)
		O	2.615(7)	0.995(36)ᶜ	1.657(33)	163.0(30)		
CuCl₂ · 2H₂O(455,1088)	(e)	Cl	3.201	0.954	2.258	170.7	110.0	97.3
CuF₂ · 2H₂O(4) (298°K)	(e)	F	2.649(6)	0.980(7)	1.687(8)	164.9(7)	115.5(4)	96.5(2)
CuF₂ · 2H₂O(2) (4.2°K)	(e)	F	2.652(5)	0.959(5)	1.707(6)	167.7(6)	110.1(4)	96.7(2)
CuSO₄ · 5H₂O(70)	H₂O(5)	O	2.830(30)	0.940(30)	1.910(30)	168.0(10)	111.0(10)	121.0(10)
		O	2.760(30)	0.960(30)	1.810(30)	171.0(10)		
	H₂O(6)	O	2.790(30)	0.960(30)	1.890(30)	154.0(10)	109.0(10)	130.0(10)
		O	2.760(30)	1.000(30)	1.760(30)	172.0(10)		
	H₂O(7)	O	2.750(30)	0.970(30)	1.790(30)	168.0(10)	114.0(10)	119.0(10)
		O	2.700(30)	0.970(30)	1.730(30)	176.0(10)		
	H₂O(8)	O	2.680(30)	0.960(30)	1.720(30)	173.0(10)	109.0(10)	105.0(10)
		O	2.720(30)	0.940(30)	1.790(30)	167.0(10)		
	H₂O(9)	O	2.790(30)	0.970(30)	1.840(30)	167.0(10)	106.0(10)	122.0(10)
		O	2.990(30)	0.960(30)	2.070(30)	161.0(10)		

TABLE II. (*Continued*)

Compound	Proton acceptor Y	Distances, Å			Angles, deg		
		O···Y	O—H	H···Y	O—H···Y	H—O—H	Y···O···Y
Fe₃(PO₄)₂ · 4H₂O[3] (4.2°K) (Ludlamite)							
H₂O(5)	O	2.546(5)	1.030(10)	1.520(9)	173.7(9)	111.3(10)	107.0(2)
	O	2.679(8)	0.970(20)	1.827(15)	145.0(12)		
H₂O(6)	O	2.644(9)	1.010(40)	1.632(40)	176.5(22)	104.4(11)	92.6(2)
	O	2.742(7)	0.990(10)	1.793(9)	158.3(8)		
FeSiF₆ · 6H₂O[624]	F	2.681(13)	0.924(50)	1.857(38)	160.7(94)	111.9(16)	113.5(5)
	F	2.720(13)	0.920(25)	1.817(18)	168.1(15)		
KAu(CN)₄ · H₂O[168]	N	2.940(1)	0.920(20)	2.083(22)	155.8(19)	112.7(23)	110.1(4)
	N	3.140(1)	0.880(30)	2.268(27)	174.6(24)		
K[B₅O₆(OH)₄] · 2H₂O[52]	O	2.878(9)	1.004(18)c	1.901(18)	177.0(2)	108.0(2)	114.0(20)
	O	2.845(18)	0.954(26)c	1.935(23)	171.0(2)		
K₂C₂O₄ · H₂O[1214]							
(e)	O	2.754(2)	0.962(3)	1.800(3)	169.7(2)	107.6(3)	118.8(1)
KCr(SO₄)₂ · 12H₂O[71] (KCr alum)							
H₂O(1)	O	2.660(30)	1.020(30)	1.650(30)	169.2(20)	107.0(30)	102.0(30)
	O	2.660(50)	1.030(40)	1.630(30)	180.0(20)		
H₂O(2)	O	2.640(50)	1.030(60)	1.610(60)	180.0(20)	103.0(40)	94.0(40)
	O	2.720(110)	0.950(60)	1.800(60)	162.8(20)		
K₂CuCl₄ · 2H₂O[290]							
(e)	Cl	3.116(5)	0.966(6)c	2.165(6)	170.0(60)	109.7(7)	100.3(2)
K₂Mn(SO₄)₂ · 4H₂O[1243] (Manganese-leonite)d							
H₂O(1)	O	2.697(6)	0.991(13)c	1.710(11)	168.1(9)	104.4(6)	92.7(4)
	O	2.711(16)	0.995(10)c	1.740(11)	165.5(10)		

Compound	Water	Atom						
LiClO$_4$ · 3H$_2$O[368]	H$_2$O(3)	O	2.738(18)	0.970(25)c	1.870(18)	147.5(9)	108.8(18)	103.8(4)
	H$_2$O(2)e	O	2.767(12)	0.931(20)c	1.840(25)	171.4(10)	106.5(15)	116.8(4)
	(e)	O	2.704(6)	0.994(13)c	1.720(13)	171.4(11)	102.5	128.0
Li$_2$SO$_4$ · H$_2$O[1232]		O	2.98(4)	0.942(4)c	2.104	158.0	110.7(21)	73.5(4)
		O	2.870(20)	0.990(30)	1.990(30)	152.9(30)		
		O	2.950(10)	1.000(50)	2.110(50)	150.8(10)		
MgSO$_4$ · 4H$_2$O[111]	H$_2$O(1)	O	2.884(5)	0.969(10)	1.919(10)	173.6(8)	110.4(8)	105.2(2)
		O	2.754(5)	0.951(13)	1.818(12)	167.6(11)		
	H$_2$O(2)	O	2.835(5)	0.968(15)	1.948(15)	151.3(11)	110.0(15)	146.8(2)
		O	3.042(5)	0.931(17)	2.388(18)	127.1(14)		
	H$_2$O(3)	O	2.860(6)	0.952(13)	2.063(12)	140.2(10)	108.7(10)	137.5(2)
		O	2.833(6)	0.989(10)	1.847(10)	173.8(9)		
	H$_2$O(4)	O	2.831(5)	0.958(13)	1.901(11)	162.9(15)	108.5(10)	114.2(2)
		O	2.734(5)	0.981(10)	1.753(10)	177.7(9)		
MnCl$_2$ · 4H$_2$O[452]	H$_2$O(1)	Cl	3.173(5)	0.964(5)	2.227(30)	165.7(4)	104.2(5)	76.9(1)
		Cl	3.281(5)b	0.947(5)	2.499(30)	140.3(3)		
	H$_2$O(2)	Cl	3.166(5)	0.971(30)	2.201(30)	174.1(4)	105.7(5)	112.3(2)
		O	2.926(6)	0.968(30)	1.967(30)	170.4(4)		
	H$_2$O(3)	Cl	3.202(7)	0.959(7)	2.294(30)	158.7(4)	112.1(7)	143.2(2)
		Cl	3.317(7)b	0.921(7)	2.495(30)	148.3(4)		
	H$_2$O(4)	Cl	3.291(5)	0.973(6)	2.374(30)	164.4(4)	108.5(5)	120.2(2)
		O	2.964(7)	0.956(6)	2.014(30)	174.4(4)		
Mn(HCOO)$_2$ · 2H$_2$O[737]	H$_2$O(1)	O	2.750(20)	1.020(40)c	1.790(40)	158.0(30)	109.0(20)	114.0(7)
		O	2.790(20)	1.010(40)c	1.800(30)	177.0(30)		
	H$_2$O(2)	O	2.800(20)	0.960(30)c	1.860(30)	172.0(30)	107.0(20)	110.4(7)
		O	2.750(20)	0.890(30)c	1.860(30)	178.0(20)		

TABLE II. (*Continued*)

Compound	Proton acceptor Y	Distances, Å			Angles, deg		
		O···Y	O–H	H··Y	O–H···Y	H–O–H	Y···O···Y
(NH$_4$)$_2$C$_2$O$_4$ · H$_2$O[1054]							
(e)	O	2.800(20)	0.970(20)	1.850(20)	168.0(20)	105.6(19)	125.0(20)
(NH$_4$)$_2$Cu(SO$_4$)$_2$ · 6H$_2$O[235]							
H$_2$O(7)	O	2.826(2)	0.982(3)c	1.871(3)	170.1(3)	109.3(3)	118.0(1)
	O	2.821(2)	0.983(3)c	1.871(3)	168.7(3)		
H$_2$O(8)	O	2.707(2)	0.993(3)c	1.730(3)	178.1(3)	105.9(3)	104.2(1)
	O	2.743(2)	0.993(3)c	1.765(3)	177.5(3)		
H$_2$O(9)	O	2.732(2)	0.991(3)c	1.760(3)	170.4(3)	105.7(3)	99.0(1)
	O	2.683(2)	1.002(3)c	1.715(3)	171.3(2)		
NaAl(SO$_4$)$_2$ · 12H$_2$O[349] (NaAl alum)							
H$_2$O(1)	O	2.747(4)	1.054(24)c	1.783(25)	162.0(30)	103.0(20)	95.7(1)
	O	2.822(3)	1.017(24)c	1.907(23)	156.0(30)		
H$_2$O(2)	O	2.623(3)	1.009(17)c	1.635(20)	174.0(20)	108.0(20)	103.2(1)
	O	2.649(3)	1.006(20)c	1.673(17)	178.0(20)		
Na$_2$Al$_2$Si$_3$O$_{10}$ · 2H$_2$O[1301] (Natrolite)							
H$_2$O(7)	O	2.840(30)	0.980(20)	1.870(20)	169.0(20)	108.1(19)	134.0(5)
	O	3.010(30)	0.940(30)	2.130(30)	154.4(18)		
Na$_3$H(CO$_3$)$_2$ · 2H$_2$O[69]	O	2.760(20)	1.030(30)	1.780(30)	167.0(10)	107.0(10)	114.0(10)
	O	2.780(20)	0.990(30)	1.780(30)	167.0(10)		
NiCl$_2$ · 6H$_2$O[765]							
H$_2$O(1)	Cl	3.210(10)	0.940(20)	2.300(20)	164.0(20)	104.0(20)	89.5(4)
	O	2.740(10)	0.940(20)	1.800(20)	174.0(20)		
H$_2$O(2)	Cl	3.170(20)	1.070(30)	2.110(30)	168.0(10)	107.0(20)	124.7(4)
	O	3.060(10)	0.960(40)b	2.270(20)	139.0(10)		

Compound / Water							
NiSO$_4$ · 6H$_2$O[1039]							
H$_2$O(1)	O	2.690(20)	0.930(20)	1.770(20)	167.3(14)	113.0(15)	*120.0(4)*
		2.810(20)	1.000(20)	1.830(20)	168.1(13)		
H$_2$O(2)	O	2.770(20)	0.970(20)	1.860(20)	154.0(18)	107.5(15)	*124.1(4)*
		2.740(20)	0.970(20)	1.790(20)	167.3(13)		
H$_2$O(3)	O	2.840(20)	0.960(20)	1.940(20)	154.1(15)	109.7(15)	*131.4(4)*
		2.750(20)	0.960(20)	1.800(20)	167.1(15)		
β-RbMnCl$_3$ · 2H$_2$O[712]							
H$_2$O(1)	Cl	3.290(30)	0.920(30)[b]	2.460(30)	145.0(20)	109.0(20)	*72.4(3)*
	Cl	3.190(30)	0.900(30)	2.310(30)	166.0(20)		
H$_2$O(2)	Cl	3.290(30)	0.850(30)	2.470(30)	159.0(20)	109.0(20)	*102.2(3)*
	Cl	3.180(30)	0.910(30)	2.280(30)	172.0(20)		
Th(NO$_3$)$_4$ · 5H$_2$O[1282]							
H$_2$O(1)[e]	O	2.698(4)	0.953(6)	1.749(6)	173.2(5)	110.0(8)	*101.2(1)*
H$_2$O(2)	O	2.697(4)	0.971(5)	1.726(5)	177.5(6)	111.0(5)	*107.6(1)*
	O	2.953(4)	0.961(6)	1.996(5)	173.6(6)		
H$_2$O(3)	O	2.946(4)	0.954(6)	1.962(6)	167.9(6)	106.8(7)	*60.0(1)*
	O	2.901(4)	0.983(7)	2.067(8)	147.9(7)		
UO$_2$(NO$_3$)$_2$ · 6H$_2$O[1281]							
H$_2$O(7)	O	2.705(9)	0.974(12)	1.737(11)	172.8(9)	106.9(5)	*112.5(2)*
	O	2.692(9)	0.970(13)	1.736(11)	168.2(10)		
H$_2$O(8)	O	2.918(6)	0.977(11)	1.942(9)	177.4(10)	106.8(9)	*109.8(3)*
	O	2.760(10)	0.932(15)	1.852(12)	164.0(10)		
H$_2$O(9)	O	2.997(7)	0.947(15)	2.179(14)	144.0(13)	114.6(12)	*80.5(2)*
	O	2.701(9)	0.872(17)	1.849(13)	164.9(12)		

[a] The italicized values of distances, angles, and the corresponding estimated standard deviations (esd's) were calculated by us from atomic coordinates and their esd's. The esd's (in parentheses) are expressed in units of the last decimal. The distances were adjusted to three-decimal precision, so that the esd's can be compared directly. Similarly, the angles were brought to one-decimal precision. Missing entries in the table could not be calculated because of incomplete original data. The numbering of water molecules here and in subsequent tables follows that in the original reference.

[b] A hydrogen bond considered bifurcated in the original work. Only the nearest of the two proton acceptors is given (see Table III).

[c] Distances corrected for anisotropic thermal motion from the "riding model."

[d] Disorder.

[e] H$_2$O molecule with two hydrogen atoms equivalent by symmetry.

Structures for which X-ray *and* neutron data are available are classified under neutron data. Structures in which disorder seriously complicates the interpretation of molecular geometry, e.g., colemanite,[620] have been omitted. Those in which disorder leads to less serious complications, e.g., cupric formate tetrahydrate,[1044] have been placed in a category of lower accuracy. Because of lack of space only the neutron diffraction results will be presented in full (Table II).*

3.2. Hydrogen Bonding

3.2.1. *The* H···Y *Distance as the Geometric Criterion of Hydrogen Bonding*

Let us consider the immediate environment of one OH group of a typical H_2O molecule in a crystalline hydrate (Fig. 4). Those nearby atoms which are chemically suitable as hydrogen-bond acceptors (O, F, Cl, Br, I, and occasionally S, Se, and Te) will be designated Y_1, Y_2, etc. in the order of increasing H···Y distances. The energy of the interaction between OH and Y depends on the nature of Y, its chemical environment, and on the O—H···Y geometry. At least in principle this energy is never zero, even at long H···Y separations and for OHY angles far removed from $180°$. We thus require an operational criterion which will permit us to decide which of the potential hydrogen-bond acceptors Y_1, Y_2, etc. is to be considered hydrogen-bonded to a given OH group. Such a criterion will of necessity be arbitrary. Clearly, a restrictive criterion will result in many OH groups being described as not hydrogen-bonded, while application of a permissive criterion will have the effect of describing many OH groups as bifurcated, trifurcated, etc., i.e., forming simultaneous hydrogen bonds to two or more acceptor atoms.

A simple geometric criterion of the presence of a hydrogen bond must be based on a single geometric parameter. Using the H···Y distance as the most suitable quantity, the following operational definition will be adopted (cf. Refs. 627 and 628):

---

* A few of the entries in Table II are for deuterated compounds. It is known that replacement of H by D lengthens short O—H···O bonds (O···O distance 2.5–2.7 Å) by as much as 0.04 Å.[540] However, for longer O—H···O bonds (O···O distance 2.7–2.9 Å), such as occur typically with a water molecule as the proton donor, the expected dimensional changes are quite small. For most hydrates these changes should be comparable with experimental error on angles and distances [cf. α-(COOH)$_2$ · 2H$_2$O and α-(COOD)$_2$ · 2D$_2$O in Table II]. In the various correlations of this chapter the effect of isotopic replacement on the geometry of the water molecule has not been taken into account.

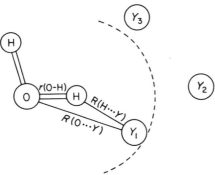

Fig. 4. Near-neighbor environment of an OH group in a crystalline hydrate. Atoms Y_1, Y_2, and Y_3 are possible hydrogen-bond acceptors. Atoms Y hydrogen-bonded to OH lie, by definition [Eq. (1)], inside a sphere centered on H and having a radius $r(H) + r(Y) - 0.2$ Å. The sphere is represented by the broken line.

A hydrogen bond is said to exist when the distance between the atoms H and Y, $R(H \cdots Y)$, is shorter by at least 0.20 Å than the sum of the accepted van der Waals contact radii $r(H)$ and $r(Y)$.

Thus

$$R(H \cdots Y)_{H-bond} < r(H) + r(Y) - 0.20 \text{ Å} \qquad (1)$$

Substituting the value of 1.20 Å for $r(H)$,[1080] we obtain

$$R(H \cdots Y)_{H-bond} < r(Y) + 1.00 \text{ Å} \qquad (2)$$

The concept of the van der Waals contact radius is based on an over-simplified, average description of molecular packing, and is inherently uncertain to about ± 0.05 Å.[627] The constant term, 0.20 Å, in eqn. (1) contains that uncertainty as well as the experimental uncertainty in the H$\cdots$Y distance. Values of up to ± 0.06 Å have been reported for the latter in the neutron diffraction determinations of Table II.

3.2.2. Neutron Diffraction Data — H$\cdots$Y Distance Known

The observed geometries of O—H$\cdots$Y contacts involving the H_2O molecule as a proton donor can now be surveyed in the light of the criterion just given. The values of H$\cdots$O and H$\cdots$Cl distances involving O and Cl atoms nearest to the particular H atom of a water molecule are shown for

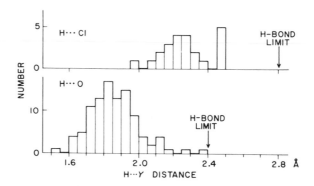

Fig. 5. Distribution of H···O and H···Cl distances in
hydrates listed in Table II.

hydrates in Table II and in Fig. 5. These distances are distributed over a
wide range but they all lie below the limit set by eqn. (1); only one H···O
value is borderline. A more restrictive criterion would not have altered the
situation significantly. For example, changing the value of the constant
term in eqn. (1) from 0.2 to 0.4 Å would reclassify only two OH groups
out of the 129 as "nonbonded."

The distribution of OHY angles for hydrates in Table II is shown in Fig.
6. The angles span the wide range from 180 to 127°, with a mean value of
169° and a standard deviation of 7°. The deviations of the O—H···Y
groups from the linear configuration which ought to correspond to lowest

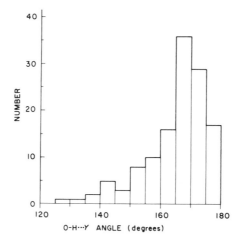

Fig. 6. Distribution of OHY angles in hydrates listed in Table II.

energy may be viewed as the result of a compromise between the structural requirements of hydrogen bonding and those of other forces to which water molecules in the crystal are subjected.

Hamilton[625] has observed that shorter hydrogen bonds are on the average less bent than are longer bonds. Figure 7 verifies this for the data on hydrates. The regression lines have slopes of -41.3 ± 9.2 deg Å^{-1} for the O—H$\cdots$O data and -33.7 ± 17.3 deg Å^{-1} for the less numerous O—H$\cdots$Cl data, indicating a correlation between angle and hydrogen-bond distance. This relationship is similar to that observed by Hamilton[625] and Chidambaram and Sikka[292] for a variety of O—H$\cdots$O bonds in crystals, some of them not involving H_2O. An explanation for the increasing amount of bending which can be tolerated in weaker and longer hydrogen bonds has been offered by Chidambaram[289,292] on the basis of an empirical potential function for the hydrogen bond.

When an O—H$\cdots$Y group in a crystal is highly bent additional acceptor atoms are likely to come within hydrogen-bonding range and multiple (i.e., bifurcated, trifurcated, etc.) hydrogen bonds may occur. Whether the situation about a given OH group (cf. Fig. 4) is to be described as a multiple hydrogen bond, a simple hydrogen bond, or no hydrogen bond depends on what criterion of hydrogen bonding is adopted. As most authors have not applied a uniform, standard criterion of hydrogen bonding when reporting the results of their crystal structure determinations, the descriptions in the literature of situations around OH groups are inconsistent.

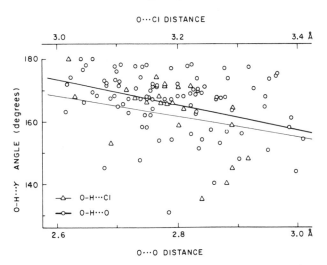

Fig. 7. Plot of OHY angle against O$\cdots$Y distance. Thick line, regression for Y = O; thin line, regression for Y = Cl.

TABLE III. Highly Bent Hydrogen Bonds in Hydrates Listed in Table II

Compound	H$\cdots$Y, Å	OHY angle, deg	Description of H bonding	
			Original	Present
NiCl$_2 \cdot$ 6H$_2$O$^{(765)}$ O(2)—H(3)⟨$\cdots$O(1) / $\cdots$O(1)⟩	2.27 / 2.27	139 / 139	Bifurcated	Bifurcated
C$_4$N$_3$O$_4$D$_3 \cdot$ D$_2$O$^{(346)}$ (Violuric acid) O(w)—D(w2)⟨$\cdots$O(6) / $\cdots$O(5)⟩	2.07 / 2.10	131 / 150	Bifurcated	Bifurcated
MnCl$_2 \cdot$ 4H$_2$O$^{(452)}$ O(1)—H(12)⟨$\cdots$Cl(1) / $\cdots$Cl(2)⟩	2.50 / 2.64	140 / 127	Bifurcated	Bifurcated
BaCl$_2 \cdot$ 2H$_2$O$^{(1063)}$ O(2)—H(4)⟨$\cdots$Cl(2) / $\cdots$Cl(1)⟩	2.49 / 2.66	135 / 125	Bifurcated	Bifurcated
MnCl$_2 \cdot$ 4H$_2$O$^{(452)}$ O(3)—H(32)⟨$\cdots$Cl(2) / $\cdots$Cl(1)⟩	2.50 / 2.95	148 / 113	Bifurcated	Single

Compound	Bond	Distance	Angle		
β-RbMnCl₃ · 2H₂O[712]	O(1)—H(1)···Cl(1)	2.46	145	Bifurcated	Nearly bifurcated
	···Cl(3)	2.84	127		
MgSO₄ · 4H₂O[111]	O(w3)—H(3a)···O(4)	2.06	140	Single	Nearly bifurcated
	···O(w4)	2.42	126		
	O(w2)—H(2b)···O(2′)	2.39	127	No bond	Borderline
	···O(2)	2.59	132		
UO₂(NO₃)₂ · 6H₂O[1281]	O(9)—H(5)···O(1)	2.18	144	Single	Single
	···O(6)	2.89	116		
	···O(4)	3.04	117		
K₂Mn(SO₄)₂ · 4H₂O[1243]	O(3)—H(4)···O(2)	1.89	147	Single	Single
	···O(2)	2.57	144		
Th(NO₃)₄ · 5H₂O[1282]	O(13)···O(23)	2.07 / 2.41	148 / 126	Single	Nearly bifurcated
	O(3)—H(5)···O(13)	2.66	112		
	···O(2)	3.19	53		

If the criterion of eqns. (1) and (2) is applied to those hydrates in Table II in which the OHY angle is smaller than 150° (11 cases), four cases of bifurcation and three cases of near-bifurcation are found (Table III). In $NiCl_2 \cdot 6H_2O$ the bond involving hydrogen H(3) is symmetrically bifurcated. The two acceptor atoms are related by a mirror plane passing through the O and H atoms of the water molecule and must participate equally in the bonding. Violuric acid monodeuterate contains one nearly symmetrically bifurcated bond, the two $D \cdots Y$ distances differing by only 0.03 Å. The hydrogen bonds involving H(12) in $MnCl_2 \cdot 4H_2O$ and H(4) in $BaCl_2 \cdot 2H_2O$ are bifurcated asymmetrically: one of the two $H \cdots Y$ distances is longer than the other by 0.14 and 0.17 Å, respectively, but both lie well within the hydrogen-bonding range.

The hydrogen bonding about H(32) in $MnCl_2 \cdot 4H_2O$ has been interpreted as bifurcated,[452] but according to our criterion the second acceptor lies outside of hydrogen bond range. Hydrogen bonding about H(1) in β-$RbMnCl_3 \cdot 2H_2O$ has also been interpreted as bifurcated.[712] In this case the second acceptor lies only 0.04 Å outside the hydrogen bond range, and the bonding can be described as nearly bifurcated. The hydrogen atoms H(5) in $Th(NO_3)_4 \cdot 5H_2O$ and H(3a) in $MgSO_4 \cdot 4H_2O$ are also engaged in nearly bifurcated hydrogen bonding, the second-nearest acceptor atoms being only 0.01 and 0.02 Å, respectively, outside the range. The second-nearest acceptor neighbors of the atoms H(4) in $K_2Mn(SO_4)_2 \cdot 4H_2O$ and H(5) in $UO_2(NO_3)_2 \cdot 6H_2O$ are outside the range by 0.17 and 0.49 Å, respectively. Finally, H(2b) in $MgSO_4 \cdot 4H_2O$ is the most weakly bonded hydrogen atom in hydrates so far examined by neutron diffraction. Not only is the $O-H \cdots O$ bond severely bent (the OHO angle is 127°), but in addition the $H \cdots O$ distance of 2.39 Å is only just within the adopted hydrogen bond limit of 2.40 Å. There is a second acceptor atom nearby, but it corresponds to an almost equally bent OHO angle and to an $H \cdots O$ distance well outside the range for hydrogen bonding.

Trifurcated hydrogen bonding has not been established in any of the hydrates examined by neutron diffraction, but such bonds have been postulated in $NiCl_2 \cdot 2H_2O$ on X-ray evidence;[997] the environment of H(5) in $Th(NO_3)_4 \cdot 5H_2O$ (cf. Table III) appears to approach trifurcation.

3.2.3. X-Ray Diffraction Data — $H \cdots Y$ Distance Unknown

The positions of hydrogen atoms have been determined by neutron diffraction for only 41 hydrate structures. In a few additional cases the hydrogen atoms have been located by X-ray diffraction with an accuracy sufficient to establish the hydrogen-bonding scheme. For several more hydrates the positions of hydrogen atoms have been derived, on certain assumptions, from the magnitudes and directions of the H—H vectors as determined by nuclear magnetic resonance,[451] or calculated by methods as discussed at the end of this section. In the majority of the 151 hydrate structures determined by X-ray diffraction (category B) the positions of only the heavier atoms are known with reasonable accuracy. We shall now examine what conclusions are possible about the presence and type of hydrogen bonding from this information alone.

The relevant interatomic distances (Fig. 4) are related by

$$R(O \cdots Y) \leq R(H \cdots Y) + r(O—H) \tag{3}$$

where the equality sign applies in the special case when the atoms O, H, and Y are collinear. Combining eqns. (2) and (3) and substituting the average value of 0.98 Å for $r(O—H)$ (cf. Section 3.4.1) yields the inequality

$$R(O \cdots Y)_{H-bond} < r(Y) + 1.98 \text{ Å} \tag{4}$$

which sets an upper limit for hydrogen-bonded $O \cdots Y$ distances (Table IV). This constitutes a necessary, but not sufficient, condition for the existence of a hydrogen bond between O and Y. An $O \cdots O$ distance longer than 3.38 Å signifies that the two oxygen atoms are *not* hydrogen-bonded, while a shorter distance only implies that they *may* be hydrogen-bonded. More precisely, above 3.38 Å the criterion of eqn. (1) cannot be met for any OHO angle, while for distances less than 3.38 Å it may be met for some angles.

If a suitable proton-acceptor atom Y occurs within the range of $O \cdots Y$ distances that qualify for hydrogen bonding, one must next enquire whether the OH groups are suitably oriented for the $H \cdots Y$ distance to satisfy the criterion of eqn. (1). The hydrogen atoms of the water molecule would be expected to orient away from the electron-acceptor neighbors X (metal ions, OH or NH groups), and so the XOY angles provide an indication of how likely the hydrogen atom is to meet the requirement for hydrogen bonding. For the XOY angle to be consistent with an $O—H \cdots Y$ bond, $70°^{[679]}$ and $95°^{[117]}$ have been proposed as minimum values. The requirement that the XOY angle be greater than a minimum within this range usually precludes the possibility of a hydrogen bond along an edge of a coordination polyhedron about a cation. If O and Y are both coordinated to the same metal ion M, the MOY angle would tend to be 45–60°, depending on the type of polyhedron and on the distortion from regular geometry.

TABLE IV. Observed O⋯Y Distances (in Å) in Crystalline Hydrates

Y	$r(Y)^a$	Maximum $R(O⋯Y)$ for H bond[b]	Observed $R(O⋯Y)$							
			Neutron data[c]				X-ray data[d]			
			Cases	Minimum	Average	Maximum	Cases	Minimum	Average	Maximum
F	1.35	3.32	4	2.65	2.68	2.72	12	2.52	2.71	2.78
O	1.40	3.37	100	2.55	2.79	3.06	647	2.60	2.81	3.33
N	1.5	3.47	2	2.94	3.04	3.14	11	2.78	3.03	3.63
Cl	1.80	3.77	23	3.02	3.19	3.32	64	2.87	3.20	3.75
S	1.85	3.82	—	—	—	—	18	3.24	3.35	3.49
Br	1.95	3.92	—	—	—	—	12	3.32	3.38	3.58
Se	2.00	3.97	—	—	—	—	7	3.31	3.42	3.56
I	2.15	4.12	—	—	—	—	5	3.59	3.63	3.67

[a] Contact radius, in Å.[1080]
[b] From eqn. (4).
[c] Neutron data on 41 hydrates (cf. Table II).
[d] High-quality X-ray data on 151 hydrates.

Another guideline in the choice of the acceptor atoms most likely to be hydrogen-bonded is the angle subtended by the two hydrogen-bond acceptor atoms at the same water molecule, $Y_1 \cdots O \cdots Y_2$. It is usually assumed that this angle must not deviate by more than $35°$ from the tetrahedral angle, $109.5°$.[117] In a majority of cases such considerations lead to a unique, though provisional, hydrogen bond scheme. For some hydrates the number of suitable acceptor atoms is greater than two for a particular water molecule, and it may then not be possible to propose an unambiguous bonding scheme.

There are very few cases of hydrates in which water molecules have no suitable acceptor atoms within the limiting hydrogen bond distance. Twelve hydrates in category B of our survey contain OH groups described as nonhydrogen-bonded on the basis of X-ray diffraction results.[115,350,468,480,613,830,855,944,1287,1440,1452,1453] For four of these hydrates approximate positions of the hydrogen atoms are available;[480,613,830,1452] the distances concerned are within the hydrogen bond range as defined by eqn. (1). In four additional compounds[115,350,468,1287] the $O \cdots O$ distances involved are all less than 3.38 Å and hence *may* correspond to hydrogen bonds. In three compounds[855,944,1440] the situation is not clear, as the $O \cdots Y$ distances involved have not been stated. In only one of the 12 hydrates, $Na_2[Fe(CN)_5NO] \cdot 2H_2O$, is there evidence for an OH group being outside the hydrogen bond range.[679,944] The nearest neighbor in this case is a nitrogen atom at 3.63 Å from the water oxygen.

Figure 8 shows the distributions of $O \cdots O$ and $O \cdots Cl$ distances that correspond to the presumed hydrogen bonding donated by the 412 distinct water molecules in the 151 hydrates of category B. The distributions include $O \cdots Y$ distances described in the original papers as "too long" for a hydrogen bond, and also the shorter of the pairs of $O \cdots Y$ distances in bonds that have been reported as bifurcated. The distributions are similar to the corresponding distributions based on neutron diffraction results, also shown in Fig. 8. Hence it appears likely that the assignments of hydrogen bonds involving the water molecules are on the whole correct.

A theoretical estimate of the H positions can be made by calculating the orientation of the H_2O molecule that gives a minimum of electrostatic lattice energy of the hydrate crystal. Baur[114] assumed a fixed geometry of the water molecule (HOH angle of $109.5°$ and O—H distance of 0.97 Å) and varied the point charges on the O atom (-2ε) and on each H atom $(+\varepsilon)$ of the water molecules. For $BaCl_2 \cdot 2H_2O$ the calculated H positions agreed within a mean deviation of 0.05 Å with those obtained from neutron diffraction data and were not very sensitive to the assumed values of ε.

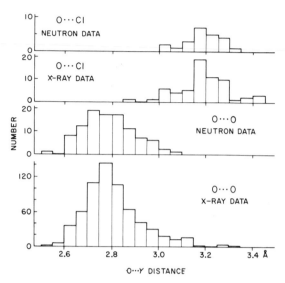

Fig. 8. Distribution of O···O and O···Cl nearest-
neighbor distances in hydrates.

The agreement was almost as good for several other hydrates, though it
was more dependent on the assumed effective charges placed on the different atoms. Independently, Ladd[818] treated the problem by using a
rigid dipole model of the H_2O molecule with HOH angle of 104.5°, O—H
bond distance of 0.99 Å, and the vapor-phase value of the dipole moment,
1.84×10^{-18} esu. For $BaCl_2 \cdot 2H_2O$ the agreement with the experimentally
determined atomic coordinates was significantly poorer than for Baur's
calculation.

3.3. Electron-Acceptor Coordination

The electron-acceptor neighbors of a water molecule are either metal
cations or proton-donor groups, NH or OH. In classifying the 663 H_2O
molecules in our survey according to the number and type of electron-
acceptor nearest neighbors (Fig. 9), we have relied mostly on the authors'
own descriptions in the original papers of what constitutes the immediate
environment of a water molecule. The bulk of the molecules (630 of 663)
have one or two electron acceptors as nearest neighbors (coordinations
within the heavy frame in Fig. 9). Only 33 molecules have coordinations
involving three or four electron acceptors.

When a water molecule is coordinated by one electron acceptor X and
two proton acceptors Y_1 and Y_2 the angles XOY_1, XOY_2, and Y_1OY_2

should average 120° for a coplanar arrangement and 109.5° for a regular tetrahedral arrangement. For most of the three-coordinated molecules the sum of these three angles deviates in fact very little from 360°, i.e., the H_2O surroundings are very nearly planar. In very few cases does the sum of the three angles fall substantially below 360°; two such molecules occur in $Zn(NO_3)_2 \cdot 6H_2O$.[479]

For a water molecule to have a regular coordination tetrahedron of two electron acceptors X_1 and X_2 and two proton acceptors Y_1 and Y_2, the six angles X_1OX_2, X_1OY_1, X_1OY_2, X_2OY_1, X_2OY_2, and Y_1OY_2 should all be 109.5°. The observed individual angles in an accidentally encountered subsample of ten H_2O molecules vary widely, from 74° to 141°, with a mean of 108.7° and a standard deviation of 14.7°. For most of these molecules the average of the six angles is within 2° of the tetrahedral value and the overall mean within 1°. An exceptional case of four-coordination (not included in the above subsample) is the water molecule in $K_2C_2O_4 \cdot H_2O$, discussed below.

Most (62 of 70) of the water molecules with only OH or NH as electron-acceptor neighbors have tetrahedral coordination. Of the 8 of 70 cases of trigonal coordination several involve acidic OH proton-donor groups and very short hydrogen bonds of about 2.55 Å. By contrast, about half

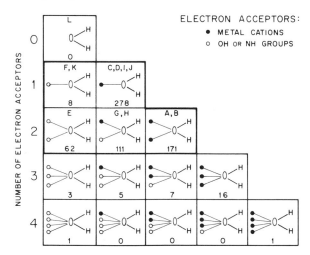

Fig. 9. Classification of H_2O molecules according to the number and type of electron-acceptor nearest neighbors. The number of cases observed is shown in each diagram. Letters refer to coordination types according to Chidambaram *et al.*[291] About 95% of the reported cases is accounted for by coordinations shown inside the heavier frame.

the water molecules associated with a cation as a nearest neighbor are trigonally coordinated. Trigonal coordination is almost always found when the cation carries a high formal charge. This is true without exception for the M^{4+} ions in our sample (Zr, Hf, Th; 23 cases in all) and in 34/43 cases for the M^{3+} ions (Al, Cr, Fe, Y, Ga, Gd, Nd). Monovalent (alkali) cations favor tetrahedral two-cation coordination of the water molecule: In 179 of 195 cases the molecule is clearly four-coordinated, while in only 1 of 195 cases is the coordination number clearly three. The total sample is too small to assess the effect of ionic size, if any.

Cations of typical (Be, Mg, Ca, Sr, Ba) or post-transition (Zn, Cd, Hg) divalent elements tend to be associated with trigonally coordinated water molecules (112 of 175 cases compared with 24 of 175 cases of tetrahedral two-cation coordination). This tendency is even more pronounced for divalent cations of the transition elements (V, Mn, Fe, Co, Ni, Cu), where trigonal coordination is found in 101 of 142 cases, while two-cation four-coordination occurs in only 12 of 142. It is possible that ionic size is a factor with Mg and the alkaline earths, but our sample is not large or varied enough to allow definite conclusions to be drawn. Nevertheless, the incidence of trigonal coordination for Mg^{2+}, Ca^{2+}, and Ba^{2+} (44 of 53, 22 of 39, and 11 of 38 cases, respectively) seems to decrease with the size of the ion.*

A further consequence of a high formal charge on a nearest-neighbor cation is the shortening of the $O—H \cdots Y$ hydrogen bonds donated by the water molecule associated with the cation. This may be considered a special case of a rule stated by Baur[110,116,1292]: The length of a hydrogen bond varies inversely with the electrostatic bond strength received by the donor atom.† The selection given in Table V of average $O \cdots O$ and $O \cdots Cl$ distances illustrates this effect. The effect of charge on the shortening of the distance is more pronounced for the $O \cdots Cl$ distances; for the $O \cdots O$ distances the effect becomes clearly noticeable only when the monovalent and divalent cations are contrasted with the trivalent and tetravalent cations.

The 33 reported cases of unusual coordination consist of 31 instances of water molecules with three electron-acceptor neighbors and two instances

---

* Taylor[1283] analyzed, as early as 1934, the coordination of water molecules in a number of zeolites and arrived at similar conclusions. In situations he described (natrolite, analcite) Na^+ is associated with tetrahedral two-cation coordination, while Ca^{2+} (scolecite) and Ba^{2+} (eddingtonite) are found with trigonally coordinated water molecules.
† The concept of electrostatic bond strength has been discussed by Pauling (Ref. 1080, p. 547).

TABLE V[a]

	$R(O\cdots O)$, Å	$R(O\cdots Cl)$, Å
Monovalent cations (Li, Na, K, Cs)	2.84 (217)	3.38 (6)
Divalent cations (Ba, Ca, Cd, Co, Cu, Fe, Mg, Mn, Ni, V, Zn)	2.82 (30)	3.23 (27)
Trivalent cations (Al, Cr, Gd)	2.68 (18)	3.12 (13)
Tetravalent cations (Hf, Zr)	2.68 (34)	—

[a] Numbers of cases are given in parentheses.

of water molecules with four such neighbors. A closer examination of the environments of these water molecules shows that almost always some of the $M\cdots OH_2$ or $OH\cdots OH_2$ distances involved fall outside the range of distances normally observed (or expected), so that such cations or proton-donor groups belong only marginally to the proper nearest-neighbor environment of the water molecule. If allowance is made for the abnormally long approaches, all but two cases can be reclassified under trigonal or tetrahedral coordination. The exceptions are $K_4Na_2Te_2O_8(OH)_2\cdot 14H_2O$,[855] which contains a water molecule coordinated with Na, K, and K at 2.48, 2.83, and 2.94 Å, respectively, and $Na_5P_3O_8\cdot 14H_2O$,[993] which contains a water molecule coordinated with Na, Na, and —OH at distances 2.38, 2.40, and 3.15 Å, respectively. These interatomic distances are in the "normal" range, but the chemical and structural complexity of the two compounds makes verification of the crystallographic results desirable.

On the view that the orbitals of the water oxygen are trigonally (sp^2) or tetrahedrally (sp^3) hybridized, the cation environment can be classified according to the relative orientation of the lone pairs on the oxygen and the nearest-neighbor cations. Such a classification has been attempted by Chidambaram et al.[291] (see Fig. 9), who also took into account the type of electron acceptor and proposed, on the data then available, 12 classes, A–L.

Although in principle Chidambaram's classification scheme has practical usefulness, in some ways it is unnecessarily complicated, while at the same time it does not accommodate several new types of lone-pair coordination. Specifically, distinction between monovalent and polyvalent cations (types A and B, C and D, G and H, I and J) has little structural significance. Distinction between C and I, or D and J, depends on the knowledge of the whereabouts of the lone pairs. This can only be deduced, from the positions of the hydrogen atoms, by making certain assumptions, and so the direction of the bisector of the lone pairs is known unambiguously only in the relatively rare symmetric case. Even if one could determine the direction of the lone pairs, geometries intermediate between types C and I, or D and J, could not be accommodated. Water molecules of at least one of the types, I, have not been observed, while type L is limited to the one example discussed below; 33 cases have been described of lone-pair coordination higher than two, and five cases of two-coordination involve simultaneously a monovalent and a polyvalent cation, a type not provided for by the classification of Ref. 291.

The only water molecule in crystalline hydrates which has been described as being without lone-pair ligands (the solitary example of class L) occurs in $K_2C_2O_4 \cdot H_2O$.[291,1214] The coordination of this water molecule involves two K^+ at a distance of 2.93 Å, i.e., in the normal $H_2O \cdots K^+$ range, but these two cations lie almost in the same plane as the two hydrogen atoms: The angle between the $K \cdots O \cdots K$ and H—O—H planes is only 12.5°, compared to 90° in a regular tetrahedron. The average value of the six ligand–H_2O–ligand angles is 114.5°. As the lone-pair orbitals would be expected to lie in a plane perpendicular to the H—O—H plane, they have been described as "not specifically directed."[291] However, in view of their proximity, the two K^+ ions ought to be counted as ligands regardless of their apparently unfavorable orientation.

3.4. Internal Geometry of the H_2O Molecule

3.4.1. *The O—H Bond Length*

The equilibrium bond length in the free H_2O molecule, as determined from spectroscopic studies, is 0.9572 ± 0.0003 Å.[137] Hydrogen bonding can be expected to lengthen the O—H bond.[61,1099] For strong hydrogen bonds, such as those involving acid hydrogens ($O \cdots O$ distances of about 2.5 Å) the lengthening is of the order of 0.1 Å, as shown in Fig. 2.6 of ref. 627. For hydrogen bonds involving water in hydrates (typical $O \cdots O$ distances of 2.7–2.9 Å) the lengthening expected, 0.01–0.02 Å, is small compared with the experimental error in the individual determination of the O—H distance, which is about 0.03 Å.

The 30 O—H bond lengths in hydrates of Table II that have been corrected for the effect of thermal motion range from 0.89 to 1.05 Å, with a mean of 0.980 Å. The standard error of this mean is 0.006 Å, assuming

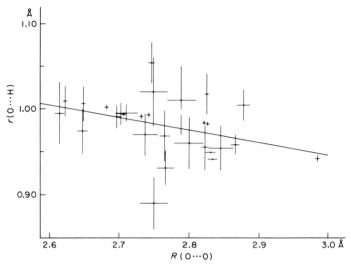

Fig. 10. Plot of the O—H bond length (corrected for thermal motion) against the O$\cdots$O distance for hydrates of Table II.

normal distribution. The conclusion that, on the average, the O—H bond is elongated by 0.023 $\pm$ 0.006 Å relative to the free molecule is subject to the assumption that the determination of the O—H bond length, including the thermal correction, is free from systematic error.

A plot of the corrected O—H distance r against the O$\cdots$O distance R is shown in Fig. 10. Regressing r on R yields a straight line defined by $r = (1.376 \pm 0.193) - (0.143 \pm 0.069)R$. The regression is significant (the correlation coefficient is 0.37), but as it explains only 14% of the sum of squared deviations, it has little predictive value. While this straight line cannot be taken to represent a functional relationship between r and R, the trend it shows would not disappear for any nonlinear monotonic function $r = f(R)$ having the property that r approaches asymptotically the value of 0.957 Å for large values of R.

Assuming that deviations of individual points in Fig. 10 from the regression line are due to random error in the determination of the O—H distance, the standard error on this distance is found to be 0.03 Å. This is comparable with a realistic estimate of the degree of accuracy which is attainable in the determination of the O—H bond length.[627]

3.4.2. *Asymmetry*

The hydrates in Table II contain 54 crystallographically distinct H_2O molecules for which the two hydrogen-bonded acceptor atoms Y_1 and Y_2

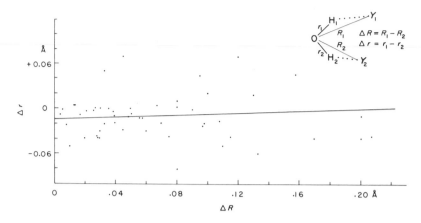

Fig. 11. Plot of the measure of internal asymmetry Δr of the water molecule against the measure of asymmetry of its environment ΔR for hydrates of Table II.

are chemically the same (both O, Cl, F, or N) and the two hydrogen-bonded distances $O \cdots Y_1$ and $O \cdots Y_2$ are not equal by symmetry or by accident. We shall number the Y atoms and the corresponding H atoms in such a way that the $O \cdots Y_1$ distance R_1 is longer than the $O \cdots Y_2$ distance R_2 (Fig. 11). The asymmetry of the external environment of the water molecule may then be represented by $\Delta R = R_1 - R_2$, where ΔR is positive by definition. The internal asymmetry of the H_2O molecule is measured by $\Delta r = r_1 - r_2$, where r_1 and r_2 are the respective bond lengths. The quantity Δr may be negative (the longer O—H bond corresponds to the shorter $O \cdots Y$ bond), positive (the longer O—H bond corresponds to the longer $O \cdots Y$ bond), or zero. Both corrected and uncorrected OH distances have been used in the plot of Δr against ΔR (Fig. 11). A least-squares fit of the data to a straight line $\Delta r = a + b(\Delta R)$ leads to the regression parameters $a = -0.015 \pm 0.032$ Å and $b = -0.069 \pm 0.067$. Both a and b are within one standard deviation from zero and the corresponding correlation coefficient is small, 0.14. Hence Δr seems to be scattered at random about zero and uncorrelated with ΔR. The expected relation between Δr and ΔR is completely obscured by experimental error.

3.4.3. *The* HOH *Angle*

The equilibrium HOH angle of the free water molecule is 104.52 $\pm 0.05°$.[137] In crystalline hydrates investigated by neutron diffraction the HOH angles range from 102.5 to 115.5°, with a mean of 108.0. The standard error of this mean is 0.4°, assuming normal distribution of error. This amounts to an average enlargement of the angle by $3.5 \pm 0.4°$, again subject to the assumption that the determination of the HOH angle by neutron diffraction is free from systematic error. The enlargement can be viewed as a consequence of changes in the hybridization of the oxygen orbitals. Tetrahedrally coordinated H_2O molecules would tend to have increased sp^3 hybridization and angles of up to $109.5°$. In trigonally co-ordinated H_2O molecules the hybridization would tend toward sp^2, yielding angles which might approach $120°$.[291,627] The observed HOH angle, on the average, is enlarged for both of the major classes of coordination, its mean value being $106.9 \pm 0.6°$ for tetrahedrally coordinated H_2O and $109.0 \pm 0.5°$ for trigonally coordinated H_2O.

As has been pointed out by Chidambaram,[288] Baur,[114] and Hamilton and Ibers,[627] the variation in the HOH angle is much smaller than that in the angle subtended by the acceptor atoms, Y_1OY_2. This is clearly shown in Fig. 12. In fact, the HOH angle is essentially independent of the Y_1OY_2 angle, the correlation coefficient for linear regression being only 0.18 (Fig. 13). Moreover, the HOH and the Y_1OY_2 planes usually do not coincide and often deviate considerably.[111] Qualitative considerations indeed show that the HOH angle should be stiff and hence remain close to its respective

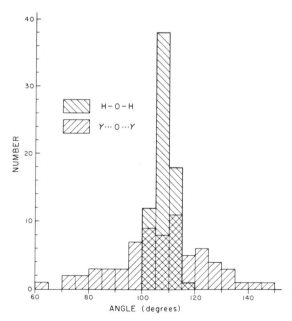

Fig. 12. Distribution of the angles HOH and of the acceptor angles YOY for hydrates of Table II.

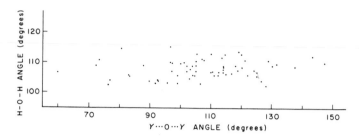

Fig. 13. Plot of the angle HOH against the acceptor angle YOY.

mean value for tetrahedrally or trigonally coordinated water molecules, while the much more flexible O—H $\cdots$ Y hydrogen bonds may bend considerably.[288,627]

4. INFORMATION FROM VIBRATIONAL SPECTROSCOPY

4.1. Vibrations of Water Molecules in Crystals

4.1.1. Isolated Molecules at Sites of C_{2v} Symmetry

On incorporating an H_2O (or D_2O, or HDO) molecule into a hydrate crystal, all nine degrees of freedom the molecule possesses become vibrational: two stretching vibrations, one bending vibration, three hindered rotations (librations), and three hindered translations. The fundamental frequencies corresponding to motions of each type fall in distinct regions of the spectrum.

An H_2O (or D_2O) molecule of symmetry C_{2v} has a total vibrational representation $3A_1 + A_2 + 2B_1 + 3B_2$ (Table VI). Assuming that motions of different type do not mix, the form of all nine normal coordinates is completely determined by symmetry. The only difference between the normal coordinates of H_2O and D_2O is the greater relative participation of the oxygen atom in the vibrations of D_2O. With the exception of the infrared-inactive twisting libration R_z, all nine fundamentals are Raman and infrared active.

The total vibrational representation of the HDO molecule, which has the lower symmetry C_s, is $6A_1 + 3A_2$ (Table VII). Both stretching vibrations now belong to the same irreducible representation. In fact, there are only two irreducible representations for the nine fundamentals. The form of the normal modes is not determined by symmetry, even if it is assumed that different types of motion do not mix, and detailed normal coordinate

TABLE VI. Normal Vibrations of H₂O in Sites of C_{2v} Symmetry

Type of motion	Symbol	Description	Irreducible representation[a]	Transition moment[b]	Frequency (cm⁻¹)	
					Vapor[c]	Solid[d]
Stretching	ν_3 or ν_{as}	Antisymmetric stretch	B_2	y	3755.8	2850–3625
	ν_1 or ν_s	Symmetric stretch	A_1	z	3656.7	
Bending	ν_2 or δ	Bend	A_1	z	1594.6	1498–1732
Rotation (libration in crystals)	ν_R $\begin{cases} R_x \text{ or } r \\ R_y \text{ or } w \\ R_z \text{ or } t \end{cases}$	Rotation about x (rock)	B_2	y	0	335–1080
		Rotation about y (wag)	B_1	x		
		Rotation about z (twist)	A_2	None		
Translation (hindered in crystals)	ν_T $\begin{cases} T_x \\ T_y \\ T_z \end{cases}$	Motion along x	B_1	x	0	200–490
		Motion along y	B_2	y		
		Motion along z	A_1	z		

[a] Irreducible representation under which the normal coordinate transforms (symmetry species).

[b] Direction of the transition moment. The z axis is taken along the H—O—H bisector, y axis is perpendicular to z and in the plane of the molecule, x axis is perpendicular to the plane of the molecule.

[c] From Ref. 137.

[d] Range of values for hydrates covered by our survey; the site symmetry of H₂O is usually lower than C_{2v}.

TABLE VII. Normal Vibrations of HDO in Sites of C_s Symmetry

Type of motion	Symbol	Approximate description	Irreducible representation[a]	Transition moment[b]	Frequency, cm^{-1} Vapor[c]	Frequency, cm^{-1} Solid[d]
Stretching	ν_3 or ν(OH)	OH stretch	A_1	yz	3707.5	3608–3170
	ν_1 or ν(OD)	OD stretch	A_1	yz	2726.7	2657–2421
Bending	ν_2 or δ	Bend	A_1	yz	1402.2	1391–1465
Rotation (libration in crystal)	ν_R { R_x	H and D motion in plane (rock)	A_1	yz	0	[e]
	R_{yz}	H motion out of plane	A_2	x		
	R'_{yz}	D motion out of plane	A_2	x		
Translation (hindered in crystal)	ν_T { T_x	Motion along x	A_2	x	0	[e]
	T_{yz}	Motion in yz plane	A_1	yz		
	T'_{yz}	Motion in yz plane ($\perp$ to T_{yz})	A_1	yz		

[a] Irreducible representation under which the normal coordinate transforms (symmetry species).
[b] Direction of the transition moment; yz signifies a direction in the yz plane, which is the plane of the molecule, x signifies direction perpendicular to the plane of the molecule.
[c] From Ref. 137.
[d] Range of values for hydrates covered by our survey.
[e] Few values available.

analysis is required for each HDO molecule in the crystal. In a first approximation, the normal vibrations will tend to involve the motion of either H or D, and this allows us to visualize roughly the form of the normal coordinates. In particular, the two stretching vibrations and the two out-of-plane librations in H_2O (or D_2O) involve both H (or D) atoms equally, while in HDO they may involve largely the motion of either H or D. All HDO fundamentals are in principle infrared and Raman active.

For example, $K_2CuCl_4 \cdot 2H_2O$[1291] contains only one crystallographically distinct type of water molecule. This molecule is located at sites of symmetry C_{2v}:

$$\text{Cu}\cdots\text{O}\underset{\text{H}\cdots\text{Cl}}{\overset{\text{H}\cdots\text{Cl}}{\Big\langle}}$$

The spectrum is nearly free from complications arising from vibrational coupling (see below).

Only one translational mode appears above 300 cm^{-1} in the infrared spectrum (Table VIII). It corresponds most likely to the T_z mode, which for trigonally coordinated water molecules can be described as cation–water stretching. The other translational modes are either below 300 cm^{-1} or too weak to be detected. The one librational mode observed for H_2O (or D_2O) is probably the R_y libration. The rocking mode is usually of much lower intensity, and the twisting mode is as a rule not observed, even when the site symmetry is lower than C_{2v} and the mode is no longer inactive in the infrared. The wagging libration shows up as the fundamental, overtone, and in combination with the bending vibration.

For HDO two librational bands are observed. Both occur as fundamentals, first overtones, and in combination with HOD bending. Their frequencies are very nearly the same as for H_2O and D_2O; evidently one of them corresponds predominantly to H motion and the other to D motion. The third HDO libration (rocking) is not observed. It would have a frequency between those of H_2O and D_2O.

4.1.2. Effect of Asymmetry

When an H_2O (or D_2O) molecule occupies a site of symmetry lower than C_{2v}, the form of the nine normal vibrations may be significantly altered and the description of the normal modes in Table VI may no longer be appropriate.

TABLE VIII. Infrared Absorption Frequencies (cm^{-1}) of Water in K$_2$CuCl$_4$ · 2H$_2$O at Different Degrees of Deuteration[a]

		H$_2$O	HDO	D$_2$O
OH Stretch		$\begin{cases} 3350 \text{ (anti)} \\ 3260^b \text{ (sym)} \end{cases}$	*3262*	—
OD Stretch		—	*2428*	$\begin{cases} 2480 \text{ (anti)} \\ 2425^b \text{ (sym)} \end{cases}$
2 × Bend		3170^b	*2840*	2347^b
Bend + Libration	H motion	2260	*1995*	—
	D motion	—	*1850*	1680
Bending		*1616*	*1431*	*1195*
		1600	—	1184
2 × Libration	H motion	1075	*1050*	—
	D motion	—	*780*	775
Libration	H motion	573	*563*	—
	D motion	—	*416*	414
Translation		440	*427^c*	417^c

[a] One crystallographically distinct water molecule on C_{2v} site. Room-temperature data from Ref. 1291; spectral range 4000–300 cm^{-1}. Frequencies in italics refer to isotopically dilute water species.
[b] Fermi resonance.
[c] Calculated values (region blocked by more intense bands).

    If the site has symmetry C_1 or C_s (the plane of symmetry coinciding with the plane of the molecule), the two H atoms become nonequivalent. An HDO molecule at such a site can assume either of the two possible configurations $Y_1 \cdots$ H—O—D$\cdots Y_2$ and $Y_1 \cdots$ D—O—H$\cdots Y_2$. The two configurations have equal abundances in the crystal and are distributed at random. The fundamental vibrations corresponding to the two HDO configurations are in principle different, and so the number of HDO fundamentals will *double*, while the number of H$_2$O or D$_2$O fundamentals is not affected by the asymmetry. The doublet splittings will vary depending on the sensitivity of the fundamental to environmental changes. If the main source of asymmetry is the unequal strength of hydrogen bonds to the two H atoms, the splitting will be particularly large for the stretching and librational modes, which undergo the greatest frequency shift with hydrogen bonding. However, all HDO fundamentals will be doubled, thus making

TABLE IX. Infrared Absorption Frequencies (cm^{-1}) of Water in
Na$_2$[Fe(CN)$_5$NO] · 2H$_2$O at Different Degrees of Deuteration[a]

		H$_2$O	HDO	D$_2$O
OH Stretch	OH$_1$	3625	*3603*	—
	OH$_2$	3545	*3568*	—
OD Stretch	OD$_1$	—	*2654*	2690
	OD$_2$	—	*2630*	2594
2 × Bend		3202	—	2358[b]
Bending		1624	—	1198[b]
		1617	—	1193[b]
		1611	—	1187[b]
		1616.5	$\left\{ \begin{array}{l} 1419.0 \\ 1427.0 \end{array} \right\}$	*1191.2*
Libration (H motion)[c]		519[b]	493[b]	([c])
		—	454[b]	—

[a] One crystallographically distinct water molecule on C_1 site. Data from Ref. 679; liquid nitrogen temperature. Frequencies in italics refer to isotopically dilute water species.
[b] Data from Ref. 754.
[c] Assignments in the region of librational frequencies are uncertain and translational frequencies are not available.

the HDO spectrum a very sensitive probe of local asymmetry, especially at low temperatures when the bands are sharp.

An example of the vibrational spectrum of a single, asymmetric, and essentially isolated water molecule is afforded by the infrared spectrum of sodium nitroprusside dihydrate, Na$_2$[Fe(CN)$_5$NO] · 2H$_2$O[679] (Table IX). The water molecules in this hydrate are all crystallographically equivalent and occupy sites of symmetry C_1.[944] The spectrum shows that appreciable vibrational coupling occurs only for bending vibrations. The doubling of the bending and stretching HDO fundamentals is clearly observed.

4.1.3. Several Types of Water Molecules Present in the Crystal

In this case the observed spectrum consists of a superposition of the vibrational bands from each water molecule. Every band of H$_2$O, D$_2$O, or HDO will be split into several components. The extent of the splitting will be different for each band, depending on the difference between the environments of the individual water molecules and on the sensitivity of the particular vibration frequency to environmental effects. Low-temperature spectra have narrow bandwidths and can be examined to particular advantage for the presence of closely spaced multiplets.

4.1.4. *Vibrational Coupling and the Isotopic Dilution Technique*

The vibrations of equivalent water molecules in a crystal may engage in dynamic coupling. The occurrence of such coupling presents a major complication in the interpretation of spectra of crystalline hydrates. The coupling causes each vibration of the isolated molecule to split into as many components as there are equivalent molecules in the unit cell. The phenomenon is often referred to as *correlation field splitting* or *factor group splitting*.[163a,1334] The number of infrared-active and Raman-active components can be derived from a correlation of the irreducible representations for (a) the point group symmetry of the isolated molecule, (b) the site group symmetry, and (c) the unit cell group which is isomorphic with the factor group.

For example, the correlation table for water molecules in sodium nitroprusside dihydrate (Fig. 14) shows that every vibration of the isolated water molecule will give rise to four bands in the Raman and three bands in the infrared spectrum, with no coincidences. The predicted number of components has in fact been observed for the bending vibrations at liquid nitrogen temperature.[679,754] The magnitude of the frequency splitting due to dynamic coupling varies widely: it can be as high as 100 cm^{-1} for the stretching fundamentals. The complications arising from vibrational coupling are an increased number of bands, increased width due to unresolved components, and frequency shifts.

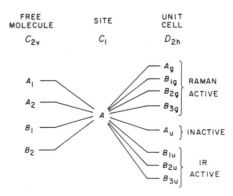

Fig. 14. Correlation table for H_2O fundamentals in sodium nitroprusside dihydrate, $Na_2[Fe(CN)_5NO] \cdot 2H_2O$.

Interpretation of vibrational spectra of water in hydrates is much simplified by the use of the isotopic dilution technique. In this technique,[666,682] one studies spectra of partially deuterated specimens and determines the number and frequencies of bands due to H_2O, D_2O, and HDO at low isotopic concentration of each. For such bands vibrational coupling is virtually eliminated: it occurs to an appreciable extent only for vibrations of identical frequency occurring in neighboring molecules in the crystal. The technique has been successfully applied to ices I–VI, by Bertie and Whalley[164,166,167] and to hydrates by Schiffer[482–484,766,767, 1191,1192] and others.[225,227,228,296,498,607,679,1210]

The most valuable recent spectroscopic results on hydrates come from isotopic dilution studies (Table X). They provide the basis of discussion in the following sections. Curiously, several entries in Table X originate in "unintended" spectra of isotopically dilute HDO, the authors' real aim having been complete deuteration of the specimen; in some cases the HDO bands were not even identified as such. The high quality of the experimental data in Table X is borne out by the good agreement of frequencies reported independently by different workers. The standard deviation is about 2 cm⁻¹ (between results from different laboratories) and the largest difference is 6 cm⁻¹.

4.2. Site Symmetry of the Water Molecules and Their Equivalence

Van der Elsken and Robinson[1325] attempted to deduce the number of structurally different water molecules from the number of bands in the librational region. The method assumes that each type of water gives two bands. It appears to work for the series of hydrates examined in Ref. 1325 (NaBr · $2H_2O$, NaI · $2H_2O$, and $BaCl_2$ · $2H_2O$ each contain two types of water molecule and give four bands, while $BaBr_2$ · $2H_2O$, $BaCl_2$ · $2H_2O$, and $SrCl_2$ · $2H_2O$ each contain one type and give two bands) but it does not seem to be capable of extension to unrelated hydrates: the number of infrared bands observed in the librational region may be larger (if vibrational coupling is appreciable) or smaller (as in K_2CuCl_4 · $2H_2O$, Table VIII, where the rocking libration is not detectable) than two per water molecule.

Information concerning the number of crystallographically distinct water molecules n and the number of distinct OH groups m can be more easily derived from the stretching and bending fundamentals. A single HOH (or DOD) bending band in the spectrum of an undeuterated (or fully deuterated) hydrate implies $n = 1$, while a single band for any HDO

TABLE X. Fundamental Frequencies (cm⁻¹) of Uncoupled HDO in Crystals

No.	Compound	t, °C	ν(OH)	ν(OD)	Bend	Ref.
1	$BaCl_2 \cdot 2H_2O$	−195	—	2521 2465 2447 2440	—	1192
2	$Ba(ClO_4)_2 \cdot 3H_2O$	−165	—	2632.9 2621.8	1428.3[a]	225
		28	—	2626[a]	1428.8[a]	225
3	$Ba(NO_2)_2 \cdot H_2O$	−170	3509 3324	2592 2459	—	228
		25	3498 3350	2584 2477	—	228
4	$CaSO_4 \cdot \frac{1}{2}H_2O$	28	—	2640	1427	([b])
5	$CaSO_4 \cdot 2H_2O$	−175	—	2577 2516	—	1210
		30	3494 3404	2582 2517	1465 1451	1210
		−195	3494 3411	2574 2517	—	766
		33	3500 3404	2583 2516	—	766
6	$CoCl_2 \cdot 2H_2O$	−188	3437	2540	1422	483
7	$CuCl_2 \cdot 2H_2O$	−188	3377	2503	1419	482–484
		37	3385	2511	1419	482–484
8	Ice I hexagonal	−173	3277[a]	2421[a]	—	166
		0	3304[a]	2444[a]	—	497
9	Ice II	−173	3373 3357 3323[a]	2493 2481 2460 2455	—	167
10	Ice III[h]	−173	3318[c]	2461[a] 2450[a]	—	167

TABLE X. (*Continued*)

No.	Compound	t, °C	ν(OH)	ν(OD)	Bend	Ref.
11	Ice V	−173	3350[c]	2461[c]	—	167
12	Ice VI	−173	3338[c]	2464[c]	—	164
13	$FeCl_2 \cdot 2H_2O$	−188	3423	2530	1421	483
14	$K_2CuCl_4 \cdot 2H_2O$	−135	—	2425	1428	1291
		28	~3262	2428	1431	1291
15	$K_2FeCl_5 \cdot H_2O$	−88	—	2494	1401	1291
		28	—	2499	1402	1291
16	$K_4[Fe(CN)_6] \cdot 3H_2O$	−165	—	2623 2607 2573 2540 2490[a]	—	[d]
		28	—	2605 2500 2480	—	[d]
17	$K_2HgCl_4 \cdot H_2O$	−147	—	2543	1418	1291
		28	~3430	2545	1411	1291
18	$K_4[Mo(CN)_8] \cdot 2H_2O$	(*f*)	—	2640[e] 2614[e]	—	639
19	$K_4[Ru(CN)_6] \cdot 3H_2O$	−165	—	2630 2614 2590 2553	—	[d]
		28	~3540[c] ~3470[c]	2620[c] 2570[c]	—	[d]
20	$K_2SnCl_4 \cdot H_2O$	−111	—	2596 2541	1429 1419	1291
		28	3527 3442	2595 2545	1426 1418	1291
21	$K_4[W(CN)_8] \cdot 2H_2O$	(*f*)	—	2641[e] 2614[e]	—	639

TABLE X. (*Continued*)

No.	Compound	t, °C	ν(OH)	ν(OD)	Bend	Ref.
22	$LiClO_4 \cdot 3H_2O$	−165	—	2618.4	1437.8	225
		28	—	2618.5	1441.8	225
23	$LiI \cdot 3H_2O$	−175	—	$\begin{cases} 2642 \\ 2529 \end{cases}$	$\begin{matrix} 1417 \\ 1397 \end{matrix}$	227
		25	—	$\begin{cases} 2600 \\ 2540 \end{cases}$	1409^a	227
24	$LiMnO_4 \cdot 3H_2O$	—	—	2570	—	$(^g)$
25	$MgCl_2 \cdot H_2O$	−180	$\begin{cases} 3491 \\ 3474 \end{cases}$	$\begin{matrix} 2577 \\ 2565 \end{matrix}$	1416^a	607
		25	3490^a	2580^a	1419^a	607
26	$MgCl_2 \cdot 2H_2O$	−180	3462	2557	1433	607
		25	3471	2564	1433	607
27	$MgCl_2 \cdot 4H_2O$	−180	$\begin{cases} 3445 \\ 3415 \end{cases}$	$\begin{matrix} 2542 \\ 2521 \end{matrix}$	—	607
		25	$\begin{cases} 3443 \\ 3413 \end{cases}$	2545^a	$\begin{cases} 1445 \\ 1433 \end{cases}$	607
28	$MgCl_2 \cdot 6H_2O$	−180	$\begin{cases} 3527 \\ 3500 \\ 3393 \\ 3358 \end{cases}$	$\begin{matrix} 2602 \\ 2582 \\ 2513 \\ 2486 \end{matrix}$	$\begin{matrix} 1423 \\ 1407 \\ 1391^a \\ — \end{matrix}$	607
		25	$\begin{cases} 3500^a \\ 3460^a \end{cases}$	—	1407^c	607
29	$MnCl_2 \cdot 2H_2O$	−188	3464	2558	1424	483
30	$Mg_3La_2(NO_3)_{12} \cdot$ $24H_2O$	−190	$\begin{cases} 3590^{c,e} \\ 3460^{c,e} \\ 3370^{c,e} \end{cases}$	—	—	405
31	$NaBr \cdot 2H_2O$	−195	—	$\begin{cases} 2608 \\ 2544 \\ 2532 \\ 2522 \end{cases}$	$\begin{matrix} — \\ 1451^a \\ 1435^a \\ — \end{matrix}$	1191

TABLE X. (*Continued*)

No.	Compound	t, °C	ν(OH)	ν(OD)	Bend	Ref.
31	NaBr · 2H$_2$O	−78	3442 ∼3430[c]	2610 2546 2534 2525	— 1447[a] 1433[a] —	1191
32	NaCl · 2H$_2$O	−195	3530 3434 3426 3416	2607 2536 2531 2523	— 1460[a] 1443[a] —	1191
		−78	3535 3425[c]	2611 2532[c]	1456[a] 1442[a]	1191
		−170	3531.2 3432.7 3423.2 3413.3	2610.4 2537.9 2532.6 2524.6	—	498
		−102	3535 3431 3424 3416	2612 2538 2533 2526	—	498
		−26	3537 3425[c]	2613 2534[c]	—	498
33	NaClO$_4$ · H$_2$O	−165	3584 3541	2641 2610	1427 1416	225
		28	3562[a]	2636[a]	1432[a]	225
34	Na$_2$[Fe(CN)$_5$NO] · 2H$_2$O	−173	3608 3564	2657 2626	1431 1418	754
		27	3606 3570	2656 2632	— —	754
		−165	3607.0 3562.5	2654.4 2624.5	1429.3 1417.6	679
		30	3603.0 3568.0	2653.5 2629.5	1427.4 1419.0	679
		(f)	3604 3570	2656 2630	1428 1419	1066

TABLE X. (*Continued*)

No.	Compound	t, °C	ν(OH)	ν(OD)	Bend	Ref.
35	NaI · 2H$_2$O			$\begin{cases} 2579 \\ 2546 \\ 2541^a \end{cases}$		
		−195	—		—	1191
		23	—	$\begin{cases} 2587 \\ 2553^c \end{cases}$	1425^c	1191
36	UO$_2$(NO$_3$)$_2$ · 2H$_2$O	(^{f})	3440^{c,e}		1429^{c,e}	296
37	UO$_2$(NO$_3$)$_2$ · 3H$_2$O	(^{f})	$\begin{cases} 3540^{c,e} \\ 3500^{c,e} \\ 3290^{c,e} \\ 3170^{c,e} \end{cases}$		1412^{c,e}	296
38	UO$_2$(NO$_3$)$_2$ · 6H$_2$O	(^{f})	$\begin{cases} 3588^{c,e} \\ 3520^{c,e} \\ 3280^{c,e} \\ 3180^{c,e} \end{cases}$		—	296

^{a} Probably an unresolved doublet.
^{b} V. Seidl, O. Knop, and M. Falk, unpublished results.
^{c} Probably an unresolved multiplet.
^{d} M. Holzbecher, O. Knop, and M. Falk, unpublished results.
^{e} "Unintended" HDO bands, occurring in the reported spectrum of deuterated hydrate because of incomplete deuteration.
^{f} Temperature unspecified, presumably room temperature.
^{g} G. Brink and M. Falk, unpublished results.
^{h} These is now evidence[1401] that ice III at liquid nitrogen temperatures is ordered. It is this ordered form (named ice IX) which was studied in Ref. 167.

fundamental implies $m = 1$. In such cases isotopic dilution is not required. With multiple bands, splitting caused by dynamic coupling has to be distinguished from that due to $n > 1$ or $m > 1$; isotopic dilution is then indicated. The number of uncoupled H$_2$O or D$_2$O bending fundamentals yields the value of n, while the number of HDO fundamentals yields the value of m. Low-temperature spectra and instrumental resolution of 1 cm^{-1} or better are essential for optimizing peak resolution.

Table XI lists hydrates for which n and m can be unambiguously derived from the spectra available to date (for references see Table X). It is interesting to note that in two of the cases, LiI · 3H$_2$O and Ba(NO$_2$)$_2$ · H$_2$O, the values of m derived spectroscopically imply that the water molecules occupy sites of symmetries lower than derived from X-ray evidence.[227,228]

TABLE XI. Derivation of the Number of Crystallographically Nonequivalent Water Molecules n and OH Groups m from the Number of Uncoupled H_2O, D_2O, and HDO Fundamentals

	$n = 1, m = 1$	$n = 1, m = 2$	$n = 2, m = 3$	$n = 2, m = 4$
Result compatible with crystal structure determination	$CuCl_2 \cdot 2H_2O$ $CoCl_2 \cdot 2H_2O$ $FeCl_2 \cdot 2H_2O$ $MnCl_2 \cdot 2H_2O$ $K_2CuCl_4 \cdot 2H_2O$ $K_2HgCl_4 \cdot H_2O$ $LiClO_4 \cdot 3H_2O$ $LiMnO_4 \cdot 3H_2O$	$CaSO_4 \cdot 2H_2O$ $K_2FeCl_5 \cdot H_2O$ $K_2SnCl_4 \cdot H_2O$ $K_4[Mo(CN)_8] \cdot 2H_2O$ $Na_2[Fe(CN)_5NO] \cdot 2H_2O$	$K_4[Fe(CN)_6] \cdot 3H_2O$ (28°)	$BaCl_2 \cdot 2H_2O$ $NaBr \cdot 2H_2O$ $MgCl_2 \cdot 6H_2O$ Ice II
Structural ambiguity removed	—	$Ba(ClO_4)_2 \cdot 3H_2O$	—	—
Prediction (crystal structure not determined)	$CaSO_4 \cdot \tfrac{1}{2}H_2O$	$NaClO_4 \cdot H_2O$ $K_4[W(CN)_8] \cdot 2H_2O$	$K_4[Ru(CN)_6] \cdot 3H_2O$ (28°)	$NaCl \cdot 2H_2O$ $NaI \cdot 2H_2O$
Result conflicts with crystal structure determination	—	$LiI \cdot 3H_2O$ $Ba(NO_2)_2 \cdot H_2O$	—	—

4.3. OH and OD Stretching Frequencies

4.3.1. *Correlation of ν(OH) with the Strength of Hydrogen Bonding*

The lowering of OH stretching frequency due to hydrogen bonding is well known. It is caused by a reduction of the OH stretching force constant brought about by a redistribution of electron density within the OH bond.[1450] The OH frequency may also be slightly affected by cation–water interactions (apart from the indirect effect of such interactions on hydrogen bonding discussed in Section 3.3), but this effect is likely to be very small compared with the effect of hydrogen bonding. Thus the lowering of the stretching frequency from the vapor value may serve as a measure of the strength of the hydrogen bond in which the particular OH group participates.

An approximately linear relationship between the OH frequency and the hydrogen bond strength has been established experimentally[433,713] and justified theoretically[1123] by Drago and co-workers. The negative slope of the straight line corresponds to a lowering of the OH stretching frequency by about 60 cm^{-1} for each kcal mol^{-1}.[713] Accepting this relationship, the range of ν(OH) in crystalline hydrates,* 3608 to 3170 cm^{-1} (Table X), corresponds to hydrogen bond strengths ranging widely from 1.7 to 6.7 kcal mol^{-1}.

4.3.2. *ν(OH) as a Spectroscopic Criterion of Hydrogen Bonding*

Figure 15 shows the values of ν(OD) observed for ices I to VI and for hydrates. The wide range of frequencies corresponds to the wide range of hydrogen bond strengths in which water molecules in crystals participate. It is of interest to note the occurrence of both weakly bonded and strongly bonded OH groups in many of the crystals. The highest stretching frequency so far recorded is that of the more weakly bonded of the two OH groups in $Na_2[Fe(CN)_5NO] \cdot 2H_2O$. While neutron diffraction data are not available for this hydrate, consideration of the geometry of the nonhydrogen atoms in its structure leads one to conclude that this OH has no neighbor nearer than a nitrogen atom at 3.63 Å from the O atom,[679] i.e., outside the hydrogen bond range defined in Section 3.2.1. While on the geometric criterion of hydrogen bonding this OH group must be considered non-bonded, its stretching frequency, 3603 cm^{-1}, is 104 cm^{-1} below the vapor

---

* The symbols ν(OH) and ν(OD) will be used to refer to the OH and OD stretching frequencies, respectively, of isotopically dilute HDO.

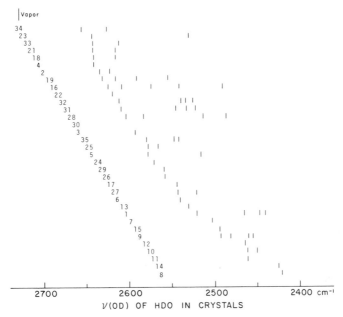

Fig. 15. Stretching frequencies $\nu(OD)$ of isotopically dilute HDO
in hydrates and ices. The identification numbers refer to Table X.

value. A frequency lowering of this magnitude is characteristic of hydrogen
bonding. For example, an almost identical lowering of $\nu(OH)$ is observed
for HDO dissolved in nitrobenzene or ethyl acetate,[760] solvents with which
the water molecule interacts through what would normally be considered
hydrogen bonds.

Clearly the lowering of $\nu(OH)$ is a much more sensitive probe of
hydrogen bonding than the comparison of H$\cdots$Y distance with the sum
of H and Y contact radii. If we use the lowering of $\nu(OH)$ or $\nu(OD)$ from
the vapor value as a spectroscopic criterion of hydrogen bonding, it will
tend to be more permissive than the geometric criterion of Section 3.2.1.
The two criteria can be matched by specifying a frequency which would
correspond to the limiting length of hydrogen bond [eqn. (1), Section 3.2.1]:
An OH group with $\nu(OH)$ above this value would not be considered hydrogen-bonded. This arbitrary cutoff frequency should lie, provisionally,
between that of the "nonhydrogen-bonded" OH group in sodium nitroprusside dihydrate, 3603 cm^{-1} for $\nu(OH)$ and 2654 cm^{-1} for $\nu(OD)$, and
that of the "hydrogen-bonded" OH group in LiClO$_4$ · 3H$_2$O, about 3560
cm^{-1} for $\nu(OH)$ and 2618 cm^{-1} for $\nu(OD)$. Because of the paucity of combined spectroscopic and crystallographic (preferably neutron diffraction)

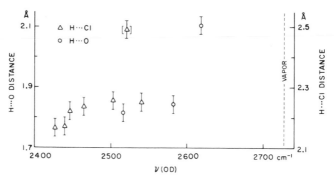

Fig. 16. Plot of H···O and H···Cl distances against the stretching frequency $\nu(OD)$ of isotopically dilute HDO in hydrates. The experimental point in brackets is for $BaCl_2 \cdot 2H_2O$ and probably corresponds to the bifurcated hydrogen bond reported for this hydrate. Bars on the experimental points indicate estimated standard error.

data on very weakly hydrogen-bonded OH groups, the exact frequency to be adopted is uncertain. An appropriate value to choose, when it becomes available, might be the frequency of the least strongly bonded OH group in $FeSO_4 \cdot 4H_2O$, which has an H···O distance of 2.39 Å, that is, on the borderline of hydrogen bonding according to the geometric criterion.

The extent to which the geometric criterion of hydrogen bonding (based on the H···Y distance) can be made to match a spectroscopic criterion [based on $\nu(OH)$ or $\nu(OD)$] depends on the degree to which $R(H···Y)$ correlates with $\nu(OH)$ or $\nu(OD)$. In view of the lack of sufficient data at the time of writing (Fig. 16), the extent of this correlation is uncertain.

4.3.3. Spectroscopic Measure of Distortion of the Water Molecule

The splitting of $\nu(OH)$ or $\nu(OD)$ gives a spectroscopic measure of the asymmetry in the internal force field of the water molecule. Owing to the relatively large standard error implicit in the determination of interatomic distances by diffraction methods, the spectroscopic method often yields the only direct demonstration of the existence of molecular asymmetry. For example, in gypsum[1210] the two O—H distances, as determined by neutron diffraction,[61] are 1.002 ± 0.030 and 0.981 ± 0.024 Å, differing by 0.021 ± 0.038 Å, while the two O···O distances are 2.816 ± 0.014 and 2.824 ± 0.015 Å, the difference being 0.008 ± 0.021 Å. The differences of the two O—H distances and the two O···O distances are absorbed in the stated uncertainty intervals and are not statistically significant, so that a distortion of the water molecule in gypsum cannot be regarded as proven from the crystal structure determination. On the other hand, the large splitting of the HDO frequencies [90 cm^{-1} for ν(OH), 65 cm^{-1} for ν(OD), and 14 cm^{-1} for the bending] demonstrates that the water molecule is indeed distorted.[1210]

4.3.4. Correlation of ν(OH) with $R(O \cdots Y)$ and $r(O—H)$

An empirical correlation of the shift of the OH stretching frequency from its vapor value $\Delta\nu$ and $R(O \cdots Y)$ has been proposed by Rundle and Parasol[1164] and Lord and Merrifield;[869] it has since been confirmed by many authors. The correlation, which appears to be approximately linear except for very long hydrogen bonds, has also been derived from simple potential functions for the hydrogen bond.[132,858] It has been reviewed recently by Efimov and Naberukhin.[441]

While the existence of the correlation can be considered proven, the detailed shape of the correlation line and the extent of scatter of the values of $\Delta\nu$ and $R(O \cdots Y)$ about that line are quite uncertain. The existing plots are based largely on mixed data, including acid OH groups and OH$^-$ ions, and, as has been pointed out,[441] many of the spectroscopic or crystallographic values used are doubtful or even outright incorrect. The correlation lines proposed by different authors differ by as much as 0.2 Å and 400 cm^{-1} in certain regions. For prediction of $O \cdots Y$ distances in hydrates from ν(OH) or ν(OD) of isotopically dilute HDO (which are accurate to about 2 cm^{-1}), a correlation based entirely on data from water molecules should be superior to correlations based on mixed data, in spite of its more limited range. A correlation based on the data for HDO in ices I–VI has, in fact, been presented by Kamb,[723] while a correlation based on H$_2$O data for hydrates has been proposed by Glemser and Hartert.[560]

The currently available ν(OD) values of HDO for both hydrates and ices whose crystal structures have been determined by X-ray or neutron diffraction are plotted in Fig. 17 against the $O \cdots O$ or $O \cdots Cl$ distances. In plotting the $O \cdots Cl$ values the "equivalent" $O \cdots O$ distances were used, which are equal to the $O \cdots Cl$ distances less 0.40 Å, the difference between the van der Waals contact radii of Cl and O. In hydrates containing several nonequivalent hydrogen bonds the frequencies were assumed to increase in the same sequence as the $O \cdots Y$ distances.

The $R(O \cdots O)$ and $R(O \cdots Cl)$ data are represented by the least-squares straight line $R = a + b(\Delta\nu)$, where $a = 0.707$ Å, $b = (8.40 \pm 1.31)$

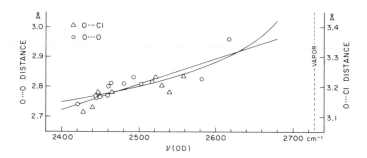

Fig. 17. Plot of O···O and O···Cl distances against the stretching frequency $\nu(OD)$ of isotopically dilute HDO in hydrates and ices.

$\times 10^{-4}$ Å $(cm^{-1})^{-1}$, and $\Delta\nu = \nu(OD) - 2727$ cm^{-1}. The regression has predictive value, as it explains 67% of the sum of squared deviations. The correlation coefficient is 0.82 and the standard error of R is 0.0314 Å. The same data are fitted marginally better by the function $R = c \ln(d/\Delta\nu)$, which has the correct asymptotic behavior at high values of R. In this case $c = 0.1391$ Å, $\ln d = 25.56$; the standard error of R is 0.0308 Å. The deviations of some of the points from the correlation lines are significant, as the experimental standard error is about 0.01 Å for $R(O\cdots Y)$ and 2 cm^{-1} for $\nu(OD)$. In its current form the correlation can be used to estimate $R(O\cdots O)$ from observed $\nu(OD)$ with $\sigma(O\cdots O) = 0.03$ Å.

The correlation of $\nu(OH)$ or $\nu(OD)$ with the O—H distance r based on the scanty data available at present is of no value. However, as neutron diffraction results of high quality become more abundant a meaningful correlation eventually may emerge. In the meantime a very rough estimate of r can be calculated from the $\nu(OH)$ value by using Badger's rule.[72] The latter provides an empirical relationship between the force constant for bond stretching k (in dyn cm^{-1}) and the bond length r (in Å):

$$r = 0.335 + (1.86 \times 10^5/k)^{1/3}$$

Under the harmonic diatomic approximation the force constant is given by

$$k = 4\pi^2\mu c^2 [\nu(OH)]^2$$

where $c = 2.998 \times 10^{10}$ cm sec^{-1}, $\mu = 1.562 \times 10^{-24}$ g (the reduced mass of the OH group), and $\nu(OH)$ is in cm^{-1}. For gypsum one obtains $r_1 = 0.985$ Å and $r_2 = 0.997$ Å, the difference being 0.012 Å. The standard error on this difference is difficult to estimate.

4.4. Bending Frequencies

Sartori *et al.*[(1182)] have shown that cation–water interactions increase the bending frequency by as much as 50 cm^{-1}. Hydrogen bonds involving the water hydrogens are also said to increase the bending frequency and by a similar order of magnitude.[(1450)] If the effect of both of the major environmental influences is in the same direction, one would expect the bending fundamental in hydrates always to have a frequency higher than in the vapor. However, this expectation is not fulfilled for about 12% of the frequencies reported for the hydrates in our survey.

Figure 18 shows the distribution of values of the bending frequencies of H_2O, D_2O, and HDO in the hydrates surveyed. The mean frequencies are 1618, 1194, and 1428 cm^{-1}, respectively, about 20 cm^{-1} above the corresponding vapor values. However, many of the individual bending frequencies are *below* the vapor values: 16 of 114 for H_2O, 6 of 51 for D_2O, and 3 of 33 for HDO. For some of the H_2O bending frequencies an incorrect assignment may be suspected, particularly where spectra of the deuterated hydrate were not recorded. For H_2O and D_2O bending the wide range of frequencies reported may be in part the result of strong dynamic coupling,

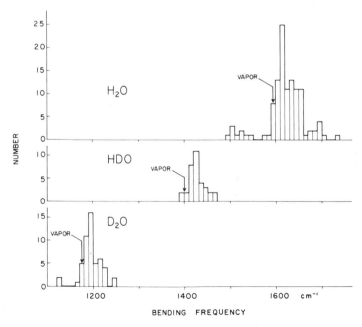

Fig. 18. Distribution of bending frequencies ν_2 of H_2O, HDO (isotopically dilute), and D_2O for hydrates and ices.

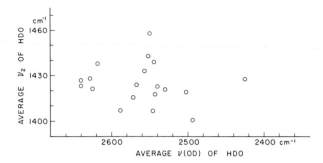

Fig. 19. Plot of bending frequency ν_2 against stretching frequency $\nu(OD)$ of isotopically dilute HDO for hydrates of Table X. When several frequencies occur in the spectrum of a hydrate their average is plotted.

with most of the infrared intensity concentrated in the lowest- or the highest-frequency component. For HDO, on the other hand, the assignments of the bending frequency are undoubtedly correct, and dynamic coupling does not enter the picture, since the HDO frequencies have been recorded at high isotopic dilution. Thus the occurrence of three HDO bending frequencies several cm^{-1} below the vapor value shows that there are some as yet unrecognized effects which lower the bending frequency in the crystal.

The bending frequency does not appear to correlate with the stretching frequency (Fig. 19), hence it must depend on factors other than hydrogen bonding. Clearly, more thorough analysis of the frequency shifts of the water bending fundamentals will be required before structurally useful information can be extracted from them.

4.5. Librational Frequencies

Since the main moving masses in the librations are the H atoms, the principal force constant is that opposing H motion at right angle to the O—H bond; it should be a direct function of hydrogen bond strength. One would thus expect the librational frequencies to be an index of hydrogen bonding and to correlate with the lowering of the OH stretching frequency. Oswald[1057] has, in fact, shown that ν_R increases almost linearly with a decrease of the OH stretching frequency (presumably unresolved ν_1 and ν_3) for undeuterated monohydrates of the kieserite type. However, results for other hydrates hardly satisfy this relationship. The most obvious factors obscuring the expected relation between librational and stretching frequencies are (a) occurrence of three normal modes of libration for each distinct water molecule and (b) dynamic coupling, which leads to broad bands of complex profiles. The three librations are usually assigned on the basis of relative intensity and frequency. The intensity is expected to vary in the infrared spectrum in the sequence wag > rock > twist,[987] while the sequence expected for the frequency is rock > twist > wag.[1325] There may be exceptions to these rough rules[976] and it is suspected that some of the assignments of the librational frequencies in the literature are incorrect.[239] Owing to dynamic coupling, the maximum of the librational band in the infrared spectrum may not correspond to the fundamental frequency. Also, the large widths and complex shapes of these bands make the assessment of the band maxima difficult. How much such assessments may vary between authors is seen for the librational frequencies of H_2O and D_2O in $BaCl_2 \cdot 2H_2O$ (Table XII).

Because of the difficulties of band assignment and location of the maxima, the H_2O librational frequencies cannot in general be used for comparisons and structural conclusions. Valid comparisons can, however, be made for closely related compounds.

4.6. Translational Frequencies and the Cation–Water Interaction

Hindered translations of water molecules which are not involved in water–cation coordination occur typically below 300 cm⁻¹. Cation-coordinated water molecules have one or more hindered translations which involve cation–water stretching and commonly occur in the range 300–500

TABLE XII. Bands in the 300–700-cm⁻¹ Region in the Spectrum of $BaCl_2 \cdot 2H_2O$

$H_2O^{(976)}$	$H_2O^{(1325)}$	$D_2O^{(1325)}$	$H_2O^{(615)}$	$D_2O^{(615)}$	$H_2O^{(537)}$	$D_2O^{(537)}$
~700	690	515	—	—	{ 710	500
					684	488
—	—	—	604	495	—	—
568	560	405	—	—	555 }	380
~530	520	370	—	—	525 }	
—	—	—	481	404	—	—
417	410	—	419	302	405	—

cm^{-1}. For trigonally bonded water molecules there is only one such frequency. It can usually be identified without ambiguity in the infrared, Raman, or neutron inelastic scattering spectrum provided that spectrum of the deuterated hydrate is also available.

Infrared data of Nakagawa and Shimanouchi[1013] and neutron inelastic scattering data of Prask and Boutin[1120] show the cation–water stretching frequencies to lie in the 490–440-cm^{-1} range for trivalent cations (Al and Cr) and in the 440–310-cm^{-1} range for divalent cations (Ni, Mn, Fe, Cu, Zn, and Mg). Frequencies involving monovalent cations are expected to be still lower, but few have been clearly identified. Only a handful of the cation–water stretching force constants have been calculated.

It appears that the cation–water stretching frequencies and the corresponding force constants offer a useful spectroscopic measure of the strength of the cation–water interaction. More work on the experimental determinations of these frequencies would be desirable.

4.7. Band Intensities

The large increase in the infrared intensity of OH stretching bands resulting from hydrogen bonding is well known. The intensity change is caused by a redistribution of electron density within the OH bond, which leads to greater dipole moment changes accompanying the motion of the hydrogen atom. The phenomenon is understood only qualitatively, but the increased intensity of OH stretching bands has been used empirically, together with the frequency shift, to diagnose hydrogen bonds and assess their relative strengths. Intensities of other infrared bands may be similarly useful. Good infrared absorptivity data are currently becoming available for ice I and its high-pressure modifications.[165] Such data for crystalline hydrates would be of considerable interest.

4.8. Bandwidths and Orientational Disorder

The OH stretching bands in many hydrates are extremely broad, the half-widths being often 100 cm^{-1} or more. The many explanations of the large bandwidth that have been offered (for a recent review see Ref. 627, pp. 86–87) assume the band broadening to be the property of an individual hydrogen-bonded OH group. That this is not correct is shown by the remarkably sharp OH and OD stretching bands of HDO in fully ordered ice II[167] and in many of the hydrates listed in Table X. The major contribution to the large bandwidth is clearly from vibrational coupling.

The residual width of the ν(OH) and ν(OD) bands in crystalline hydrates may be due to (a) proton tunneling, which shortens the lifetime of the vibrational upper state, (b) occurrence of sum and difference bands and of "hot" bands involving low-frequency vibrations, or (c) orientational disorder. Since the contribution from (a) would lead to anomalously high OH/OD bandwidth ratio and that from (b) would show characteristic changes with temperature, HDO bandwidths can be used diagnostically to demonstrate the presence or absence of orientational disorder. An interesting illustration of the effect of orientational disorder on HDO bandwidths is provided by the two closely related hydrates $LiClO_4 \cdot 3H_2O$[225] and $LiI \cdot 3H_2O$.[227] In the perchlorate the narrow bands indicate that the water molecules are fully ordered, whereas in the iodide the much broader bands are attributed to disorder.

4.9. Single-Crystal Spectra and the Orientation of Water Molecules in the Crystal

Infrared spectra of suitably oriented single crystals, when recorded with linearly polarized radiation, can be used to determine the dichroic ratio of a band. From this ratio one can derive the direction of the transition moment vector for the vibration.[1268] Similarly, Raman spectra of single crystals can be used to determine the relative magnitudes of the components of the polarizability derivative tensor.[953] If the direction of the transition moment vector or of the principal axes of the polarizability derivative tensor is imposed by symmetry, as for H_2O under C_{2v} site symmetry, then the orientation of the water molecule in the crystal can be determined from the infrared or Raman spectrum. Unfortunately, the method is not strictly applicable for water molecules in sites of symmetry lower than C_{2v}, which account for the majority of all hydrates. Nevertheless, the approximate determinations of H_2O orientation may be useful when neutron diffraction or NMR data are not available.

5. PSEUDOHYDRATES

While the chemical formula may indicate that a compound contains a stoichiometric number of water molecules, this water need not be present as H_2O molecules. Some or all of it may exist in the structure as H_3O^+ or OH^- ions, or as —H and —OH groups attached to some ion or radical. Water present in this form is referred to as *water of constitution* and hydrates containing it as *pseudohydrates*.

Until the crystal structure of a hydrate has been determined or its spectrum examined it is not known whether any of its water content is water of constitution: all hydrates liberate H_2O upon decomposition. Many compounds which had been formulated as true hydrates have turned out to be pseudohydrates. For example, the acid salt hydrate $2ZnCl_2 \cdot HCl \cdot 2H_2O$ has the structure $H_5O_2^+Zn_2Cl_5^-$,[495] $As_2O_5 \cdot 4H_2O$ is in reality $2H_3AsO_4 \cdot H_2O$,[1434] $3CaO \cdot Al_2O_3 \cdot 6H_2O$ has the structural formula $Ca_3Al_2(OH)_{12}$,[324,325] and $SrB_2O_4 \cdot 4H_2O$ is $Sr[B(OH)_4]_2$.[753] Among hydrates of inorganic acids the monohydrates of hydrochloric, hydrobromic, perchloric, nitric, and sulfuric acids are structurally $H_3O^+Cl^-$,[1449] $H_3O^+Br^-$,[900] $H_3O^+ClO_4^-$,[835] $H_3O^+NO_3^-$,[904] and $H_3O^+HSO_4^-$.[1273] Dihydrates of these acids contain either $H_5O_2^+$ ions, as in $H_5O_2^+Cl^-$,[901] $H_5O_2^+Br^-$,[900] and $H_5O_2^+ClO_4^-$,[1045] or two H_3O^+ ions, as in $(H_3O^+)_2$-SO_4^{2-}.[1274] Higher hydrates may contain various aquated oxonium ions and water of crystallization, as in the formal trihydrates $H_5O_2^+Cl^- \cdot H_2O$,[902] $H_5O_2^+Br^- \cdot H_2O$,[900] and $H_3O^+NO_3^- \cdot 2H_2O$,[905] and in the formal tetrahydrates $H_7O_3^+H_9O_4^+2Br^- \cdot H_2O$[903] and $H_5O_2^+AuCl_4^- \cdot 2H_2O$.[1409] Similar variety is encountered among hydrates of certain inorganic oxides and hydroxides. These compounds, sometimes termed "aquoxides," have been recently reviewed by Schwarzmann.[1209]

Infrared spectroscopy can often confirm the presence of water of constitution in a crystalline compound, and in favorable circumstances can unambiguously distinguish between a true hydrate and a pseudohydrate. Recent examples are the infrared studies of the "hydrates" of HCl and HBr by Gilbert and Sheppard,[553] the "dihydrates" of $GaPO_4$ and $GaAsO_4$ by Pâques-Ledent and Tarte,[1071] and the "hemihydrate" of $SnHPO_4$ by Yellin and Cilley.[1448] Absence of the H_2O bending band in the 1600–1700-cm^{-1} region is usually a clear proof of the absence of ordinary water of hydration, while broad bands near 1700 cm^{-1} (bending) and near 3000 cm^{-1} (stretching) show the presence of H_3O^+, $H_5O_2^+$, and higher oxonium ions. The presence of OH^- ions is invariably indicated by sharp OH stretching bands in the region of 3600 cm^{-1}.

6. SUMMARY

Compared with water or aqueous solutions, the H_2O molecule in a crystalline hydrate is considerably more constrained, both by the rigidity of the crystal structure and by the restrictions imposed by crystallographic symmetry. However, this very fact facilitates the study of the interaction of the water molecule with its environment and of the changes the molecule undergoes as a consequence of being incorporated in the hydrate.

The internal geometry of the water molecule in crystals differs very little from that in the vapor. Every H_2O molecule in a stoichiometric hydrate has at least one electron-acceptor and two proton-acceptor nearest neighbors. The distance between the water hydrogen and the proton-acceptor atom Y is shorter, with very few exceptions, by at least 0.2 Å than the sum of the van der Waals contact radii for H and for the acceptor atom. In this sense, nearly all water molecules in such hydrates must be considered hydrogen-bonded. The shorter hydrogen bonds tend to be nearly linear, but longer bonds show wide deviations from linearity. Severely bent O—H $\cdots$ Y bonds tend to have additional proton-acceptor atoms nearby. In some cases the hydrogen bonds are bifurcated or nearly bifurcated.

Most water molecules in crystalline hydrates have one of two clear-cut coordinations: approximately planar trigonal, with a single electron acceptor, or approximately tetrahedral, with two electron acceptors. Large deviations from these limiting configurations appear to be rare. Unusual coordinations which involve three or even four electron acceptors occasionally occur. Almost all of them involve some long and clearly marginal intermolecular approaches.

Valuable complementary information concerning the water molecule in hydrates is obtainable from infrared and Raman spectra. The lowering of the stretching frequency $\nu(OH)$ or $\nu(OD)$ of HDO, for example, can be used as a criterion of hydrogen bonding. Such a spectroscopic criterion is more sensitive than a geometric criterion based on the distance between a water hydrogen and a proton-acceptor atom. The correlation which exists between $\nu(OH)$ or $\nu(OD)$ and the distance between the water oxygen and the proton acceptor makes it possible to estimate that distance with a standard error of about 0.03 Å. The O—H bond lengths can be estimated roughly from $\nu(OH)$ or $\nu(OD)$ by Badger's rule. The splitting of $\nu(OH)$ and of $\nu(OD)$ provides a spectroscopic measure of the distortion of the water molecule in the crystal.

ACKNOWLEDGMENT

We are indebted to Mr. P. F. Seto for his untiring and many-faceted assistance; to Mr. Alfred Black for statistical computations; to Dr. F. R. Ahmed and Mrs. M. E. Pippy for making available a program for computing bond distances and angles; and to Dr. John A. Walter for constructive comments on the manuscript.

Clathrate Hydrates

D. W. Davidson

Division of Chemistry
National Research Council of Canada
Ottawa, Canada

1. INTRODUCTION

1.1. Gas Hydrates: Historical Survey

The term "gas hydrate" has been applied for nearly a century to the *solids* which are formed by the combination of many gases and volatile liquids with a large excess of water. Their study goes back at least to 1810 when Humphry Davy, in the Bakerian lecture[384] to the Royal Society, reported that an aqueous solution of oxymuriatic acid gas (which he suggested be renamed chlorine) froze more readily than water itself. In 1823 Faraday found[470] the composition of the solid formed to be roughly $Cl_2 \cdot 10H_2O$ but recognized the possibility that incomplete drying of the crystals or loss of gas during the analysis could have led to underestimation of the chlorine content. This was indeed the case, and Faraday's work illustrates the difficulties inherent in direct analysis of the gas hydrates which, along with the general assumption of simple whole-number stoichiometry, led to considerable controversy about the compositions of the 40 or so gas hydrates reported in the ensuing 100 years. The period up to 1925 is the subject of a detailed review by Schroeder.[1206] There was much speculation about the nature of the forces uniting the gas and water molecules. These were early recognized to be much weaker than those of primary chemical bonds. By 1880 it was clear that the vapor pressure of the hydrate system was independent of its water content over a wide range and depended on temperature alone.[697,1361] Pressure–temperature measurements along the three-phase equilibrium lines played an important role in Roozeboom's elaboration[1157] of the phase rule for heterogeneous equilibria during the period between 1884 and 1887.

Figure 1, based mainly on Roozeboom's classical study[1155,1156] of SO_2 hydrate, is fairly typical of the projections on the P–T plane of the phase diagrams of gas hydrate systems. The region of stability of the hydrate is bounded by the h-I-g, h-l_1-g, and h-l_1-l_2 equilibrium lines, where h stands for hydrate, I for ice I, g for gas, and l_1 and l_2 for the liquid phases rich in water and hydrate-former, respectively. Two invariant points Q_1 and Q_2 mark the coexistence of h, I, l_1, and g and of h, l_1, l_2, and g. A third quadruple point (not shown) occurs at the eutectic temperature where l_2 freezes in the presence of h and g.

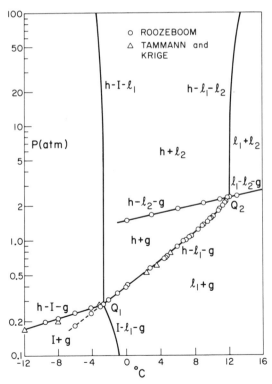

Fig. 1. Phase diagram of the SO_2–H_2O system. The h-I-l_1 and h-l_1-l_2 lines are based on the high-pressure measurements.[1279]

De Forcrand was the first to suggest[393] that the composition of a hydrate $M \cdot nH_2O$ could be determined indirectly from a knowledge of the heats of the reactions $h \rightarrow l_1 + g$ and $h \rightarrow I + g$, since the difference is merely the heat of fusion of n moles of ice. With the aid of an empirical relationship between the heat of the second reaction and the temperature at which the hydrate decomposition pressure reached 1 atm, and by application of the Clapeyron equation to the h-l_1-g equilibrium line, de Forcrand obtained[395] values of n which ranged mostly between 5.7 and 8.3 for a large number of simple gas hydrates. These and other results contradicted part of Villard's rule,[1345] according to which "the dissociable compounds formed by combination of water with various gases and existing only in the solid state are isomorphous with one another, crystallize in the cubic system, and have a constitution expressed by $M \cdot 6H_2O$." The observation that the hydrates almost invariably crystallized in the regular system, that is, had no effect on polarized light, provided a convenient method of distinguishing them from hexagonal ice.

In their 1919 study of H_2S hydrate Scheffer and Meyer[1188] refined the indirect method of analysis by applying the Clapeyron equation to the dependence of $\ln P$ on $1/T$ along both the h-I-g and h-l_1-g curves to determine the hydrate composition at Q_1, the point of intersection. Appropriate corrections were made for the vapor pressure of water, the solubility of H_2S in water, and the difference in heat capacities of hydrate and decomposition products. This became the most common and reliable method of determining the compositions of simple gas hydrates. Despite the accumulation of data of increased accuracy which pointed to nonintegral values of n in $M \cdot nH_2O$ for many hydrates, the tendency to assign the nearest integer persisted. As recently as 1950, an experimental value of 6.71 for ethylene hydrate led the authors[427] to conclude that the formula was probably $C_2H_4 \cdot 7H_2O$.

The first of many studies of mixed or double hydrates, that of chloroform and H_2S, was reported[863] in 1852. For this hydrate de Forcrand found[392] $CHCl_3 \cdot 2H_2S \cdot 23H_2O$ by direct analysis, and 20 years later,[397] $CHCl_3 \cdot 2H_2S \cdot 19H_2O$ by the thermodynamic method. Meanwhile, a simple hydrate of chloroform, $CHCl_3 \cdot 18H_2O$, had been reported.[280] This and a number of other "liquid hydrates" of halogenated hydrocarbons displayed,[1345] except in their relatively large water contents, the same general properties as gas hydrates. These included the formation of regular crystals which could serve as nuclei for crystallization of gas hydrates and a sensitivity to the presence of "help gases" like nitrogen and oxygen which increased hydrate stability. Villard[1345] did not consider that help gases entered into the hydrate structure and proposed a thermodynamically untenable explanation of their effect.

In the 1930's Nikitin[1029–1031] prepared by "isomorphous coprecipitation" the mixed hydrates of the rare gases with SO_2 and developed a method of separating the rare gases based on the different partition coefficients between the gas and hydrate phases. The rare gas hydrates were regarded as prototypes of many isomorphous hydrates formed by molecules of suitable size and van der Waals character, all, from argon to chloroform, representable, as he thought;[1032] by $M \cdot 6H_2O$, where M could be a mixture of components.

Meanwhile von Stackelberg[217,1356] had initiated a series of thermodynamic, composition, and X-ray studies which extended over 20 years and saw the proper nature of the gas hydrates gradually emerge. The first X-ray results,[536,1093] for single crystals of SO_2 hydrate and of the double hydrate of $CHCl_3$ and H_2S and for ten other polycrystalline hydrates, were only published[1350] after the war. A cubic unit cell with a lattice constant of 12 Å was found, the space group being O_h^3 ($Pm3n$) except for the double hydrates, which were "probably" O^2 ($P4_232$). Proper intensity measurement of the diffraction lines was prevented by loss of X-ray film during the war, but a structure was proposed[1350] which had some relationship to the structure previously found by von Stackelberg[1360] for CaB_6. The hydrate structure was a kind of ice in which each water molecule had four water neighbors and which contained holes or voids of two different geometries large enough to hold gas molecules. There were six times as many water molecules as holes. Variability of composition was ascribed to variable occupancy of the holes and the increased stability produced by help gases to their incorporation in otherwise empty holes.

Although this first structure provided an explanation for the general properties of gas hydrates, the O- -O bond distances (2.42–2.60 Å) and the O- -O- -O angles (61–145°) departed so greatly from their values in hexagonal ice (2.76 Å and essentially tetrahedral angles) as to be unrealistic for a modification of ice. Starting from the notion that the pentagonal dodecahedron should provide a suitable spherical cavity in which the angles are not appreciably different from tetrahedral, Claussen[307] constructed a cubic unit cell in which 136 water molecules were arranged to form 16 dodecahedra as well as eight larger hexakaidecahedra consisting of four hexagonal and 12 pentagonal faces of hydrogen-bonded water molecules. This structure was immediately confirmed by von Stackelberg and Müller[1357,1358] for the "liquid hydrates" of large molecules like chloroform and ethyl chloride and for their double hydrates with H_2S. The space group was O_h^7 (*Fd3m*), the unit cell dimension 17.2 Å, and the ideal composition $CHCl_3 \cdot 17H_2O$ for all the large holes filled and $CHCl_3 \cdot 2H_2S \cdot 17H_2O$ for the small cages filled with H_2S. The distinct 12-Å structure formed by smaller molecules remained to be determined. This was done almost simultaneously by Claussen,[308] Müller and von Stackelberg,[1009] and Pauling and Marsh.[1083] The 12-Å unit cell was found to contain 46 water molecules arranged to form two pentagonal dodecahedra and six tetrakaidecahedra, each made up of 12 pentagonal and two parallel hexagonal faces. Ideal compositions are $M \cdot 5\frac{3}{4}H_2O$ if all cages are occupied and $M \cdot 7\frac{2}{3}H_2O$ if only the larger 14-hedra are filled.

Further publications of von Stackelberg and his students[1351,1353–1356,1359] present the results of extensive studies of the properties of hydrates of structure I (12-Å cell) and structure II (17-Å cell). The great majority of gas hydrates are now known to conform to one or the other of these structures.

1.2. Gas Hydrates as Clathrates

Most gas hydrates thus constitute a class of solids in which small molecules of many types occupy almost spherical holes in icelike lattices made up of hydrogen-bonded water molecules. They clearly qualify[1083] as "clathrates," the term introduced by Powell[1115] for "the structural combinations of two substances which remain associated not through strong attractive forces but because strong mutual binding of the molecules of one sort makes possible the firm enclosure of the other." Dielectric and nuclear magnetic resonance measurements (see Section 5) have demonstrated a remarkable degree of rotational mobility on the part of molecules encaged in clathrate hydrate lattices of structures I and II and the absence of any great directional dependence of the interaction with the water molecules of the cage. This interaction appears to be primarily of the van der Waals type, even for polar "guest" molecules (Section 5.6).

The lattices of some amine and salt hydrates resemble the type I and II lattices in some respects (Sections 2.5 and 2.6). They differ from typical clathrate hydrates, however, in the presence of hydrogen bonds or strong ionic interactions between guest and host molecules. Although it is sometimes useful for structural reasons to regard these hydrates as clathrates,[708] a more appropriate designation is "semiclathrate hydrates."[705]

A considerable number of other clathrate systems are known.[174,617,943,1116,1117] The properties of β-quinol clathrates,[1067] in particular, have been extensively studied[1246] and have some relevance here in so far as the

β-quinol cage is roughly the same size as, if less spherical than, the 12-hedral cages of the clathrate ices.

The gas hydrates are also classed among inclusion compounds,[343, 344,1352] a more general grouping which includes lamellar and channel as well as cage structures.[100] In so far as the constraints imposed by the lattice are predominately one- or two-dimensional, in layer or channel structures the included molecules generally have much freer access to the surface of the solid than when incorporated in cages.

In common with other clathrates, ice clathrates thermodynamically are solid solutions of guest molecules in a host lattice,[1328] a concept already implicit in the work of Nikitin.[1029–1032] As such, they are nonstoichiometric. The variability of composition is, however, confined between the minimum occupancy of cages necessary for thermodynamic stability of the host lattice with respect to ice or aqueous solution[1328] and complete filling of the cages which the guest molecules are capable of entering. In simple structure II hydrates this range is narrow and the composition is close to the stoichiometric $M \cdot 17H_2O$ (Section 3.2).

1.3. Classification of Clathrate Hydrates

The most fundamental system of classification is according to the structure of the host lattice. As shown in Table I, the known clathrate hydrates, with two exceptions, appear to be all of von Stackelberg's structures I or II. These and some related semiclathrate hydrate structures are the subject of Section 2.

Another classification is based on the nature of the guest molecules, which vary in chemical constitution (Table I) from rare gases to ketones and amines. Although size of the guest molecule is the single most significant parameter in determining whether structure I or II is formed, the molecular structure and polarity determine the details of the guest–host interactions and the mobility of the guest molecule (Section 5).

The term "simple hydrate" seems appropriate to the presence of only one guest species. In the case of several guest components it is useful to preserve a somewhat arbitrary distinction between a "double hydrate," in which one species occupies only the larger cages and the other primarily the smaller cages, and a "mixed hydrate," in which cages of the same kind are occupied by two or more species. This conforms roughly to von Stackelberg's original usage, although his conclusion[1352] that double hydrates are stoichiometric is no longer likely (see Section 3.4).

An additional classification scheme—a thermodynamic one based on the nature of the phase diagram of the M–H$_2$O system—has been used in Table I as a convenient way of describing the conditions under which a hydrate is stable. Systems with the basic phase diagram of Fig. 1 are labeled i_2 and are characterized by the temperatures and (total) pressures of the quadruple points Q_1 and Q_2. This is the most common system and reflects the sparing solubility of most hydrate-formers in water. The Q_2 point is absent for light gases whose h-l_2-g equilibrium line ends (at the l_2-g critical point in the presence of hydrate) before it intersects the h-l_1-g line. In such i_1 systems hydrate may exist to indefinitely high temperatures as long as a pressure at least equal to its steeply rising decomposition pressure is maintained. In general the h-l_2-g line tends to run nearly parallel to the h-l-g line and there appears to be no reported case of a h-l-l_2-g point among simple hydrates of water-insoluble species.

The 40 authentic simple gas hydrates mentioned in Schroeder's 1926 review[1206] included no hydrates of water-soluble species. The reluctance to consider solubility consistent with gas hydrate formation was only overcome by von Stackelberg's identification[1359] of a structure II hydrate of dimethyl ether. If M is sufficiently soluble, there is only one liquid phase and Q_2 is absent. For some hydrates it is replaced (as the maximum tem-

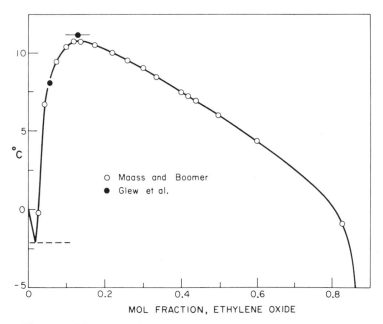

Fig. 2. Melting points in the ethylene oxide–water system.[568,575,907]

perature of coexistence of hydrate, liquid, and gas) by the congruent melting point. The phase relationships in such a case, identified as s_c in Table I, are most conveniently shown in the temperature–composition projection of the phase diagram. This is illustrated in Fig. 2 for the ethylene oxide–water system,[907,959] in which the hydrate formed has proved to be structure I.[927,1356] Some hydrates, like that of acetone,[1124] melt incongruently and are labeled s_i. The entries under Q_2 for s_c and s_i systems are the temperatures of congruent or incongruent melting. In s_c systems the water-rich eutectic temperature is given under Q_1.

1.4. Tabulation of Clathrate Hydrate Properties

Table I lists structural and stability characteristics of the simple clathrate hydrates, along with literature references.

Column 3 identifies the hydrate structure and the cubic unit cell dimension a (in Å), if known. Unless a reference number is shown the cell dimension is the $0°C$ value given by von Stackelberg and Jahns.[1354] Where no X-ray data exist the assignment of structure I or II is based on published composition studies and/or the size of the guest molecule.

Column 4 shows the type of phase diagram of the $M–H_2O$ system in the nomenclature of the last section and column 5 gives the year the hydrate was first reported.

In columns 6 and 7 are given the temperature ($°C$) and total pressures (atm) of the invariant points already discussed. Two hydrates are formed by cyclopropane and additional quadruple points occur at $-16.01°C$, 0.295 atm (h_I-h_{II}-I-g) and $1.46°C$, 0.846 atm (h_I-h_{II}-l_1-g). Both hydrates of trimethylene oxide melt incongruently.

Column 8 shows the temperature at which the hydrate dissociation pressure reaches 1 atm.

An attempt has been made to include in Table I all hydrates considered which can be properly characterized as clathrates, but some have undoubtedly been overlooked. Hydrates which, like that of CCl_4,[1355] appear to be stable only in the presence of air or other help gas have been omitted. Some reports of gas hydrate formation have proved erroneous. Villard's early description[1343] of a CF_4 hydrate with a decomposition pressure at $0°C$ of about 1 atm is clearly inconsistent with the pressure of 41.5 atm recently found[981] and must be attributed to the encagement of other gases of the CCl_nF_{4-n} type formed during the thermal reaction[1343] of CCl_4 and AgF. Decomposition of SO_2Cl_2 may have been responsible for the gas hydrate

TABLE I. Structural and Stability Properties of Simple Clathrate Hydrates

Type of guest	Guest molecule	Structure, value of a	System type	Year reported	T, P at Q_1	T, P at Q_2	T at $P=1$ atm	Additional references
Rare gas	Ar	I	i_1	1896[1344]	$-0.8°$, 87[946]	No Q_2	-124[101]	399
	Kr	I	i_1	1923[399]	$-0.1°$, 14.3	No Q_2	-49.8[101]	580
	Xe	I, 12.0	i_1	1925[400]	$0°$, 1.5[1350]	No Q_2	-10.4[101]	580
Diatomic molecule	N_2	I	i_1	1960[1320]	$-1.3°$, 141.5[1322]	No Q_2	—	946
	O_2	I	i_1	1960[1320]	$-1.0°$, 109.2[1322]	No Q_2	—	—
	Cl_2	I, 12.03, 11.82 ($4°$)[1083]	i_2	1811[384]	$-0.22°$, 0.316[1414]	$28.3°$, 8.4[782]	9.7	470, 575
	Br_2	$4/mmm$[19]	i_2	1828[872,873]	$-0.30°$, 0.0555[1007]	$5.80°$, 0.127	✓	1359, 1455
	BrCl	I, 12.02[1359]	i_2	1828[872,873]	$0°$, 0.165[1350]	$25°$, 2.5	18[38]	569
Triatomic molecule	CO_2	I, 12.07	i_2	1882[1361]	$0°$, 12.4[651,1279]	$9.9°$, 44.4[1815]	-55[979]	831, 982, 1345
	N_2O	I, 12.03	i_2	1888[1340]	$0°$, 9.7[1345]	$12°$, 41	—	—
	H_2S	I, 12.02	i_2	1840[1422]	$-0.4°$, 0.918[1211]	$29.5°$, 22.1	0.4	781, 1188
	H_2Se	I, 12.06	i_2	1882[391]	$0°$, 0.455[398]	$30°$, 11[401]	8	—
	SO_2	I, 11.97	i_2	1829[404]	$-2.6°$, 0.274[1279]	$12.1°$, 2.33[224]	6.8	1155, 1156, 1345
	ClO_2	I(?)	i_2	1843[224]	$-0.79°$[218]	$18.2°$	15	175
	COS	I, 12.14	i_2	1954[1359]	—	—	—	—

TABLE I. (*Continued*)

Type of guest	Guest molecule	Structure, value of a	System type	Year reported	T, P at Q_1	T, P at Q_2	T at $P=1$ atm	Additional references
Inorganic polyatomic molecule	PH_3	I	i_2	1882[1455]	0°, 2.3	28°	—	398
	AsH_3	I	i_2	1915[398]	0°, 0.81	28.3°, 17.5	1.6	—
	SF_6	II, 17.21	i_2	1954[1354]	0°, 0.80[1240]	14.0°, 19.9	1.0	981, 1328
Hydrocarbon	CH_4	I	i_1	1888[1340]	−0.2°, 25.3[386]	No Q_2	−78.7[979]	395, 541, 564, 946, 1173
	C_2H_2	I	i_2	1878[262]	0°, 5.75[1345]	15.0°, 33	−40.2[979]	394
	C_2H_4	I	i_1	1888[1340]	−0.1°, 5.44[427]	No Q_2	−36.9[979]	782, 1134, 1321, 1323, 1341, 1345
	C_2H_6	I	i_2	1888[1340]	−0.03°, 5.23[1147]	14.7°, 33.5	−32.0[979]	386, 541, 1134
	Propylene	II	i_2	1952[1134]	−0.134°, 4.60[305]	0.958°, 5.93	—	1059
	Cyclopropane	I, 12.14(5°)[616] II	i_4	1960[86] 1969[616]	— −0.05°, 0.619	16.21°, 5.59[616]	2.8	87, 936, 978
	C_3H_8	II, 17.40	i_2	1890[1343]	0°, 1.74[386]	5.7°, 5.45[87]	−11.6	275, 770, 1134, 1406
	iso-Butane	II, 17.57	i_2	1954[1359]	0.00°, 1.12[85]	1.88°, 1.653	−2.8	224, 386, 357, 629, 923, 1203

Cyclopentene	II	i_2	$1950^{(1069)}$	—	3.2°	—	857
Cyclopentane	II	i_2	$1950^{(1069)}$	—	$7.7°^{(1394)}$	—	857
Fluorinated hydrocarbon CH_3F	I	i_2	$1890^{(1342)}$	0°, 2.1	18.8°, 32	—	—
CH_2F_2	I	i_2	$1890^{(1343)}$	—	17.6°	—	—
CHF_3	I, 12.05	i_2	$1890^{(1343)}$	—	21.8°	—	—
CF_4	I	i_2	$1969^{(981)}$	0°, 41.5	—	—	—
C_2H_3F	I, 12.11	i_2	$1954^{(1354)}$	—	—	—	—
C_2H_5F	I	i_2	$1890^{(1342)}$	0°, 0.7	22.8°, 8	3.7	396
C_2F_4	II (?)	i_2	$1890^{(1343)}$	—	10.5°	—	—
CH_3CHF_2	I, 12.12	i_2	$1954^{(1354)}$	0°, 0.54	$14.9°, 4.30^{(84)}$	4.3	86, 87
$(CH_3)_3CF$	II	i_2	$1969^{(1394)}$	—	—	—	—
Chlorinated hydrocarbon CH_3Cl	I, 12.00	i_2	$1856^{(162)}$	$0°, 0.41^{(1347)}$	20.5°, 4.9	7.5	75, 395, 574, 1345
CH_2Cl_2	II, 17.33	i_2	$1897^{(1345)}$	$0°, 0.153^{(971)}$	1.7°, 0.211	—	1350, 1351, 1355
$CHCl_3$	II, 17.33	i_2	$1885^{(280)}$	$-0.09°, 0.065^{(1420)}$	$1.7°, 0.090^{(1069)}$	—	103, 104, 857, 590, 1279, 1350, 1351, 1359
C_2H_3Cl	II	i_2	$1897^{(1345)}$	—	$1.15°, 1.80^{(86)}$	—	87
C_2H_5Cl	II, 17.30	i_2	$1890^{(1342)}$	$0°, 0.265^{(971)}$	$4.8°, 0.77^{(224)}$	—	1345, 1350, 1355, 1359
CH_3CHCl_2	II	i_2	$1897^{(1345)}$	$0°, 0.072^{(971)}$	1.5°, 0.092	—	392, 1350, 1355

TABLE I. (*Continued*)

Type of guest	Guest molecule	Structure, value of a	System type	Year reported	T, P at Q_1	T, P at Q_2	T at $P=1$ atm	Additional references
Chloro-fluorinated hydrocarbon	CH_2ClF	I	i_2	1960[86]	$-0.2°$, 0.222[87]	$17.88°$, 2.825	9.83	224
	$CHClF_2$	I, 11.97(2°)[1421]	i_2	1947[294]	$-0.2°$, 0.84[1421]	$16.3°$, 7.6	0.9	—
	$CHCl_2F$	II	i_2	1954[80]	$-0.13°$, 0.145	$8.61°$, 0.998	—	86, 87, 224
	CCl_2F_2	II, 17.37, 17.13(2°)[1421]	i_2	1947[294]	$-0.1°$, 0.36[1421]	$12.1°$, 4.27	5.2	—
	CCl_3F	II, 17.29	i_2	1947[294]	$-0.1°$, 0.080[1421]	8.5, 0.65	—	—
	CH_3CClF_2	II, 17.29	i_2	1954[1354]	$-0.04°$, 0.136[224]	$13.09°$, 2.294[87]	9.1	86, 223
Alkyl bromide	CH_3Br	I, 12.09	i_2	1856[162]	$-0.24°$, 0.238[87]	$14.73°$, 1.51	11.3	86, 224, 971, 1350
	C_2H_5Br	II	i_2	1948[971]	$0°$, 0.2	$1.4°$, 0.22	—	1350, 1355, 1359
Alkyl iodide	CH_3I	II, 17.14	i_2	1880[390]	$0°$, 0.097[971]	$4.3°$, 0.23	—	1342, 1345, 1350, 1359
Bromofluoro-methane	$CHBrF_2$	II (?)	i_2	1960[86]	—	$9.87°$, 2.65[87]	—	—
	$CBrF_3$	II	i_2	1961[1421]	$-0.1°$, 0.88	$11°$	0.5	—

Type	Compound	Structure						References
	CBr_2F_2	II	i_2	1960[86]	—	4.9°, 0.501[87]	—	—
	$CBrClF_2$	II	i_2	1960[562]	0.00°, 0.189	9.96°, 1.673	7.6	86, 87, 223, 1272
Mercaptan	CH_3SH	I, 12.12	i_2	1887[763]	0°, 0.31[1359]	12.0°, 1.25	10.0	1353
Ether	Ethylene oxide	I, 12.1 (−10°)[1356]; 12.03 (−25°)[917,927]	s_c	1863[1441]	−2.1°[568]	11.1[575]	—	570, 907, 959
	Dimethyl ether	II, 17.47	—	1954[1359]	—	—	—	—
	Propylene oxide	II, 17.124 (−138°)[1178]	s_c	1952[1405]	−4.7°	−3.5°[1394]	—	647
	Trimethylene oxide	II, 17.095 (−138°)[1178]	s_i	1966[647]	—	−13.1°	—	—
		I	s_i	1966[647]	—	−20.8°	—	—
	1,3-Dioxolane	II, 17.118 (−138°)[1178]	s_c	1966[1178]	—	−3.0°[1335]	—	—
	Furan	II, 17.3 (−10°)[1356]	i_2	1950[1069]	—	4.6°	—	—
	Dihydrofuran	II, 17.166 (−138°)[1178]	s_c	1966[647]	−3.3°	−1.2°	—	—
	Tetrahydrofuran	II, 17.18 (−10°)[1356]; 17.170 (−138°)[1178]	s_c	1950[1069]	−1.0[458]	4.4°	—	590, 647
Ketone	Acetone	II, 17.16 (−38°)[1124]	s_i	1961[1124]	—	−19.8°[1000]	—	141, 322, 1416
	Cyclobutanone	II, 17.161 (−138°)[1178]	s_c	1966[371]	—	0.0°[1000]	—	—
Amine	t-Butylamine	$I\bar{4}3d$[1928]	s_c	1967[706]	—	−1°[930]	—	—

reported[76] for this molecule, which is too large to form a normal clathrate hydrate. Other clathrate hydrates remain to be discovered.

No attempt has been made to make the list of references complete, except for the X-ray results. Many studies of hydrates such as those of Cl_2, H_2S, and $CHCl_3$ are mainly of historic interest.

Table I includes only hydrates with a single encaged species. References to the early studies of mixed and double hydrates may be found in Schroeder.[1206] Lattice parameters of structure II double hydrates with H_2S have been reported by von Stackelberg[1354,1356,1359] and a detailed structure of the tetrahydrofuran–H_2S hydrate by Mak and McMullan.[938] Decomposition temperatures of a considerable number of new double hydrates of relatively large molecules with H_2S are given by Glew *et al.*[573]

2. STRUCTURES OF THE LATTICES

Most of the structural information presented in this section is taken from the comprehensive X-ray studies of clathrate and clathratelike hydrates by Professor G. A. Jeffrey and his co-workers. These were reviewed in 1967 by Jeffrey and McMullan.[708]

2.1. Hydrates of Structures I and II

The positions of the oxygen atoms of the 46 water molecules of the unit cell of structure I have been best defined by McMullan and Jeffrey[927] for the ethylene oxide hydrate. The coordinates of the 136 water molecules in the unit cell of structure II have been similarly defined for the double hydrate of tetrahydrofuran and H_2S by Mak and McMullan.[938] Of the four hydrogen bonds formed to the nearest neighbors of each water molecule, three form edges of a specific cage (Fig. 3) and the fourth (not shown) is directed outward from the cage. Every bond forms an edge common to three planar or almost planar pentagonal or hexagonal rings, each shared between two cages, and every water molecule is a common element of six rings and four cages. In each structure there are three crystallographically nonequivalent oxygen sites, depicted by different symbols in Fig. 3. The geometries of the 14- and 16-hedra are discussed by Allen.[18]

The positions of the hydrogen atoms are disordered. That is, each water molecule has six orientations which are consistent with full hydrogen bonding as in ice Ih and Ic and other disordered ice polymorphs (see Volume 1, Chapter 4). This is shown more convincingly by the dielectric properties

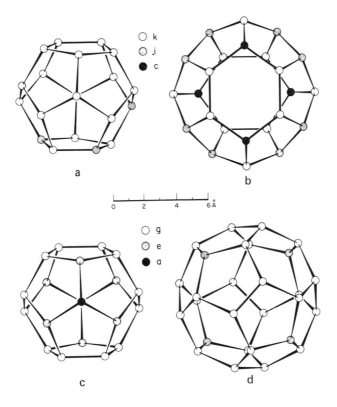

Fig. 3. Cages of structure I (a, b) and structure II (c, d) clathrate hydrates viewed along axes of highest symmetry. Literal
identification of the sites follows the notation of *The International Tables for X-Ray Crystallography.*

(Section 4) than by the small "half-hydrogen" electron density peaks of the
X-ray studies. Moreover, the symmetry of some sites [e.g., the tetrahedral
symmetry of the (a) sites in structure II] is clearly inconsistent with
ordering.

Some geometric parameters[927,938] are given in Table II. In terms of
cage radius the most spherical cage is the large cage of structure II, in which
the distances from oxygen atoms to cage center vary by only 2%. This is
also the cage of highest symmetry. The variation in radius is 14% in the
large structure I cage, which resembles an oblate spheroid with the
shorter axis perpendicular to the plane of Fig. 3(b). The small cages of
the two structures differ little in size; that in type I is somewhat more
spherical. Each is slightly distorted from a regular pentagonal dodecahedron.

TABLE II. Geometry of Unit Cells and Cages

	Structure I	Structure II
Space group, cell parameter	$Pm3n$, 12.03 Å	$Fd3m$, 17.31 Å
Number of H_2O molecules	46	136
Nonequivalent O sites	6 (c), 16 (i), 24 (k)	8 (a), 32 (e), 96 (g)
Oxygen site symmetry	$\bar{4}2m$, $3m$, m	$\bar{4}3m$, $3m$, m
Departure of O- -O- -O angle from tetrahedral:		
Average at O site	1.2°, 1.2°, 5.1°	0°, 2.1°, 3.6°
Average in structure	3.7°	3.0°
Average O- -O length, Å	2.793	2.790
Small cages:		
Number, symmetry	2, $m3$	16, $\bar{3}m$
Oxygen sites	8 (i), 12 (k)	2 (a), 6 (e), 12 (g)
Distance to center, Å	3.83, 3.96	3.748, 3.845, 3.956
Average cage radius, Å	3.91	3.902
Large cages:		
Number, symmetry	6, $\bar{4}2m$	8, $\bar{4}3m$
Oxygen sites	4 (c), 8 (i), 8 (k), 4 (k)	4 (e), 12 (g), 12 (g)
Distance to center, Å	4.25, 4.47, 4.06, 4.645	4.729, 4.715, 4.635
Average cage radius, Å	4.33	4.683

    The relationship of the particular lattice structure to the size of the guest molecule is illustrated schematically in Fig. 4. The size shown is the maximum molecular diameter as estimated by von Stackelberg[1354,1356,1359] or as roughly evaluated from molecular geometry and van der Waals radii.

    Structure I is exclusively formed by molecules of some 5.3 Å or less in diameter, structure II by molecules with diameters in the range of 5.6–6.6 Å. If the radius of a water molecule is taken as 1.4 Å (the van der Waals radius of oxygen), the mean "free diameters" of the small cages are about 5.0 Å and of the larger cages about 5.8 and 6.6 Å, respectively. It is therefore possible for guest molecules of up to 5.0 Å in size to fit easily into both the small and large type I cages and the composition to approach $M \cdot 5\frac{3}{4}H_2O$. Somewhat larger molecules fit with some difficulty into the smaller cages, as in the case of ethylene oxide, which forms a hydrate of composition

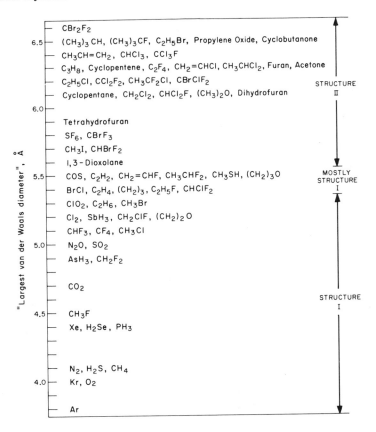

Fig. 4. Molecules which form simple clathrate hydrates, arranged in order of size.

M · 6.89H$_2$O at the congruent melting point.[575] Molecules like CH$_3$SH and CH$_2$=CHF are expected to occupy the larger cages only, with M contents which do not exceed M · 7$\frac{2}{3}$H$_2$O. Presumably because of the shape of the 14-hedra, molecules as large as 5.8 Å are not known to form type I hydrates. The transition to structure II occurs at about 5.5 Å. At least two molecules near this size, trimethylene oxide[647] and cyclopropane,[616] form hydrates of both types. For three others, COS, CH$_3$Br, and CH$_3$CHF$_2$, the hydrates formed change from structure I to structure II in the presence of H$_2$S.[1354] Molecules between 5.6 and 6.6 Å in size form structure II hydrates, with compositions near M · 17H$_2$O. Compositions are discussed in Section 3.2.

The lattice parameter of the simple isostructural hydrates shows no substantial dependence on the nature of the guest species. Von Stackelberg and Jahns[1354] found a slight lattice expansion for guest molecules near the maximum size consistent with the structure formed. All structure I values of a at 0°C for guest molecules less than 5.5 Å in diameter were 12.03 ± 0.06 Å; for COS, CH_3SH, $CH_2=CHF$, and CH_3CHF_2 hydrates (Fig. 4) they were 12.12 ± 0.02 Å, i.e., $\frac{3}{4}\%$ higher.[1354] With the exception of the unusually low lattice parameter of CH_3I hydrate and the high value for $(CH_3)_2O$, all simple type II hydrates and double hydrates with H_2S of molecules less than 6.5 Å in diameter gave 17.32 ± 0.11 Å. Molecules of 6.5 Å or greater in size formed hydrates with $a = 17.51 \pm 0.09$ Å. These included, besides the simple hydrate of isobutane, the H_2S double hydrates of molecules too large to form simply hydrates [CCl_4, CCl_3NO_2, $(CH_3)_2S$, n-C_3H_8Br, CCl_3Br, and C_6H_6] and which ranged up to about 6.9 Å in size.

Sargent and Calvert's[1178] precise lattice parameters at $-138°C$ of six structure II hydrates of molecules ranging from trimethylene oxide to tetrahydrofuran in size, average 17.133 Å, from which the small variation (± 0.037 Å) shows no correlation with molecular size.

McIntyre and Petersen[917] examined the dependence of the lattice parameter of ethylene oxide hydrate on temperature and on the composition of the solution from which hydrate was prepared. They found a at $-20°C$ to be about 0.5% higher and the thermal expansion coefficient larger by a factor of two for hydrate formed from a 12.8 mole % solution of ethylene oxide than from solutions containing less than about 5 mole %. These differences probably arise from sensitivity to the degree of occupancy of the small cages[575] and the outward expansion of these cages with increased thermal motion of encaged ethylene oxide. For the smallest cage occupancy the thermal expansivity is only slightly larger than that of ice.[917] An average expansivity between -163 and 0°C twice as great as for ice may be estimated from the structure II data of Sargent and Calvert.[1178]

The small degree of dimensional variation among isostructural hydrates contrasts with the β-quinol clathrates, where the cages are more flexible. Thus the c dimension of the β-quinol hexagonal unit cell varies from some 5.49 Å for such small guest molecules as HCl, H_2S, and CH_3OH to 6.24 Å for CH_3CN.[1068]

2.2. Hydrates I and II as Ices

These hydrates are properly ices in the sense that each water molecule is hydrogen-bonded to four nearest neighbors, as in all the known forms of ice, the guest molecules playing a stabilizing and space-filling role which depends little on their identity (Section 3.1). The average departure of

O- -O- -O angles from the tetrahedral values found in ices Ih and Ic is only 3.7° and 3.0° in structures I and II, respectively (Table II), much less than in the high-pressure ices II, III, V, and VI.[723] The O- -O bond lengths on average exceed those in hexagonal ice (2.76 Å) by only 1%, and are comparable to those in most of the other forms of ice.

Aside from the cages, the most obvious structural feature which distinguishes the clathrate ices from other forms of ice is the predominance of planar (structure II) and almost planar (structure I) five-membered rings. The less numerous pentagons which occur in ice IX[724] (and III) are far from planar. The hexagonal rings are likewise planar (structure I) or nearly so (structure II), in contrast to the puckered hexagons found in other forms of ice. Such planarity means that the three water molecules hydrogen-bonded to the molecule at each end of a hydrogen bond are eclipsed when viewed along that bond, or, in Bjerrum's terminology,[177] that the arrangement about the bond is mirror symmetric (ms). In cubic ice Ic and probably in ices VII and VIII the orientations about all bonds are staggered or trans in a center symmetric (cs) arrangement; this is true also in Ih except for one-fourth of the bonds which lie parallel to the c axis, about which the arrangement is ms.

Bjerrum considered[177] that the ms arrangement was more stable than the cs but that it was probably not possible to construct a crystal in which more than one-fourth of the bonds were of the ms type. Strictly speaking, the requirements of ring and cage closure produce some distortion of most bonds in the clathrate hydrates from the ideal tetrahedral ms arrangement found around c bonds in Ih. It is this distortion of angles and bond lengths and the weaker attractive forces between nonbonded neighbors, rather than any fundamental instability of ms with respect to cs bonding, which accounts for the low stability of the empty clathrate hydrate lattices with respect to Ih (Section 3.1).

Pentagonal rings outnumber hexagonal rings in a ratio of 8:1 in structure I and 9:1 in structure II, both close to the ratio of 8.62:1 which would make the average angle exactly tetrahedral if all faces were exactly planar. The departures from planarity which occur are in the direction of equalization of hydrogen bond lengths throughout the structures.[938]

2.3. Bromine Hydrate

This, the second gas hydrate to be discovered, was the subject of many early studies. Löwig[872,873] described the crystals as octahedral, but other early observations of crystal shape and the effect on polarized light appear

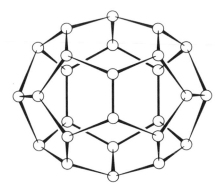

Fig. 5. The pentakaidecahedral cage which
occurs in bromine hydrate.[19]

not to have been reported. Estimates of the hydration number ranged
from six to ten. The most accurate composition determination was that of
Mulders,[1007] who obtained $Br_2 \cdot 8.36H_2O$ and $Br_2 \cdot 8.47H_2O$ at Q_1 from
vapor pressure measurements along the h-l_1-g and h-I-g equilibrium lines.
Von Stackelberg and Müller[1359] assigned bromine hydrate to structure I
on the basis of rather unsatisfactory diffraction patterns and associated the
relatively high water content with the inability of the large Br_2 molecule
to enter the small cages.

Allen and Jeffrey[19] found a tetragonal unit cell of Laue symmetry
$4/mmm$ with $a = 23.8$ and $c = 12.2$ Å at $-10°$. A detailed structure has
not been presented but, on the basis of the similarity in symmetry and cell
size to the tetragonal tetra n-butyl ammonium salt hydrates (Section 2.6),
the structure is presumed to consist of the "ideal" tetragonal cell containing
172 water molecules, ten 12-hedra, sixteen 14-hedra, and four 15-hedra (Fig.
5). With the Br_2 molecules encaged in the two larger polyhedra the ideal
composition is $Br_2 \cdot 8.6H_2O$. The Br_2 molecule forms the only known hydrate in which the guest molecules occupy undistorted 15-hedra.

2.4. "Ice VIII"

In 1896 Barendrecht[88] concluded that the cubic crystals formed at
low temperatures in aqueous solutions of more than 50% ethanol and in
solutions of other alcohols and acetaldehyde were a form of ice. Cohen and
van der Horst[322] came to the same conclusion in 1938 with respect to the
solids formed in solutions of acetone, acetaldehyde, and other organic
liquids, and proposed the name "ice VIII." Rau[1132] found that water

droplets, after many cycles of freezing and thawing, froze consistently as cubic crystals at $-72°C$. These observations could not be repeated for pure water[220,356] but similar behavior was induced by the presence of ether or acetone vapor[356] and by ethanol.[220] It was inferred[220] that Rau's samples had gradually taken up ethanol from the cooling bath and that his cubic ice, like the solid found to form below $-72°C$ in 70% ethanol solution, was probably ethanol monohydrate.

The X-ray study by Quist and Frank[1124] showed the "ice VIII" formed in concentrated acetone solutions to be a structure II hydrate, as did subsequent dielectric studies,[1000,1416] which also showed the absence of hydrogen bonding between the acetone and water molecules. Acetone hydrate has a peritectic melting point at $-19.8°$.[1000]

Figure 6 shows a phase diagram of the ethanol–water system constructed from thermal analysis and dielectric studies.[1114] An ethanol hydrate was found to exist below $-73.5°C$ and to have a composition close to that expected for a structure II hydrate. The ethanol molecules are not rotationally mobile, however, and are probably hydrogen-bonded to water. A somewhat earlier thermal analysis study by Vuillard and Satragno[1362] put the hydrate decomposition temperature at $-74°C$ and the composition at ~ 5 moles of water per mole of ethanol. Recent X-ray work by Calvert

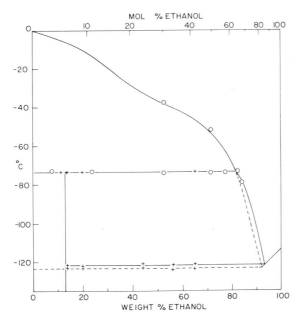

Fig. 6. Phase diagram of the ethanol–water system.[1114]

and Srivastava[265] has shown the formation of both a type I ($Pm3m$, $a = 11.88$ Å at $-165°C$) and a "modified" type II ($F4_132$, $a = 17.25$ Å) ethanol hydrate, the former possibly only stable when a large proportion of its cages are occupied by air.* Detailed structures have not been published.

There is no evidence about the structures of the solids formed in aqueous solutions of the other liquids once considered to promote the formation of "ice VIII," a name which has now been assigned[1400] to an authentic high-pressure form of ice. A clathrate hydrate may well be formed, at least temporarily, by dropwise addition of water to acetaldehyde at $-20°C$,[322] but fail to form during cooling of a prepared solution because of rapid hydrolysis to $CH_3CH(OH)_2$.[128,129]

2.5. Amine Hydrates

Pickering's[1089] early study of the melting points in amine–water systems established the existence of numerous alkylamine hydrates which melted at congruent compositions ranging from $\frac{1}{2}$ to about 37 mol of water for each mole of amine. Some are listed in Table III. Melting curves[1089,1239] also showed the presence of a number of incongruently melting hydrates of more uncertain composition. Glew's analysis[565,566] of the earlier data[1089,1239] gave two higher hydrates of diethylamine and led him to suggest that one of these likely had the composition $(C_2H_5)_2NH \cdot 6\frac{2}{3}H_2O$ expected for a structure I clathrate in which the NH group replaced a water molecule in the lattice. Moreover, $(CH_3)_3N \cdot 10H_2O$ was probably[565,566] the ideal hexagonal clathrate structure related to the known structure of (iso-$C_5H_{11})_4NF \cdot 38H_2O$ (see Section 2.6).

An extensive crystallographic study of structures of higher amine hydrates by Jeffrey's group has revealed an interesting variety of "semi-clathrate" structures in which the amino groups are generally linked by hydrogen bonds to water molecules. To date, only the *tert*-butylamine hydrate has been found[928] to lack such bonds and consequently to be a true clathrate. Some of the other structures may be recognized as modifications, with degraded symmetry, of the archetypal clathrate lattices of cubic or hexagonal symmetry. Thus $(CH_3)_3N \cdot 10\frac{1}{4}H_2O$ is found[1070] to differ from the "ideal" hexagonal structure suggested[565,566] by Glew in the presence of hydrogen bonding of water to amine molecules contained in two kinds of irregular cages, one (Fig. 7a) related to a double 14-hedron and the second (Fig. 7b) very much distorted from the 15-hedron of the ideal structure. Hexagonal symmetry is maintained by sixfold or threefold

---

* L. D. Calvert, private communication.

TABLE III. Higher Hydrates of Amines[a]

Amine (A)	Melting point, °C, and hydration number from mp curve[1089]	Space group, unit cell composition	Unit cell dimensions, Å	Range of O--O lengths, Å, O--O--O angles, deg	Polyhedra
1 $(CH_3)_2NH$	−16.9, 6.9	$P23$ or $Pm3$[930]	12.55	—	$3\times12, 2\times d14^*, 2\times d15^*$
2 $(CH_3)_3N$	5.9, 10.22[566]	$P6/mmm$[1070], $4A \cdot 41H_2O$	12.378, 12.480	2.74–2.85, 96.5–137.3	$3\times12, 2\times d14^*, 2\times d15^*$
3 $C_2H_5NH_2$	−7.5, 5.45	$P\bar{4}3n$ or $Pm3n$[930]	12.17	—	—
4 $(C_2H_5)_2NH$	−6.57, 6.80[566]	$P2_1/c$[930]	13.86, 8.44, 10.93, $\beta = 97.5°$	—	—
5 —	−7.28, 8.10[566]	$Pbcn$[718], $12A \cdot 104H_2O$	13.44, 11.77, 27.91	2.76–2.96, 84.1–123.3	$4\times18^*, 8\times$irreg*
6 n-$C_3H_7NH_2$	−20 (incong.)	$P2_1/n$[1221], $16A \cdot 104H_2O$	12.43, 20.73, 27.28, $\beta = 89.3°$	—	11, 14*, $d16^*$, 2 kinds of irreg*
7 —	−13.5, 7.96	$(6/mmm)$[930]	12.20, 12.38	—	—
8 i-$C_3H_7NH_2$	−4.2, 7.53	$P6_3/mmc$[929], $10A \cdot 80H_2O$	12.30, 24.85	2.68–2.88, 88.9–144.2	$2\times8, 2\times12, 6\times14^*, 4\times16^*$
9 $(CH_3)_3CNH_2$	−1[928]	$I\bar{4}3d$[928], $16A \cdot 156H_2O$	18.81	2.77–2.85, 87.9–133.2	$12\times8, 16\times17^*$
10 $(CH_2)_6N_4$	13[462]	$R3m$[937], $A \cdot 6H_2O$	7.30, $\alpha = 105.4°$	2.764–2.766, 108.5–126.6	$1\times$irreg*

[a] Under Polyhedra, 3×12 means three dodecahedra per unit cell, d a cage derived from a polyhedron, irreg a cage not easily identified with a polyhedron. Cages marked with an asterisk are occupied. Structures 1 and 3 are probably variants of structure 1; 2, 8, and probably 7 are variants of the ideal hexagonal structure. Symmetry is maintained in 2 and 8 by disorder of the positions of the amine and some of the H_2O molecules, while there is no positional disorder of H_2O molecules in 5, 6, 9, or 10. All H_2O molecules are four-coordinated, except for four in the unit cell of 2, eight in 8, and three in 10, which are three-bonded.

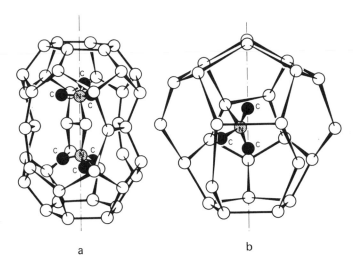

Fig. 7. The occupied cages in trimethylamine hydrate, showing one
of the disordered arrangements of guest molecules.

disorder of the positions of the amine molecules and of seven of the 41
water molecules of the unit cell.

Some additional structural information is given in Table III. There is
considerably more variability in O- -O length and O- -O- -O angle than is
found in the simpler structures I and II. In some hydrates disorder in the
amine configurations is accompanied by disorder in the sites of some water
molecules and, in several, hydrogen bond balance is achieved by three-
coordination of some water molecules. In all of the alkylamine hydrates,
however, the great majority of the water molecules are four-coordinated.
This is not true for the rather special case of hexamethylenetetramine
hexahydrate, whose relatively simple structure is discussed below.

2.5.1. $(CH_3)_3CNH_2 \cdot 9\frac{3}{4}H_2O$

The unhydrogen-bonded *tert*-butylamine molecules occupy 17-hedral
cages (Fig. 8), which are so far unique to this hydrate.[928] The C—N axis
normally coincides with the threefold cage axis, although some statistical
interchange of the NH_2 and CH_3 groups is possible. The C—C bonds are
not aligned with the centers of the large heptagonal rings in the equilibrium
orientations about the threefold axis but rather toward their edges, presumably to maximize the van der Waals attractive energy. Large-amplitude
rotational oscillations (through rms angles of 18°) were, however, observed.[928] It is possible that the barrier to rotation about the CN axis is
relatively small. (See Section 4.9.1).

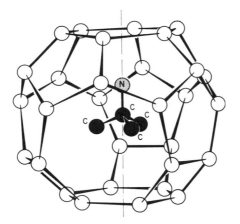

Fig. 8. The heptakaidecahedron of *tert*-bu-
tylamine hydrate.

2.5.2. $(CH_2)_6N_4 \cdot 6H_2O$

In the hydrate of hexamethylenetetramine the guest molecule of tetrahedral symmetry is located at a site of threefold symmetry (Fig. 9) in a
rhombohedral water structure.[710,937] Three N atoms act as proton acceptors
in hydrogen bonds with one (labeled 2 in Fig. 9) of the two nonequivalent
kinds of water molecules which occupy alternate positions in hexagonal

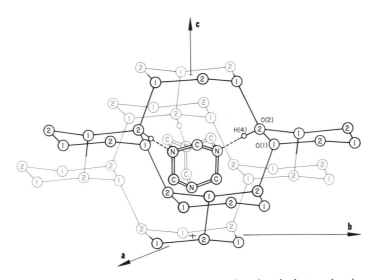

Fig. 9. Structure of hexamethylenetetramine hexahydrate, showing
the hydrogen bonding to three of the four nitrogen atoms.

rings. The framework formed by these rings is similar to one of the two interpenetrating lattices which define the cages in β-quinol clathrates.[1067] In the hydrate case the relatively open cage is bounded by eight hexagons. An electron density difference map showed protons in the hydrogen bond to nitrogen and at the O(1) end of the bond between rings. A slight modification to the conclusion[937] that the protons are disordered only within the bonds forming the hexagonal rings was suggested by the dielectric results (Section 4.9.4).

2.6. Semiclathrate Hydrates of Peralkyl Ammonium and Other Salts

2.6.1. Structures Containing Pentagonal Dodecahedra

The formation of hydrates containing from 18 to about 60 molecules of water per molecule of tetra n-butyl or tetra iso-amyl ammonium salt was shown in Kraus's laboratory in 1940.[499] McMullan and Jeffrey[926] found that isomorphous tetragonal crystals were formed by nine tetra n-butyl ammonium salts with monovalent and divalent anions and about 32 water molecules for every cation present. A second isomorphous series of crystals was formed by the orthorhombic hydrates of four tetra iso-amyl ammonium salts with about 40 water molecules per cation. It was suggested[926] that these were clathratelike compounds similar in general character to the gas hydrates. This observation was verified by later detailed structural studies of hydrates of tetra iso-amyl ammonium fluoride,[476] tetra n-butyl ammonium benzoate,[208] tri n-butyl sulfonium fluoride,[170,707] and tetra n-butyl ammonium fluoride[925] (Table IV).

Figure 10 gives some indication of the adjustments in the ideal hexagonal clathrate lattice, whose unit cell consists of six 12-, four 14-, and four 15-hedra constructed of 80 water molecules, which are made to accommodate the tetra iso-amyl ammonium cation.[476] A water molecule (labeled 1 in Fig. 10) located at the junction of two orthogonal hexagonal rings of the ideal structure and common to two 14-hedra (one behind the other) and two 15-hedra is replaced by the N atom. Neighboring water molecules are pushed apart to permit each of the four iso-amyl groups to fit quite snugly into a 14- or 15-hedral cage. The fluoride anion is hydrogen-bonded within the lattice, where it probably replaces water molecule 2. The removal of four water protons from the ideal structure for each salt molecule present is accompanied by the disappearance of the four original hydrogen bonds to molecule 1 and the number of hydrogen bonds remains equal to the number of available protons. At the same time the number of stable orientations of the water molecules is reduced from the six of the ideal framework to three at sites neighboring the N^+ and F^- sites.

Table IV gives the relationship between the structures of other salt hydrates and the ideal frameworks present in bromine hydrate and in structure I hydrates. In these cases the structures are complicated by the presence of positional disorder of some of the lattice water molecules and in the orientations of the cations. Hydrogen-bonded water molecules were found to occupy some distorted dodecahedra. In the tetra n-butyl ammonium benzoate hydrate the phenyl rings occupy 14-hedra unoccupied by the cations.

The salts listed in the last row of Table IV are thought,[172,708] largely on the basis of isomorphism, to have related structures derived from the same ideal clathrate framework. Variation in the detailed structures with the size and valence of the anions is to be expected.

The structure of a second hydrate of tri n-butyl sulfonium fluoride, monoclinic $(n\text{-}C_4H_9)_3SF \cdot 23H_2O$, has been found[170] not to be simply related to any of the ideal clathrate lattices, although pentagonal dodecahedra still occur. Pairs of cations occupy large irregular 60-hedra, formed of rhombs as well as pentagons and hexagons, which are so far unique to this structure. A number of other onium salt hydrates[708] have structures which are unknown and probably unrelated to those of Table IV.

The cations of the salts which form high hydrates appear to be almost exclusively restricted to tetra n-butyl ammonium and phosphonium, tetra *iso*-amyl ammonium, and tri *iso*-amyl sulfonium ions. Close van der Waals fit of the guest species in the cages seems to contribute a necessary element to their stability. The nature of the anion is less restricted, although the multiplicity of structural types displayed by fluorides probably reflects the ability of F^- to substitute for H_2O with a minimum of distortion of the host lattice.

2.6.2. Structures Based on Truncated-Octahedra

A rather different clathrate framework of four-coordinated water molecules may be constructed of close-packed truncated-octahedra.[1385] In undistorted form these consist (Fig. 11a) of eight regular hexagonal and six square faces. They fit together to occupy all the space within a cubic *Im3m* lattice. The ideal unit cell composition is $2M \cdot 12H_2O$. The structure is unique among clathrate and semiclathrate hydrates in the equivalence of all the water molecules. The only hydrate known to form this ideal framework is the hexahydrate of hexafluorophosphoric acid,[206] in which

TABLE IV. Tetraalkyl Ammonium Salt and Related Hydrates[a]

Unit cell of ideal framework

System	Hexagonal	Tetragonal	Cubic
Space group	$P6/mmm$	$P4_2/mmm$	$Pm3n$
Polyhedra	$6\times12, 4\times14, 4\times15$	$10\times12, 16\times14, 4\times15$	$2\times12, 6\times14$
Composition (12-hedra empty)	$8M \cdot 80H_2O$	$20M \cdot 172H_2O$ (Br_2 hydrate)	$6M \cdot 46H_2O$ (structure I)

Actual unit cell

Salt S	$R_4'NF$	$R_4NC_6H_5CO_2$	R_4NF	R_3SF
MP, °C	31	6	25	~5
System	Orthorhombic	Tetragonal	Tetragonal	Cubic
Dimensions, Å	12.08, 12.61	23.57, 12.45	23.52, 12.30	12.34
Space group	$Pbmm$	$P4_2/mmm$	$P4_2/m$	$Pm3n$
Compound cages enclosing cations	$2\times(2\times14+2\times15)$	$4\times(3\times14+1\times15)$	$4\times(3\times14+1\times15),$ $1\times(4\times14)$	$2\times(3\times14)$

Anion	Replaces H_2O	Two O's of benzoate replace H_2O's	Replaces H_2O	Probably replaces H_2O
Number of H_2O molecules absent from ideal cell	4	16^b	10^b	Probably 8^b
Composition	$2S \cdot 76H_2O$	$4S \cdot 158H_2O$	$5S \cdot 164H_2O$	$2S \cdot 40H_2O$
O--O bond lengths, Å	2.75-2.86	2.62-2.95	2.70-2.87	—
O--O angles, deg	91-133	91-135	98-126	—
Related hydrates	$R_4'N$-Cl, -OH, -$C_6H_5CO_2$, -HCO_2; $(R_4'N)_2$-CrO_4, -WO_4; $R_2R_2'NF$	R_4N-n-$C_3H_7CO_2$, -RCO_2, -$R'CO_2$	R_4N-Cl, -Br, -HCO_3, -CH_3CO_2, -n-$ClC_6H_5CO_2$, -OH, -NO_3, -HCO_2; $(R_4N)_2$-CrO_4, -WO_4, -C_2O_4, -HPO_4, -p-$C_6H_4(CO_2)_2$; $R_3R'NF$; R_4PF; $(R_4P)_2C_2O_4$	R_4N-n-$C_3H_7CO_2$, -RCO_2; R_3NH-n-$C_3H_7CO_2$; $(R_4N)_2$-C_2O_4,^{c} -WO_4;^{c} $(R_4P)_2C_2O_4^c$

[a] R and R' refer to n-butyl and iso-amyl groups, respectively. $4 \times (3 \times 14 + 1 \times 15)$ means that each of four cations occupies a compound cage derived from three tetrakaidecahedra and one pentakaidecahedron.

[b] Two additional water molecules occupy distorted 12-hedra.

[c] These salts form cubic cells of 24.6 Å with compositions of about $6S \cdot 344H_2O$.

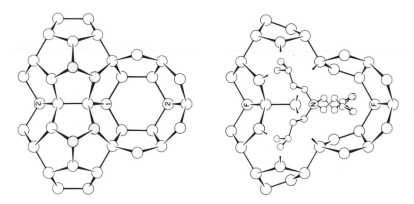

Fig. 10. Distortion of part of the ideal hexagonal structure (above) to accommodate (*iso*-C$_5$H$_{11}$)$_4$NF (below). The cation occupies two 15-hedra and two 14-hedra, of which one is above and one below the hexagon in the plane of the upper figure. The N and F atoms replace water molecules 1 and 2 of the undistorted structure. The shaded water molecules are omitted from the lower figure for clarity.

the PF$_6^-$ anion is the guest species. The 14-hedral cages are similar in general size to the 14-hedral cages of structure I, but all 24 cage water molecules are equidistant from the cage center. A general lack of stability associated with considerable distortion of the O- -O- -O angles from tetrahedral values may be overcome by the ionic nature of HPF$_6$. The position of H$^+$ is unknown but possibly is distributed as H$_3$O$^+$ over all the water molecules.

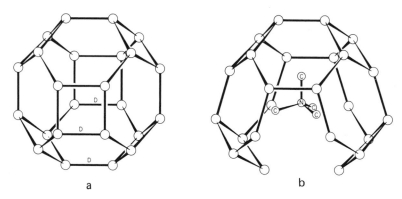

Fig. 11. (a) The ideal truncated-octahedral cage found in HPF$_6$ · 6H$_2$O[206] and (b) the distorted form found in (CH$_3$)$_4$NOH · 5H$_2$O.[931] In the latter, three of the CH$_3$ groups display orientational disorder about the vertical C—N axis.

This basic framework is considerably distorted in $(CH_3)_4NOH$ · $5H_2O$,[931] in which the substitution of OH^- ions for lattice water molecules is accompanied by the disappearance of three parallel hydrogen bonds (D in Fig. 11a) of the cage and expansion to accommodate the relatively large tetramethyl ammonium ion (Fig. 11b). There are now three non-equivalent O atom sites in an orthorhombic (Cm3m) unit cell of 20 water molecules. The distributions of the four OH^- ions over the O sites and of the numerically deficit protons over the bonds are uncertain, but both are probably disordered.

2.7. Comparison of Clathrate and Semiclathrate Structures

Although the "ideal" clathrate structures of cubic symmetry are of quite general occurrence, the ideal tetragonal and hexagonal structures, except in the single instance of bromine hydrate, are not known to occur in undistorted form. Given the variety of size and shape of guest molecules available, a somewhat lower intrinsic stability on the part of these non-cubic structures is suggested, possibly as a result of an element of strain introduced into the hydrogen-bonded framework by the presence of 15-hedra. The distortion of O- -O- -O angles from the tetrahedral angle is particularly significant in these structures at the numerous sites which are common to two orthogonal hexagons. These are the sites at which water molecules are partially replaced by N, F, or other atoms of guest species in the distorted salt hydrate structures.

The salt hydrates of Table IV owe a considerable part of their stability to the closeness of fit in the cages of the *n*-butyl and *iso*-amyl groups and some of the cations. The structures of hydrates of organic salts of other geometries, like those of the larger amines, are expected to show little relationship to the ideal clathrates.

With the possible exception of the highly disordered cubic hydrate of tri *n*-butyl sulfonium fluoride, the semiclathrate hydrates must be considered as stoichiometric, in contrast to hydrates of the gas hydrate type. Little information is available about other than structural properties. The presence of hetero-atoms in the hydrogen-bonded water lattice, of some three-bonded (but apparently never less than three-bonded) water molecules, and of wide distributions of O- -O bond lengths and O- -O- -O angles (cf. entries in Tables III and IV with ranges of 2.77–2.81 Å, 105.7–119.9° and 2.77–2.84 Å, 105.5–124.3° in structures II and I, respectively) should be reflected in the width of the infrared bands and distributions of re-orientation rates of the water molecules. (See Sections 4.9, 4.10, and 4.11.4.)

Significant differences in relative stability of different orientations of some
water molecules are to be expected. No exceptionally high overall molecular
reorientation rates of guest species are to be expected (Section 5.8.5),
except possibly for PF_6^- and $(CH_3)_4N^+$ in the structures mentioned in
Section 2.6.2. However, large-amplitude thermal oscillations of parts of
guest molecules not directly incorporated in the lattice framework were
generally observed in the X-ray studies.

3. THERMODYNAMIC PROPERTIES

3.1. The Ideal Solution Model

Once the clathrate and nonstoichiometric nature of the gas hydrates
was clear, it was natural to consider a thermodynamic model in which the
guest molecules are dissolved or absorbed in the host water lattice.[1326]
The theory was first applied by van der Waals[1327] to β-quinol clathrates,
in which the cages are all of one kind, and was later extended to gas hydrates
with guest molecules of one[1105] or more kinds.[104] The most elegant treatment is that by van der Waals and Platteeuw.[1328]

The results are most directly derived from the partition function (PF)
written in the form[1328]

$$\Xi = \exp\left(-\frac{F^\circ}{kT}\right) \sum_{N_{Mi}} \prod_i \left[\frac{N_i!}{(N_i - N_{Mi})!\, N_{Mi}!} (b_{Mi}\lambda_M)^{N_{Mi}} \right] \tag{1}$$

when only a single guest species M is considered. F° is the free energy of the
water molecules in the empty lattice, N_1 and N_2 are the numbers of small
and large cages, respectively, and N_{M1} and N_{M2} the numbers of M-type
molecules in the respective cages. The absolute activity λ_M is related to the
μ potential of M by $\mu_M = kT \ln \lambda_M$, and b_{Mi} is the molecular PF of M in
cage i.

Equation (1) is the product of an ordinary PF of the empty host
lattice, which is a function of T, V, and the number of water molecules
N_w, and a grand PF of guest molecules, which depends also on λ_M. The
latter is precisely the grand PF for ideal localized adsorption,[500] the factor
$N_i!/[(N_i - N_{Mi})!\, N_{Mi}!]$ giving the number of distinguishable ways of dis-
tributing N_{Mi} molecules over N_i adsorption sites.

Equation (1) depends on the validity of two principal assumptions.
The first is that the contribution of the H_2O molecules specified by F° is
independent of the degree of occupancy of the cages and therefore of the

nature of M, the interaction between M and H_2O molecules being entirely given by the grand PF of M molecules. The second assumption is that b_{Mi}, the PF of a guest molecule which can only singly occupy a cage, is the same for each cage of the ith type and is independent of the occupancy of other cages. The first assumption is not strictly valid for large guest molecules which distort the lattice, although there is no evidence of serious distortion of the lattice dimensions (Section 2.1). The second assumption is a much better approximation for the clathrate hydrates than for adsorption generally, where the sites are defined and localized less well. Strong polarity of guest molecules may be expected to affect the validity of both assumptions.

The summation in eqn. (1) over all possible values of N_{M1} and N_{M2} may be evaluated by the multinomial theorem[1166] to give

$$\Xi = \exp(-F^\circ/kT) \prod_i (1 + b_{Mi}\lambda_M)^{\nu_i N_w} \tag{2}$$

where N_i has been replaced by $\nu_i N_w$. In structure I, for example, $\nu_1 = 1/23$ and $\nu_2 = 3/23$.

The familiar thermodynamic functions are related to Ξ by equating the sum in eqn. (1) to its most probable term, that corresponding to the equilibrium distribution of M over the cages. The result is[1328]

$$kT \, d(\ln \Xi) = (U/T) \, dT + P \, dV + N_M kT \, d(\ln \lambda_M) - \mu_w \, dN_w \tag{3}$$

in which N_M is the equilibrium number of encaged molecules of activity λ_M, that is,

$$N_M = N_{M1} + N_{M2} = [\partial(\ln \Xi)/\partial(\ln \lambda_M)]_{T,V,N_w}$$
$$= \sum_i [\nu_i N_w b_{Mi}\lambda_M/(1 + b_{Mi}\lambda_M)]$$

from eqn. (2). Thus

$$\theta_i = N_{Mi}/(\nu_i N_w) = b_{Mi}\lambda_M/(1 + b_{Mi}\lambda_M) \tag{4}$$

gives the degree of occupancy of the cages. Similarly,

$$\mu_w(h) = -kT[\partial(\ln \Xi)/\partial N_w]_{T,V,\lambda_M} = \mu_w^\circ - kT \sum_i \nu_i \ln(1 + b_{Mi}\lambda_M)$$
$$= \mu_w^\circ + kT \sum_i \nu_i \ln(1 - \theta_i) \tag{5}$$

in which μ_w° is the chemical potential of the empty lattice. This equation expresses the effect of cage occupancy on the stability of the lattice and may be recognized as a form of Raoult's law.

The activity of encaged molecules in equilibrium with gaseous M at partial pressure (strictly, fugacity) p_M is

$$\lambda_M = \frac{p_M}{kT\phi_M} = \frac{p_M}{kT(2\pi m k T/h^2)^{3/2} j_M} \tag{6}$$

with j_M the molecular PF for vibration and rotation of M in the gas. The product $b_{Mi}\lambda_M$ may be written as $C_{Mi}p_M$, in which C_{Mi} depends on T and only slightly (through the effect of a finite compressibility on b_{Mi}) on pressure. Equation (4) now becomes Langmuir's isotherm

$$\theta_i = C_{Mi}p_M/(1 + C_{Mi}p_M) \tag{7}$$

For stability of the hydrate with respect to ice or liquid water $\mu_w(h) \leq \mu_w(I)$ or $\mu_w(l_1)$. At equilibrium with ice at a temperature along the h-I-g line the equality of μ potentials leads from eqn. (5) to*

$$\Delta\mu_w = \mu_w{}^\circ - \mu_w(I) = -kT \sum_i \nu_i \ln(1 - \theta_i) \tag{8}$$

which determines the minimum cage occupancy and, from eqn. (7), the minimum pressure of M required for hydrate stability. At higher temperatures $\mu_w(I)$ in eqn. (8) is replaced by $\mu_w(l_1)$ corresponding to the h-l_1-g equilibrium.

The constancy of $\sum \nu_i \ln(1 - \theta_i)$ for isostructural hydrates at a fixed temperature of equilibrium does not mean in general that the composition at that temperature is the same for different M molecules, since the individual values of θ_1 and θ_2 are not fixed. For the many hydrates, however, in which M is too large to be readily accommodated in the small cages (Section 2.1), b_{M1} and θ_1 approach zero and θ_2 is determined by eqn. (8). Thus, except for the small effect of pressure, the model predicts that all simple structure II hydrates have the same composition $\theta_2 M \cdot 17 H_2O$ at the same temperature along the h-I-g or h-l_1-g† equilibrium line. The corresponding composition of structure I hydrates for which $\theta_1 = 0$ is $\theta_2 M \cdot 7\frac{2}{3} H_2O$.

A value of $\theta_2 = 0.905$ for structure I at 0°C was originally taken[1105,1328] from an experimental composition[1007] of Br_2 hydrate at Q_1 and used to give $\Delta\mu_w = 167$ cal mol^{-1}. This much-quoted value must be discarded since Br_2 is now known not to form a structure I hydrate (Section 2.3).

---

* The variation of $\Delta\mu_w$ with pressure is given by $P \Delta V_w$, which is normally small ($\sim$0.08 P/atm in cal/mol) for I $\rightarrow$ h but becomes significant at high pressures, especially for $l_1 \rightarrow$ h.

† Along h-l_1-g this is only true for M negligibly soluble in water.

From eqns. (2) and (3) the internal energy is

$$\frac{U}{T} = kT\left[\frac{\partial(\ln \Xi)}{\partial T}\right]_{V,N_w,\lambda_M} = \frac{U^\circ}{T} + kT\sum_i \frac{\nu_i N_w \lambda_M\, \partial b_{Mi}/\partial T}{1 + b_{Mi}\lambda_M}$$

or, from eqn. (4),

$$U - U^\circ = N_w kT^2 \sum_i \nu_i \theta_i\, \partial(\ln b_{Mi})/\partial T \tag{9}$$

gives the contribution to the energy of the hydrate from encagement of $N_w \sum \nu_i \theta_i$ molecules of M.

3.2. Determination of Hydrate Compositions

The composition is the only property of hydrates which can serve as a test of the solution model without recourse to calculation of partition functions of the guest molecules. In Tables V and VI experimental information about the compositions of hydrates of structures I and II is collected, mostly from application of the de Forcrand method already mentioned (Section 1.1). Accurate application of this method requires adequate definition of $d(\ln P)/d(1/T)$ along the h-I-g and H-l_1-g equilibrium lines at their point of intersection (Q_1) and correction of the resulting enthalpies for nonideality of the gas and solubility of M in liquid water. Results of this method are given in column 4 of Tables V and VI. An attempt has been made to express the relative accuracy of difference measurements by weighting them on a scale of zero to three, where zero indicates dubious significance.

The method of Miller and Strong[975] has been comparably recently used to determine the compositions in column 6. It depends on the lowering effect of added solute (usually NaCl) on the activity of water in the liquid phase and a consequent increase in hydrate decomposition pressure. At constant temperature the equilibrium constant of the reaction

$$M \cdot nH_2O(h) \rightleftharpoons M(g) + nH_2O(l_1)$$

is assumed to be unaffected by the addition of solute and n is found from

$$n = \frac{\ln[p_M(S)/p_M(O)]}{\ln[a_w(O)/a_w(S)]} \tag{10}$$

where $p_M(S)$ and $p_M(O)$ are the equilibrium partial pressures (fugacities) of M in the presence and absence of added solute and $a_w(S)$ and $a_w(O)$ the

TABLE V. Compositions of Structure I Hydrates

Hydrate	A, $\Delta H(\mathrm{h} \rightarrow l_1+\mathrm{g})$, kcal mol⁻¹	B, $\Delta H(\mathrm{h} \rightarrow \mathrm{I}+\mathrm{g})$, kcal mol⁻¹	From $A - B$		From effect of NaCl		From analysis
			n	w	n	w	n
Ar	11.72[946]	2.94[101]	6.15	2	—	—	—
Kr	13.9[1350]	3.98[101]	6.90	1	—	—	6.05–6.25[a][101]
Xe	16.7[1350]	5.77[101]	7.61	0	—	—	
N₂	12.38	3.81	6.01[1322]	3	6.21[1319]	2	—
O₂	11.84	3.19	6.06[1322]	3	6.11[1319]	2	—
Cl₂	17.34	7.22	7.05[1414]	3	6.90[1414]	2	7.27[17]
			6.86[b][1414] (0°)	1	(4°)		7.04[1414] (0°)
BrCl	—	—	7.28[569]	1	—	—	7.77[569]
CO₂	14.4	5.68	6.07[831]	2	—	—	—
			5.98[651]	2			
H₂S	14.98	6.10	6.06[781]	3	—	—	—
SO₂	16.6	7.9	6.20[1350]	1	—	—	6.0[1279]

Compound							
ClO$_2$	16.5[218]	—	—	—	—	—	5.92[175]
CH$_4$	12.83	4.55	5.765[564]	2	—	—	6.0[541]
	—	4.34^a[979]	5.92	2	—	—	(16°)
C$_2$H$_2$	15.0[1350]	5.33^a[979]	6.74	1	—	—	—
C$_2$H$_4$	14.87	5.275	6.68[427]	2	6.9[1323]	2	—
	16.27	6.27	6.99[782]	3	—	—	—
C$_2$H$_6$	17.98	6.13	8.25[386]	2	—	—	8.24[541]
	—	6.33^a[979]	8.11	1	—	—	(4°)
	16.3[1147]	6.33	6.94	1	—	—	—
Cyclopropane (deuterate)	19.06	7.405	8.12[616]	2	7.87[616]	3	—
	—	—	7.76[616]	2	—	—	—
CH$_3$Cl	18.25	(6.5)	8.2[574]	1	—	—	—
CH$_2$ClF	19.70[87]	—	—	—	7.98[87]	2	—
CHClF$_2$	25.1	7.0	12.6[1421]	0	—	—	—
CH$_3$Br	19.34	8.01	7.89[86]	2	7.89[224]	2	—
	19.5	8.1	7.94[1350]	1	—	—	—
Ethylene oxide	—	—	—	—	—	—	6.89[575] (11.1°)

^{a} From low-temperature measurements.
^{b} From density and lattice parameter.

TABLE VI. Compositions of Structure II Hydrates

Hydrate	A, $\Delta H(\text{h} \to l_1+\text{g})$, kcal mol⁻¹	B, $\Delta H(\text{h} \to l+\text{g})$, kcal mol⁻¹	From $A - B$		From effect of NaCl		From analysis
			n	w	n	w	n
SF₆	29.57	5.14	17.02[1240]	3	—	—	—
Cyclopropane (deuterate)	29.2	6.42	15.87[616]	1	17.05[616]	3	—
	32.37	6.44	17.18[616]	3	—	—	—
n-Propane	32.1	6.34	17.94[386]	1	17.9[87]	1	19.7[275]
iso-Butane	30.5	5.45	17.45[85]	2	17.50[85]	2	17.1[1314]
CH₃I	31.4	7.3	16.79[1350]	1	—	—	—
CHCl₃	30.7[590]	—	17.25ᵃ[590]	1	—	—	18,[280] 17.7[103]
CHCl₂F	32.72	8.51	16.86[80]	2	16.80[1421]	2	—
CCl₃F	35.45	11.57	16.63[1421]	0	—	—	—
CCl₂F₂	30.14	7.79	15.57[1421]	0	—	—	—
CBrClF₂	32.567 31.86[1272]	8.254	16.94[562]	4	16.57[1272]	2	—
CBrF₃	29.42	6.99	15.62[1421]	0	—	—	—
CH₃CH₂Cl	31.9	8.7	16.16[1350]	1	—	—	—
CH₃CClF₂	31.11	7.49	16.45[224]	1	17.18[224]	2	—
C₄H₈O	27.1[590]	—	16.86ᵃ[590]	3	—	—	—

ᵃ From density and lattice parameter.

corresponding activities of water in the liquid. In so far as the composition of the hydrate is assumed not to be changed by the change of pressure [eqn. (7) is therefore violated], this method is not strictly valid except in the extrapolation to infinite dilution. However the error from this source does not appear to be large for the systems to which the method has been applied, namely hydrates in which $\theta_2 \approx 1$ and $\theta_1 \approx 0$. [The error may be estimated by using the value of n obtained to find the approximate value of C_{M2} from eqn. (7) and thence the change of n for a change of partial pressure from $p_M(0)$ to $p_M(S)$.] The method of Pieron,[1092] in which the effect of added solute on the h-l_1-g equilibrium temperature is determined at constant pressure, requires a knowledge of $\Delta H(h \rightarrow l_1 + g)$. It has been used,[1323] with added ethanol, to determine the composition of ethylene hydrate.*

As pointed out by von Stackelberg,[1350] comparison of measured hydrate density with the X-ray lattice parameter yields information about the degree of occupancy of the cages. Glew and Rath[575] determined the densities of ethylene oxide hydrate under a number of equilibrium conditions by flotation and demonstrated its variable stoichiometry. The compositions of tetrahydrofuran and chloroform hydrates were determined[590] from measurements[590,1350] of the volume changes when hydrate decomposes and of the densities of the resulting liquid phases. The largest uncertainty in the density method is at present contributed by the uncertainty in the lattice parameter at the required temperature.

The difficulties of direct chemical analysis are well known. Among indirect techniques which have been used to determine the amount of excess water in hydrate systems are Tammann and Krige's measurements[1279] of the volume change at the melting point of ice present in SO_2 hydrate and various applications of the Schreinemaker technique, in which the changes in concentration of an inert solute are used to determine the amount of water which has reacted. Barrer and Ruzicka's application,[103] using Cs_2SO_4, to $CHCl_3$ hydrate in the presence of help gases was more successful than other applications to Br_2 hydrate[637] with KBr, to BrCl hydrate[38] with $CuSO_4$, and to Cl_2 hydrate[1414] with $CaCl_2$. A difficulty in determining hydrate composition by simple measurement of the quantity of gas consumed by reaction with a known amount of water stems from the slowness of the reaction in its later stages, presumably because the unreacted water is encrusted with hydrate. Barrer and Edge have shown[101] that relatively rapid reaction, even with ice, is promoted by the presence of small steel

---

* In an earlier study[426] propylene was falsely identified as ethylene.

balls and violent agitation of the reaction vessel. In measurements to as low as 90°K they found values of n between 6.05 and 6.25 for the hydrates of Ar, Kr, and Xe.

Determination of hydrate composition from accurate determination of the concentration at which congruent melting occurs appears to have been confined to ethylene oxide hydrate,[575] although Glew has shown[565,566] how the curves of decomposition temperature versus concentration of hydrates of water-soluble species may be used to yield hydrate compositions. The method depends on a relationship similar to that given by eqn. (10) with the partial pressures replaced by the activities of M in the liquid. It may be applied to hydrates which melt incongruently.

3.3. Free Energies of the Lattices of Structures I and II

Table V demonstrates a considerable variation in composition of structure I hydrates. We examine now the possibility that this variability may reasonably be accounted for by a common value of $\Delta\mu_w$ at 0°C, as defined in Section 3.1. In doing so, we discard as unlikely the exceptionally high value of n reported[1421] for $CHClF_2$ hydrate.

Among molecules too large to occupy appreciably the small cages (see Fig. 4) the most accurate compositions have been determined for cyclopropane and CH_3Br hydrates, for each of which n is about 7.9. This value corresponds to $\theta_2 = 0.97$, considerably in excess of the value 0.905 previously indicated for $\Delta\mu_w = 167$ cal mol^{-1}. Moreover, there is a clear indication from relatively accurate data that for a considerable number of hydrates of small molecules the value of n is about 6.1. Values of n smaller than 6.9 are inconsistent with $\Delta\mu_w = 167$ cal mol^{-1}. This is shown in Fig. 12, where n [$= 5.75/(\frac{1}{4}\theta_1 + \frac{3}{4}\theta_2)$] is plotted against $\log(1 - \theta_2)$ for a number of values of $\Delta\mu_w$. The minimum value of n for $\Delta\mu_w = 250$ cal mol^{-1}, the value which permits $\theta_2 = 0.97$, $\theta_1 = 0$, is 6.19, while a value as low as 6.07 is given by $\Delta\mu_w = 278$ cal mol^{-1}. The latter corresponds to $\theta_2 = 0.98$ at $\theta_1 = 0$, for which $n = 7.8$. A fair compromise between the extreme values of n, with due regard for the greater accuracy of the lowest values, is given by $\Delta\mu_w = 265 \pm 15$ cal mol^{-1} at 0°C, which defines the minimum n value as 6.13 ± 0.07 and, for $\theta_1 = 0$, $n = 7.85 \pm 0.05$. The frequency with which values of n near the minimum are observed for small guest molecules is to be expected, since such compositions correspond to approximately equal occupancy of both kinds of cage (about 94%) and therefore to similar Langmuir constants and partition functions for M in the two cages.

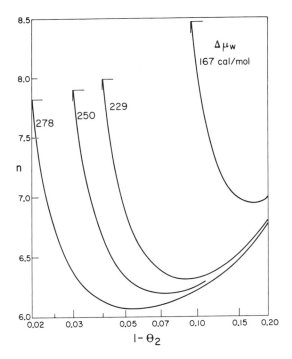

Fig. 12. Dependence of hydration number on degree of occupancy of large cages of structure I for various values of $\Delta\mu_w$. The upper limits of n shown correspond to $\theta_1 = 0$.

We conclude that the presently available composition data for structure I hydrates are consistent with the ideal solution model and that departures from ideality may be detected from composition measurements only if the compositions, in selected cases, can be determined with considerably better accuracy.

The compositions of structure II hydrates have, generally speaking, been measured with somewhat greater accuracy. Here, however, the extent of occupancy of the large cages is so near unity that no experimentally significant departure from the ideal stoichiometry of $M \cdot 17H_2O$ appears to have been detected for any simple structure II hydrate. Moreover, statistical analysis of the composition data in Table VI gives[590] for the unweighted mean, with cases of $w = 0$ omitted, $n = 16.99 \pm 0.54$, and, with weighting according to w^2, $n = 17.01 \pm 0.31$. The conclusion at about the 85% confidence level is that more than 98% of the large cages are occupied, and therefore that $\Delta\mu_w > 125$ cal mol^{-1} for structure II.

A value considerably larger than this minimum value is expected from the close similarity between such properties as O- -O bond lengths, departures of O- -O- -O angles from tetrahedral values, and *ms* bonding in structures I and II (Section 2.2). Since structure II is slightly more like ice I, it is a fair inference that $\Delta\mu_w$ for structure II lies between 125 and 265 cal mol^{-1} at 0°C and probably nearer to the larger value, which corresponds to $\theta_2 = 0.99975$.

3.4. Mixed Hydrates

The equations of Section 3.1 for the ideal solution model may readily be generalized to more than one encaged species M.[1328] Not surprisingly, since μ_w(h) depends not on the nature of the encaged species but only on the degree of occupancy of the cages, eqn. (7) becomes

$$\theta_i = \sum_M \theta_{Mi} = \sum_M \frac{C_{Mi}p_M}{1 + \sum_M C_{Mi}p_M} \tag{7'}$$

The partial pressure of a particular M for equilibrium of hydrate with ice or liquid water,

$$p_M = \frac{\theta_{Mi}}{C_{Mi}(1 - \theta_i)} \tag{11}$$

like θ_{Mi} itself, is now less than its value in the presence of M alone.

The stabilizing effect of a second encageable component is particularly evident for structure II hydrates, in which a help gas may occupy the otherwise empty small cages. This is illustrated in Table VII, based on the work of Villard[1345] and von Stackelberg and Meinhold.[1355] In the absence of help gas the decomposition temperature shown is that of the h-l_1-l_2-g invariant point. Since addition of a help gas as a third component to the system makes the h-l_1-l_2-g equilibrium univariant, the decomposition temperature becomes dependent on the pressure of help gas. In principle it is possible to calculate the rise in decomposition temperature from a knowledge of how $\Delta\mu_w$ and the required values of C_{Mi} vary with temperature (see Section 3.5) but not in the simple manner which has been suggested.[104] The evolution of air has been observed during the decomposition of many clathrate hydrates[590,1350,1355] and of some related amine[930] and salt[925] hydrates which were prepared in the presence of air.

The hydrates formed only in the presence of help gases by the large molecules listed at the bottom of Table VII are probably of structure II,

TABLE VII. Stabilizing Effects of Help Gases

Hydrate former	Decomposition temperature, no help gas, °C	Ref. 1355 Help gas, pressure, atm	Ref. 1355 Decomp. temp., °C	Ref. 1345 Help gas, pressure, atm	Ref. 1345 Decomp. temp., °C
C_2H_5Cl	4.8	CO_2 5	6.8	N_2 12	8.0
		15	10.8	O_2 8	
		N_2 5	6.0	N_2 55	15.5
		15	8.6	O_2 38	
				H_2 23	5.5
				40	6.0
$CHCl_3$	1.7	CO_2 5	5.9	—	—
		15	10.5		
		N_2 5	3.6		
		15	6.5		
CH_2Cl_2	1.7	CO_2 5	5.0	N_2 23	8.0
		15	8.4	O_2 15	
		N_2 5	3.1	N_2 75	15.5
		15	6.2	O_2 55	
C_2H_5Br	1.4	CO_2 5	4.9	—	—
		15	9.1		
		N_2 5	2.8		
		15	5.6		
CCl_4	[a]	CO_2 5	7.3	N_2 45	14
		15	12.6		
		N_2 5	3.1		
		15	7.0		
C_2H_5I	[a]	—	—	N_2 100	8
CS_2	[a]	—	—	N_2 50	0
				O_2 32	
				O_2 80	7
CH_2ClCH_2Cl	[a]	—	—	Air 100	18
I_2	[a]			O_2[1346] 300	8

[a] No hydrate found.

except for I_2, where Villard[1346] may well have prepared O_2 hydrate itself for the first time.

Waller reported[1372] the decomposition temperatures of the hydrates of acetone, CH_2Cl_2, $CHCl_3$, and CCl_4 to be 3.0, 8.6, 10.9, and 13.7°C, respectively, with 1 atm of Xe as help gas. The significance of the results is less clear for Ar and Kr,[1372] since the hydrates appear to have been formed under pressures of help gas larger than those necessary to form simple Ar and Kr hydrates.

Von Stackelberg[1352] distinguished between "double hydrates" of structure II with H_2S or H_2Se as one component, in which the vapor pressure apparently remained constant under intermittent pumping at constant temperature until all hydrate disappeared,[1353] and "mixed hydrates," whose pressures gradually decreased when the hydrates were subjected to similar pumping.[1355] The anomalous behavior of such "double hydrates," namely a univariant equilibrium where a bivariant one is expected, was attributed[1353] to a fixed, stoichiometric composition, e.g., $CHCl_3 \cdot 2H_2S \cdot 17H_2O$. Such a violation of the solution theory is, however, much less likely than the presence of a fourth phase, e.g., liquid $CHCl_3$. Another explanation, applicable in at least some cases, was given by Platteeuw and van der Waals on the basis of their studies of the H_2S–C_3H_8[1106] and CH_4–C_3H_8[1328] hydrates. At -3°C the former hydrate exhibits a minimum vapor pressure at a point where the composition of the hydrate is the same as that of the gas, about 3 mol of H_2S per mole of C_3H_8. Once this "azeotropic" composition is reached hydrate decomposition proceeds at a constant pressure. In the CH_4–C_3H_8 hydrate, however, the minimum vapor pressure occurs for pure propane and at all other compositions the hydrate is richer in propane than the gas in equilibrium with it. For both hydrates the results are readily explained by the solution theory, in terms of which the Langmuir constant for incorporation in the small cages is some 25–50 times as large for H_2S as for CH_4. An azeotropic hydrate has also been reported[496] to be formed by C_2H_4 and C_3H_8.

The validity of the Langmuir isotherm [Eq. (7)] has been experimentally confirmed[1104] for Ar in β-quinol clathrate, where there is only one type of cage. Comparable measurements of the simple gas hydrates are considerably more difficult, even when only the larger cages may be occupied, because of the high degree of occupancy of the cages at the minimum gas pressure necessary for stability. Barrer and Edge[101] avoided this problem by examining the dependence on pressure of the absorption of Ar and Kr by $CHCl_3$ hydrate at temperatures below 0°C. Since all the large cages are effectively occupied by $CHCl_3$, little error is introduced by the assump-

tion that gas molecules enter only the small cages. The Langmuir isotherm
may then be written

$$\frac{P}{V} = \frac{1}{C_{M1} V_m} + \frac{P}{V_m}$$
(12)

where V is the volume of gas absorbed at pressure P of rare gas and V_m
is the limiting volume absorbed at very high pressures where all the cages
are occupied. Plots of P/V versus P are shown in Fig. 13, where P is in
cm Hg and V in ml at STP per gram of water in the hydrate. The Langmuir
isotherm is seen to be confirmed.

Values of C_{M1} obtained in this way[101] may be extrapolated to give
0.11 and 0.75 atm^{-1} for Ar and Kr, respectively, at 0°C. If these values are
assumed applicable to the small cages of structure I, which are geometrically
similar to those of structure II, one finds $\theta_1 = 0.91$ for both argon and
krypton hydrates at their respective dissociation pressures. In view of the
accumulated uncertainties, there is satisfactory agreement with values of
0.93–0.94 obtained from the compositions of type I hydrates of small
guest molecules (Section 3.3).

Determination of the compositions of double hydrates appears potentially to be one of the best methods of defining the poorly known value
of $\Delta\mu_w$ for structure II hydrates. Thus the rather inaccurate composition[1353]

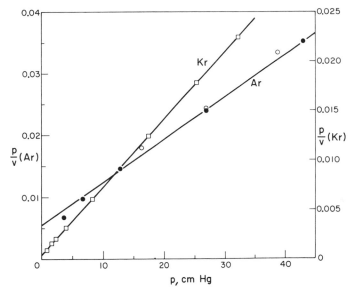

Fig. 13. Langmuir isotherms [Eq. (12)] for absorption of argon and
krypton in chloroform hydrate at −78°C.[101]

$(0.85\pm0.05)CCl_4 \cdot (1.90\pm0.088)H_2S \cdot 17H_2O$ defines $\Delta\mu_w = 250$ cal mol^{-1} to within about 60 cal mol^{-1}. For $\theta_1 = 0$ and the same value of $\Delta\mu_w$, $\theta_2 = 0.9996$ would need to be defined between about 0.9974 and 0.9999 to give comparable accuracy.

The following mixed and double hydrates have been studied in varying detail: Ar–N$_2$,[946,1173] Ar–CH$_4$,[548,946,1173] Ar–SiH$_4$,[548,777] CH$_4$–H$_2$S,[1033] CH$_4$–CO$_2$,[1315] $\overset{.}{C}$H$_4$–C$_2$H$_4$,[1237] CH$_4$–C$_2$H$_6$,[386,923] CH$_4$–SF$_6$,[1240] CH$_4$–CH$_3$CH=CH$_2$,[1059] CH$_4$–C$_3$H$_8$,[270,386,923,1328] CH$_4$–iso-C$_4$H$_{10}$,[386,923] CH$_4$–n-C$_4$H$_{10}$,[386,923] H$_2$S–C$_3$H$_8$,[1106] H$_2$S–CHCl$_3$,[1353] H$_2$S–CCl$_4$,[1353] C$_2$H$_4$–C$_3$H$_8$,[496] CH$_3$Cl–(CH$_3$)$_2$C=CH$_2$,[783] ethylene oxide–acetone,[857] O$_3$–CCl$_4$,[933] C$_3$H$_8$–CH$_3$CH=CH$_2$,[1134] CHCl$_3$–C$_2$H$_5$Cl,[1355] C$_2$H$_5$Cl–CH$_3$-CHCl$_2$,[1355] CHCl$_3$–acetone,[857] CH$_3$I–cyclopentane,[857] and cyclopentane–cyclopentene.[857]

3.5. Enthalpies of Encagement and of the Empty Lattices

Since C_{Mi} is a measure of the equilibrium constant which describes the distribution of M molecules between the gas and cages i, the heat of encagement or intercalation is $\Delta H_{Mi} = -R\,d(\ln C_{Mi})/d(1/T)$. Barrer and Edge[101] found -6.1 and -6.7 kcal mol^{-1} for encagement of Ar and Kr, respectively, in the small cages of chloroform hydrate, and, with somewhat less extensive data, -7.9 kcal mol^{-1} for Xe.

Although these appear to be the only direct measurements of heats of encagement, there have been many measurements (cf. Tables V and VI) of the heat of the reaction between gaseous M and ice to form hydrate, which, in terms of the solution theory, is

$$\Delta H(I + g \to h) = n\,\Delta H_w(I \to h^\circ) + \Delta H(h^\circ + g \to h) \qquad (13)$$

per mole of M encaged, with

$$\Delta H(h^\circ + g \to h) = n\sum_i \nu_i\theta_i\,\Delta H_{Mi} \qquad (14)$$

The heat of formation of empty lattice from ice is expected to be small and positive for structural reasons and is sometimes taken to be zero.[1350,1351] Because of the factor of n in eqn. (13), however, $\Delta H_w(I \to h^\circ)$ may well make an important contribution to $\Delta H(I + g \to h)$.

For a temperature change along the h-I-g equilibrium line

$$\frac{d(\Delta\mu_w/T)}{d(1/T)} = -k\,\frac{d\sum \nu_i \ln(1 - \theta_i)}{d(1/T)}$$

$$= \Delta H_w(I \to h^\circ) + \frac{\Delta V_w(I \to h^\circ)}{T}\,\frac{dP}{d(1/T)} \qquad (15)$$

and a small positive value of $\Delta H_w(I \rightarrow h^\circ)$ means that the extent of cage occupancy under equilibrium conditions decreases slowly with rise of temperature. Only in exceptional cases of very high dissociation pressures could predominance of the second term reverse this variation. Similarly, along the h-l_1-g line the variation in composition is determined by the sum of a relatively large negative value of $\Delta H_w(l_1 \rightarrow h^\circ)$ and a pressure-dependent term of the same sign. The consequence is clearly that for practically all hydrates the degree of cage occupancy necessary for stability is least, and the value of n greatest, at the h-I-l_1-g quadruple point. A pronounced fall in n with rise in temperature along the h-l_1-g line has been experimentally shown for Cl_2[1414] and ethylene oxide[575] hydrates.

3.6. Spherical Cell Models

3.6.1. Application of the Lennard-Jones and Devonshire Theory

This model of liquids and compressed gases has been applied by Barrer and Stuart[104] and van der Waals and Platteeuw[1105,1328] to the clathrate hydrates. The development[1328] is similar to that for liquids, with the average radius of the cell formed by the nearest neighbors (usually taken as 12 for liquids) replaced by the average radius a_i of the actual cage formed by z_i water molecules. The interaction energy between a guest and a water molecule is taken as

$$4\varepsilon[(\sigma/r)^{12} - (\sigma/r)^6] \qquad (16)$$

where ε is the minimum energy which corresponds to a separation (the van der Waals radius) of $r = 2^{1/6}\sigma$ and σ is the separation at which the attractive and repulsive energies are equal. The required values of ε and σ are found from

$$\varepsilon = (\varepsilon_w \varepsilon_M)^{1/2}, \qquad \sigma = (\sigma_w + \sigma_M)/2 \qquad (17)$$

where ε_M and σ_M are known for some gases from fit of second virial coefficients or viscosities by the 12–6 potential for M–M pair interaction. If the van der Waals radius of water is assumed to be 1.40 Å, $\sigma_w = 1.25$ Å. The value of ε_w may be determined empirically by fit of the theory to the dissociation pressure of a particular hydrate.[1328] Alternatively, this degree of empiricism may be avoided by estimation of ε from various equations for the London dispersion constant,[103,104] of which the equation of Slater and Kirkwood is the most satisfactory.[101]

The molecular partition function is given by

$$b_{Mi} = \phi_M \{\exp[-w_i(0)/kT]\} 2\pi a_i{}^3 g_i \qquad (18)$$

in which the energy at the cage center is

$$w_i(0) = z_i \varepsilon (\alpha_i^{-4} - 2\alpha_i^{-2}), \qquad \alpha_i = a_i^3/(\sigma^3 \sqrt{2}) \tag{19}$$

and

$$g_i = \int \exp\left\{ \frac{z_i \varepsilon}{kT} \left[-\frac{l(y)}{\alpha_i^4} + \frac{2m(y)}{\alpha_i^2} \right] \right\} y^{1/2} \, dy \tag{20}$$

ϕ_M [see eqn. (6)] is the product of the translational, rotational, and internal partition functions of gaseous M. The integral in eqn. (20), where $l(y)$ and $m(y)$ are[500,1327] algebraic functions of $y = r^2/a_i^2$, with r the distance from the center of the cage, may be readily evaluated numerically for each distinct guest species in each cage.

From eqns. (6) and (18) the Langmuir constant for a specific M is

$$C_{Mi} = (2\pi a_i^3 g_{Mi}/kT) \exp[-w_{Mi}(0)/kT] \tag{21a}$$

Rearrangement of the Langmuir equation and replacement of kT/p_M by v_M, the molecular volume of M in the gas phase, gives

$$C_{Mi} p_M = \frac{\theta_{Mi}}{1 - \theta_{Mi}} = \frac{2\pi a_i^3 g_{Mi}}{v_M} \exp\left[-\frac{w_{Mi}(0)}{kT} \right] \tag{21b}$$

which shows that the degree of occupancy of the cage is determined by the ratio of the "free volume" of M in its cage to the molecular volume in the gas and by the Boltzmann factor in the energy of M at the cage center with respect to the ideal gas.

The partition function given by eqn. (18) strictly applies only to monatomic guest molecules. Its application to polyatomic guests requires the assumption that the rotational and internal vibrational contributions are the same as in the free gas. The assumption of free rotation is expected to be a reasonable one for molecules which are close to spherical in shape.

The dissociation pressure of the gas hydrates may be calculated from eqns. (8) and (7) with values of C_{Mi} given by eqn. (21). This is most simply done by calculating θ_{M1} and θ_{M2} for assumed values of the pressure until a pressure is found at which the cage occupancy factors satisfy eqn. (8). Van der Waals and Platteeuw took $\Delta\mu_w = 167$ cal mol^{-1} at 0°C for structure I (Section 3.1) and used an experimental value for the dissociation pressure of argon hydrate to find $\varepsilon_w/k = 166.9°$. The dissociation pressures found[1328] for other hydrates are given in Table VIII, along with the ratio of the occupancy factors of the small and large cages and n in M · nH$_2$O. The hydrates of oxygen and nitrogen were unknown at the time of this work,[1328] as was the proper value[981] of the dissociation pressure of CF$_4$

TABLE VIII. Dissociation Pressures at 0°C from the Lennard-Jones and Devonshire Model[a]

Hydrate	ε_M/k, deg	σ_M, Å	Van der Waals–Platteeuw[b]			Revised[c]			Experimental p_M, atm
			p_M, atm	θ_1/θ_2	n	p_M, atm	θ_1/θ_2	n	
Ar	119.5	3.408	95.5	0.981	6.87	112	1.012	6.12,	90
Kr	166.7	3.679	15.4	1.002	6.92	14.2	1.028	6.13	14.3
Xe	225.3	4.069	1.0	0.974	6.93	0.57	1.022	6.12	1.5
CH₄	142.7	3.810	19.0	0.978	6.92	17.3	1.018	6.12	25.5
CF₄	152.5	4.70	1.6	0.315	7.76	1.01	0.414	6.93	41.5
N₂	95.05	3.698	90	0.958	6.88	108	1.003	6.12	150
O₂	117.5	3.58	63	0.979	6.89	73	1.013	6.12	112
AsH₃*	281	4.06	—	—	—	0.137	1.035	6.13	0.80
CO₂	205	4.07	0.71	0.913	6.83	1.00	1.017	6.12	12.4
*	190	3.996	—	—	—	2.16	1.019	6.12	—
C₂H₆	243	3.954	1.1	1.012	6.93	0.62	1.037	6.13	5.23
C₂H₄	199.2	4.523	0.5	0.596	7.28	0.23	0.777	6.33	5.45
*	205	4.232	—	—	—	0.51	0.985	6.12	—
C₂H₂	185	4.221	—	—	—	1.0	0.983	6.12	5.74
COS*	335	4.13	—	—	—	0.25	1.036	6.13	—
Cl₂*	257	4.40	—	—	—	0.51	0.918	6.16	0.33
SO₂*	252	4.29	—	—	—	0.095	0.977	6.13	0.29
SF₆	200.9	5.51	—	—	—	0.48[d]	—	17.004	0.80
C₃H₈	242	5.637	—	—	—	0.046[d]	—	—	1.73
CCl₄*	327	5.881	—	—	—	0.0016[d]	—	—	—

[a] Lennard-Jones parameters[674,1328] are from second virial coefficients or viscosities (*) of gases.
[b] $a_1 = 3.95$, $a_2 = 4.30$ Å, $\Delta\mu_w = 167$ cal mol^{-1}, $\varepsilon_w/k = 166.9°$.
[c] $a_1 = 3.91$, $a_2 = 4.33$ Å, $\Delta\mu_w = 265$ cal mol^{-1}, $\varepsilon_w/k = 230.4°$.
[d] $a_2 = 4.683$ Å, $\theta_1 = 0$, $\theta_2 = 0.99975$.

hydrate. The agreement between the calculated and observed dissociation pressures was regarded as satisfactory for the rare gas and methane hydrates and the disagreement for ethane and ethylene hydrates attributed to hindered rotation and possibly to the inadequacy of the central force field [eqn. (16)] for large polyatomic molecules. Mazo[958] has pointed out, however, that inclusion of hindered rotation would *reduce* the calculated dissociation pressures. There is a consistent drop in the ratio $p_M(\text{calc})/p_M(\text{obs})$ with increasing size of M which suggests that the "free volume" of relatively large M molecules is overestimated. The disagreement for CF_4 hydrate indicates that molecular asphericity is not the principal source of inadequacy of the model.

As already indicated (Section 3.2) the high values of n given by $\Delta\mu_w = 167$ cal mol^{-1} are clearly inconsistent with the known compositions of the hydrates of small guest molecules. Dissociation pressures have therefore been recalculated with $\Delta\mu_w = 265$ cal mol^{-1}, with minor adjustments of mean cage radii to conform to the values given (Table II) by the most recent X-ray study.[927] The value $\varepsilon_w/k = 230.4°$ was chosen to give agreement with the observed dissociation pressure of Kr hydrate. The dissociation pressures obtained for other hydrates (Table VIII) fit the experimental values slightly less well than before. The compositions found are much more realistic for small M but less so for larger M. It may be pointed out that for small molecules for which $C_{M1} \approx C_{M2}$, the compositions are practically determined by $\Delta\mu_w$ and are nearly independent of the Langmuir constants. The values of $p_M(\text{calc})/p_M(\text{obs})$ show the same trend as before toward increasing inadequacy of the Lennard-Jones and Devonshire model with increasing σ_M.

Van der Waals and Platteeuw[1328] used $\varepsilon_w/k = 166.9°$ to find C_{M2} for the structure II hydrate of SF_6, and from the observed dissociation pressure found $\Delta\mu_w = 196$ cal mol^{-1} at $-3°C$. The use of $\varepsilon_w/k = 230.4°$ leads in a similar manner to $\Delta\mu_w = 288$ cal mol^{-1}. For structural reasons, however, this value is unlikely to be larger for structure II than the value of 265 cal mol^{-1} recommended (Section 3.3) for structure I. The latter value has been assumed in the calculations of the dissociation pressures of a number of structure II hydrates shown in Table VIII. These pressures may be considered as maximum pressures consistent with the Lennard-Jones and Devonshire model as presented above. The inadequacy is pronounced for the large CCl_4 molecule for which a hydrate with a dissociation pressure only 4% as great as the vapor pressure of liquid CCl_4 at $0°C$ is predicted, in contradiction to the finding that a hydrate of CCl_4 is only formed in the presence of help gases.[1355]

Calculation of heats of encagement subjects the Lennard-Jones and Devonshire model to a less stringent test than dissociation pressures since the "free volume" of the encaged molecule contributes little to the result. According to eqns. (9) and (18),

$$U - U^\circ = N_w k T^2 \sum_i \nu_i \theta_i \left[\frac{\partial(\ln \phi_M)}{\partial T} + \frac{w_{Mi}(0)}{kT^2} + \frac{\partial(\ln g_{Mi})}{\partial T} \right] \quad (22)$$

and since the sum in $\partial(\ln \phi_M)/\partial T$ may be shown to be the energy of $1/n$ moles of perfect gas,

$$\Delta U(h^\circ + g \to h) = nR \sum_i \nu_i \theta_i \{[w_{Mi}(0)/k] + T^2[\partial(\ln g_{Mi})/\partial T]\} \quad (23)$$

is the energy of encagement of 1 mol of gas. Since $H_M(g) = U_M(g) + RT$ and the difference between $H_M(h)$ and $U_M(h)$ may be neglected, the heat of encagement may be found by subtracting RT from the rhs of eqn. (23).

The resulting values of $\Delta H(h^\circ + g \to h)$ are given in Table IX for van der Waals and Platteeuw's choice of parameters (column 2) and for the revised parameters (column 3) identified in Table VIII. The heats of formation of hydrate from ice and gas given in column 4 are experimental values from Tables V and VI.

The close numerical agreement of the heats of encagement obtained by van der Waals and Platteeuw with the heats of hydrate formation from ice was taken[1105] as evidence that $\Delta H(I \to h^\circ)$ is effectively zero. With the revised value of ε_w consistently positive values (last column of Table IX) are obtained for this quantity. For structure I the average value is 0.19 kcal mol^{-1} and there is no pronounced trend with molecular size. The somewhat larger value obtained for structure II would be reduced by choice of $\Delta\mu_w < 265$ cal mol^{-1}.

The model is satisfactory in the sense that it accounts properly for the relative magnitudes of the heats of hydrate formation from ice and gas and yields a nearly constant positive value of $\Delta H(I \to h^\circ)$. It may be noted, however, that the calculated heats ΔH_{M1} of encagement in the *small* structure I cages [the entries in Table IX include contributions from both cages, as defined in eqn. (14)] are -4.33, -5.89, and -8.18 kcal mol^{-1} for Ar, Kr, and Xe, respectively, which, except for Xe, are appreciably less negative than the corresponding experimental values for the small cages of structure II (Section 3.5).

By arguments based on comparisons of square-well-potential free volumes [i.e., $4\pi(a_i - \sigma_M)^3/3$] with those of β-quinol clathrates, Child[293] has deduced $\Delta S(I \to h^\circ) = -0.42 \pm 0.23$ and $+0.05 \pm 0.08$ eu and

TABLE IX. Comparison of Calculated Heats of Encagement with Experimental Heats of Formation from Ice (kcal mol⁻¹)

Hydrate[a]	$-\Delta H(h^\circ + g \rightarrow h)^{(1105)}$	$-\Delta H(h^\circ + g \rightarrow h)^b$	$-\Delta H(I + g \rightarrow h)$	$\Delta H(I \rightarrow h^\circ)$
Ar	3.29	3.80	2.94	0.14
Kr	4.45	5.16	3.98	0.19
Xe	6.33	7.33	5.77	0.25
CH_4	4.45	5.69	4.55	0.19
CF_4	6.74	7.75	—	—
N_2	3.46	4.01	3.81	0.03
O_2	3.59	4.14	3.19	0.16
AsH_3*	—	8.13	—	—
CO_2	—	7.08	5.68	0.23
C_2H_6	6.17	7.17	6.13	0.17
C_2H_4	—	8.33	6.27	0.29
*	—	7.64	—	0.20
C_2H_2*	—	7.15	5.33	0.27
COS*	—	9.17	—	—
Cl_2*	—	9.05	7.22	0.26
SO_2*	—	8.58	7.9	0.11
SF_6	—	11.24	5.14	0.35
C_3H_8	—	12.65	6.34	0.37

[a] Lennard-Jones parameters from second virial coefficients or viscosities (*).
[b] This work.

$\Delta H(I \rightarrow h^\circ) = 0.18$ and 0.20 kcal mol⁻¹, respectively for hydrates of types I and II. Although the numerical significance of this method is questionable, it is interesting that it yields ΔH (and $\Delta \mu_w$) values similar to those obtained above.

3.6.2. Cell Model with the Kihara Potential

As a result of an examination[922] of the reasons for the failure of the Lennard-Jones and Devonshire cell model to account satisfactorily for dissociation pressures of the gas hydrates of large molecules, McKoy and Sinanoglu made use of the Kihara potential as a more adequate representation than the L-J 12–6 potential of the interaction between water and polyatomic guest molecules. The size and shape of the guest molecules were represented by a spherical (CH_4, CF_4) or linear (N_2, C_2H_6) core, the Kihara potential function averaged over all orientations of the guest molecule with respect to a fixed water molecule, and the result incorporated in the spherical cell model. Parameters were chosen to yield results identical with those of van der Waals and Platteeuw for monatomic gases, i.e., for consistency with $\varepsilon_w/k = 166.9°K$.

The dissociation pressures found[922] for hydrates of spherical molecules (13 atm for CH_4, 0.6 atm for CF_4) were somewhat less satisfactory than those given by the L-J 12–6 potential (see Table VIII, column 4). There was, however, an improvement in the pressures obtained for "linear" molecules (N_2, 115 atm; O_2, 120 atm; CO_2, 9.0 atm; C_2H_6, 8.4 atm). As is evident from Fig. 14, the reason for the higher dissociation pressures in these cases is the less negative energy of encagement given by the Kihara potential. Although heats of encagement were not specifically evaluated, it is clear that they will generally be consistent only with negative values of $\Delta H_w(I \rightarrow h°)$. Another unsatisfactory result is the low equilibrium extent of occupancy of the cages. In recent applications of the Kihara potential[1012] $\Delta\mu_w = 167$ cal mol^{-1} and $\Delta H(I \rightarrow h°) = 0$ are again assumed.

Since in any cell model the dissociation pressures are more sensitive to the parameters of the intermolecular potentials than are the second virial coefficients from which these parameters are determined,[922] it is possible that suitable variation of the parameters of the Lennard-Jones or Kihara potentials will result in satisfactory values for dissociation pressures, heats

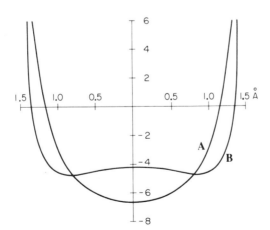

Fig. 14. Spherical cell potential energies (kcal mole^{-1}) of ethane as a function of distance from the center of the 14-hedron.[922] Curve A is from the 12–6 and curve B from the Kihara potential.

of encagement, and compositions. It appears, however, that the better the calculated pressures, the less satisfactory the heats and compositions become.

3.7. Lennard-Jones Model with Discrete Water Molecules

Barrer and Edge[101] summed the L-J 12–6 potential of pairwise interaction between an encaged Ar, Kr, or Xe molecule and all the water molecules in the nearest 27 unit cells of structure I or all 136 water molecules of the unit cell of structure II. Small contributions from neighboring guest molecules were also included. For a given cage the heat of encagement was taken as $E_{min} + \frac{1}{2}RT$, where E_{min} is the minimum energy within the cage and $RT/2$ arises from the difference between the classical contribution of a three-dimensional oscillator $(3RT)$ and that of a monatomic gas $(5RT/2)$. The Lennard-Jones force constant for each type of guest was chosen to give agreement with the experimental heats of encagement (Section 3.5) in the small cages of structure II chloroform hydrate. Values of $\Delta H(h^\circ + g \rightarrow h)$ obtained[101] were -5.54, -6.02, and -7.38 kcal mol^{-1} for the structure I hydrates of Ar, Kr, and Xe, respectively. From the corresponding values of $\Delta H(I + g \rightarrow h)$ and $n = 6.12$, $\Delta H_w(I \rightarrow h^\circ)$ was found to be 0.42, 0.33, and 0.26 kcal mol^{-1}, respectively, or 0.34 kcal mol^{-1} on average. This is perhaps the best value available for this quantity. It provides some evidence that $\Delta H_w(I \rightarrow h^\circ) > \Delta \mu_w$ and therefore that $\Delta S_w(I \rightarrow h^\circ) \lesssim 0$. A small positive entropy difference is to be expected if the spectrum of lattice vibrations of ice is shifted somewhat to lower frequencies in the hydrate because of somewhat weaker hydrogen bonding (Section 4.12).

An unexpected feature of Barrer and Edge's results should be mentioned: that the experimental values of their A constant $[A = 4\varepsilon\sigma^6$, see eqn. (16)] increases less rapidly than σ^6 in the sequence Ar, Kr, and Xe.

The energy of small encaged molecules (Ar, Kr) was found[101] to be a minimum not at the cage center but at displaced positions which bring the guest molecule more nearly in van der Waals contact with some of the water molecules (Fig. 15). This result is also given by the spherical cell model with the L-J 12–6 potential[103,104] and for "diatomic" molecules with the Kihara potential[922] (cf. Fig. 14). For direct pairwise summation of potentials the energy also depends on the direction of displacement from the center (Fig. 15). It is apparent that at low temperatures most guest molecules occupy off-center positions and that there are finite barriers to rotation of polyatomic guest molecules.

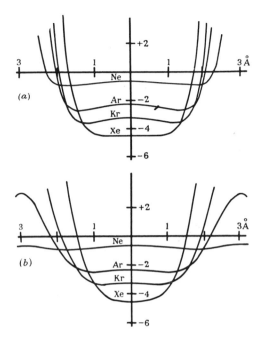

Fig. 15. Dependence of potential energy (kcal mol^{-1}) of rare gases on displacement along a diameter of the equatorial plane (a) and along the polar $\bar{4}$ axis (b) of the 14-hedron.[101]

In their early X-ray study of Cl_2 hydrate, Pauling and Marsh[1083] found that the scattering intensities at 4°C could be best represented by orientations of the Cl_2 which were weighted by the square of the cosine of the angle between the Cl—Cl bond and the equatorial plane of the large cage. Successively poorer results were given for Cl_2 molecules confined to the equatorial plane, with orientations distributed uniformly over the surface of a sphere, or aligned with the $\bar{4}$ axis of the cage. An attempt has been made[373] to apply an "atom–atom" L-J 12–6 potential to this hydrate. The interaction of each Cl atom with each water molecule was summed over all the water molecules of the large cage, the appropriate value of ε being determined by $\Delta H(I + g \rightarrow h) = -6.5$ kcal mol^{-1},[564] $\Delta H(I \rightarrow h^\circ) = 0.35$ kcal mol^{-1}, and $\Delta H(h^\circ + g \rightarrow h) = E_{min} + \frac{3}{2}RT$. It was assumed that the heat of intercalation of the $\sim$14% of molecules in the small cages is the same as in the large ones.

The results[373] showed an absolute minimum in energy with the center of mass located in the equatorial plane at $x = 0$, $y = z = 0.065a$, $a = 12.1$ Å, and the Cl—Cl axis inclined by 1.5° to the normal to the diagonal. The

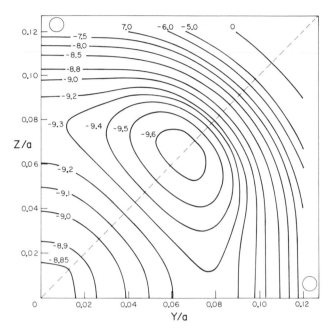

Fig. 16. Dependence of potential energy (kcal mole^{-1}) on position of center of mass of chlorine within one quadrant of the equatorial plane of the 14-hedron.[373]

energy is scarcely affected by the slight rotation necessary to bring the molecule into the equatorial plane, for which the equipotential contours are given in Fig. 16. The coordinates define the position of the center of mass for the particular orientation which gives the minimum energy at that position. The centers of the Cl atoms at least energy are defined by circles. The energy contours in the other three quadrants are the same. From the dependence of energy on displacement from the configuration of least energy, harmonic frequencies between 18 and 40 cm^{-1} were found for the three translational and two rotational oscillations. These frequencies are of the same order as those found with the spherical cell theory[103] for the triply degenerate translational oscillations of Ar, Kr, Xe, and CH$_4$ in the small cages of structure II and appear to be of the right order for oscillations of encaged molecules generally in hydrates at very low temperatures. With increase of temperature the motions become increasingly anharmonic and Cl$_2$ molecules increasingly able to cross the ~0.3 kcal mol^{-1} barrier separating the regions of potential minima.

At the temperature of the X-ray study, $RT = 0.55$ kcal mol^{-1} and the model suggests that the Cl electron density should be smeared into a diffuse circular ring extending somewhat above and below the equatorial plane, in reasonable agreement with the X-ray results. However, the model shows that the reorientation is accompanied by large-scale translation of the center of mass not envisaged in the X-ray models. The axial configuration is found to correspond to a relative minimum of energy, which, however, is less stable by about 1 kcal mol^{-1} than the minimum in the equatorial plane.

Similar atom–atom Lennard-Jones calculations have been attempted for the hydrates of ethylene oxide and trimethylene oxide. These are polar molecules, however, and further discussion is postponed until after the electrostatic fields of the water molecules are considered (Section 5.6).

3.8. Effect of Pressure on Hydrate Stability

The decomposition temperatures of the hydrates of SO_2 and $CHCl_3$ along the h-l_1-l_2 equilibrium line (see Fig. 1) were studied by Tammann and Krige[1279] at mechanical pressures to about 2 kbar. For SO_2 hydrate the decomposition temperature rises to a maximum value at ~2 kbar, where $\Delta V(\text{h} \rightarrow l_1 + l_2) = 0$. The effect of pressure on $CHCl_3$ hydrate is to lower the decomposition temperature, since the hydrate is less dense than its liquid decomposition products. Similar behavior has recently been found[590] for tetrahydrofuran hydrate (Fig. 17), whose congruent melting point (along

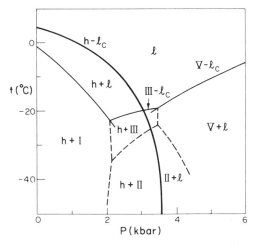

Fig. 17. Effect of pressure on the decomposition temperature (heavy line) of tetrahydrofuran hydrate.[590] Lighter lines are melting points of ices I, III, and V and dashed lines separate the regions of stability of different ices.

the h-l_c line, where l_c is the liquid of the same composition as the hydrate) becomes incongruent at the point ($-19.8°$, 3.05 kbar) of intersection with the freezing curve III-l_c of ice III. The decomposition temperature thereafter falls steeply along the h-III-l and h-II-l lines. Other congruently melting structure II hydrates are expected to show qualitatively the same behavior. Acetone hydrate,[1000] on the other hand, melts incongruently (h $\rightarrow$ I $+$ l) at normal pressure with an increase of 5–6% in volume, and an initial rise of decomposition temperature with increase of pressure is expected.

For the common i_2 systems (Section 1.3) it may be generally anticipated that, barring solidification of l_2, the h-l_1-l_2 curve will eventually intersect the melting curve of one of the high-pressure forms of ice, most likely the h-VI-l_1 line.

3.9. Lower Critical Decomposition Temperature

For CO_2 hydrate Miller and Smythe find[982] that the I-h-g equilibrium line, extrapolated from measurements between 193 and 152°K, should intersect the vapor pressure curve of solid CO_2 (s_2) at about 121°K, the temperature of the I-s_2-h-g quadruple point below which CO_2 hydrate is unstable with respect to solid CO_2 and ice. The existence of such a lower critical decomposition temperature requires that $\Delta H(s_2 \rightarrow g)$ be greater than $\Delta H(h \rightarrow I + g)$, i.e., that $\Delta H(s_2 + I \rightarrow h)$ be positive. The latter quantity is approximately zero for many hydrates.[1351] It may be considered as the sum of the slightly positive $\Delta H(I \rightarrow h°)$ (Section 3.6) and the heat of transfer of M from s_2 to hydrate cages, probably negative unless there are relatively strong s_2 lattice forces. Equilibrium pressures have not been measured to sufficiently low temperatures to show a lower critical point for other hydrates, but comparison of heats of sublimation of M with $\Delta H(h \rightarrow I + g)$ (Tables V and VI) suggests the possibility of the phenomenon in the hydrates of, among others, Cl_2, SO_2, $CHCl_3$, and SF_6, and its absence in the rare gas, N_2, O_2, and simple hydrocarbon hydrates.

4. KINETIC PROPERTIES OF THE WATER MOLECULES

The orientational disorder of the water molecules in clathrate hydrates, just as in ice Ih and the other disordered forms of ice, manifests itself in a rotational mobility which affects the dielectric and nuclear magnetic resonance properties at relatively high temperatures. Reorientations of the water molecules result in interchanges among the $\sim(3/2)^N$ lattice configurations until the temperature becomes so low that the rate of interchange is inappreciable and the configurational disorder and corresponding residual entropy are "frozen-in."

The present section deals mainly with the reorientation of water molecules with some reference to self-diffusion and infrared properties. Infrared studies provide independent evidence of the presence of an orientational disorder which persists to low temperatures.

4.1. Dielectric Properties of Ice Ih

The dielectric behavior of hexagonal ice, described in Volume 1, Chapter 4, serves as a convenient standard with which to compare the clathrate hydrate properties. For ice Ih the Cole–Cole complex permittivity locus is accurately a semicircle, as first shown by Auty and Cole,[62] that is, reorientation of the water molecules proceeds with only a single relaxation time. At $-10.5°$ the static permittivity $\varepsilon_0 = 99$, the high-frequency permittivity $\varepsilon_\infty = 3.1$, and the relaxation time (obtained as the reciprocal of the angular frequency of maximum dielectric loss ε'') is 5.4×10^{-5} sec. The variation of relaxation time τ with temperature is given by

$$\tau = Ae^{E_A/RT} \tag{24}$$

with $A = 5.3 \times 10^{-16}$ sec and the Arrhenius "activation" energy is 13.2 kcal mol^{-1}.

Examination of a fully four-coordinated hydrogen-bonded network shows that reorientation of the water dipole in any manner that maintains the hydrogen bonding is impossible without the cooperation of neighboring water molecules. The most successful mechanism proposed to account for reorientation of the water molecules in ice is that of Bjerrum.[177,178] The occasional thermally activated misorientation of a water molecule about one of its hydrogen bonds is accompanied by the formation of two defect bonds, a D bond bearing two protons and an L bond bearing none, which may then become separated by rotation of defect-bearing molecules about one of their normal hydrogen bonds. The process of diffusion of these orientational defects leaves molecules along the diffusion path reorientated and fully hydrogen-bonded. A single defect may effect reorientation of many water molecules before it is annihilated by encounter with a defect of the complementary kind. In ice Ih at $-10°$ the ratio of defects to normal water molecules is of the order of 10^{-7} and becomes rapidly smaller with decrease of temperature.[597] In the ideal crystal, defects are generated

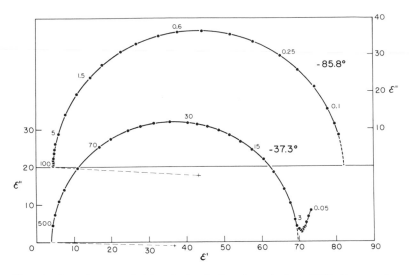

Fig. 18. Complex permittivity loci of dioxolane hydrate.[1335] Frequencies
are in kHz.

intrinsically by thermal excitation of normal water molecules. It is easily
appreciated, however, that disruption or distortion of the hydrogen bonding
near impurities and crystal imperfections may introduce extrinsic defects
into the lattice in sufficient numbers as to considerably accelerate the relaxation rate even of molecules which are many molecular diameters removed
from the injection sites.

4.2. Dielectric Properties of Structure II Hydrates

The complex permittivity plots of 1,3-dioxolane hydrate[1335] given in
Fig. 18 are typical of all the structure II hydrates which have been adequately
studied.* In shape these loci may reasonably well be described as circular
arcs with centers lying below the ε' axis, as given by the equation of Cole
and Cole[326]

$$\varepsilon^* = \varepsilon' - i\varepsilon'' = \varepsilon_{\infty 1} + (\varepsilon_0 - \varepsilon_{\infty 1})/[1 + (i\omega\tau_0)^{1-\alpha}] \qquad (25)$$

for the complex permittivity in terms of the limiting values of the permit-
tivities ε_0 and $\varepsilon_{\infty 1}$ reached at low and high frequencies, the most probable
relaxation time τ_0, and the distribution parameter α. The latter formally

---

* The theory of the dielectric permittivity of aqueous solutions is discussed in detail
  in Chapter 7.

specifies the width of a continuous distribution of relaxation times which is logarithmically symmetric about τ_0. In the case of ice Ih, $\alpha = 0$ and the distribution function collapses to a delta function.

Figure 19 shows the dependence on temperature of the dielectric parameters of dioxolane hydrate.[1335] The failure of the measured static permittivity to follow the linear dependence on $1/T$ to indefinitely low temperatures is due to the development of cracks in the sample or its contraction away from the electrodes.[591] Large permittivities are particularly sensitive to this kind of inhomogeneity: A parallel air gap of thickness less than 0.001 of the sample thickness is sufficient, for example, to reduce a real permittivity of 90 to a measured value of 84. The presence of such gaps also reduces somewhat the measured relaxation times from their true values, though not to the extent necessary to account for the pronounced departure observed at low temperatures (see Fig. 19) from linearity of the $\log \tau_0$ versus $1/T$ plots. Other explanations must be invoked,[1000] of which the most likely is the increasing predominance at low temperatures of relaxation controlled by Bjerrum defects of extrinsic origin (Section 4.1).

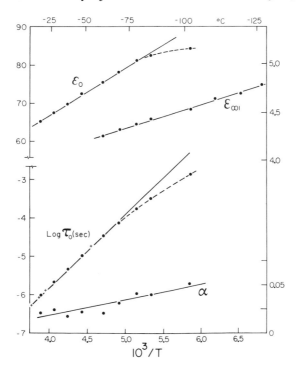

Fig. 19. Temperature dependence of dielectric parameters
of dioxolane hydrate.[1335]

The departures of the dispersion-absorption loci of structure II hydrates from full Debye semicircles are small, and accurate measurement of their shape is hindered by the presence of liquid rich in hydrate-former and, especially, by the presence of ice.[647,1000] The many difficulties attendant upon the formation of relatively large homogeneous samples of hydrates of water-insoluble species explain why most of published measurements have been of hydrates of molecules soluble in water.

The values of α which provide the "best fit" do not depend on the guest molecule. They change with temperature from 0.020 at $-25°C$ to 0.050 at $-105°C$ and 0.07 at $-130°C$. However, consistent departures from the Cole–Cole arcs observed for all the structure II hydrates take the form of a perpendicular approach at high frequencies to $\varepsilon_{\infty 1}$, as illustrated in Fig. 20,[1000] rather than at the angle of $(1 - \alpha)\pi/2$ predicted by eqn. (25). A similar approach to ε_0 at low frequencies has been suggested by some results, but is normally obscured by the additional low-frequency absorption and polarization associated with ionic conductance and space-

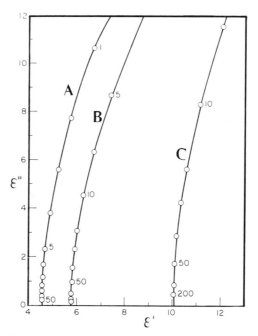

Fig. 20. Plots illustrating the perpendicular approach to $\varepsilon_{\infty 1}$ for structure II hydrates of (A) dioxolane, (B) trimethylene oxide, and (C) cyclobutanone.[1000] Temperatures are near $-110°C$.

charge or electrode effects. The implication of these results is that the density of reorientational relaxation times falls off considerably more rapidly for times respectively small and large in comparison with τ_0 than the distribution function

$$F(s)\,ds = \frac{1}{2\pi}\,\frac{\sin \alpha\pi}{\cosh[(1-\alpha)s]-\cos \alpha\pi}\,ds \qquad (26)$$

with $s = \ln(\tau/\tau_0)$, to which the Cole–Cole equation formally leads.[326]

Examination of the measured static permittivities of the hydrates listed in Table X shows them to be very similar to each other. Due consideration of the likely effects of sample inhomogeneity, where necessary, suggests that such differences as are experimentally significant merely reflect the variability of $\varepsilon_{\infty 1}$. To within two or three permittivity units the amplitude of the dispersion associated with reorientation of water molecules, $\varepsilon_0 - \varepsilon_{\infty 1}$, in all cases rises in proportion to $1/T$, from a value of 58 at $-15°C$ to 70 at $-60°C$ and to about 86 at $-105°C$. Thus neither the amplitude nor the shape of the water dispersion in structure II hydrates is sensitive to the nature of the guest molecules. In these respects the dielectric behavior is characteristic of the lattice structure and was in fact used[371] to first show the formation of some of the structure II hydrates given in Table X.

Consideration of $\varepsilon_{\infty 1}$, which is mainly determined by the dipole moment of the guest molecule, is postponed to Section 5.1.

TABLE X. Dielectric Parameters of Structure II Hydrates

Hydrate	ε_0 $(-40°)$	τ_0 $(-40°)$, μsec	E_A, kcal mol^{-1}	$\varepsilon_{\infty 1}$ $(-105°)$	μ(gas), D
SF$_6$[935]	63	780	12.3	2.9	0
1,3-Dioxolane[1335]	69	5.4	8.7	4.57	1.47[a]
Propylene oxide[647]	$\sim$70	2.0	8.0	5.94	2.00
2,5-Dihydrofuran[647]	68	1.5	7.5	5.03	1.54[b]
Tetrahydrofuran[647]	67	1.0	7.4	5.06	1.63
Trimethylene oxide[647]	65	0.48	7.0	5.63	1.93
Cyclobutanone[1000]	71	0.49	6.5	9.85	2.89
Acetone[1000]	$\geq$47	0.57	6.5	$\geq$8.6	2.88

[a] From solutions in cyclohexane.
[b] From solutions in benzene.

4.3. Relaxation Times of the Water Molecules in Structure II

Table X shows the relaxation times τ_0 of the Cole–Cole representation at $-40°C$ and activation energies E_A which describe their dependence on temperature according to eqn. (24). The relaxation times at $-40°C$ range over three orders of magnitude; the largest activation energy is almost twice the smallest. The rate of reorientation of water molecules is obviously highly sensitive to the nature of the guest species present.

There is no apparent correlation of the relaxation rate (or activation energy) with the maximum van der Waals diameter of the guest (Fig. 4) or with the small variations shown in unit cell dimension (Table I).

The relatively high degree of stability of the SF_6 hydrate lattice is, however, pronounced, as is its resemblance in relaxation behavior to ice Ih, for which the relaxation time is 1500 μsec at $-40°C$ and $E_A = 13.2$ kcal mol^{-1}. The lowest nonzero multipole moment of the highly symmetric SF_6 molecule is the hexadecapole moment and the properties of its hydrate may be taken as close to the properties of the ideal hydrate lattice, unperturbed by electrostatic interaction with the guest molecules. As Table X shows, there is some tendency for the relaxation rate to increase with the dipole moment of the guest. The correlation is not close, however, and it has been suggested[591,1000] that the relaxation rates of ether and ketone hydrates may be controlled by the occasional formation of hydrogen bonds between guest and host molecules so as to inject Bjerrum defects of the L type into the lattice. The geometry of the guest molecule should favor the formation of such bonds by ketones more than by ethers and by trimethylene oxide more than by propylene oxide, in agreement with the experimental data. Studies of reorientation of the guest molecules (Section 5.3) have yielded no evidence of hydrogen bonding, but the extent of such bonding would presumably have to exceed several per cent to be readily detectable. To account for the relaxation behavior, the number of hydrogen-bonded guest molecules need represent only a fraction $\sim 2 \times 10^{-6}$ of those present.[591]

Relaxation times at a few relatively high temperatures have been reported[222] for hydrates of dichlorofluoromethane ($\mu = 1.3$ D) and trichlorofluoromethane (0.5 D), measured in the form of powder covered by liquid hydrate-former. The activation energies found (13.9 and 13.3 kcal mol^{-1}, respectively) were close to that of ice, although the relaxation times (130 and 450 μsec, extrapolated to $-40°C$) were considerably shorter. The dispersion loci were much broader than those characteristic of structure II hydrates and possibly reflect the presence of considerable quantities of ice. Nevertheless, the observation that these relaxation times are nearer to that of SF_6 hydrate than to those of the cyclic ether hydrates lends support to the possibility of a hydrogen-bonding mechanism in the ether hydrates.

Relaxation of the water molecules in type II hydrates accords well with the Bjerrum model. The pronounced variation from hydrate to hydrate in mean relaxation time while the narrow distribution of relaxation times about the mean remains substantially invariant has a ready explanation in a variability of the number of orientational defects which leaves unaffected the relative preference for different defect diffusion paths as determined by the lattice structure. In common with the departures from simple Debye behavior found in the orientationally disordered high-pressure ices, the distribution of relaxation times reflects the nonequivalence of sites and orientations of water molecules and therefore excludes relaxation mechanisms such as local melting which involve the simultaneous reorientation of more than a very few water molecules.[371]

4.4. Dielectric Properties of Structure I Hydrates

In all cases the absorption and dispersion curves of structure I hydrates are decidedly broader than for structure II, although additional experimental problems with the generally more volatile type I hydrates make less certain the identification of a shape of permittivity locus which is characteristic of the type I lattice. Some examples of experimental loci are given in Fig. 21. Of the hydrates given in Table XI, the best-defined loci have been obtained for cyclopropane[936] and ethylene oxide[376,542] hydrates. These are again approximately Cole–Cole arcs but with larger values of α, which make the departures from the arc which occur at high frequencies more pronounced than in hydrates of type II. The best samples of ethylene oxide gave values of α which rose uniformly from about 0.07 at $-40°C$ to 0.14 at $-130°C$,[542] values which agree fairly well with those found for cyclopropane hydrate (see Fig. 21B) at relatively high temperatures and for trimethylene oxide hydrate[647] (Fig. 21C) at relatively low temperatures. The broader loci measured[591] for argon (Fig. 21A) and nitrogen hydrates prepared under gas pressures of 1–2 kbar are possibly the result of some phase and composition inhomogeneity (see Section 4.5). On balance it appears likely that a shape of permittivity locus characteristic of "pure" structure I hydrates takes the form of a circular arc with values of α close to those given above for ethylene oxide hydrate, but, as for the case of structure II hydrates, with the absorption depressed at frequencies high and low compared with the frequency of maximum absorption.

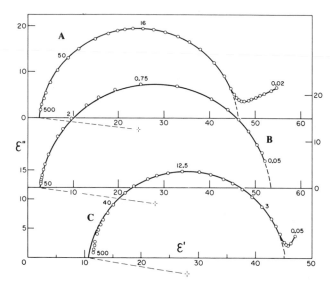

Fig. 21. Complex permittivity plots of structure I hydrates of
(A) argon at 8.0°C, (B) cyclopropane at −39.3°C, and (C) trimethylene oxide at −115.5°C. Frequencies are in kHz.

Although structure I hydrates are clearly distinguishable from those of
structure II by the shapes of their absorption and dispersion curves, this
does not appear to be true of the contributions of $\varepsilon_0 - \varepsilon_{\infty 1}$. For the best
defined cases, e.g., ethylene oxide hydrate between $+11$ and $-60°$[542]
and argon and nitrogen hydrates at relatively high temperatures,[591] these

TABLE XI. Dielectric Parameters of Structure I Hydrates

Hydrate	ε_0 (−40°C)	τ_0 (−40°C), μsec	E_A, kcal mol^{-1}	$\varepsilon_{\infty 1}$ (−40°C)	μ(gas), D
Ar	62[a]	96(2000b)[a]	5.7	2.85	0
N$_2$	61	180(1200b)	7.9	2.85	0
Cyclopropane	≥53	280	10	2.9	0
Ethylene oxide	62	0.33	7.7	10.8 (−105°C)	1.90
Trimethylene oxide	≥49	0.03[b]	5.8	≥11.1 (−105°C)	1.93

[a] Extrapolated from higher temperatures.
[b] Extrapolated from lower temperatures.

contributions are very similar to those found for structure II. In many cases conditions favorable to the development of relatively large gaps and voids existed, which led to low measured static permittivities, especially at low temperatures. This is true of Fig. 21C, for example, which illustrates the difficulty of obtaining adequate cell-filling factors for incongruently-forming hydrates prepared *in situ*. The effect of this is more serious for the structure I hydrate of trimethylene oxide than for the incongruent structure II hydrate of the same molecule, presumably because of the further contraction which occurs when structure II reacts with trimethylene oxide-rich liquid to form structure I. The occurrence of hydrates of both these structures was eventually[647] recognized by the dielectric behavior, after initial difficulties associated with the slow achievement of equilibrium between the solid phases. Fig. 22 illustrates the difference between the dielectric absorption of these two hydrates for samples substantially free of contamination by the second hydrate.

4.5. Relaxation Times of Water Molecules in Structure I

The relaxation times of the hydrates of nonpolar molecules (Table XI) are again much longer than those of the ether hydrates.

The activation energies present a different picture, however, and the low values for Ar and N_2 hydrates warrant further attention. At 0°C the

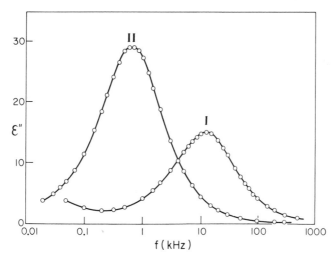

Fig. 22. Dielectric absorption at −115°C by water molecules in trimethylene oxide hydrates of types I and II. Corresponding values of $\varepsilon_{\infty 1}$ are 11.6 and 5.76, respectively.

relaxation times of all the hydrates with nonpolar guests are very close to the value of 21.5 μsec found for ice itself. These values are 16, 15, 15, and 18 μsec for the hydrates of argon (at 2000 bar), nitrogen (at 1200 bar), cyclopropane, and sulfur hexafluoride, respectively. Since the energies of the empty clathrate lattices are certainly less negative than that of ice (Section 3.6), it is a fair inference that the τ_0 values for nonpolar guests will be no larger than for ice and therefore that these high-temperature values are not sensibly too low for intrinsic relaxation. For ice the relaxation times become increasingly affected by impurities with decrease of temperature, so that by $-60°C$ the relaxation behavior characteristic of pure ice is only found in samples which have been carefully purified, as by zone-refining.[329,1163] The relaxation times of ice samples of ordinary purity show a considerable reduction in measured activation energy below about $-40°C$[589] a behavior almost certainly due to impurity-related extrinsic Bjerrum defects. The slower the intrinsic relaxation rate, the more sensitive is the relaxation to impurities.

It is therefore not unlikely that the slowly relaxing hydrates of Ar and N_2 will show such sensitivity, even at relatively high temperatures, and that their low activation energies may be ascribed to impurities. It was suggested[591] that, under formation pressures in excess of 1000 bar of gas, Ar and N_2 molecules are able to replace water molecules in the lattice in sufficient numbers to provide the principal source of defects, except at the highest temperatures of measurement. The smaller size of Ar would enable it to enter the lattice more easily and account for its lower activation energy, while a nonuniform distribution of substitutional impurities throughout the sample would account for the somewhat atypical breadth of the dielectric absorption.

Considerable gas pressures are of course necessary for the formation of Ar and N_2 hydrates above 0°C (see Table I) and even greater pressures were used[591] to increase the solubility in water and promote the very slow formation of hydrate under static conditions.

4.6. Comparison with Dielectric Relaxation of the Ices Proper

In Table XII dielectric parameters of the disordered (D) and ordered (O) polymorphs of ice are compared. The Kirkwood g factor has been calculated from

$$\varepsilon_0 - \varepsilon_{\infty 1} = 2\pi N(n^2 + 2)^2 g\mu^2/9kT \qquad (27)$$

where N is the number of water molecules per cm³, n has been taken as the sodium D refractive index as determined by the Lorentz–Lorenz equation

TABLE XII. Dielectric Parameters of Various Forms of Ice at $-30°C$[a]

"Ice"	Space group	Orientation	Conditions	Water molecules cm^{-3} $\times 10^{-22}$	n_D^2	$\varepsilon_{\infty 1}$	ε_0	α	g
Hydrate I	$Pm3n$	D	—	2.64	1.60	2.9	61	0.06	2.5
Hydrate II	$Fd3m$	D	—	2.62	1.59	2.9	63	0.021	2.6
Ih	$P6_3/mmc$	D	—	3.08	1.72	3.1	107	0.0	3.4
Ic	$F\bar{4}3m$	D	$-100°C$	3.10	1.73	~3.0	~159	—	~3.7
II	$R\bar{3}$	O	$-30°C$, 3 kbar	3.98	2.00	3.7	—	—	—
III	$P4_12_12$	D	$-30°C$, 3 kbar	3.88	1.97	4.1	117	0.04	2.6
IX	$P4_12_12$	O	$-175°C$, 1 bar	3.94	1.99	3.6[b]	—	—	—
V	$A2/a$	D	$-30°C$, 5 kbar	4.22	2.08	4.6	144	0.015	2.8
VI	$P4_2/nmc$	D	$-30°C$, 8 kbar	4.52	2.19	5.1	193	0.05	3.4
VII	$Im3m(?)$	D	25°C, 25 kbar	~5.52	~2.59	—	Large	—	~3
VIII	$Im3m(?)$	O	$-50°C$, 25 kbar	~5.53	~2.59	—	—	—	—

[a] Structural information is from Ref. 444, dielectric data from Refs. 62, 1400, 1401, 1415.
[b] $-120°C$, 2.3 kbar.

with $n_D^2 = 1.72$ for ice Ih, and $\mu = 1.84$ D. The value of g is an empirical measure of the average degree of local correlation of dipole directions, the equation having been derived[762] for a point dipole immersed in a sphere of isotropic polarizability expressed in terms of n^2. The assumption that all hydrogen-bonded configurations have equal probability led Powles[1118] to derive a value of 2.7 from the structure of ice Ih. With respect to the g values listed in Table XII, the clathrate hydrates are typical ices. Values of g for cubic ice Ic were found[589] to be indistinguishable from those of Ih near $-100°C$. Accurate quantitative measurements of the interesting cubic ice VII have not been made, but g appears to be smaller than for ice VI.[1400]

It is noteworthy that for all the high-pressure disordered polymorphs of ice the relaxation times[1415] extrapolated to $0°C$ and 1 bar are some 200 times smaller than the relaxation time of ice Ih, whereas, as already noted, the relaxation times of the clathrate hydrates of nonpolar guests at $0°C$ are hardly distinguishable from that of ice Ih. This clearly implies that when the relaxation is intrinsic in origin and not subject to perturbation by polar guest molecules the relatively small distortion from tetrahedral bonding in the hydrates produces no significant effect on the reorientation rate and therefore on the "strength" of the hydrogen bonding. The effect is much more pronounced for the appreciable angular distortions which exist in the high-pressure ices.

Departures from Cole–Cole arcs like those noted for the clathrate hydrates are found for ices III and VI, the perpendicular approach to ε_0 being particularly pronounced.[1415] For these two ices the dielectric relaxation may be more accurately represented by the superposition of two discrete relaxation times than by the continuous distribution of eqn. (26).

4.7. Outline of a Molecular Model of Relaxation of Ices and Hydrates

The reorientation of water molecules by diffusion of Bjerrum defects is a stochastic process, the details of which have been treated in a number of ways for ices Ic and Ih.[612,702,1048–1050,1165] In order to indicate the likely origin of the distributions of relaxation times which exist in the clathrate hydrates and the high-pressure polymorphs of ice most concisely, we here use an approach based on Hoffman's model[675a] of rotator-phase solids. Hoffman's model, which refers to a single-axis rotator in which jumps to only two adjacent rotational sites are permitted, is modified to permit reorientational jumps to four sites by rotation about any of the four O- -O bonds. It is also necessary to consider the correlations between neighboring orientations required for maintenance of full hydrogen bonding.

4.7.1. Ice Ic

In the case of ice Ic all water molecule sites are crystallographically equivalent, on average, and all six orientations at any site are equally probable in the absence of perturbation. After removal of a perturbation (e.g., by an applied electric field) of the equilibrium, the population of orientation 1 reverts to equilibrium according to

$$dN_1/dt = -4kN_1 + k(N_2 + N_3 + N_4 + N_5)$$

where N_2, N_3, N_4, and N_5 are the populations of the four orientations realizable from orientation 1 by a single jump and k is the common transition rate constant for any rotational jump. Simultaneous solution of the six similar rate equations, which may be written as

$$\begin{vmatrix} -(D+4k) & k & k & k & k & 0 \\ k & -(D+4k) & k & k & 0 & k \\ k & k & -(D+4k) & 0 & k & k \\ k & k & 0 & -(D+4k) & k & k \\ k & 0 & k & k & -(D+4k) & k \\ 0 & k & k & k & k & -(D+4k) \end{vmatrix} = 0$$

gives $D(D + 4k)^3(D + 6k)^2 = 0$ and

$$N_i = \tfrac{1}{6}N + \tfrac{1}{2}[N_i(0) - N_{7-i}(0)]e^{-4kt} + \{\tfrac{1}{2}[N_i(0) + N_{7-i}(0)] - \tfrac{1}{6}N\}e^{-6kt} \quad (28)$$

where the $N_i(0)$ are the populations of the six orientations at zero time and N is the total number of water molecules. Since the electric polarization is proportional to sums of terms which contain as factors differences in populations of orientations which are opposite to one another (e.g., $N_1 - N_6$), the relaxation term in e^{-6kt} is dielectrically inactive for the orthogonal set of orientations possible in cubic ice. The polarization takes the simple form $\Pi = \Pi(0)e^{-4kt}$. In compressional elastic relaxation, on the other hand, the strain relaxes as e^{-6kt}.[(1050)]

4.7.2. Ice Ih

In ice Ih the six orientations at any site form two subsets which consist, respectively, of the three orientations in which a proton lies in (or near) the O- -O bond in the c direction and of those which bear only protons in oblique directions. Since the correlation between orientations in the c

direction requires that half of the water molecules belong to each set, the equilibrium populations of all six orientations are again the same. Reorientation rates within the first set (k_1) may be distinguished from those within the second (k_2) and from those which take a molecule from one set to the other (k_3). The value of k_3 is the same for a transition in either direction since for every molecule which jumps from the first set to the second a neighboring molecule must jump from the second to the first. The characteristic determinant is now

$$
\begin{vmatrix}
-(D+2k_1 \\ +2k_3) & k_1 & k_1 & k_3 & k_3 & 0 \\
k_1 & -(D+2k_1 \\ +2k_3) & k_1 & k_3 & 0 & k_3 \\
k_1 & k_1 & -(D+2k_1 \\ +2k_3) & 0 & k_3 & k_3 \\
k_3 & k_3 & 0 & -(D+2k_2 \\ +2k_3) & k_2 & k_2 \\
k_3 & 0 & k_3 & k_2 & -(D+2k_2 \\ +2k_3) & k_2 \\
0 & k_3 & k_3 & k_2 & k_2 & -(D+2k_2 \\ +2k_3)
\end{vmatrix}
$$

The factors are D, $D + 4k_3$, $D + \frac{1}{2}(S + R)$ (twice), and $D + \frac{1}{2}(S - R)$ (twice), where $S = 3k_1 + 3k_2 + 4k_3$ and $R = [9(k_1 - k_2)^2 + 4k_3^2]^{1/2}$. For orthogonal orientations and a polycrystalline sample the polarization becomes

$$ \Pi = \tfrac{1}{3}\Pi(0)\{e^{-4k_3 t} + [(R + Q)/R]e^{-(S+R)t/2} + [(R - Q)/R]e^{-(S-R)t/2}\} \quad (29) $$

with $Q = 3k_1 - 3k_2 - 2k_3$. Thus as many as three relaxation times are possible, although the amplitude of the second process is relatively small unless the values of k differ greatly.

The experimental observation of a single dielectric relaxation time for ice Ih shows the rate constants to be indistinguishable. This is not surprising in light of the similarity in bond lengths and the close approach ($\pm 0.2°$) of the O--O--O angles to tetrahedral.[866] It should be emphasized that the k's are not rate constants in the usual sense since they depend on the product of the probability that a defect will appear in a particular bond and the probability that it will "jump" to produce the specified reorientation. In the case of cubic ice, Onsager and Runnels[1050] find that the relaxation rate is given approximately by

$$1/\tau = 4k = (N_D \nu_D + N_L \nu_L)/4N \tag{30}$$

where N_D and N_L are the numbers of D and L defects, respectively, the ν's specify their jump rates, and the factor of four in the denominator is an approximate measure of back-tracking during diffusion.[1165] In cases where the jump rates depend on the orientations the defects are no longer uniformly distributed over the bonds and differences in k may arise from the tendency of the defects to follow preferred diffusion paths. The fact that each effective reorientation of a molecule is accompanied by reorientation of molecules behind it and ahead of it in the diffusion chain will, for most geometries, tend to prevent the individual values of k from becoming greatly different.

4.7.3. Structure II Hydrate

In the most general case, each nonequivalent set of water molecules requires five relaxation times to describe the approach to equilibrium among the six orientations. The sixth root of the characteristic equation corresponds to the time independence of the equilibrium populations. Thus, structure I and structure II hydrates, each of which has three sets of sites, may have at most 15 discrete dielectric relaxation times.

The site symmetries in structure II are shown in Fig. 23. The symmetry of the eight (a) sites in the unit cell is the same as in cubic ice and dielectric relaxation proceeds as

$$\Pi(a) = \Pi(0, a)e^{-4k(a)t}$$

The symmetry of the 32 (e) sites is the same as in hexagonal ice, with the (e)–(a) bond taking the place of the bond in the direction of the c axis. The bonding to (a) requires that all orientations be equally populated at equilibrium, and again that $k_3(e)$ be the common rate constant for exchange between the two subsets. The result is eqn. (29). Moreover, since the transition described by $k_3(e)$ must be accompanied by reorientation at (a), $k_3(e) = k(a)$.

The remaining 96 molecules at (g) sites form the hexagonal rings of the structure II framework. Each is bonded (Fig. 23) to two (g) molecules in the same hexagonal ring as well as to one (g') molecule of an adjacent hexagon and to one (e) molecule. The correlation with the orientations at (e) and the equivalence of both ends of the (g)–(g) bonds require that the

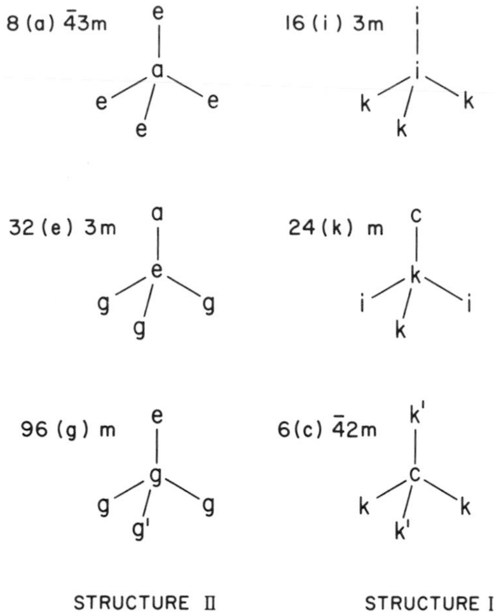

Fig. 23. Symmetry and bonding at the nonequivalent sites of the water molecules.

populations of orientations (e)–(g)–(g′) and (g)–(g)–(g) be equal and that
the remaining four orientations be also equally populated in the absence of
perturbation. The symmetry leads to 12 distinct values of k at the (g) sites
and to five relaxation times which are complicated functions of these
k values.

Summation of the contributions of the three sites thus leads to as
many as eight discrete relaxation times for structure II. Without any detailed
knowledge of the relative values of the 15 values of k required, further
analysis seems at present fruitless. The contributions to the permittivity
from some of the eight processes will be much mole important than from
others.

4.7.4. Structure I Hydrate

Within structure I, 16 (i) sites and 24 (k) sites have symmetries identical
to the (e) and (g) sites of structure II (Fig. 23) and together contribute as
many as eight discrete relaxation times. In addition, six (c) sites, located at
the intersections of the hexagonal rings, have two equivalent and opposite
orientations in which both protons lie in the same hexagon and four mutually
equivalent orientations in which the protons lie in different hexagons. If

k_1 and k_2 refer to transitions from the first to the second set and the reverse, and k_3 to transitions within the second set, the characteristic determinant is

$$\begin{vmatrix} -(D+4k_1) & k_2 & k_2 & k_2 & k_2 & 0 \\ k_1 & \begin{matrix}-(D+2k_2\\+2k_3)\end{matrix} & k_3 & k_3 & 0 & k_1 \\ k_1 & k_3 & \begin{matrix}-(D+2k_2\\+2k_3)\end{matrix} & 0 & k_3 & k_1 \\ k_1 & k_3 & 0 & \begin{matrix}-(D+2k_2\\+2k_3)\end{matrix} & k_3 & k_1 \\ k_1 & 0 & k_3 & k_3 & \begin{matrix}-(D+2k_2\\+2k_3)\end{matrix} & k_1 \\ 0 & k_2 & k_2 & k_2 & k_2 & -(D+4k_1) \end{vmatrix}$$

Of the four relaxation times which describe the approach of the populations to equilibrium after a perturbation, only two are dielectrically active and we have

$$\Pi(c) = [\Pi(c, 0)/(2k_1 + k_2)][k_2 e^{-4k_1 t} + 2k_1 e^{-(2k_2+2k_3)t}]$$

A total of ten relaxation times is thus predicted for structure I.

Since at any site in either of these unperturbed structures every orientation must remain equal in population to at least one other orientation, unique orientational ordering is not possible without a change in unit cell symmetry.

4.8. Resolution of Structure II Dielectric Spectra into Discrete Relaxation Times

A well-known characteristic of relaxation spectra generally is the difficulty associated with their resolution into individual relaxation processes. This is particularly true in the present instance where the possible presence of a considerable number of relaxation times is anticipated and where the overall departure from Debye behavior is relatively small. The results depend critically on the accuracy of the data and may be misleading in the presence of small systematic instrumental errors, of relaxing impurities (like ice), and of mobile ions which can contribute a space-charge polarization.

An analysis into discrete relaxation times has been attempted for a number of structure II hydrates for which the data are best defined. The results are illustrated for a sample of dioxolane hydrate of composition $C_3H_6O_2 \cdot 17.0H_2O$ at $-69.7°C$, the measurements having been made at 37 frequencies between 0.05 and 200 kHz. Values of ε'' were corrected for a small spurious low-frequency absorption by subtracting a dc conductivity of 3.6×10^{-11} ohm^{-1} cm^{-1}. The best-fit Cole–Cole arc [eqn. (25)] gave $\varepsilon_0 = 78.04$, $\varepsilon_{\infty 1} = 4.05$, $\tau_0 = 0.0714$ msec, and $\alpha = 0.0369$. This arc represented the experimental data (37 values each of ε' and ε'') with a standard deviation (σ) of 0.192 permittivity units, with systematic errors in the sense already discussed (Section 4.2) at relatively low and again at relatively high frequencies, where the experimental data clearly define the larger $\varepsilon_{\infty 1} = 4.32$.

The analysis into discrete relaxation times was made by adjustment of the $2m$ parameters in the relation

$$\varepsilon^* = 4.32 + \sum_{i=1}^{m} \frac{\varDelta \varepsilon_i}{i + i\omega\tau_i} \tag{31}$$

to minimize the sums of the 74 squares of differences between experimental and computed values by a program for estimation of nonlinear parameters which made use of Marquardt's algorithm[945] to reduce the number of reiterations required. The relaxation times resulting from this analysis are shown in Fig. 24 as solid line segments whose heights are proportional to the corresponding values of $\varDelta \varepsilon_i$. There is a substantial fall in σ as the number of relaxation times increases from one to three, a further fall at four, but no further change at five. The minor contribution associated with a relaxation time of about 1 msec for $m \geq 3$ almost certainly arises from space-charge polarization: It is contributed principally by the ε' data which have not been "corrected" for polarization associated with dc conductivity. It is inferred that the data justify resolution into three proper reorientational relaxation times, of which the one with a value close to τ_0 of the Cole–Cole representation makes a dominant contribution to permittivity. More than three relaxation times may be present, but this cannot be shown by the data. Already some elements of the correlation coefficient matrix are large and values of the individual parameters are consequently rather poorly defined.

The pattern of four relaxation times found from less extensive data for cyclobutanone hydrate at $-76.4°C$ (dashed lines in Fig. 24) is similar to that found for dioxolane hydrate. The agreement, except for the larger contribution of low-frequency polarization to the longest time, is well within the 90% nonlinear confidence estimates of the parameters.

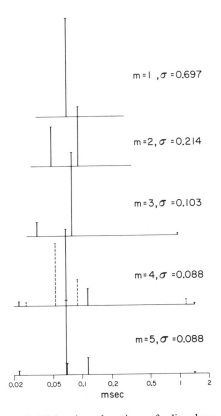

$m = 1, \sigma = 0.697$

$m = 2, \sigma = 0.214$

$m = 3, \sigma = 0.103$

$m = 4, \sigma = 0.088$

$m = 5, \sigma = 0.088$

0.02   0.05   0.1   0.2   0.5   1   2

msec

Fig. 24. Dielectric relaxation of dioxolane hydrate at $-69.7°C$ represented by superposition of a variable number of discrete Debye relaxation times. The four relaxation times of cyclobutanone hydrate at $-76.4°C$ represented by the dashed lines have been multiplied by ten.

4.9. Dielectric Properties of Amine Hydrates

Exploratory measurements which have been made of a number of amine hydrates suggest that the diversity of these structures is accompanied by a considerable variability in their dielectric properties. Effects associated with relatively high ionic conductance frequently interfere with a proper definition of the dielectric behavior.

4.9.1. $(CH_3)_3CNH_2 \cdot 9\frac{3}{4}H_2O$

The hydrate of *tert*-butylamine (Section 2.5.1) is the only known true clathrate among the amine hydrates. The dielectric absorption shown in

Fig. 25A is very broad and obscured by ionic conductance at low frequencies. Two large overlapping regions of absorption, of roughly equal amplitude, may be distinguished. The activation energy associated with each is about 9 kcal mol^{-1}. These may probably be assigned to relaxation of two groups of water molecules, both undergoing reorientation at rates considerably faster than found in hydrates of types I and II. It may be noted that although the 20 unlike O- -O- -O angles in this structure range over very wide values (Table III), half of these angles lie within $\pm 5°$ of the tetrahedral angle.[928] The origin of the small tail at the high-frequency end is uncertain, although it may be associated with some orientational disorder of the amine molecule. This tail extends over many decades of frequency at 77°K, where it extrapolates to $\varepsilon_{\infty 1} = 3.1_3$. There is, however, no such extraordinary reorientational mobility on the part of the guest molecule as exhibited in the other clathrate hydrates (Section 5).

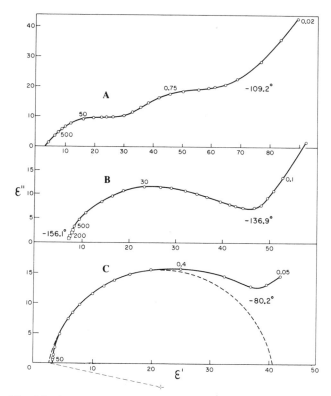

Fig. 25. Complex permittivity curves from exploratory studies by Y. A. Majid and J. Easterfield of the hydrates of (A) tert-butylamine, (B) trimethylamine, and (C) diethylamine.

4.9.2. $(CH_3)_3N \cdot 10\frac{1}{4}H_2O$

Figure 25B illustrates the qualitative behavior of this hydrate at low temperatures. Only a single, very broad absorption region is found at audio and radio frequencies down to 77°K. The mean reorientation rate is the fastest yet found in the amine hydrates and the activation energy is only about 4 kcal mol^{-1}. The high-frequency tail is less broad than in *tert*-butylamine hydrate. An interesting feature is the relatively large high-frequency intercept (>6) which appears to be too high to be ascribable to rapid reorientation of the amine molecules ($\mu = 0.65$ D) alone. Of the 41 water molecules in the unit cell, seven have been assigned to sites which are disordered.[1070] It is possible that some of these water molecules undergo rapid reorientation by processes which are accompanied by changes in the orientations of amine molecules to which they are hydrogen-bonded. This could also account, by injection of defects, for the comparatively rapid relaxation rate of the other water molecules which are more "normally" bonded to their neighbors.

4.9.3. $(C_2H_5)_2NH \cdot 8\frac{2}{3}H_2O$

Figure 25C illustrates the dielectric behavior found for this hydrate. The reorientational relaxation is relatively slow and corresponds to an activation energy of about 8 kcal mol^{-1}. The value of $\varepsilon_{\infty 1}$ (3.2) appears to include no contribution from rapidly rotating molecules in agreement with the structure[718] of this hydrate, which precludes such reorientation of the amine. Indeed, in marked contrast to the trimethylamine case, hydrogen bonding of diethylamine to water molecules appears to exert a stabilizing effect on the lattice. Since the secondary amino group both donates and accepts a hydrogen atom in forming hydrogen bonds to two adjacent water molecules, there is no net injection of Bjerrum defects into the framework. Moreover, the rotational immobility of the amine molecules prevents exchange of the donator and acceptor bonds (except possibly by an ionization mechanism) and blocks the diffusion of Bjerrum defects across the amine molecules. The 24 (of 104) water molecules which are directly hydrogen-bonded to the 12 amine molecules in the unit cell can undergo reorientation between only three sites instead of the six normally available. The result is a mean relaxation rate which is slower than that of tetrahydrofuran hydrate, for example, despite the considerable distortion of many of the O--O--O angles from tetrahedral values. The effect of this angular spread may be seen in the considerable width of the dielectric absorption locus.

4.9.4. $(CH_2)_6N_4 \cdot 6H_2O$

The observation of a relatively large absorption (Fig. 26) apparently attributable to the relaxation of water molecules in the hexahydrate of nonpolar hexamethylenetetramine indicated[372] that the partial orientational ordering of the water molecules is not as great as the structural determination suggested.[937] The original assignment of a full hydrogen atom to the O(1) end of the bond between hexagonal rings (see Fig. 9) is consistent only with little or no dielectric polarization. The difficulty is removed if the three-coordinated O(1) molecules are permitted three possible orientations with two-thirds of a proton, on average, at the O(1) end of each of its bonds. It is interesting that the increased relaxation rate expected from the presence of water molecules [O(1)] which are hydrogen-bonded to only three neighbors appears to be in large part compensated for by the less effective defect diffusion possible in a three-coordinated network.

Simple analysis of the kinetics of reorientation among the three orientational sites of each of the water molecules by the method described in Section 4.7 leads to the prediction of only two relaxation times for this hydrate. This is at least consistent with the observed behavior, which is obscured by ionic conductance at low frequencies and an absorption apparently associated with impure ice at high frequencies (Fig. 26). More definitive study of this hydrate is warranted also by the possibility that the orientations of the water molecules become more ordered at low temperatures.

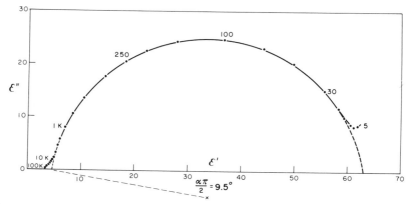

Fig. 26. Complex permittivity locus of hexamethylenetetramine hexahydrate at $-53.3°C$.[372] The high-frequency inflection appears to be contributed by ice.

4.10. Dielectric Properties of Tetraalkyl Ammonium Salt Hydrates

The work of Carstensen and Koide[775] showed the dielectric absorption in the tetra-*n*-butyl and tetra-*iso*-amyl ammonium bromide, chloride, and hydroxide hydrates to be broad and complex. To the extent that it is presently possible to properly identify the contribution of reorientation of the water molecules, this reorientation appears to be most rapid (≥ 100 MHz at $0°$C) in the hydroxide hydrates, somewhat less rapid in tetra-*n*-butyl ammonium halide hydrates, and slowest (0.1–1 MHz at $0°$C) in the tetra-*iso*-amyl ammonium halide hydrates. Such a sequence would suggest that reorientation of water molecules is accelerated by the presence of OH^- in the lattice frameworks and by the disorder of the tetragonal frameworks of tetra-*n*-butyl ammonium salt hydrates (Section 2.6.1).

4.11. NMR Spectra of the Protons of the Water Lattices

4.11.1. *Second Moments*

The wide-line proton magnetic resonance spectra of the clathrate hydrates of proton-containing guest molecules consist, at most temperatures, of a narrow line from the guest species superimposed on a broad line from the water protons, as shown for structure I cyclopropane hydrate in Fig. 27.[936] The narrowness of the component originating with the encaged molecules is due to their rapid reorientation. This component is considered in Section 5.8.

The broad component, as is shown most clearly when the narrow component is absent (e.g., in Xe and SF_6 hydrates), is far from Gaussian in shape at low temperatures, with less absorption in the wings, and shows evidence of an unresolved doublet structure.[935] This shape is probably the result of the broadening effect of intermolecular proton spin–spin interaction on the Pake doublet[1065] associated with the dominant interaction between protons of the same water molecule.

Second moments of the total proton absorption signals for hydrates of structures I and II are plotted against temperature in Figs. 28 and 29. With the exception of the data shown for Cl_2[7] and SF_6[935] hydrates, the results are from unpublished work of S. K. Garg and for some hydrates are merely exploratory in nature.

Where the only protons present are those of water, the rigid-lattice second moments (32.5 ± 0.6, 32.6 ± 0.5, and 33.4 ± 1.5 G² for Xe, SF_6, and tetrahydrofuran-d_8 hydrates, respectively, at $100°$K) are very similar to that of hexagonal ice (32.4 ± 1.1 G² at $75°$K[1125]). This is to be ex-

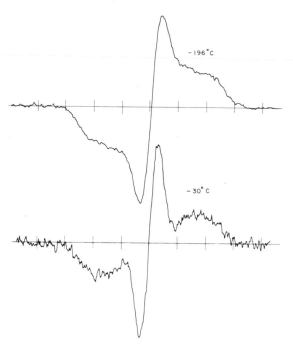

Fig. 27. Derivative curves of proton magnetic resonance absorption in cyclopropane hydrate.[936] Scale divisions are 5 G apart.

pected from the similarities in the local structures, as detailed calculations[935] have shown. These hydrates clearly show an increase of second moment with increase of temperature until the temperatures of ultimate line narrowing are reached. Such an increase is opposite to the trend expected from the usual effects of density and vibrational changes. It is possible that the mean H—O—H angle and/or the O—H bond length decrease somewhat with rising temperature, although this explanation becomes implausible if the gas-phase angle is preserved[287] in ice at all temperatures. The question of the geometric distortion of the water molecule in ices[1399] remains unsettled. Saturation difficulties (see below) may have prevented definite observation of a similar trend with temperature in the second moment of ice Ih, although an increase with rising temperature in the linewidth measured at right angles to the c axis by Kume and Hoshino[815] has been pointed out.[319]

When the guest molecules contain protons, similar low-temperature second moments of the water proton absorption are obtained if the observed second moments are "corrected" for the reduction produced by the presence of the narrow component. The low-temperature behavior of the second moment for the type I hydrate of trimethylene oxide (Fig. 28) appears to reflect the temperature dependence of the contribution of this guest species (Section 5.8.2).

4.11.2. Line Narrowing and Diffusion

The linewidths and second moments begin to decline at temperatures where the reorientation or diffusion rate of the water molecules becomes sufficiently fast. It is apparent from Figs. 28 and 29 that the decline begins when the dielectric relaxation time has reached 10^{-4} sec and is substantial at $\tau_0 = 10^{-5}$ sec. The correlation is sufficiently good as to suggest that the simpler NMR linewidth measurements may be used to estimate dielectric relaxation times of other hydrates.

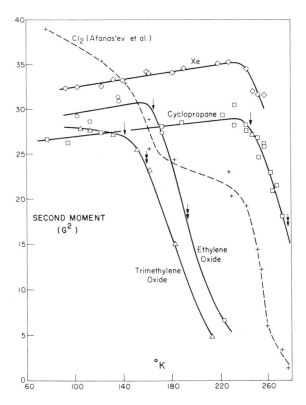

Fig. 28. Second moments of proton resonance absorption bands of structure I hydrates. Single- and double-headed arrows mark temperatures at which the dielectric relaxation times are 10^{-4} and 10^{-5} sec, respectively.

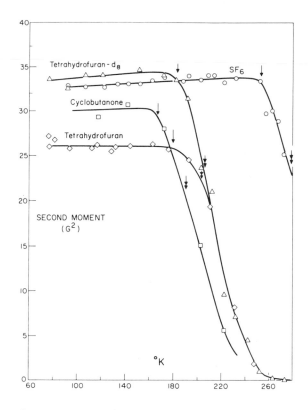

Fig. 29. Second moments of proton absorption curves
of structure II hydrates.

At high temperatures the water component of the spectrum merges
with the guest component (if any) and may, as in the ethylene oxide and
tetrahydrofuran cases,[237] become the narrower of the two before the
decomposition temperature is reached.

Although the initial narrowing shows a well-defined correlation with
reorientation of the water molecules, diffusion of water molecules (or at
least of their protons) appears to be required to account for the low second
moments reached at higher temperatures. Reorientation of water molecules
around fixed sites can only lower the second moment to about 7 G^2. The
lack of a definite plateau near this value suggests that rotational and diffusional narrowing become effective in roughly the same temperature range.
The role of diffusion is less clear in the behavior of the hydrates of nonpolar
molecules, where the second moments do not become small enough to
show definite diffusional effects before decomposition occurs.

Afanas'ev et $al.$[7] suggested that the two linewidth transition regions shown by their second moments of Cl_2 hydrate (Fig. 28) corresponded respectively to rotational and diffusional narrowing. It appears more likely that one of the transitions (possibly the one at higher temperatures) is associated with the presence of ice and that the exceptionally high second moment of 39 G^2 measured at $-196°C$ resulted from partial saturation of the ice component. The possible role played by hydrolysis of Cl_2 is unclear. It seems probable that hydrolysis of SO_2 contributes to the exceptionally fast dielectric relaxation of SO_2 hydrate, whose absorption covers a broad range of audio and radio frequencies at $-196°C$. The second moment of SO_2 hydrate at $-53°C$ is no more than about 0.7 G^2 and therefore smaller than for any of the hydrates of Figs. 28 and 29.

Onsager and Runnels[1050] have pointed out that there is closer agreement of the local magnetic field correlation times[92] in ice at relatively high temperatures with the jump times of translational diffusion than with the dielectric relaxation times, which are about an order of magnitude longer. This suggests that the line narrowing in ice, which begins at $-40°C$ according to some results[237,814] and is not yet appreciable at $-35°C$ according to others,[92] is mainly the result of the self-diffusion of water molecules.

Line narrowing begins at a higher temperature in SF_6 hydrate than in ice and at at least as high a temperature in Xe and cyclopropane hydrates, despite the somewhat faster reorientation rates shown by the hydrates. Diffusion therefore appears to be slower in these hydrates than in ice. On the other hand, it must be a great deal faster in the hydrates of polar guest molecules given in Figs. 28 and 29. The empirical relationship noted above between initial line narrowing and an average dielectric relaxation time of 10^{-4} sec suggests, if indeed diffusion is invariably more effective than reorientation in causing line narrowing in the clathrate hydrates, that there is a close correlation between the dielectric and diffusion rates. It should be noted that the distribution of dielectric relaxation times will lead to narrowing at a lower temperature than that expected from the average reorientation rate.

It may be concluded that the diffusion step times (given by $d^2/6D$, where d is the interwater distance and D the self-diffusion coefficient) and the dielectric relaxation times are of roughly equal magnitude in the clathrate hydrates at relatively high temperatures and that they depend in a similar way on the nature of the guest molecule. Some support is thus provided for the suggestion of Haas[612] that the diffusion of Bjerrum defects is necessarily accompanied by the translation of interstitial water molecules or of vacancies. In the case of ice, strong arguments have been presented[598,1050] that diffusion and dielectric relaxation are unrelated processes. The details of any interstitial defect diffusion mechanism will of course be different for the hydrates, which lack the long hexagonal channels present in ice. Interstitial diffusion in the hydrates might be expected to be facilitated by the presence of unoccupied small cages but Fig. 28 provides no support for this. It would appear that the question of the mechanism or mechanisms of diffusion in "ices" is not yet settled.

4.11.3. *Spin–Lattice Relaxation Times*

Values of T_1 calculated[1050] for ice from the molecular reorientation rate are larger by a factor of about five than the experimental values. There is better agreement with values calculated from the diffusion coefficient provided that the water molecule moves several molecular diameters in the diffusion step.[1050] No quantitative T_1 measurements of the water protons of clathrate hydrates appear to have been reported. There is a significant difference, however, between ice and most clathrate hydrates in the level of saturation of the absorption signal found at low temperatures. Thus the value of T_1 measured at 30 MHz in ice has reached 100 sec at $-38°C$ and saturation was only avoided (if then) at $-183°C$ by working with an rf field of 40 μG or less.[814] Rabideau *et al.*[1125] found that similarly low fields are required to avoid saturation in ices Ih, Ic, II, IX, V, and VI at 60 MHz and $-200°C$. None of the measurements of the clathrate hydrates of Figs. 28 and 29 show saturation at 16 MHz and rf fields of the order of 5 mG, which implies that T_1 is never larger than the order of seconds. Argon hydrate, however, shows saturation comparable to ice.

It does not seem possible to attribute the small low-temperature T_1 values of clathrate hydrates to diffusion or reorientation of water molecules. A more likely origin is in the interaction between the spins of the reorientating guest molecules and the stationary proton spins of the lattice. Such a spin–rotation mechanism has been proposed, for example, to account for the low values of T_1 in the low-temperature solid phase of SF_6 itself.[204] The contrast between the saturation behavior of argon and xenon hydrates may reflect the coupling between the lattice and the large-amplitude "rattling" motion of encaged ^{129}Xe and ^{131}Xe.

4.11.4. *Amine Hydrates*

Definitive studies of the relationship of the NMR spectra to the interesting structures (Section 2.5) of the amine hydrates have not been made. Cursory study of $(CH_3)_3N \cdot 10\frac{1}{4}H_2O$ showed a monotonic fall in total second moment from 29 G^2 at 100°K to about 2 G^2 at 200°K. The relaxation time which corresponds to the maximum of the very broad dielectric absorption (Section 4.9.2) reaches 10^{-4} sec at about 120°K. The hexamethylenetetramine hexahydrate spectrum began to show narrowing at 250°K, where $\tau_0 = 10^{-4}$ sec. Serious saturation occurred at lower temperatures.

4.12. Infrared Spectra

In general features the infrared spectra of the water lattices of clathrate hydrates resemble those of ices Ih and Ic, with, however, significant differences in some details. In SO_2 hydrate the maximum absorption of the librational band at 780 cm^{-1} is shifted downward by some 30 cm^{-1} with respect to ice Ic[640] and the maximum of the first overtone OH stretching band is shifted upward by 70 cm^{-1} with respect to ice Ih.[894] Both shifts are consistent with somewhat weaker hydrogen bonding between the water molecules of the hydrate. Bertie and Othen* find that the OD stretching band of isotopically dilute HOD in ethylene oxide hydrate is broader ($\sim$85 cm^{-1}) than in ice, with a flat-topped, structureless shape which accords with the presence of (four) crystallographically distinct O--O bonds in which the positions of the protons are disordered. Disorder also contributes to the width of the translational vibration band of ethylene oxide hydrate which shows maximum absorption at 225 cm^{-1} and differs considerably in shape from the corresponding band of ice Ih or Ic.

Falk[464] found the center of gravity of the OD stretching band of dilute HOD in $(CH_3)_3N \cdot 10\frac{1}{4}H_2O$ to be 10 cm^{-1} higher than in ice Ih and the band to be twice as wide, a result consistent with the occurrence of ten distinct O--O and two O--N bonds in this disordered structure (Section 2.5). An IR study of hexamethylenetetramine hydrate by Solinas and Bertie now in progress* may elucidate the degree of order present. (See Section 4.9.4.)

5. MOTION OF THE GUEST MOLECULES

In this section the available experimental information about the motion of encaged molecules is reviewed and some models relevant to this motion considered. Perhaps the feature which most distinguishes gas hydrates,

* J. E. Bertie, private communication.

along with some other clathrates, from other solid systems is the remarkable degree of rotational mobility of the guest species. If the guest is polar, its rotational motion is most directly studied by dielectric measurements, the results of which are presented first.

5.1. Contribution of Guest Molecules to the Static Permittivities (ε_{02})

As noted in Section 4.2, the permittivity measured on the high-frequency side of the dispersion region associated with reorientation of the water molecules reaches a limiting value $\varepsilon_{\infty 1}$ which depends on the nature of the guest molecule as well as on the temperature. As Fig. 30 shows, this limiting value, as measured for structure II hydrates at a fixed, relatively high temperature, is very nearly proportional to the square of the dipole moment of the guest molecule. Moreover, above about 150°K for most hydrates the value varies linearly with $1/T$. Taken together, these two observations indicate that the guest molecules are undergoing reorientation at rates considerably in excess of the frequencies of measurement, say 1 MHz, either isotropically or between two or more preferred orientations of similar energy. To emphasize that $\varepsilon_{\infty 1}$ is a static permittivity of the dispersion associated with reorientation of guest molecules, it will be labeled ε_{02} in this section.

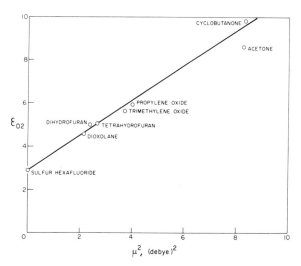

Fig. 30. Dependence on dipole moment of the reorientational contribution of guest molecules to the experimental permittivities of structure II hydrates at 168K.

Since the cages are almost spherical, these results provide an opportunity for application of the Onsager cavity model[1047] in which a molecule is taken to be a point dipole at the center of a spherical void of radius a immersed in a dielectric continuum. For the present case the Onsager result may be written

$$\frac{(\varepsilon_{02} - 1)(2\varepsilon_{02} + 1)}{3\varepsilon_{02}} = \frac{(\varepsilon^{\circ} - 1)(2\varepsilon^{\circ} + 1)}{3\varepsilon^{\circ}}$$

$$+ \frac{4\pi N_M}{1 - f_M \alpha_M} \left[\alpha_M + \frac{\mu_M^2}{3kT(1 - f_M \alpha_M)} \right] \qquad (32)$$

where ε_{02} is the static permittivity, ε° is the permittivity of the empty rigid lattice, and N_M, μ_M, and α_M are, respectively, the number density, dipole moment, and polarizability of encaged M molecules. The reaction field factor f_M is $2(\varepsilon_{02} - 1)/a^3(2\varepsilon_{02} + 1)$. In the usual application of this model to liquids[213] a is an effective molecular radius given by $4\pi Na^3/3 = V$, where N is Avogadro's number and V the molar volume. For clathrate hydrates a must be close to the "free radius" (Section 2.1) of the real cage, which in the case of the large structure II cage lies between 3.2 and 3.3 Å.

The value of ε° may be estimated (from the similarity in lattice structure to ice Ih) by setting the first term of the rhs of eqn. (32) equal to $(\varepsilon_{\infty} - 1)(2\varepsilon_{\infty} + 1) d_r/3\varepsilon_{\infty}$, where $\varepsilon_{\infty} = 3.10$ for ice and $d_r = 0.888$ is the ratio of the density of the empty lattice to that of ice. This yields $\varepsilon^{\circ} = 2.81$.

TABLE XIII. Structure II Hydrates at −105°C

Guest	α, Å³	ε_{02}(exp)	ε_{02} [eqn. (32)]
SF$_6$	6.46	2.9	3.04
1,3-Dioxolane	6.72	4.57	4.27
2,5-Dihydrofuran	7.74	5.03	4.51
Tetrahydrofuran	7.93	5.06	4.72
Trimethylene oxide	6.23	5.63	5.14
Propylene oxide	6.18	5.94	5.29
Cyclobutanone	7.57	9.84	8.58
Acetone	6.41	≥8.6	8.02

In Table XIII values of ε_{02} given by eqn. (32) with $a = 3.2\,\text{Å}$ are compared with the experimental values. The calculated contributions of the polar guest molecules to ε_{02} are seen to represent only $81 \pm 4\%$ of the experimental contributions, except for acetone hydrate, for which the experimental value[1000] is poorly defined. A similar underestimate is found for the structure I hydrates of ethylene oxide[542] and trimethylene oxide.[647] Formally, the calculated contributions may be made to agree closely with the experimental ones for structure II hydrates by allowing a to decrease to $2.8\,\text{Å}$. A more realistic modification of the model, however, is given in Section 5.4.

5.2. Behavior of ε_{02} at Low Temperatures

At low temperatures ε_{02} ceases to increase as $1/T$ increases. This behavior is illustrated for cyclobutanone hydrate in Fig. 31, where the departure occurs at temperatures below 150°K. A broad maximum in the region of 45°K (heavy line) is followed by a decrease at lower temperatures. Below about 30°K the static permittivity could not be measured at frequencies above 1 Hz because the reorientation of cyclobutanone molecules became too slow. Dispersion of ε' at temperatures between 15 and 50°K is reflected by the differences in the permittivity curves at the fixed fre-

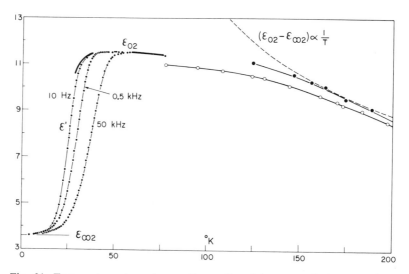

Fig. 31. Temperature dependence of ε_{02} of cyclobutanone hydrate (various samples).[1000] Values of ε' measured at a number of fixed frequencies are also shown.

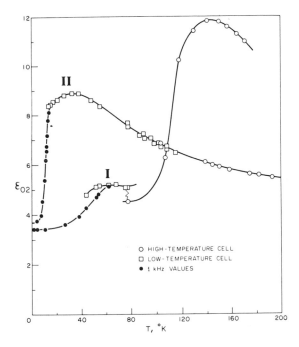

Fig. 32. Temperature dependence of ε_{02} of trimethylene oxide hydrates of types I and II. The mismatch at 77°K is probably a result of contraction of I in the "high-temperature" cell.

quencies shown in Fig. 31. Qualitatively the same behavior has been found for the hydrates of tetrahydrofuran, acetone, and ethylene oxide.[542] An interesting contrast between the behavior of trimethylene oxide in its two hydrates is shown in Fig. 32. The plateau shown by trimethylene oxide in its structure I lattice in the region of 50–90°K is unique among the hydrates so far examined.

The symmetry defined by the positions of the oxygen atoms of the water molecules (Fig. 3) is sufficiently high as to require a symmetric distribution of preferred orientations of equal energy within any cage type. Failure of ε_{02} to increase with $1/T$ at low temperatures can only mean a degradation of this symmetry. Orientations which are statistically equivalent when averaged over many cages acquire within individual cages differences in energy which are significant multiples of kT, leading to an increasing depopulation of less stable orientations as the temperature falls. Such behavior is to be expected from the orientational disorder of the water molecules, as discussed in Section 5.5.

5.3. Dielectric Relaxation of Guest Molecules

Davies and Williams[375,380] measured the dielectric absorption associated with reorientation of the guest molecules in the hydrates of ethylene oxide, tetrahydrofuran, and acetone at frequencies up to 8.5 GHz and temperatures down to 90°K. Their estimated relaxation times were of the order of 10^{-11} sec at 90°K. The Eyring activation energies ($= E_A - RT$) were about 0.5 kcal mol^{-1} for ethylene oxide hydrate and 0.3 kcal mol^{-1} for the type II hydrates. Since all measurements (except possibly one for ethylene oxide hydrate) were made to the low-frequency side of maximum absorption, the shapes of the absorption curves were only partly defined. They appeared to vary from close to the ideal single-relaxation-time Debye shape for ethylene oxide hydrate to much broader absorption for acetone hydrate.

More recently dielectric measurements have been extended to lower temperatures where the absorption may be more adequately defined at lower frequencies (1 Hz to 1 MHz). The absorption curve[373] shown in Fig. 33 for tetrahydrofuran hydrate is fairly typical of those obtained. The absorption is invariably so much broader than the ideal Debye absorption as to make difficult its characterization by measurements at a fixed tem-

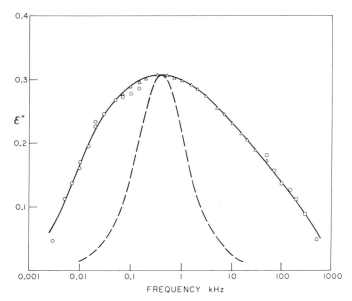

Fig. 33. Dielectric absorption associated with reorientation of tetrahydrofuran molecules in the hydrate at 20.4°K.[373] A Debye absorption curve is shown for comparison.

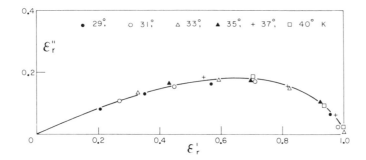

Fig. 34. Reduced permittivity locus of cyclobutanone hydrate from
data at a number of temperatures.[1000]

perature over the available frequency range. The absorption falls much
more slowly on the high-frequency than on the low-frequency side of the
maximum and the Cole–Cole plots (cf. Fig. 34) have some resemblance to
flat, skewed arcs.[374] Analysis of the shape of the locus in the case
of ethylene oxide hydrate, however, has shown[542] that the absorption
falls rather less rapidly at relatively low frequencies than for the skewed
arc. In all cases the absorption becomes broader with decrease of temperature.

Figures 35 and 36 show the dependence on temperature of the absorption at 1 kHz for a number of guest molecules. Such plots provide a clear
indication of the variation of relaxation rate with the size of the guest
molecule but tend to obscure the real asymmetry in the frequency dependence of the absorption.

Some dielectric properties associated with reorientation of guest molecules are given in Table XIV. These include values of ε_{02} at the broad
maximum (cf. Fig. 31), temperatures of maximum absorption at a number
of frequencies in the kHz range, and frequencies of maximum absorption
estimated[380] near 90°K from measurements in the GHz range. Values of
E_A have been derived from the low-temperature measurements at frequencies below 1 MHz; values labeled E_A* are averages over the range
between the low-temperature measurements and those at 90°K. There is a
tendency for the apparent activation energy to undergo some decrease with
increase of temperature, although the change is small for such a large
range of relaxation rates. Within each structure the relaxation rate decreases
and the activation energy increases with increase in size of guest molecule.
There is a particularly noteworthy contrast between structures I and II in
the reorientational mobility of trimethylene oxide.

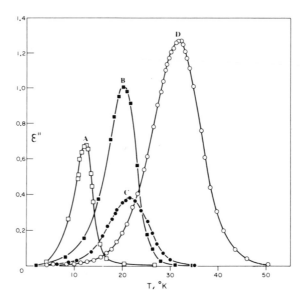

Fig. 35. Dielectric absorption at 1 kHz by guest molecules in structure II hydrates of (A) trimethylene oxide,
(B) acetone, (C) tetrahydrofuran, and (D) cyclobutanone.

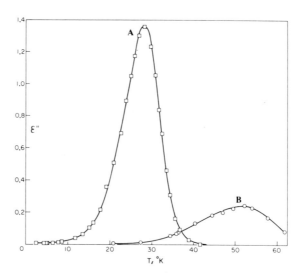

Fig. 36. Absorption at 1 kHz by (A) ethylene oxide and
(B) trimethylene oxide in their structure I hydrates.

TABLE XIV. Dielectric Parameters Associated with Guest Molecules

Hydrate	Max. ε_{02}, at T °K	T, °K, of max. ε'' at						Frequency of max. ε'' at 88°K [380]	E_A, kcal mol^{-1}	E_A^*, kcal mol^{-1}
		10 Hz	100 Hz	1 kHz	10 kHz	100 kHz	1 MHz			
Ethylene oxide	14, 50°K	—	25.7	28.0	30.8	—	—	6.6 GHz	1.42	1.26
Trimethylene oxide (I)	11.9, 145°K	—	45.0	51.9	56.7	63.1	71	—	2.43	—
Trimethylene oxide (II)	8.9, 35°K	—	10.9	12.0	13.2	14.4	—	—	0.60	—
Acetone	12.5, 45°K	17.8	18.9	20.2	22.2	24.4	27.6	37–140 GHz (93°K)	1.08	0.82–0.94
Tetrahydrofuran	6.3, 50°K	18.3	19.8	21.6	23.8	—	—	14 GHz	1.06	0.92
Cyclobutanone	11.6, 45°K	25.8	—	31.8	35.4	40.0	—	—	1.44	—

5.4. Limiting Permittivities at Very Low Temperatures

At temperatures near $4°K$ the measured permittivities approach the values listed as $\varepsilon_{\infty 2}$ in the third column of Table XV, the error estimates being based on cumulative uncertainty in the measured permittivities (mainly in the low-temperature cell constants) and possible inaccuracy in the estimation of the proper limiting high-frequency permittivity from the measurements at kHz frequencies. In all cases there is still detectable absorption and dispersion even at $4°K$.

These experimental values of $\varepsilon_{\infty 2}$ are significantly higher than the ε_{∞} values found for nonpolar guest molecules, which are 2.9 ± 0.1 (cf. Tables X and XI) at relatively high temperatures and are expected to be no higher at $4°K$. There appears therefore to be a contribution to the permittivity associated with dipole moments of guest molecules in addition to that contributed by reorientation. The magnitude of this contribution has been estimated for a number of hydrates in Table XV, which shows values of $\varepsilon_{\infty 2}$ calculated from eqn. (32) with the dipolar term removed, ε_{02} replaced by $\varepsilon_{\infty 2}$, and polarizabilities given in Table XIII. For these polar guest molecules $\varepsilon_{\infty 2}(exp)$ is seen to significantly exceed $\varepsilon_{\infty 2}(calc)$.

The most likely source of this difference is the rotational oscillations of the guest molecules about their preferred orientations within the cages.[1000] These vibrations are expected to be of large amplitude and to occur in the far-infrared region. Formally, for agreement with the Onsager

TABLE XV. Low-Temperature Permittivities and Related Parameters[a]

Hydrate	$\varepsilon_{\infty 2}$ (calc)	$\varepsilon_{\infty 2}$ (exp, $4°K$)	$\alpha_M{}^*$, $Å^3$	I_C	I_A	I_B	ν_i, cm^{-1}
Ethylene oxide	3.16^b	4.0 ± 0.2	12	43.8	24.8	—	32, 43
Trimethylene oxide (I)	3.27	3.4 ± 0.1	8	75.1	—	43.1	58, 71
Trimethylene oxide (II)	3.03	3.7 ± 0.2	19	75.1	—	43.1	17, 23
Tetrahydrofuran	3.11	3.5 ± 0.1	16	126.1	—	72.5	15, 20
Acetone	3.04	3.9 ± 0.1	21	103.0	49.7	—	20, 29
Cyclobutanone	3.08	3.6 ± 0.2	18	142.1	—	105.2	22, 25

[a] I values in amu $Å^2$.
[b] With $\alpha_M = 4.52$ $Å^3$.

equation, they must raise the effective polarizability of the guest molecule to a value ($\alpha_M{}^*$, Table XV) which is some two to three times larger than the "normal" polarizability.

It is possible to roughly estimate the frequency region in which the rotational oscillations should occur by relating the increment in refractive index $[\varepsilon_{\infty 2}(\exp)]^{1/2} - [\varepsilon_{\infty 2}(\text{calc})]^{1/2}$ to the integrated absorption in the far infrared.[165] For simplicity it is assumed that there are two rotational oscillations of frequencies ν_i which give rise to bands of negligible width. The absorption intensity of each is proportional to

$$(\partial \mu/\partial Q)^2 = \mu^2(\sin^2 \theta)/I_i\theta^2 \approx \mu^2 I_i,$$

where I_i is the moment of inertia about the i axis and θ the rotation about this axis from the equilibrium orientation. The corresponding contribution to the refractive index is proportional to $\mu^2/I_i\nu_i^2$, the proportionality factor being $N_M/6\pi$.[165] With the i axes taken to be the principal inertial axes other than the polar axis of the molecule, the frequencies shown in the last column of Table XV are obtained. Except for the significantly higher frequencies for trimethylene oxide in structure I, the model is too crude to suggest more than the presence of intense absorption at low temperatures at frequencies near 20 cm^{-1} and above. The electrostatic fields of the water molecules (Section 5.5) are expected to result in *broad* absorption.

It seems likely that the enhanced polarizability responsible for the large values of $\varepsilon_{\infty 2}$ is also responsible for the relatively large values of ε_{02} which, as noted in Section 5.1, considerably exceed the values given by the Onsager equation with "normal" polarizabilities and $a = 3.2$ Å. To take account of the contributions of rotational oscillations, $\alpha_M{}^*$ is substituted for α_M in the second term of Eq. (32) and one of the factors $1/(1 - f_M\alpha_M)$ in the third term is replaced by $[1 + f_M(\alpha_M{}^* - \alpha_M)]/(1 - f_M\alpha_M{}^*)$.[1000] With $\alpha_M{}^*$ (Table XV) derived from the low-temperature values of $\varepsilon_{\infty 2}$, the calculated values of ε_{02} now exceed the experimental ones at 168°K. For formal agreement a must be permitted values in the range 3.4–3.6 Å. Although some increase in effective cage radius may be expected from the hydrate structure (the innermost shell of neighbors consists of water molecules whose effective permittivity is less than the macroscopic ε_{02}), the values of $\alpha_M{}^*$ employed are undoubtedly too large to be applicable at 168°K. One can proceed no further without a knowledge of $\varepsilon_{\infty 2}$ at relatively high temperatures, but it may be concluded that the modified Onsager model accounts at least reasonably well for the contribution to the high-temperature permittivity from the reorientation of guest molecules.

5.5. Electrostatic Fields of the Water Molecules

The treatments of Sections 3.6 and 3.7 of interactions between guest and host molecules are applicable only to nonpolar guest species. Since the energy of alignment of the dipoles of ethylene oxide and of tetrahydrofuran in the field of one water dipole one cage radius away amount to about -1.0 and -0.6 kcal mol^{-1}, respectively, the total field of the 24 or 28 water dipoles of the cage might be expected to establish a well-defined preferred orientation and to produce a considerable hindrance to reorientation of such polar guest molecules. As already noted in Section 5.1, however, the value of ε_{02} remains linear in $1/T$ down to temperatures below 150°K and there is therefore no great distortion in the symmetry of the orientation-dependent energy of the guest molecule from that determined by the positions of the water O atoms. Moreover, the data of Table XIV suggest that the average activation energies for molecular reorientation in the same temperature range are only of the order of 1 kcal mol^{-1}.

The origin of the relative smallness of the perturbation produced by the electrostatic fields is the high degree of mutual cancellation of the individual fields at the center of the cage due to the water dipoles. It has been shown[373] that if all the water dipoles of the cage are uniformly distributed on the surface of a sphere of radius r and the molecular dipole moments are resolved into components m_{OH} along the O- -O bonds to which the water molecules contribute hydrogen atoms, the resulting field at the cage center is

$$\mathbf{E}(D) = \frac{(1 - \cos\theta)(1 + 3\cos\theta)m_{OH}}{r^3 \cos\theta} \sum_k \mathbf{r}_k \tag{33}$$

where θ is the angle between the radius vector of a water molecule with respect to the cage center and one of the O- -O bonds to neighboring molecules in the surface. The unit vector $\mathbf{r}_k$ specifies the radial direction of a water molecule which has an m_{OH} component pointing outward from the cage and the summation is over all water molecules with such "external" OH groups. Full hydrogen bonding requires that half the water molecules of the cage bear external OH groups.

If θ is tetrahedral, the resultant dipolar field at the cage center vanishes. In real cages θ cannot be exactly tetrahedral and in fact varies slightly with the water molecules and bonds considered. Small resultant dipolar fields are then obtained.

The small cages of structures I and II are somewhat distorted forms of the regular dodecahedron, for which $\theta = 110°54.3'$. For this angle eqn. (33) gives

$$\mathbf{E}(D) = 0.268(m_{OH}/r^3) \sum_{k=1}^{10} \mathbf{r}_k$$

from which the field may be evaluated for any assumed distribution of orientations of water molecules consistent with full hydrogen-bonding. Figure 37 shows the distribution of resultant dipolar fields for 4000 randomly generated cage configurations. The median field is only $0.44m_{OH}/r^3$, that is, considerably less than the field of only one water dipole.

The lower histogram of Fig. 38 shows the distribution of dipole fields for a limited number of random configurations of the large structure II cage. The water molecules were located at the sites given by an X-ray study,[938] the dipole moments were assumed to bisect the O- -O- -O proton donor angles, and the fields were determined by direct summation of the fields of the 28 water dipoles.[373] The resultant fields are again found to be generally smaller than the field of an individual water molecule. This was also the case for similar calculations for the 14-hedra of structure I.[373]

The upper histogram of Fig. 38 gives the resultant field distribution when the fields contributed by the quadrupole and octupole moments of the

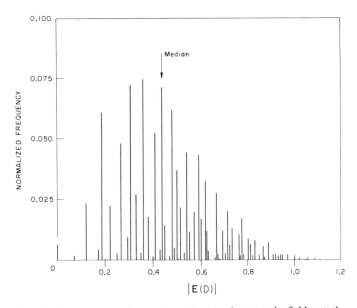

Fig. 37. Distribution of resultant dipolar electrostatic fields at the center of a 12-hedral cage taken as a regular pentagonal dodecahedron.[373] Fields are in units of m_{OH}/r^3 (~19 kesV cm⁻¹).

water molecules are added to those contributed by the dipole moments. The particular elements of the higher moment tensors employed[373] were those calculated by Glaeser and Coulson[556] for the gas-phase water molecule. The mean central field is seen to be raised by a factor of three and the distribution of fields considerably broadened. This is largely the effect of the quadrupolar fields, which do not tend to mutually compensate one another as do the dipolar fields. The use of more recent gas-phase quadrupole moment elements[373] than those of Glaeser and Coulson increases the fields still further. The resulting fields now appear to be too high to be entirely consistent with the relatively small electrostatic perturbations suggested, as outlined above, by the dielectric behavior.

The difficulty appears to be an artificial one in the sense that the use of gas-phase quadrupole moment elements is unjustified for hydrogen-bonded water molecules in an icelike environment. The gas-phase dipole moment of water increases from 1.84 D to perhaps 2.6 D[339] when water is incorporated in ice, and the second moment of the charge distribution must be more profoundly affected. It may be inferred from the dielectric properties

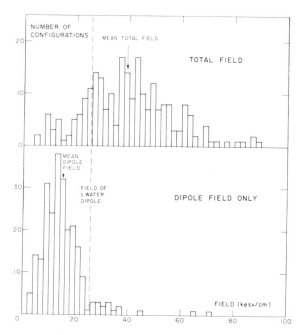

Fig. 38. Distribution of electrostatic fields at the center of the 16-hedron as given by 241 random configurations of the water molecules of the cage.[373]

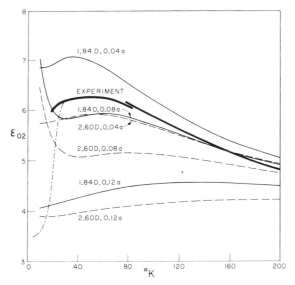

Fig. 39. Comparison of experimental values of ε_{02} of tetrahydrofuran hydrate with those calculated from a dipole–dipole model of interaction between guest and host molecules, averaged over 70 configurations of the cage.[373]

of the clathrate hydrates of polar guest molecules that this second moment must be considerably less anisotropic than in the gas.

The effect of the electrostatic fields is clearly seen in the behavior of ε_{02} at low temperatures. This is shown in Fig. 39 for a crude model[373] of tetrahydrofuran hydrate in which each guest molecule is assumed to occupy one of four tetrahedrally disposed preferred orientations whose populations are determined by Boltzmann factors $\exp(-\mu \cdot E/kT)$, E being the resultant field due to dipole moments of water molecules at the site of the guest dipole μ. The site of the dipole in each of the four orientations was displaced in turn a distance $0.04a$, $0.08a$, and $0.12a$ from the cage center along the three-fold symmetry axes of the cage to allow for the off-center position of the dipole moment of tetrahydrofuran both within the molecule and within the cage.* The experimental shape of the ε_{02}–temperature curve is seen (Fig. 39) to be consistent with the model for water dipole moments of 1.84–2.6 D and for displacements of $0.04a$ to $0.08a$. (The rise in calculated ε_{02} at very low temperatures appears to be an artifice of the limited number of cage configurations over which the result is averaged.)

* a is the unit cell dimension.

5.6. Effect of Electrostatic Fields on Reorientation of Polar Guest Molecules

In the absence of electrostatic fields dielectric absorption by the guest molecules is expected to be relatively narrow. The details will depend on the shape of the guest molecule, which will determine the energy differences between nonequivalent preferred orientations, and the variability of the potential energy barriers hindering interchange between orientations. At relatively low temperatures only the most stable of the preferred orientational sites will be occupied and since these will on the whole be equivalent sites, no wide distribution of reorientation rates is to be expected.

It seems necessary therefore to ascribe much of the width, as well as the asymmetry, of the dielectric absorption observed to the effect of electrostatic fields. Even if one allows only four tetrahedrally disposed preferred orientations within the large structure II cage, the single relaxation time given by the tetrahedral symmetry is in general replaced by three relaxation times when an appreciable electrostatic field is present. It is clear that variation in magnitude and direction of the electrostatic field from cage to cage will result in a broad continuum of relaxation times.

To the extent that the electrostatic fields $\mathbf{E}$ determine the distribution of barrier heights through contributions of the form $-\boldsymbol{\mu} \cdot \mathbf{E}$, the logarithmic width of the relaxation time distribution within a given cage type should be proportional to μ/T and the shape of the absorption curves should be similar for different guest molecules at the same value of μ/T. This has been approximately confirmed by comparison of the absorption shown by tetrahydrofuran and cyclobutanone hydrates.[1000]

The asymmetry of the absorption appears to be mainly attributable to the effect of the fields in lowering the barriers to reorientation. In terms of the model of tetrahedral preferred orientations, for example, the electrostatic fields will tend to lower the barriers between the two orientations most favored by the field and to raise the barriers between those less favored. The effect is to increase the absorption at high frequencies more than at low frequencies.

It is apparent that the molecular interpretation of the activation energies (Table XIV) derived from the temperature dependence of the frequency of maximum absorption is far from straightforward. Its low sensitivity to temperature for values of this frequency which range from $\sim$0.1 kHz to 10 GHz, however, suggests that it provides a fair estimate of the average geometric barrier to molecular reorientation.

5.7. Dipole–Dipole Interactions between Guest Molecules

Within structure II hydrates these interactions appear to have a minor effect on the motions of the guest molecules. The energy of alignment of the dipole of tetrahydrofuran, for example, in the dipole field of a neighboring tetrahydrofuran molecule is only about 0.1 kcal mol^{-1}[373] and the tetrahedral arrangement of neighbors seems to preclude general ordering of dipole directions even in the absence of overriding electrostatic fields of the water molecules. Some contribution is, however, probably made to the width of the dielectric absorption.

In structure I, however, parallel ordering of the dipole directions along the lines of centers of neighboring large cages is potentially possible, at least for the linear Ising model. Matsuo et al.[955] have recently presented evidence of a dipole order–disorder transition of HCN in its β-quinol clathrate in which the centers of neighboring cages lie 5.5 Å apart in the c direction. A maximum in static permittivity occurs at the temperature (178°K) at which a heat capacity anomaly is most pronounced. The allowed orientations of HCN are those in which the molecular axis coincides with the c axis, that is, only parallel or antiparallel alignment of neighboring dipoles is possible. Parallel alignment becomes dominant below 178°K and leads to a rather slow decline of ε_0 with decrease of temperature. (Both above and below the "transition" temperature the dielectric absorption is relatively narrow,[380] in agreement with the expectation that the lattice electrostatic fields are less important than in the hydrates.)

It is possible to explain the qualitative differences between the dielectric and NMR (see Section 5.8.2) properties of the structure I hydrate of trimethylene oxide and the other hydrates studied by a similar ordering of trimethylene oxide dipoles at low temperatures. A relatively narrow maximum in ε_{02} (Fig. 32) occurs at 145°K, a temperature some three times as high as for the broad maxima found in the other hydrates (Table XIV). Rough Lennard-Jones calculations suggest that orientations in which the polar axis of trimethylene oxide coincides with the $\bar{4}$ cage axis, and in which the center of mass is displaced from the cage center to bring the oxygen atom nearer to the adjacent hexagon of water molecules, are preferred by cage geometry. The occurrence of well-defined preferred orientations of trimethylene oxide is also suggested by the relatively high rotational oscillation frequencies estimated in Table XV. The energy of a trimethylene oxide dipole in an infinite chain of dipoles 6.0 Å apart and aligned along the chain is -1.2 kcal mol^{-1}, if allowance is made for the shielding effect of the polarizabilities of the water molecules.[373] The idealized one-dimensional model predicts an order–disorder transition at about 300°K, twice as high as the temperature observed, just as for HCN β-quinol.[955]

The electrostatic fields of the water molecules are expected to interfere with the long-range ordering both by causing frequent antiparallel correlations between neighboring pairs of molecules and by occasionally establishing preferred directions much removed from the cage axis. The plateau where $\varepsilon_{02} \sim 5$ at temperatures near 70°K (Fig. 32) is possibly attributable to such disordering effects.

X-ray indications[927] that even at 250°K ethylene oxide molecules in the 14-hedra may have strongly preferred orientations with the polar axis coinciding with the cage axis are not supported by dielectric or NMR results nor by consideration of the short-range interactions between guest and host molecules.[542] The possibility that ordering similar to that of trimethylene oxide occurs at very low temperatures cannot be excluded, however.

5.8. NMR Spectra of Guest Molecules

The NMR spectra of guest molecules which contain magnetic nuclei confirm the high degree of rotational mobility shown by the dielectric results. At relatively high temperatures the proton resonance line of the guest molecule, as shown for cyclopropane in Fig. 27, is much narrower than that originating from the water protons. It gradually broadens with fall of temperature until it merges with the water line.

The superposition of the lattice proton line may be avoided by measurement in a D_2O lattice. Most NMR studies of proton-containing guest molecules have made use of "deuterates" rather than of hydrates. In these the proton lines have widths between *extrema* of the derivative curves of the order of 1 G at high temperatures and maintain a featureless Gaussian-like shape normally down to about 50°K as the width gradually increases. At lower temperatures fine structure becomes apparent.

5.8.1. Second Moments

The second moments shown in Fig. 40 and Table XVIA tend to vary over a wide range of temperature, especially for the deuterates of trimethylene oxide (structure I) and cyclobutanone, in which the "linewidth" transition occurs over most of the range between 4 and 250°K. In other cases the second moment is almost constant between about 150 and 200°K. In general the rigid lattice value is only reached near 4°K, a further illustration of the exceptional rotational mobility of the enclathrated molecules.

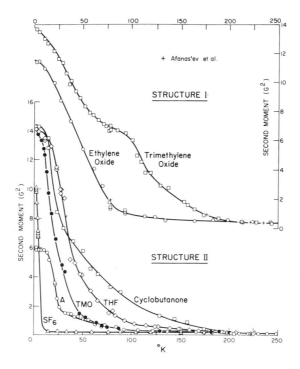

Fig. 40. Second moments of ^1H absorption (^{19}F absorption in the case of SF_6) in clathrate deuterates. (A) acetone, (TMO) trimethylene oxide, (THF) tetrahydrofuran.

TABLE XVIA. Proton Second Moments of Clathrate Deuterates: Experimental Values (G^2)

Deuterate	Temperature, °K						
	250	200	150	100	50	20	4.2
Ethylene oxide	0.30	0.51	0.56	0.96	5.38	9.93	11.4
Trimethylene oxide (I)	0.28	0.57	1.89	6.07	8.55	12.1	13.6
SF_6(^{19}F resonance)	0.19	0.20	0.21	0.22	0.22	0.23	6.9
Trimethylene oxide (II)	—	(0.18)	0.25	0.37	1.35	7.3	13.7
Tetrahydrofuran	0.11	0.25	0.42	0.86	3.95	12.3	14.0
Acetone	0.10	0.19	0.26	0.71	1.16	3.87	5.87
Cyclobutanone	—	0.18	1.01	2.33	5.4	9.7	14.3

The low second moment of acetone deuterate arises from the rotational tunneling of the CH_3 group which still persists below 2°K.

Second moments calculated for conditions appropriate to the high- and low-temperature regions are given in Table XVIB. The second column includes *inter*-guest contributions only, evaluated from the assumption that the guest molecules are undergoing effectively isotropic rotation about centers of mass which coincide with the cage centers. For this condition the *intra*-guest contribution vanishes and the *inter*-guest second moment is given[801] by $358.1n \sum_i n_i r_i^{-6}$, where n is the number of H atoms in a guest molecule, n_i is the number of neighboring cages whose centers are r_i away from the center of a given cage, and the summation is over all sets of neighboring cages which contribute appreciably. Columns 3 and 4 include as well the respective contributions from magnetic interactions with the deuterons for the cases of isotropic rotation of the D_2O molecules about the O atoms and of D_2O molecules in fixed but random configurations determined by the Bernal–Fowler rules of hydrogen bonding. Reorientation of the D_2O molecules produces only a minor reduction in the contribution from H—D interaction.

TABLE XVIB. Proton Second Moments of Clathrate Deuterates: Calculated Values (G^2)

M	M rotating			Rigid lattice	Geometry of M
	D_2O rotating and diffusing	D_2O rotating	D_2O rigid		
Ethylene oxide	0.17	0.26	0.29	11.6	(355)
Trimethylene oxide (I)	0.24	0.33	0.35	14.7	(279)
SF_6(^{19}F resonance)	0.064	0.13	0.14	10.7	$r_{SF} = 1.58$ Å
Trimethylene oxide (II)	0.072	0.13	0.15	14.0	(279)
Tetrahydrofuran	0.096	0.16	0.17	15.1	(120)
					$r_{CH} = 1.10$ Å $\angle HCH = 112°$
Acetone	0.072	0.13	0.15	$\sim 7^a$	(1022)
Cyclobutanone	0.072	0.13	0.15	13.8	(1187)
					$r_{CH} = 1.096$ Å $\angle HCH = 110.7°$

^{a} Rotating CH_3 groups.

The fifth column of Table XVIB gives the rigid lattice second moments calculated from the guest-molecule geometries given in the references cited in the last column. In all cases more than 93% of the total rigid lattice second moment calculated is contributed by interactions within the guest molecule, which illustrates the high degree of magnetic isolation of molecules within the deuterate cages. This is especially true for structure II, where the intra-guest contribution is about 97% of the total. Since the interaction between the protons of CH_2 groups contributes some 9–10 G^2, the result depends crucially on the length of the CH bond and on the H—C—H angle. In general an uncertainty of about 0.5 G^2 in the calculated moments is expected, mainly from this source. This is also the uncertainty in the experimental second moments reported at 4°K. For the particular case of acetone hydrate, the proper rigid lattice second moment is $\sim$26 G^2; the estimate of 7 G^2 is for rotating methyl groups. The tunneling of CH_3 is known[20] to persist to very low temperatures in some other molecules and is very likely not a consequence of the clathrate structure of acetone hydrate.

The relatively small *inter*-guest contributions to the rigid-lattice second moment were estimated by summing terms of the form[529] $(3 - 6h^2 +8h^4)/[3R^6(1 - 4h^2)^3]$, where R is the separation of cage centers and $h = r_0/R$. The protons of each molecule were assumed to be randomly, i.e., uniformly, distributed over the surface of a sphere of radius r_0 about the cage center. This rigid lattice contribution is larger by a factor of two to three than the *inter*-guest contribution for isotropic rotation (column 2, Table XVIB).

The guest–host contributions to the rigid-lattice second moment were found by summing over the fixed deuterons the terms[529]

$$10.01(1 + h^2)/[R^6(1 - h^2)^4],$$

where R is the distance between the cage center and an individual deuteron, and again $h = r_0/R$.

There is generally good agreement between the calculated second moments for the rigid lattices and those measured at 4°K. The measured values average about 0.5 G^2 lower, possibly in consequence of the rotational oscillations already discussed (Section 5.4) and of the ring-puckering vibrations of guest molecules with four- and five-membered rings. In the case of SF_6 deuterate rigid lattice conditions have not been reached at 4.2°K and possibly not even at 1.8°K, where the measured second moment is 9.9 G^2.

With the exception of ethylene oxide and SF_6 deuterates, the second moments measured near 250°K clearly suggest that only the interaction between rotating guest molecules makes an important contribution, and that most if not all the effects of guest–host interactions are removed by diffusion of the water molecules, in agreement with the conclusions of Section 4.11.2. The relatively high experimental second moment of ethylene oxide deuterate may arise from incomplete averaging to zero of guest–host and *intra*-guest contributions. It is apparent from Fig. 29 that at 250°K the water molecules in SF_6 hydrate are neither rotating nor diffusing fast enough to appreciably affect the second moment: The excess of the experimental second moment (0.19 G²) over the calculated value (0.14 G²) is likely attributable to the presence of $\sim$4% HDO impurity in this deuterate, which was the only one prepared outside the sample tube.

The second moment of the ${}^{19}F$ line of SF_6 hydrate is found to run parallel to that of SF_6 deuterate. It falls from 12.1 G² at 1.8°K to 1.25 G² at 12.5°K and remains at 1.2 G² as the temperature rises to 220°K, in excellent agreement with the value of 1.15 G² calculated[935] for isotropic reorientation of SF_6 molecules in the rigid water lattice.

At 200°K the proton second moments of the structure II deuterates are consistent with fixed D_2O molecules, especially if allowance is made for a slight content of HDO. (The presence of 1 at% H would increase the proton second moments of Table XVIB column 4, by 0.04 G² in structure I and by 0.03 G² in structure II.) This result is in agreement with the indications of Fig. 29 for H_2O molecules. Reorientation of D_2O molecules is expected to be slower, as observed for ice.[62]

The studies outlined here do not confirm the observation[914] of a sharp linewidth transition near 150°K in the fluorine resonance of SF_6 hydrate and deuterate, nor the very large proton second moment given[755] for acetone deuterate between 77 and 172°K. Unexpectedly high second moments have been reported for the deuterates of cyclopropane[755] and propane,[916] which are smaller than some of the guest molecules of Tables XVIA and XVIB. A rise of 1.4 G² is said[755] to occur in the second moment of both the hydrate and deuterate of CF_4 as the temperature falls from 150 to 130°K. Some of these observations may be attributed to poor hydrate content of the samples studied.

5.8.2. *Molecular Reorientation Rates*

The correlation between the rotational motion responsible for the fall of second moment with increasing temperature and the dielectric reorientation rates of the guest molecules is remarkably close in view of the dependence of line narrowing on more general molecular motion than dielectric relaxation. Thus the temperatures at which the dielectric relaxation rate is 10^5 Hz [34.8, 63.1, 14.4, 24.4, 26.4, and 40.0°K for the hydrates of ethylene oxide, trimethylene oxide (I), trimethylene oxide (II), acetone, tetrahydrofuran, and cyclobutanone, respectively] run reasonably parallel with the temperatures at which the second moment reaches one-half the rigid lattice value (48, 78, 22, 23, 34, and 36°K).

It is interesting to note that the NMR "activation energies" derived from Waugh's approximation[1380] that $E_R = 37T$, with T taken as above, all lie within about 25% of the dielectric Arrhenius energies (E_A, Table XIV) of the corresponding hydrates. These values of E_R are of course only crude averages: Actual analysis of the variation of linewidth with temperature yields effective activation energies which change from relatively low to high values as the temperature rises. The wide temperature interval over which narrowing occurs has the same origin as the wide distribution of dielectric relaxation times (Section 5.6), probably with additional contributions from the temperature dependence of reorientation about the polar axis.

The second moment (Fig. 40) of the structure I deuterate of trimethylene oxide, like the value of ε_{02} (Fig. 32), has an unusual dependence on temperature. The second moment rises rapidly as the temperature falls from 160 to 100°K, the temperature range for which the dielectric behavior suggested a change from disordered to ordered orientations of most trimethylene oxide molecules (Section 5.7). Its value, at about half the rigid lattice moment, is relatively insensitive to temperature between 100 and 70°K.

5.8.3. Line Shapes at Low Temperatures

Figure 41 illustrates the development of fine structure in the proton spectrum of ethylene oxide deuterate[542] as the temperature is lowered toward the rigid lattice condition. Such fine structure is the result of transitions between the states into which the Zeeman levels are split by interactions between the fixed magnetic nuclei of the guest molecule. It is therefore characteristic of the geometric arrangement of the four protons of ethylene oxide. Broadening produced by intermolecular magnetic interactions normally obscures the fine structure in polycrystalline solids: It is observed in clathrate deuterates because of the high degree of magnetic isolation of the guest species. The secular equations for the energy levels of the isolated four-spin system cannot be solved analytically. An approximate solution,[698] based on the assumption of two weakly interacting proton pairs, was found[542] not to be applicable to ethylene oxide, in which the ratio of inter-pair to intra-pair distance is only 1.37.

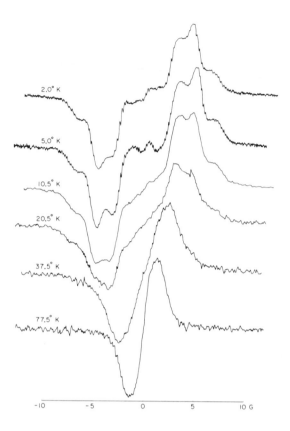

Fig. 41. Derivative line shapes of ^1H absorption in ethylene oxide deuterate at a number of low temperatures.[542]

Similar fine structure occurs in the low-temperature spectra of acetone, trimethylene oxide (both structures), cyclobutanone, and tetrahydrofuran deuterates. In the case of acetone the line shape is the triplet characteristic of methyl group rotation[37] with an additional weak pair of lines observed in the wings below about 14°K. Figure 42 shows the close resemblance between the rigid lattice line shapes at 4.2°K of the deuterates of trimethylene oxide and cyclobutanone, molecules in which the proton arrangements are similar although not identical.

The fine structure of the ^{19}F band of SF_6 deuterate at the lowest temperatures (Fig. 43) appears to consist of at least seven components. In the case of the hydrate these are not resolved. In neither case does the band display a center of symmetry, because of anisotropy of the chemical shifts.

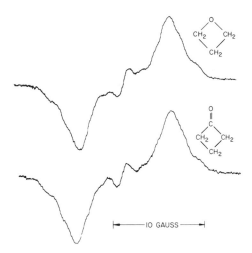

Fig. 42. Derivative rigid-lattice line shapes of
¹H absorption in structure II deuterates of trimethylene oxide and cyclobutanone.

5.8.4. *Spin–Lattice Relaxation*

The only T_1 measurements which have been reported for guest molecules are those of McDowell and Raghunathan,[916] who obtained T_1 values
at 16 MHz for propane deuterate which rose from 60 msec at 108°K to
1.6 sec at 77°K. The general lack of saturation effects on the spectra of
clathrate deuterates in rf fields at the mG level even at very low temperatures
is noteworthy. In some cases[542] some saturation is evident at this power

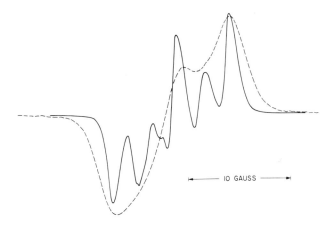

Fig. 43. Derivative line shapes of ¹⁹F absorption in SF₆ hydrate
(broken curve) and deuterate (continuous curve) at 1.8°K.

level at intermediate temperatures and it is possible that T_1 generally passes through a maximum value at some temperature between 100 and 200°K. Again (see Section 4.11.4) the relaxation mechanism at most temperatures is probably of the spin–rotational type associated with tumbling motion of the guest molecules. At the lowest temperatures rotational and translational oscillations may well be involved.

5.8.5. Amine and Alkyl Ammonium Salt Hydrates

Few results are available from NMR studies of the motion of guest molecules in semiclathrate hydrates. A cursory study of $(CH_3)_3N \cdot 10\frac{1}{4}H_2O$ showed two superimposed lines near 110°K, the second moment of the narrow component (if Gaussian) being about 2.3 G^2, i.e., less than half that expected if the only averaging motion is rotation of methyl groups. Measurements of the corresponding deuterate gave[237] a linewidth of about 1.4 G between 208 and 270°K, much like ethylene oxide deuterate, a result taken[237] to show a comparable degree of rotational mobility. Since the more recent X-ray structure (Section 2.5) shows that the amine molecules are hydrogen-bonded to positionally disordered water molecules, it appears that rapid interchange between the disordered amine configurations is made possible by rearrangement of these water molecules, as already proposed (Section 4.9.2) to account for the dielectric behavior.

No narrow component associated with reorientation of the amine molecule was observed for hexamethylenetetramine hydrate.

The proton second moment of $(i\text{-}C_5H_{11})_4NF \cdot 38D_2O$ was found[915] to fall slowly from 12.5 G^2 at 77°K to 9.5 G^2 at 240°K, values which suggest, when compared with the rigid lattice second moment of about 23 G^2, considerable narrowing from thermal motion of the iso-amyl groups as well as from methyl group reorientation. A linewidth transition was observed near 248°K, followed by a reduction of second moment to 1.85 G^2, which remained unchanged between 255 and 286°K. It is possible that concerted interchange of the carbon atoms between the twofold disordered positions shown by the X-ray study[926] is responsible for this small second moment.

Three distinct regions of line narrowing were observed in a similar study[915] of $(n\text{-}C_4H_9)_4NF \cdot 31.8D_2O$ in which the unit cell[925] (see Table IV) contains four equivalent cations with twofold disorder about the mean C chain axis and a fifth cation with fourfold disorder. Detailed analysis of the fall in second moment from 16 G^2 (rigid lattice value is $\sim$28 G^2) at 77°K to 5 G^2 at 245°K and about 1 G^2 at 280°K is difficult but it may be concluded[915] that it accompanies increasingly rapid exchange between the disordered configurations.

5.9. Infrared Spectra

For sufficiently large and spherical cages the selection rules for infrared activity of the *intra*molecular vibrational transitions of guest molecules should be the same as in the gas, with the lines of the P and R branches greatly "pressure-broadened" by collisions with the cage walls. At the other extreme of rotational mobility the breadth of the vibrational bands should be determined by the differences between the perturbing force fields of the different preferred orientations or, if this is fast enough, by the rate of interchange between preferred orientations. The second model is the more realistic for hydrates at temperatures where the reorientation rate is appreciably less than the frequencies associated with transitions between rotational levels of the gas (say 1 cm^{-1}) but at higher temperatures it is possible that gaslike rotational contours may be observed for sufficiently symmetric guest molecules.

In contrast to the β-quinol clathrates, there have been few studies of the infrared spectra of molecules encaged in clathrate hydrates. Harvey et al.[640] found the fundamental antisymmetric stretching frequency of SO_2 in its hydrate to be nearer to its value in gaseous SO_2 than in solid SO_2. Two weak shoulders observed at $120°K$ some 10 cm^{-1} above and below the main peak at 1340 cm^{-1} were possibly sum and difference bands with a low-frequency librational or translational mode, as suggested by the dependence on temperature. A similar shape is shown* by the band associated with the symmetric ring-stretching mode of ethylene oxide in its hydrate. At $90°K$ weak satellites occur at frequencies 15 cm^{-1} above and below the main peak at 1268 cm^{-1}. The latter itself is a doublet with components separated by 2 cm^{-1}, possibly a consequence of two distinct preferred orientations.*

Far-infrared measurements at low temperatures are expected (Section 5.4) to show strong absorption in the region of 20 cm^{-1} associated with rotational oscillations of the guest molecules. In view of the gradients of electrostatic fields present in the cages (Section 5.5), infrared activity of the translational ("rattling") modes is also anticipated. Bands of this kind have been found[21,254,377] in the spectra of a number of nonpolar and polar molecules encaged in β-quinol clathrates at frequencies between 30 and 80 cm^{-1}. Far-infrared studies of clathrate hydrates are now in progress in several laboratories.

---

* J. E. Bertie, private communication.

6. OTHER ASPECTS OF CLATHRATE HYDRATES

Having in the preceding sections sketched some of the structural, thermodynamic, and kinetic properties of the clathrate hydrates, we conclude this chapter by surveying some implications and applications of the unique properties of these forms of ice.

6.1. Preparation and Kinetics of Formation

Although clathrate hydrates which form congruently are easily prepared by crystallization from solution, difficulties are often encountered in the quantitative preparation of the more numerous hydrates of water-insoluble species. Direct chemical analysis of such hydrates has almost invariably shown excess water (cf. Tables V and VI) and the presence of ice may be recognized in many samples studied by X-ray,[1359] dielectric,[372,647] NMR,[7] and infrared* methods.

The conditions and kinetics of formation have been studied for, among others, the hydrates of the rare gases,[103] Cl_2,[402] C_3H_8,[402,770,779] CH_3Br,[83] ethylene oxide,[568] tetrahydrofuran,[1101] and natural gas.[1303] Ice appears to promote hydrate nucleation. In the case of water-insoluble hydrate-formers the rate-determining step in crystal growth is frequently the absorption of hydrate-former into the aqueous liquid phase, while with soluble species the removal of the heat of crystallization may be generally rate-determining. The conditions which promote rapid hydrate formation—agitation and large thermal or pressure "driving forces"—result in small crystals and difficulty in separation of crystals from adhering liquid.

Barrer and Edge[101] showed that appreciable rates of formation of rare gas hydrates from ice at low temperatures are realized by shaking with steel balls to abrade the hydrate crust which forms on the ice. Similar crusts commonly occur at the gas–aqueous liquid interface above 0°C and inhibit hydrate formation in stationary systems.

6.2. Natural Gas Hydrates

Hammerschmidt[629] pointed out in 1934 that the cause of blockages previously observed in natural gas pipelines exposed to low temperatures was more likely the formation of gas hydrates than of ice. Confirmation[386] of this

---

* J. E. Bertie, private comunication.

observation has led to regulation of the water content of natural gas and to numerous studies by petroleum engineers of the stability characteristics of simple and mixed hydrocarbon hydrates. The subject has been reviewed by Kobayashi[771] and Makogon and Sarkis'yants.[939] Production of natural gas in the Soviet Union from wells in several northern areas where temperatures fall to as low as $-60°C$[1172] has been accompanied by development of improved methods of prevention of hydrate plugs,[284,757,1172,1304,1305] including the pretrapping of water as hydrate and the injection of methanol or other hydrate inhibitor into the gas stream.

The probability also exists[757,940]* that natural gas may occur in the form of extensive gas hydrate deposits in the upper strata of wells in permafrost regions; according to preliminary estimates, the reserves of natural gas in this form in the USSR exceed 10^{13} m³.[940]* It has also been suggested that large deposits exist at higher temperatures under the sea bed, where hydrostatic pressures may amount to several hundred atmospheres.

6.3. Natural Occurrence of Clathrate Hydrates

With the exception of natural gas hydrates, the thermodynamic conditions required for stability make it unlikely that clathrate hydrates occur on the earth in any abundance. The partial pressures of CO_2 and Ar in the earth's atmosphere are too low to permit formation of hydrates of these gases even at the low temperatures of the upper atmosphere.[1279] Likewise, extrapolation of the decomposition pressures of nitrogen and oxygen hydrates[1322] indicates that these hydrates become stable at 1 atm only near $-70°C$. The existence of "air hydrate" in the Antarctic ice cover has, however, been proposed[980] by Miller to account for the presence of air found[592] in ice cores taken at depths greater than 1200 m.

Miller[979] has considered the likelihood that gas hydrates are present in abundance in other parts of the solar system. Methane hydrate may exist on Uranus, Neptune, Saturn, and Jupiter and on the satellites of Saturn and Jupiter. Delsemme et al.[407,408] suggested that hydrates of CO_2, CH_4, C_2H_6, and other molecules may exist in the heads of comets, to explain the constancy of the relative spectral intensities of different gases as comets approach the sun. This entails the presence of a mixed hydrate and excess water.[979] The existence of hydrate will obviously affect estimates of abundances based on spectroscopic observations. It is possible[982] that water is combined in CO_2 hydrate in the Martian ice cap.

---

* Press release, Soviet Embassy, Ottawa, October 16, 1971.

6.4. Uses of Clathrate Hydrates

6.4.1. *Desalination of Water*

Formation, separation, and decomposition of gas hydrates was among the methods suggested by Parker[1073] in 1942 for conversion of saline to potable water. Several processes of doing this have been patented.[430,563, 567,665,1410] Potential advantages over the simple freezing and thawing of ice are higher formation temperatures and use of direct-contact evaporation of the liquid hydrate-former to remove the heat of crystallization and its subsequent condensation to provide the heat of melting. Propane and Freon-12 (CCl_2F_2) have been used in pilot plant operations both in the United States[545,1411] and in the Soviet Union.[768] Problems have arisen[82] connected with the high porosity of the hydrate formed and inadequate separation of hydrate from adhering brine. Progress in development of the hydrate process is at present slow, although processes which rely on the freezing of ice, including one which makes use of *n*-butane as a direct-contact heat transfer medium,[727] are being actively developed in several countries.

6.4.2. *Fractionation of Gases and Liquids by Selective Enclathration*

The early work of Nikitin[1029–1031] showed that gas hydrates may be used to concentrate some components of mixtures of gases. The capacity of the host lattices to take up guest molecules is equivalent to the sorption capacity of the best activated carbons and zeolites.[102] The availability of water suggests many applications[354,692] in large-scale separations.

The enrichment factor[104] of species M with respect to species M′ is given by

$$\eta(M/M') = \left[\left(\sum_i \nu_i \theta_{Mi}\right) \bigg/ \sum_i \nu_i \theta_{M'i}\right](p_{M'}/p_M) \qquad (34)$$

where the ratio of partial pressures in the gas mixture may be written alternatively as the ratio of the activities of components of a liquid solution in equilibrium with the hydrate. If M and M′ occupy only cages of the same kind, the ideal solution theory gives

$$\eta(M/M') = C_M/C_{M'} \qquad (35)$$

the ratio of the Langmuir constants. If the cages occupied are the large cages of structure II, this result may also be written as

$$\eta(M/M') = p_{M'}(sh)/p_M(sh) \tag{36}$$

the ratio of the dissociation pressures of the corresponding hydrates of single guest species measured at the same temperature, i.e., at $\theta_M(sh) = \theta_{M'}(sh)$ [cf. eqn. (8)].

Some of the results of Barrer et al.[102] for fractionation of gases with $CHCl_3$ hydrate are given in Table XVII. If the gases are assumed to enter only the small type II cages, the experimental enrichment factor should in the ideal case be equal to the ratio of Langmuir constants for the small cages, shown in the last column as derived from the small quantities of *single* help gases absorbed during the formation of $CHCl_3$ hydrate near 0°C. The enrichment factor increases with decrease of temperature as expected, since the heat of encagement is in general more negative (cf. Table IX) for the component enriched in the hydrate. The experimental enrichment factor for O_2 in air is surprisingly high; von Stackelberg[1350] found as much O_2 as N_2 in $CHCl_3$ hydrate prepared in the presence of air.

TABLE XVII. Enrichment Factors of $CHCl_3$ Double and Mixed Hydrates

M, M'	t, °C	Mole ratio M/M' in gas	$\eta(M/M')$	$C_M/C_{M'}$
Kr, Ar	0	0.76	3.4	4.1
	−30	0.25–0.37	5.3±2.0[a]	—
	−78	0.08	10.0	—
Xe, Kr	0	1.12	2.9	5.7
	−30	0.19–0.57	4.2±1.2[a]	—
Xe, Ar	0	—	—	23
	−30	0.06	24±3[a]	—
O_2, N_2	0	0.27	1.9	1.4
N_2, H_2	0	—	—	6.9
N_2, Ne	0	—	—	1.9
$CHCl_3$, CH_3I	0	1[b]	1.87	1.52[c]
$CHCl_3$, CH_2Cl_2	0	1[b]	2.07	2.42[c]
$CHCl_3$, CCl_4	0	1[b]	1.79	—

[a] Range of several experimental values.
[b] Mole ratio in liquid.
[c] From eqn. (36).

Enrichment factors for mixed type II hydrates formed from liquid mixtures containing $CHCl_3$[103] are also given in Table XVII. From the value of $\eta(CHCl_3/CCl_4)$ and $p_M(sh, 0°C) = 0.065$ atm for $CHCl_3$ hydrate, a dissociation pressure $p_{M'}(sh, 0°C) = 0.12$ atm may be estimated for CCl_4 hydrate. This pressure is more than twice the vapor pressure of liquid CCl_4, in agreement with the observation that CCl_4 itself does not form a stable hydrate. No incorporation of benzene or toluene was observed.[103] Von Stackelberg and Meinhold[1355] inferred that such molecules as CCl_4, CH_3CCl_3, and CS_2, but not benzene, were incorporated in the hydrate of C_2H_5Cl.

For hydrates in which both sizes of cage may be occupied the enrichment factor varies with the composition of the gas in a manner predicted by the solution theory.[1328] For the structure II hydrate of C_3H_8 and CH_4 at $-3°C$, $\eta(C_3H_8/CH_4) = 27$ when the gas contains 2.5% C_3H_8 but is only 1.6 for gas with 72.5% C_3H_8.[1328] For the "azeotropic" $H_2S–C_3H_8$ system (Section 3.4) H_2S is greatly enriched in the structure II hydrate at low concentrations of H_2S in the gas, whereas C_3H_8 is enriched in the hydrate at high H_2S gas concentrations.[1106]

Several studies[270,386,546,629,1012] of the fractionation of hydrocarbon gases have shown the expected enrichment of propane and iso-butane with respect to smaller and larger molecules.

Clathrate hydrates differ from zeolites and from phenol, urea, and other clathrates in their ability to enclose spherical molecules. Thus, in contrast to hydrates, zeolites and phenol clathrates do not take up $CHCl_3$ and CCl_4, but do absorb the more extended CS_2 molecule.[1350,1352] Urea clathrates incorporate n-paraffins in preference to iso-paraffins,[481] whereas hydrates form with iso-butane and CH_3CHCl_2 but not with n-butane and CH_2ClCH_2Cl.[1350]

6.4.3. Concentration of Aqueous Solutions

A related application, described by Glew,[567] makes use of hydrate formation and separation to reduce the water content of aqueous solutions. It has been suggested[1394] that the method is most applicable to the concentration of temperature-sensitive solutions. Centrifugal separation of hydrates has been used in studies of the concentration of fruit juices,[683] coffee extract,[1394] and sucrose.[1394] It is probable that the degree of water removal may .be more easily controlled by, for example, the quantity of a water-soluble hydrate former present, than in concentration by partial freezing of ice.

6.4.4. *Storage of Gases*

Hydrates of gases with low critical temperatures (type i_1, Table I) may be used with advantage for the storage of gases at considerably lower pressures or in smaller volumes than required for the gases themselves. Thus the concentration of guest species in structure I hydrates amounts to about 7 mol liter^{-1}. The pressure required to store gas at this density is about 160 atm. Some studies[139,974,975,1072] have been made of the storage of natural gas in hydrate form.

The use of hydrate for the storage and transport of explosive or labile materials such as ClO_2[664,1412] and ozone[933,1373] has been proposed.

6.4.5. *Temperature Measurement and Far-Infrared Detection*

The sensitivity to temperature of the electrical properties of clathrate hydrates of polar molecules (Section 5.3) suggests their use as low-temperature bolometers. Thus, $d(\ln \varepsilon'')/d(\ln T)$ is about 5, and almost independent of temperature, between values of about $\varepsilon''_{max}/20$ and $\varepsilon''_{max}/2$ on the low-temperature side of ε''_{max} measured at 1 kHz for the hydrates of ethylene oxide, tetrahydrofuran, cyclobutanone, and trimethylene oxide (II). The mid-temperatures of these logarithmically linear regions are 17.5, 14.5, 21.0, and 6.6°K, respectively. Similar regions to the high-temperature side of ε''_{max} correspond to values of $d(\ln \varepsilon'')/d(\ln T)$ of between -10 and -20, around 36.0, 26.4, 41.0, and 15.0°K, respectively. Such sensitivities are considerably higher than those of resistance elements commonly used to measure low temperatures: Arsenic-doped germanium resistors, for example, have sensitivities which generally do not exceed $d(\ln R)/d(\ln T) = -2$.

The use of dielectric heating for temperature control is possible on the high-temperature side of ε''_{max}, where the tendency for a sample to cool in the presence of a liquid helium or hydrogen heat sink may be counteracted by an increase in ε'' and in heat produced.

Since broad absorption from polar guest molecules occurs in the far-infrared region at low temperatures and overlaps the absorption at somewhat higher frequencies from the water lattice, it is to be expected that some hydrates may serve as far-infrared detectors at temperatures where advantage may be taken of the large sensitivity of ε'' to temperature to measure the heating effect of the infrared radiation.

6.5. Related Structures

It is interesting to find that structures similar to the clathrate hydrates of types I and II are formed by other substances which, like water, tend to bond tetrahedrally. Kamb has concluded[722] that melanophlogite, a low-density form of silica containing organic matter, is probably the structural analog of the 12 Å hydrate. Relatively inert structures identical, except in dimensions, with the two clathrate hydrate types are formed during partial thermal degradation of alkali silicides and germanides.[730,731] These substances are relatively rich in Si or Ge, which forms the lattice, the alkali atoms occupying the cages to a variable extent. Silicon clathrates of the structure I type ($a = 10.3$ Å) have been reported[352] with Na, K, and Rb and of structure II ($a = 14.6$ Å) with Na and Cs, while for the corresponding germanium clathrates $a = 10.7$ Å and 15.4 Å. These structures apparently collapse under pressures above 100 kbar.[251]

6.6. Models of Liquid Water Structure

In 1957 Pauling[1079] proposed a clathrate hydrate model for liquid water. To account for the density, this model requires that the cages be occupied by water molecules, in contradiction to the information now available from composition, dielectric, and NMR studies of the clathrates themselves. Earlier, Glew and Moelwyn-Hughes[574] had drawn attention to the similarity between the heats of hydrate formation from ice and the heats of solution of the hydrate-forming gases in water. There has been much subsequent speculation about the formation of water cage structures around nonelectrolyte solutes. From correlations between gas hydrate dissociation pressures and the potencies of the corresponding gases as general anesthetics, Pauling suggested[272,1081,1082] that anesthesia may be attributed to the formation of microcrystals of clathrate hydrate in the brain. Miller considered[978] that anesthetic gases promote the transient "iceberg" content of liquid water and the formation of "ice-cover" on the lipid membranes so as to reduce their permeability and change their electrical characteristics.

ACKNOWLEDGMENTS

The author is indebted to J. E. Bertie, E. L. Carstensen, S. K. Garg, S. R. Gough, R. E. Hawkins, Y. A. Majid, S. L. Miller, and B. Morris for unpublished information and to J. Novak for the figures.

CHAPTER 4

Infrared Studies of Hydrogen Bonding in Pure Liquids and Solutions

W. A. P. Luck

Institut für Physikalische Chemie
Universität Marburg
Germany

1. INTRODUCTION

While many properties of nonpolar or weakly polar substances can be arranged according to the critical temperature T_c or the molecular weight M,[896] the properties of a series of strongly polar substances cannot be derived from the molecular weight or from the dipole moment μ (serving as a measure of polarity). A comparison of H_2S with H_2O illustrates this situation. The special interaction that comes into play in these substances is referred to as "hydrogen bonding" because it is found only in proton-containing substances.

An insight can be gained into the peculiar interactions of such compounds through the example of the hydrides. The clearest anomalies are displayed by H_2O, HF, and NH_3. Water might be expected to have a melting point of about $-92°C$ and a boiling point of about $-80°C$. These anomalies are also displayed by the heats of sublimation and of fusion (Fig. 1) and also by the internal heat of evaporation (Fig. 2). The ratio of the heat of evaporation to the heat of fusion is not anomalous (Fig. 3), proving that the strong forces of the hydrogen bond affect both quantities in the same way, and even for substances forming hydrogen bonds the order of magnitude is not appreciably influenced by melting.

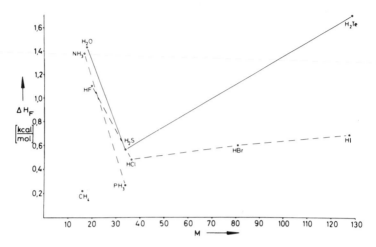

Fig. 1. Heat of fusion ΔH_F as function of the molecular weight M,
showing anomalies in molecules with H bonds.

It should be mentioned that dipole forces are proportional not only
to the dipole moment but also to the distance between two dipoles. Latimer
and Rodebush[834] already pointed out in 1920 that hydrogen atoms, having
a small electron shell, can approach electrons very closely. In this sense
they discussed, on the basis of the Lewis–Kossel model of complete eight-
electron shells, the ease with which protons could complete such shells.

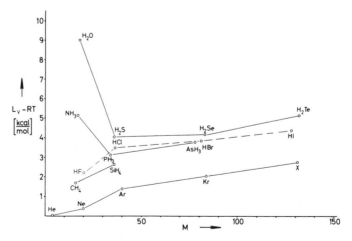

Fig. 2. The heat of internal evaporation $L_v - RT$ at 760 mm Hg
as function of molecular weight is anomalous in compounds with
H bonds.

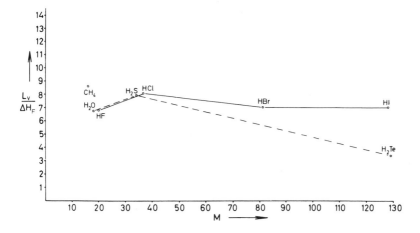

Fig. 3. The ratio of heat of vaporization to heat of fusion as function of molecular weight shows few anomalies.

One can easily interpret NH_4^+ in this sense as having the structure

$$H$$
$$..$$
$$H : N : {}^{\oplus}H$$
$$..$$
$$H$$

One may therefore expect a particularly strong interaction between groups
possessing free electron pairs (basic groups) and protons. Protons in polar
bonds (having donated the valence electron of hydrogen to a ligand)
should be particularly effective centers of positive charge. Latimer and
Rodebush discuss the case for highly associated liquids like water: "Structurally this may be represented as:

$$H$$
$$..$$
$$H : O : H : O :$$
$$..$$
$$H$$

Such combinations need not be limited to the formation of double or triple
molecules. Indeed the liquid may be made up of large aggregates of molecules continually breaking up and reforming under the influence of the
thermal agitation."

In a network of aggregates of water molecules bound to each other
in this way each oxygen atom would be surrounded by four hydrogen atoms.

Each free electron pair would have attracted one proton:

$$
\begin{array}{cc}
\text{H} & \text{H} \\
\ddot{} & \ddot{} \\
\text{H : O : H : O : H} \\
\ddot{} & \ddot{} \\
\text{H} & \text{H}
\end{array}
$$

Various molecules capable of hydrogen bond formation are capable, to varying extent, of forming such networks.

The results arising from this fact are illustrated in Fig. 4. The boiling points are given for groups of substances as a function of the number n of C, O, N, or F atoms. All the alcohols have boiling points 50° to 165° higher than those of the corresponding paraffins. Water, however, has a boiling point 260° higher than CH_4 despite their having identical molecular weights. This differential effect compared with alcohols derives from the fact that each H_2O molecule possesses two donor as well as two acceptor groups for hydrogen bonding which can all be saturated in a molecular lattice. Ammonia has a threefold possibility of functioning as a donor, but only one as an acceptor. This makes it impossible for all three protons in ammonia to form hydrogen bonds. HF possesses three acceptor functions but only one donor function per molecule. This makes it impossible to saturate all the acceptor functions.

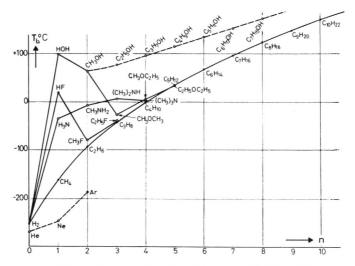

Fig. 4. Boiling points T_b as function of the sum of O, N, and C atoms clearly show the influence of H bonds.

Similar trends as apply for the boiling point are also valid for internal heats of vaporization.

The free electron pairs are restricted to a narrow range of orbitals according to quantum mechanical concepts. According to spectroscopic data, the orbitals, e.g., in water, of the free electron pairs as well as the axes of the two protons form approximately tetrahedral angles with each other. The angle between the two proton axes is found by infrared spectroscopy to be 105°3′ in the vapor, the O—H distance being 0.96 Å.[365,660,960–962,966,967] Neutron diffraction studies on ice indicate the angle to be 109.1° with a proton distance of 0.99 Å.[1087] The O-O-O angle in ice is 109.5°. We have to expect *a priori* that the hydrogen bond, being a true secondary valence force, will be orientation-dependent. This orientation dependence of hydrogen bonds leads, in the case of liquids, to the formation of orientation defects which are additional to the hole defects of nonpolar substances. In order to understand hydrogen-bond-containing liquids, it is important to determine such orientation defects. In addition to boiling points and enthalpy changes accompanying phase changes, many other properties are affected by hydrogen bonds.

The formulation of the hydrogen bond by Latimer and Rodebush did not distinguish it from a valence bond, and this gave rise, e.g., to the objection by Armstrong against the assumption of a "bigamous" hydrogen.[162]

The question has therefore to be settled whether the hydrogen bond ought to be considered as a valence effect or a secondary valence effect. This is best done from spectroscopic considerations. It can be shown that the hydrogen bond does not induce new or proper quantum states with new absorption bands, but merely a perturbation, albeit a very strong one, of the absorption bands already present in the free molecules. Furthermore, the bond energy due to hydrogen bonds can be determined spectroscopically and shown to be at the most 5 kcal mol^{-1}, i.e., less than normal chemical bond energies.

2. METHODS OF INFRARED SPECTROSCOPY FOR THE DETERMINATION OF HYDROGEN BONDS

Spectra provide the language for talking about atoms and molecules. In carefully conducted, quantitative spectroscopic investigations they answer questions about intermolecular interactions. Infrared (IR) spectroscopic investigations are particularly suitable for the detection of hydrogen bonds between protons on polar groups and free electron pairs. As far back as

1894 Paschen observed that the valence vibration of water is displaced toward longer wavelengths in the liquid as compared to the vapor.[1076,1077] Already at that time he suggested aggregation to be the cause for this effect in the liquid state.

No doubt it is to Freymann that the credit goes for being the first to observe the strong concentration and temperature dependences of the OH bands in solutions of alcohols and for interpreting the effect as being due to aggregation.[531] Following this, various frequency displacements, decreases of intensity, or increases of half-width were reported, involving OH, NH, or CO vibrations in systems capable of intermolecular or intramolecular hydrogen bond formation. Mecke and Vierling[970] found for the frequency displacement Δv of IR bands upon change of environment the relation

$$(\Delta v/v)(v_0 V)^2 \sim \text{const} \tag{1}$$

where v_0 is the fundamental vibration frequency and V is the molar volume.*

The largest displacement was found for water, in accordance with its small molar volume. Longinescu[864] suggested an association in the liquids proportional to their molar concentration. He assumed a degree of aggregation of about one-tenth of the molar concentration per liter. The work of Mecke and Vierling[970] shows the frequency displacement to increase linearly with the order of the overtone vibrations. Kempter and Mecke[745,746] computed detailed formulas for the calculation of aggregation equilibria from IR data. Mecke and his school applied this method to a large number of experimental data.

Many other workers have reported anomalous properties of IR valence vibrations, fundamental and overtone. The OH, NH, FH, and in some cases CH groups come near acceptors with free electron pairs, like O, N, F, Cl, or S. As a spectroscopist one is therefore inclined to regard the perturbation of the X—H bands as a measure of hydrogen bond formation and to use this perturbation directly as a definition of hydrogen bond interaction. The quantitative discussion of *free* OH and NH groups, not affected by hydrogen bond formation, as introduced by Mecke and Kempter is particularly fruitful in discussing the spectra of vibrational overtone bands. In these cases the perturbation through superimposition of the bands due to hydrogen bonds is much less than in the case of the fundamental vibration. The Δv is larger and the overtone bands of the hydrogen bond are much weaker.

---

* We have observed a proportionality $\Delta v \propto 1/V^2$ in some systems.[894] For the detailed theory of the medium dependence of spectra see Refs. 245, 246, 247, and 369.

In addition, for variations of concentration or temperature the experimental errors are also smaller, owing to the more favorable optical paths ranging from 1 mm to 4 cm, while in fundamental vibrational spectra optical paths as thin as about 1 μm are necessary. Accordingly the achievement of constant temperature is also much easier in the range of overtone vibrations and less subject to error, particularly as most IR instruments are arranged to have the absorption cell in front of the monochromator. This causes the light ray to heat up the sample in case of inadequate temperature control of the cell. We have undertaken wide-ranging spectroscopic investigations of hydrogen bonds in the near IR. For this reason we prefer to discuss the applicability of the method in the overtone region. For quantitative investigations of absorption, Lambert's law may be used:

$$I = I_0 \exp(-\varepsilon' cd) \tag{2}$$

with I_0 the intensity of incident light, I the intensity of light emerging from the absorbing layer of thickness d, c the concentration of the absorbing particles, and ε' the extinction coefficient. The optical density E is usually given by

$$E = \log(I_0/I) = \varepsilon cd \tag{3}$$

with ε the decimal extinction coefficient (referred to base 10). Equation (2) is only valid for monochromatic light, otherwise we have

$$I = \sum_{\lambda} I_{0\lambda} e^{-\varepsilon_\lambda cd} \tag{4}$$

When ε_λ is wavelength-dependent the sum can no longer be expressed as a single exponential function. This means that E may become concentration- or path-length-dependent.[874] However, $E = f(cd)$ is still true. The optical density remains constant even when ε_λ is wavelength-dependent, so long as the product cd (i.e., the number of molecules in the light path) remains constant. This is true for any variation of the two factors. The relationship is called Beer's law. It is valid only if ε is concentration-independent. This will be the case as long as there are no interactions. However, within the framework of the Mecke–Vierling rule [eqn. (1)] some dependence on concentration is to be expected. Figure 5 shows for one IR overtone band that only small changes in the optical density will occur with relatively large changes of concentration, so long as cd is kept constant. The same is true even for polar substances. Large deviations from Beer's law will occur, however, as soon as hydrogen bonds are involved (Fig. 6). The OH overtone band, in the case of oximes at 1.42 μm, is strongly concentration-depend-

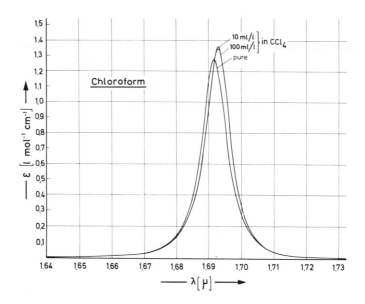

Fig. 5. The extinction coefficient ε of chloroform is almost independent of the concentration in CCl_4 solutions.

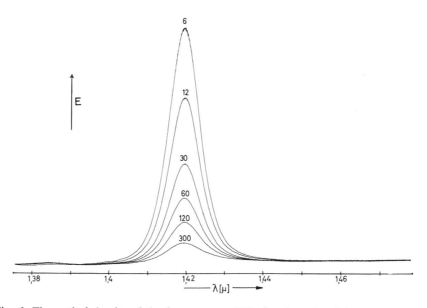

Fig. 6. The optical density of the first overtone OH vibration of cyclohexanone oxime in CCl_4 solutions at 20°C decreases markedly with concentration at (in g liter^{-1}) $cd = 60$ cm g liter^{-1} = const.[876]

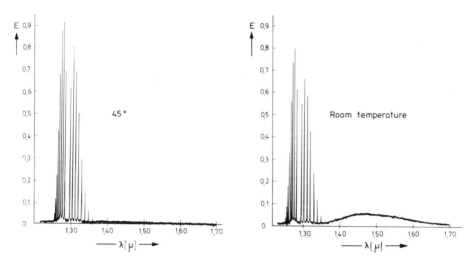

Fig. 7. In HF vapor at higher temperatures only the normal rotational structure of the first overtone band can be observed. At lower temperatures a broad band of H bonds appears at 1.5 μm.

ent.[876] At high concentrations there appears a broad absorption in the range of 1.6–2 μm. This additional absorption, which becomes very marked with increasing concentration, cannot easily be recognized as a proper band, to be ascribed to OH vibrations in a hydrogen bond. The difficulty in this case is due to superimposition of CH vibrations. This type of hydrogen bonding band is more clearly recognizable in Fig. 7. The left represents an overtone band of HF vapor at 1 atm and 45°C, clearly split into a rotational structure. This spectrum was obtained in a steel cell with thin polyethylene windows. On cooling the bulb, the intensity of the band decreases, and instead a new broad band appears with a maximum at 1.46 μm (Fig. 7).[890] This band can be attributed to the HF vibration within a hydrogen bond, since it is known that HF forms hydrogen bonds even in the vapor state. [503] The disappearance of the fine structure of the 1.5-μm band indicates that this process inhibits rotation. This disappearance of the rotational structure is common to most IR spectra of solutions and of liquids, so that normally only a broad band without any structure is observed. In the case of relatively small molecules like HCl (Fig. 8) or H_2O (Fig. 9) one can still detect the remnants of a band structure even in solutions. One may conjecture that these are remnants of a rotational structure. Barrow and Datta[106] were able to support this view for HCl and DCl by comparison of the spectra in the far IR. They found a dependence of the rotational structure on the solvent. The resolution of the structure decreases

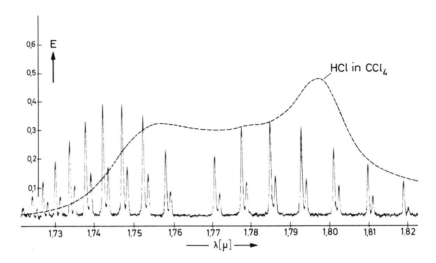

Fig. 8. The overtone band of HCl vapor consists of two branches with resolved rotation structure. HCl in CCl₄ solutions shows a broad band with remnants of the rotation structure (scale units of the solution spectrum are 10 times larger).

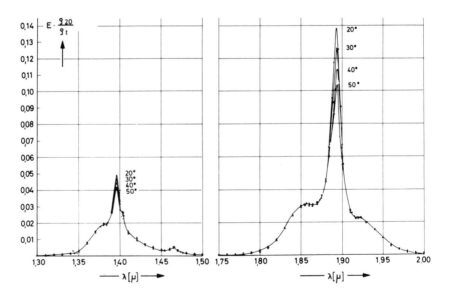

Fig. 9. The overtone spectra of small amounts of H_2O in CCl₄ show structures. (Ordinate: optical density corrected for the temperature-dependent density of CCl₄).

from CCl_4 via CS_2 to C_6F_{12}. This is capable of interpretation as perturbation caused by solvent molecules. Also, with CH_4 or C_2H_2 in CCl_4 one finds indications of a rotational structure.

2.1. Investigation of Hydrogen Bonds in Solution

2.1.1. Theoretical Considerations

The power of IR spectroscopy in the study of hydrogen bonds in solutions is most convincing, so that it appears to us necessary to discuss the method in detail. As seen in Fig. 6, it is comparatively easy to determine the fraction α to a high degree of accuracy, where $\alpha = \varepsilon_c/\varepsilon_0$ represents the ratio of the extinction coefficient ε_c at concentration C to ε_0 at infinite dilution $(C \to 0)$. Figure 10 shows corresponding α values for a series of CCl_4 solutions at 20°C.[881,963,964] Apart from its absolute magnitude the concentration dependence of α also varies greatly. It is one of the most interesting tasks in the investigation of solutions to explain these variations.

To gain an understanding of pure liquids, we are able to predict from Fig. 10 that for substances with a strong tendency to form hydrogen bonds the proportion of potential bonds not actually formed will be very small. In many systems it will be only a few per cent of the X—H groups present. On the assumption that sharp X—H bands are caused only by free X—H

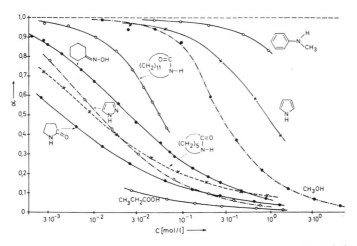

Fig. 10. Amounts of spectroscopically determined non-H-bonded X—H groups in CCl_4 solutions at 20°C.

groups but not by those bound by hydrogen bonds, we can state that

$$\alpha = \varepsilon_c/\varepsilon_0 = (\text{mono})/C \tag{5}$$

where (mono) is the concentration of monomers, i.e., molecules not forming hydrogen bonds, and C is the analytical concentration. To simplify the theory, let us consider the formation of hydrogen bonds as the formation of dimers

$$2\text{mono} \rightleftharpoons \text{di} \tag{6}$$

with

$$K_{12} = (\text{mono})^2/(\text{di}) \tag{7}$$

With eqn. (5) we have

$$(\text{di}) = \tfrac{1}{2}[C - (\text{mono})] = \tfrac{1}{2}C[1 - (\varepsilon_c/\varepsilon_0)] \tag{8}$$

From (8) and (7) we have

$$K_{12} = \frac{2C^2(\varepsilon_c^2/\varepsilon_0^2)}{C[1 - (\varepsilon_c/\varepsilon_0)]}$$

$$1 - (\varepsilon_c/\varepsilon_0) = (2/K_{12})(\varepsilon_c^2/\varepsilon_0^2)C$$

$$\varepsilon_c = \varepsilon_0 - (2/K_{12})(\varepsilon_c^2/\varepsilon_0)C \tag{9}$$

This last equation allows a more exact extrapolation of ε_0, by plotting ε_c against $\varepsilon_c^2 C$.[876]

If the assumption of exclusive dimerization is correct, a straight line should be obtained with an intercept of ε_0.

In the case of lactams with less than six carbon atoms one actually does obtain a straight line, indicating the formation of dimers exclusively. From eqn. (7) it then follows that

$$(\text{mono})^2/[C - (\text{mono})] = \tfrac{1}{2}K_{12} \tag{10}$$

As Fig. 11 shows, the left-hand side of this equation is indeed constant over an astonishingly wide range of concentrations, in fact three orders of magnitude (1:1000). Equilibria thus obtained are among the best substantiated ones known in physical chemistry. As indicated in Fig. 11, this is no longer true for lactams containing larger numbers of CH_2 groups. For steric reasons the lactams up to enanth-lactam exist in the *cis* configuration, while in the higher members the energetically more favored *trans* con-

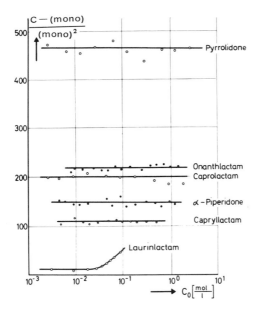

Fig. 11. Equation (10) for cyclic dimerization is valid over a wide concentration region for lactams with few carbon atoms (*cis* form of the amide group).

figurations predominate. Similarly, therefore, the simple relationship (8) does not hold for N-ethylacetamide.

Theoretically the next simplest case is the aggregation of oximes.[876] If a mean degree of association m is first determined:

$$K_m = (\text{mono})^m / [C - (\text{mono})] \qquad (11)$$

and $\log[C - (\text{mono})]$ is plotted against $\log(\text{mono})$, m is given by the slope.[876] Its magnitude, 2.3, indicates that dimerization is still the prevalent process. For a superimposition of dimer and trimer formation we have

$$3\text{mono} \rightleftharpoons \text{tri} \qquad (12)$$

$$K_{13} = (\text{mono})^3 / (\text{tri}) \qquad (13)$$

$$C = (\text{mono}) + 2(\text{di}) + 3(\text{tri}) \qquad (14)$$

$$C = (\text{mono}) + (2/K_{12})(\text{mono})^2 + (3/K_{13})(\text{mono})^3 \qquad (15)$$

$$[C - (\text{mono})]/(\text{mono})^2 = (2/K_{12}) + (3/K_{13})(\text{mono}) \qquad (16)$$

If instead of eqn. (10), we use eqn. (16) and plot $[C - (\text{mono})]/(\text{mono})^2$ against (mono), we should expect a straight line, the slope of which allows the calculation of K_{13} and the intercept that of K_{12}. As seen in Fig. 12, a straight line is in fact obtained for oximes, indicating a superimposition of dimer and trimer formation only. An increasing number of CH_2 groups favors dimer formation, a behavior also observed by Geiseler and Fruwert[547] in the case of linear oximes.

The aggregation tendency can be expressed by the mean number of monomers f forming the polymer, thus describing the average polymerization.[676,792,1426] Let us define

$$f = \left(\sum_n nC_n\right) \bigg/ \sum_n C_n = C \bigg/ \sum_n C_n \tag{17}$$

where C_n designates the concentration of the n-mer in mole liter^{-1}. f may be calculated from the formula[676]

$$f = C \bigg/ \int_0^{\varepsilon_c} (1/\varepsilon_c)\, d(\varepsilon_c C) \tag{18}$$

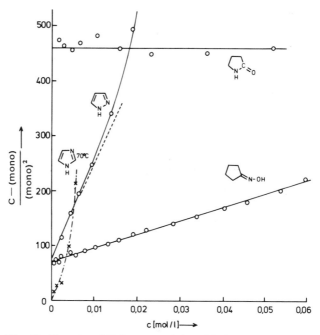

Fig. 12. Equation (16) for a coupled equilibrium of monomers, dimers, and trimers is valid for oximes and (at low concentrations) for pyrrole in CCl_4 at 20°C.

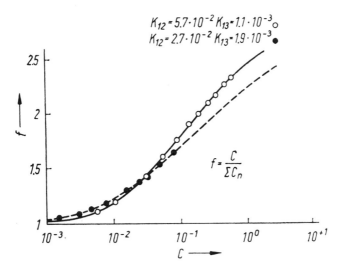

Fig. 13. Mean association number f of cyclopentanone and cyclo-dodecanone oxime in CCl$_4$ at 20°C as function of concentration.

According to eqn. (18) one obtains f by graphical integration of the plot of $1/\varepsilon_c$ against $\varepsilon_c C$. Figure 13 shows values of f for two different oximes, obtained by this method. The points are obtained from eqn. (18), the continuous curves have been calculated from spectroscopically determined equilibrium constants. The two methods show excellent agreement.

Equation (18) has the advantage that ε_0 does not appear in it. Figure 13 shows that f values may intersect. The f-value sequence of various substances at a particular standard concentration does not give an exact measure of the tendency for association. Rather, the percentage of monomers would constitute a measure for the tendency toward association. Even this may give rise to a different sequence when several substances are compared in various concentration ranges.[876]

From eqn. (16) one can approximately determine K_{12} and K_{13} for pyrrole in reasonably dilute solution (Fig. 12).

In imidazole the association tendency is so strong that one may neglect dimer formation at low concentrations ($1/K_{12} \rightarrow 0$). Similarly there is no range where trimer formation is appreciable (the results on imidazole shown in Fig. 12 refer to 70°C). For further data based on spectroscopic determinations see refs. 320, 676, 963, 964. The position is similar with 4-methylimidazole, which could be studied in a wider range of concentration because of its higher solubility (Fig. 14). On a magnifield scale one may estimate $2/K_{12}$ by extrapolation to be equal to 20–30 at 60°C. Because of

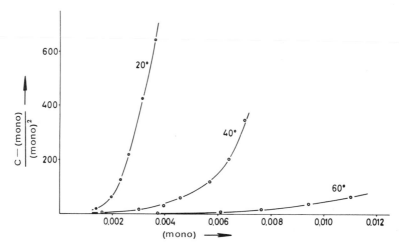

Fig. 14. The association of 4-methylimidazole in CCl₄ determined by spectroscopic methods gives indications of aggregates higher than trimers.

increasing spectroscopic errors at low concentrations this estimate is subject to some uncertainty. The shape of the graph in Fig. 14 is to be expected theoretically when aggregates larger than trimers appear. This is shown, e.g., by Fig. 15. The effect of various degrees of association on the expression $[C - (\text{mono})]/(\text{mono})^2$ has been calculated theoretically. The following equilibrium constants have been assumed for the sequence of curves beginning at the bottom and with $(\text{mono}) = 5$ mol liter^{-1} (equilibrium constants not specified are assumed to be infinity):

1. Tetramers only $K_{14} = 1000.$

2. Trimers only $K_{13} = 10.$

3. Trimers plus tetramers $K_{13} = 10, K_{14} = 1000.$

4. Dimers only $K_{12} = 1.$

5. Only dimers plus tetramers $K_{12} = 1, K_{14} = 1000.$

6. Only dimers plus trimers $K_{12} = 1, K_{13} = 10.$

7. Dimers, trimers, and tetramers $K_{12} = 1, K_{13} = 10, K_{14} = 1000.$

Higher polymers would cause even larger deviations from the straight line. The theoretical calculation of such high degrees of association is possible only approximately. One of the best known methods is the one worked out by Mecke and Kempter.[746] They assumed superimposed equilibria from dimer formation up to aggregates of n molecules:

$$C = (\text{mono}) + 2(\text{di}) + 3(\text{tri}) + 4(\text{tetra}) + \cdots + n(n\text{-mer}) \qquad (19)$$

$$\begin{aligned} C = {} & (\text{mono}) + (2/K_{12})(\text{mono})^2 + (3/K_{13})(\text{mono})^3 \\ & + (4/K_{14})(\text{mono})^4 + \cdots + (n/K_{1n})(\text{mono})^n \end{aligned} \qquad (20)$$

$$C = (\text{mono}) \sum_{i=1}^{n} (i/K_{1i})(\text{mono})^{i-1} = (\text{mono}) \sum_{i=1}^{n} (i/K_{1i})(\alpha C)^{i-1} \qquad (21)$$

with $K_{11} = 1$. The following simplifying assumption was introduced:

$$K_{12} = K_{23} = \cdots + K_{(i-1)i} = K_{MK}$$

Thus $K_{1i} = K_{MK}^{(i-1)}$ and $\alpha(\text{mono})/K_{MK} < 1$. Then for $n \to \infty$

$$\alpha = (\text{mono})/C \sim [1 - (\alpha C/K_{MK})]^2 \qquad (22)$$

or

$$\varepsilon_c^{1/2} = \varepsilon_0^{1/2} - (\varepsilon_c C/\varepsilon_0^{1/2} K_{MK}) \qquad (23)$$

By plotting $\varepsilon_c^{1/2}$ against $\varepsilon_c C$ a straight line should be obtained with an intercept of $\varepsilon_0^{1/2}$.

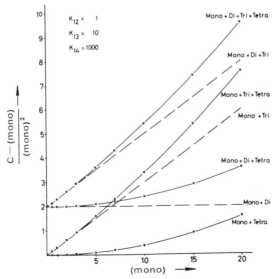

Fig. 15. Plot of $[C - (\text{mono})]/(\text{mono})^2$ as a function of the monomer concentration (mono) assuming different coupled association equilibria.

Figure 16 shows a plot according to eqn. (23) for cyclohexanol. For medium concentrations of alcohols eqn. (23) is fairly well satisfied. At low concentrations, however, clear deviations occur. The values marked for high concentrations are experimentally very uncertain owing to considerable decrease of the height of the bands. Another question which becomes acute in this range is whether the broad bands of the hydrogen bonds overlap with the region of free vibrations. At a first consideration one should not expect this to be the case, because hydrogen bonding ought to induce a displacement to longer wavelengths. The large bandwidth has, however, not yet been satisfactorily explained. If it is due to a short life of the hydrogen bonds, an overlap into the range of free vibrations may have to be considered. With this assumption one would have to consider a correction of the extinction coefficients in the range of high concentrations, and this can be done approximately. A tangent can be drawn at the minimum between the free vibration and the band due to hydrogen bonds, and the short-wavelength side of the free vibration band. (See region near the minimum at 1.43 μm in Fig. 17 and near 3600 cm^{-1} in Fig. 18.) Fortunately, only a very small correction is needed in the overtone range and it need be considered only for high concentrations. The difficulties are greater in the fundamental range, where the intensity of the bands due to hydrogen bonding is much higher and thus the superimposition effects are of quite a different order of magnitude. This is particularly true for spectra of pure liquids, where free vibration and hydrogen bond vibration bands are much less separated.

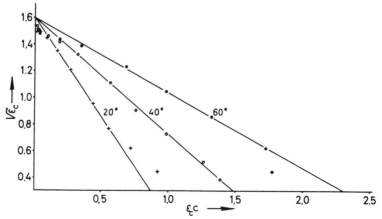

Fig. 16. The association of cyclohexanols in CCl$_4$ can be described as coupled equilibria involving chainlike aggregates; $\varepsilon_0 = (1.6)^2 = 2.56$.

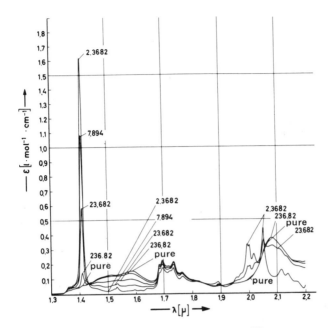

Fig. 17. Extinction coefficients of CH_3OH at different concentrations (g liter^{-1}) in CCl_4 at 20°C (overtone spectrum).

Figures 17 and 18 show two examples of the great difference between the intensity of bands due to hydrogen bonding in the fundamental and overtone regions, respectively.

This problem of background and the evaluation of ε_0 are the main sources of error in the otherwise powerful method of IR spectroscopy.

2.1.2. Determination of ε_0

We have therefore to discuss the question of the evaluation of ε_0 in some detail. We have the two complementary methods of approximation, expressed by eqns. (9) and (12), respectively. In the graphical method the temperature invariance of ε_0 for small changes of temperature is of great help. This invariance is true only if allowance is made for the volume expansion of the liquid. For pyrrolidone, at 20°C, and without the correction $\varepsilon_0 = 2.06$; at 40°C, $\varepsilon_0 = 2.03$; and at 60°C, $\varepsilon_0 = 1.96$. If the correction is made for the density change of CCl_4, then $\varepsilon_0 = 2.06$ for all temperatures. When working to high accuracy the temperature of the prepared solutions must be allowed for. Neglecting this density correction affects not only ε_0 but also α and the equilibrium constants. For pyrrolidone,

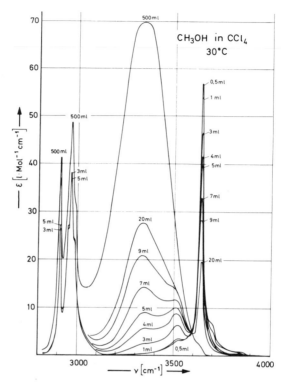

Fig. 18. Extinction coefficients of CH₃OH at different
concentrations in CCl₄ at 20°C (fundamental vibration).
In contrast to the overtone (Fig. 17), the intensity of the
H bond band is much higher than the unperturbed band.

for instance, the error due to this neglect when computing K_{12} at 60°C will
amount to 5%. The same magnitude of error will be associated with the calculation of the heat of association from measurements in the range 20–60°C.

Where dimers and trimers are superimposed an approximately straight
line [eqn. (9)] can be observed for small C values which, in a plot of ε_c
versus $\varepsilon_c^2 C$ changes at higher C into the characteristic curve of Fig. 19.
The Mecke–Kempter method (see Fig. 16) also generally leads to a straight
line[876] in the range of low concentrations, even when it is not generally valid.

With cyclooctanone oxime both methods are suitable for the verifica-
tion of a chosen value of ε_0. The slope of the straight line, valid for low C,
is determined in the case shown in Fig. 19 by K_{12}. The correctly chosen
straight line can therefore be checked by verifying the K_{12} value using the
graphical method of Fig. 12.

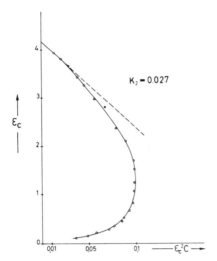

Fig. 19. Plot of $\varepsilon_c = f(\varepsilon_c^2 C)$ yields $\varepsilon_0 = \lim \varepsilon_c$ as $\varepsilon_c^2 C \to 0$. See eqn. (9); values given for cyclohexanone in CCl_4, 20°C.

A second example of the application of both graphical methods in determining ε_0 is shown in Figs. 20 and 21, representing our measurements on pyrrole. Here also both methods agree with respect to ε_0 although both are valid only in the limit $C \to 0$. In every case the use of both methods is recommended not only for the calculation of ε_0 but also because useful

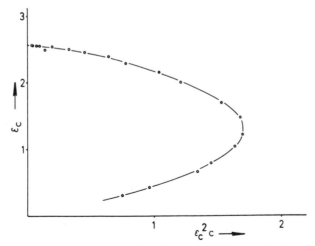

Fig. 20. The determination of ε_0 with $\varepsilon_c = f(\varepsilon_c^2 C)$ is possible for pyrrole in CCl_4, although eqn. (9) is valid only at low concentrations.

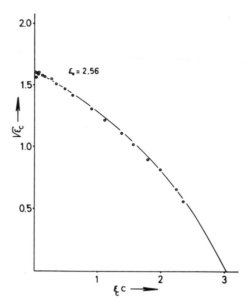

Fig. 21. The determination of ε_0 from $\varepsilon_c^{1/2}$ $= f(\varepsilon_c C)$ gives the same value for pyrrole in CCl_4 with $\varepsilon_c C \to 0$ as the method of Fig. 20.

hints are thus obtained as to the nature of the association equilibria. In the case of 4-methylimidazole no useful results can be obtained by plotting ε_c against $\varepsilon_c^2 C$ at 20°C. Here only the lower branch of the plot is observed, strongly curved toward the left. In this case, nevertheless, if the concentration is not too high, a very satisfactory straight line is obtained by plotting $\varepsilon^{1/2}$ against εC, which enables us to calculate ε_0. On the other hand, with N-ethylacetamide the concentration range where one may assume an approximate dimerization is clearly visible from the plot of ε against $\varepsilon^2 C$. In the range of higher concentrations, however, the association of N-ethylacetamide can be represented as an exclusive formation of cyclic tetramers. From the slope of the resulting straight lines K_{12} and K_{14}, respectively, will be obtained.

For propionic acid the plot of $\varepsilon^{1/2}$ against εC is completely useless; when plotting ε against $\varepsilon^2 C$ one may determine ε_0, but only approximately and only by using the assumption that ε_0 is temperature-independent. It can be seen clearly that propionic acid in CCl_4 forms not only dimers but also a considerable proportion of trimers. Wolf et al.[1425] have also found by a nonspectroscopic method a considerable proportion of trimers in the case of the lower carboxylic acids. Because Wolf and Metzger[1427] were unable to confirm this in the case of benzoic acid, they questioned the results of the former workers. Our experiments speak, however, in favor of the assumption of trimerization. One ought to investigate whether benzoic acid shows a deviation. Geisenfelder and Zimmermann[600] drew the conclusion from the X-ray structure of pure formic acid in the liquid state that chainlike aggregates are formed. This also speaks in favor of higher aggregates being formed with increasing concentration.

2.1.3. *The Validity Range of the Mecke–Kempter Relation in the Case of Chainlike Aggregates*

The Mecke–Kempter method is only an approximation. With almost every substance one observes deviations from eqn. (23).[676,963,964] The deviations occur particularly in the range of low concentrations. For example, Fig. 22 shows the constants K_{MK} determined from eqn. (22)

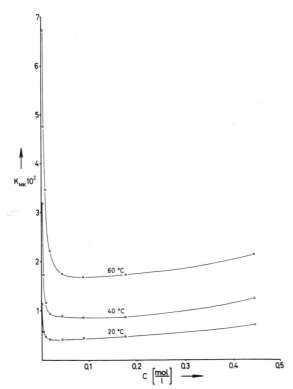

Fig. 22. Equilibrium constants K_{MK} [see eqn. (23)] for 4-methylimidazole in CCl_4. The Mecke–Kempter approximation is valid only for higher concentrations.

using the results of our measurements on 4-methylimidazole in CCl_4 solution at various temperatures. One may use a constant K_{MK} as a first approximation for concentrations above 0.02–0.1 mole liter^{-1}. At low concentrations, however, large deviations appear. This being the case, one would expect similar deviations for a large range of substances and they have in fact been observed for alcohols. Various corrections to this approximation have been suggested for small aggregates. Hoffman,[676] in the case of methanol, makes use of the assumption $K_{12} \neq K_{23} \neq K_{34}$ for $i > 4K_{(i-1)i} = K_{MK}$. His experiments yielded $K_{12} \to \infty$ or (di) $\to 0$, in agreement with our results. When determining the mean degree of association m [see eqn. (11)] it is clearly seen that $m \geq 3$. From his careful measurements of the second harmonic vibration Hoffman found the equilibrium constants shown in Table I.

Coggeshall and Saier[320] suggested the approximation

$$K_{12} = K_{23} \quad \text{for} \quad i > 3K_{(i-1)i} = K_{MK}$$

The suggestion to separate only the dimer and trimer formation from the higher aggregations is in agreement with the estimate for dipole–dipole interactions. The suggestion receives experimental support from the hint that at low concentrations all the alcohols display a band in the fundamental frequency range due to hydrogen bonding which is different from those appearing at high concentrations (3520 cm^{-1} in Fig. 18 and Refs. 123, 133, 134, 318, 884, 890, 891, 934, and 1231. However, the possibility is not ruled out that trimers, too, may have their own band due to hydrogen bonding. For further reference to dimers see Ref. 843. The method suggested by Coggeshall and Saier[320] for higher alcohols is, in our experience, less

TABLE I. Equilibrium Constants for Alcohols in CCl_4 at 21.5°C

	Methanol	t-Butanol	Phenol
K_{12}	∞	1.52	0.88
K_{13}	0.062	0.28	0.19
K_{14}	0.013	0.08	0.05
K_{23}	0.25	0.53	0.43
K_{34}	0.23	0.43	0.37
K_{MK}	0.24	0.99	0.42

suitable for methanol. From our own measurements of the first harmonic vibration one obtains equilibrium constants at 20°C displaying a systematic drift with concentration. In the concentration range of 1.5 to 32 g liter^{-1} methanol in CCl_4 it is, however, possible to fit the results of the harmonic vibration measurements into the assumption that cyclic tetramers are formed exclusively, with $K_{14} = 0.007$. This is in agreement with the experience of other authors that only small amounts of dimers are formed in solutions of alcohols. At high concentrations the mean degree of association rises steeply with increasing methanol concentration. In highly concentrated solutions the fraction of OH groups not participating in hydrogen bonds falls far below 10% in the case of methanol (see Fig. 10). From the theories of systems consisting of pure associated liquids it should follow from the above that the hydrogen bond share vastly preponderates.

Fletcher and Heller,[490] by measuring harmonic vibrations in the range of 5–100°C, have been able to show in solutions of octan-1-ol and butan-1-ol in n-decane that alcohol association can be represented under the exclusive assumption of an equilibrium between monomers and tetramers.

Dunken and Fritzsche[687] have shown in the case of alcohols that the calculation of a mean degree of association n becomes ambiguous when several association equilibria are superimposed upon each other. It is possible to represent the experimental data with the aid of various assumptions. Firm conclusions are only possible if their consistency can be demonstrated for several temperatures. Our experience shows that according to the range of concentrations, various equilibria have to be used for a successful interpretation. Dunken and Fritzsche achieved the best agreement for CCl_4 solutions with the following assumptions:

Isopropanol 20°C, $c \leq 0.1m$: $n = 1, 2$, and 3, for $c > 0.1m$, also $n > 3$.

t-Butanol: $n \leq 6$.

t-Pentanol, 15°C: $n \leq 4$.

The example of the system N-ethylacetamide/CCl_4 shows that determining the mean degree of association m according to eqn. (11) is quite suitable for verifying the assumed association mechanism. The Mecke–Kempter method does not work in this case, and plotting $\varepsilon_c^{1/2}$ against $\varepsilon_c C$ does not produce a straight line. One may try to assume an association to linear chains of n molecules, since it is known that the amide groups are predominantly in the *trans* configuration. The end groups would not form hydrogen bonds and would be almost indistinguishable from completely free groups, so that one could write

$$\frac{\varepsilon}{\varepsilon_0} = \frac{(\text{mono}) + N}{C} = \frac{(\text{mono}) + (1/n)[C - (\text{mono})]}{C}$$

where N denotes the concentration of aggregates of n molecules or, with $\alpha_n = (\text{mono})/C$,

$$\varepsilon/\varepsilon_0 = \alpha_n + (1/n)(1 - \alpha_n) = [(n-1)/n]\alpha_n + (1/n)$$

or

$$\alpha_n = [1/(n-1)][n(\varepsilon/\varepsilon_0) - 1]$$

Using

$$K_{1n} = (\text{mono})^n/N = n(\text{mono})^n/[C - (\text{mono})] = n\alpha_n^n C^{n-1}/(1 - \alpha_n) \qquad (24)$$

we obtain

$$\alpha_n = 1 - (n/K_{1n})\alpha_n^n C^{n-1} \qquad (24a)$$

According to eqn. (24a), one should obtain a straight line by plotting α_n against $\alpha_n^n C_0^{n-1}$ with a suitable choice of n. This is possible for N-ethyl-acetamide only in a very limited range of concentrations. From the range for linearity the following equilibrium constants are derived for 40°C, using the respective slopes for calculation:

K_{12}	3×10^{-1}	K_{12}	3×10^{-1}
K_{13}	1×10^{-2}	K_{23}	3.5×10^{-2}
K_{14}	3.8×10^{-4}	K_{34}	3.7×10^{-2}
K_{15}	1.6×10^{-5}	K_{45}	4.2×10^{-2}
K_{16}	5.1×10^{-7}	K_{56}	3.2×10^{-2}
K_{17}	3.9×10^{-8}	K_{67}	7.8×10^{-2}
K_{18}	2.9×10^{-9}	K_{78}	7.4×10^{-2}
K_{19}	1.9×10^{-10}	K_{89}	6.5×10^{-2}

Although the equilibrium constants $K_{n,n+1}$ are, with the exception of K_{12}, of the same order of magnitude as assumed by Mecke and Kempter, the process seems to us to be too formal. The following method of assuming cyclic associates appears more useful. In this case, according to eqn. (20) and assuming an exclusively n-fold association, we have

$$\varepsilon_c = \varepsilon_0 - (n/K_{1n})(\varepsilon_c^n/\varepsilon_0^{n-1})C^{n-1} \qquad (25)$$

From a plot of ε_c against $\varepsilon_c^n C^{n-1}$ one can then decide upon the range of validity for preponderantly n-fold association.

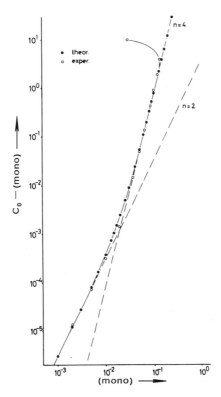

Fig. 23. Open circles: experimental values C $-$(mono) $= f[(\text{mono})]$ for N-ethylacetamide in CCl_4 at 40°C. Filled-in circles: theoretical values for a coupled equilibrium of monomers and cyclic dimers and tetramers. Broken lines: equilibria of monomers and dimers ($n = 2$) and monomers and tetramers ($n = 4$).

Figure 23 shows a plot of $\log[C - (\text{mono})]$ against $\log(\text{mono})$. We can see that there are regions of mainly cyclic dimers and others of mainly cyclic tetramers. The points shown make it clear that the whole concentration range can be excellently represented by the assumption of superimposed equilibria of dimers and tetramers. According to this, the graphically found equilibrium constants were used and, if trimers are to be included, one may put $K_{13} \approx 1.10^{-2}$.

From eqn. (20) it follows, for dimers and tetramers only, that

$$Z = [C - (\text{mono})]/(\text{mono})^2 = (2/K_{12}) + (4/K_{14})(\text{mono})^2$$

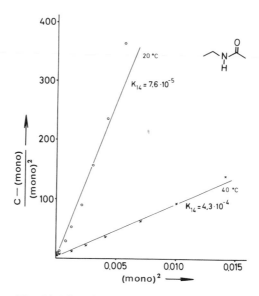

Fig. 24. Plot of Z versus (mono)^2 (see text) for a coupled equilibrium of monomers, dimers, and tetramers, seen to be valid for N-ethylacetamide in CCl$_4$.

Z plotted against (mono)^2 should give a straight line, the slope of which yields K_{14} and its intercept K_{12}. This is indeed the case, as shown in Fig. 24. The K_{14} calculated from this construction, is in good agreement with the values obtained above. The series of simple association equilibria has thus been extended by a further step. The lower lactams preferentially form dimers only ($Z = \text{const}$); oximes form dimers and trimers [$Z = $ linear function of (mono)]; and N-ethylacetamide may be represented by an equilibrium between dimers and tetramers [$Z = $ quadratic function of (mono)].

From K_{12} and K_{14}, respectively, one may try to calculate the association enthalpy by plotting ln K against $1/T$. The following values are obtained:

$$\Delta H_{12} = +4 \text{ kcal mol}^{-1} \qquad \text{or} \quad \Delta H_{12}/2 = 2 \text{ kcal mol}^{-1}$$

$$\Delta H_{14} = -14.5 \text{ kcal mol}^{-1} \quad \text{or} \quad \Delta H_{14}/4 = -3.62 \text{ kcal mol}^{-1}$$

For cyclic dimers the *cis* configuration of the amide group must be assumed; for cyclic tetramers, however, the *trans* configuration is possible. Both associates have hydrogen bond angles of $0°$. According to Huisgen and Walz,[687] the amides are present preferentially in the *trans* configuration.

Exceptions are the lower lactams, where for steric reasons the *cis* configuration is enforced.[687] Huisgen and Walz drew the conclusion from their investigations that in solution the *cis* form as well as the *trans* form is present. They estimate the heat of conversion to be roughly 1.5–2.5 kcal mol⁻¹. Our results agree with their ideas. The cyclic dimer should only be formed by the *cis* form. With the formation of dimers a *trans–cis* rearrangement is occurring. In our method of approximation to determine the association equilibria in N-ethylacetamide one would expect an energy difference between *cis* and *trans* of about 5 kcal mol⁻¹.

2.1.4. *Hydrogen Bond Entropy*

The discussion of entropy effects is necessary in order to enable us to draw conclusions about the nature of the bond from the angle dependence of the hydrogen bond energy. From infrared spectra, in cases where such calculations are possible, we get the equilibrium constants, so that values for the entropy are given by

$$R \ln K = -(\Delta H/T) + \Delta S \tag{26}$$

Table II shows some of the calculated entropy values.

In the case of exactly known equilibria, e.g., oximes and propionic acid, the drift of enthalpies is parallel to that of the entropies. In these cases all the conclusions from the angle dependence of interaction energies will hold also for the free energies. Although there is a considerable difference between ΔS_{12} and ΔS_{13} in the case of pyrazole, one should not draw conclusions from this too hastily, because the formulas used for cyclic aggregates are only approximations.

TABLE II. Entropy Changes at 20°C (in cal mol⁻¹ deg⁻¹)

	$\Delta H_{12}/T$	ΔS_{12}	$\Delta H_{13}/T$	ΔS_{13}
Propionic acid	37.5	23.3	51.1	32.5
Cyclooctanone oxime	27.3	20.7	45.4	32
Pyrazole	15	8	51.5	34
Pyrrolidone	26	15.2	—	—
Caprolactam	24.2	15	—	—
Laurin lactam	15.3	12.6	—	—
Pyrrole	12.6	13.7	—	—

2.2. The Badger–Bauer Rule

Further examples of the angle dependence of hydrogen bond energies can be derived indirectly from the frequency shift $\Delta\nu$ between free OH vibrations and those due to hydrogen bonds. It has become conventional in spectroscopy to assume a proportionality between $\Delta\nu$ and the hydrogen bond energy. This makes it possible to calculate the interaction energy ΔH directly from these experiments. These ideas have the following experimental basis. The first indications that $\Delta\nu \propto \Delta H$ were given by Badger and Bauer.[73,74] Some data for hydrogen bonds between phenol and various bases as acceptors were compiled by Pimentel and McClellan.[1099] The values for carboxylic acids used in their figure refer to the gaseous state. For phenols $\Delta\nu \simeq \Delta H$ is approximately true.

As the values in their figure originate from various sources and the calculations of hydrogen bond equilibria can only be an approximation, it is impossible to decide whether deviations from linearity are real or not. There are, however, various indications that the so-called Badger–Bauer rule, $\Delta\nu \propto \Delta H$, is true only for homologous series. The constant of proportionality may depend on parameters characteristic of the system. Figure 25 shows similar results of our own[885] with the vibrational overtone band of pyrrolidone in the presence of various hydrogen bond acceptors. It shows

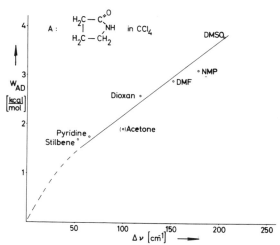

Fig. 25. The interaction energy H_{AD} of pyrrolidone in CCl_4 with different acceptors is proportional to the frequency shift $\Delta\nu$ of the H bond band (Badger–Bauer rule). Results refer to 20°C.

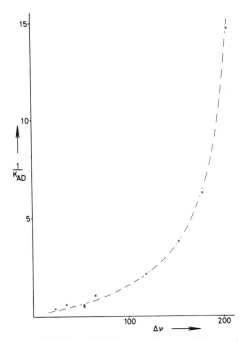

Fig. 26. Plot of $1/H_{\mathrm{AD}}$ for the association of pyrrolidone with different acceptors in CCl_4 solution at 20°C as function of the frequency shift $\Delta\nu$.

$\Delta\nu$ of the fundamental vibrational band as a function of ΔH_{AD}, the donor–acceptor interaction energy, calculated from the equilibrium constant, the latter derived by approximation from the first overtone vibration. It is to be stressed that these approximations do not approach the accuracy of the procedure discussed above for the calculation of aggregations of oximes, lactams, or carboxylic acids. The frequency values in brackets (acetone) are quite uncertain owing to the superimposition of the small shift and the free OH bands. Figure 26 shows a correlation between $1/K_{\mathrm{AD}}$ and $\Delta\nu$ for the same system. The K_{AD} are the equilibrium constants of hydrogen bond formation between pyrrolidone as donor and various acceptors.

In Fig. 27 can be seen the concentration dependence of the fundamental frequency band of cyclopentanone oxime, discussed above. The free vibration occurs at 3600 cm^{-1}. The maximum at 3300 cm^{-1} probably corresponds to the dimer, that at 3235 cm^{-1} to the trimer.

This concentration dependence of the dimer band highlights the great difficulty in a discussion of the bands due to hydrogen bonds, owing to

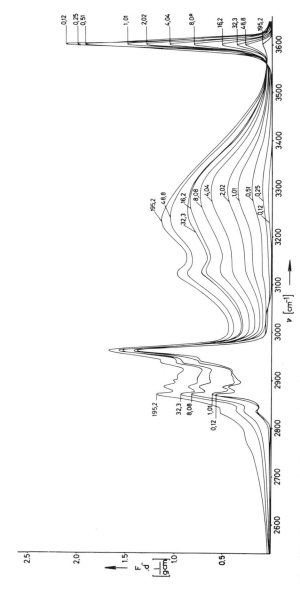

Fig. 27. Fundamental vibrations of cyclopentanone oxime solutions in CCl_4 (concentration of oxime in g liter^{-1}). There is a monomer maximum (3600 cm^{-1}) and two H-bond maxima (3300 and 3235 cm^{-1}). These maxima correspond to the concentration-dependent percentage of monomers, dimers, and trimers.

TABLE III. Test of the Badger–Bauer Rule for Aggregation of Cyclopentanone Oxime

Species	ν, cm^{-1}	$\Delta\nu$, cm^{-1}	ΔH_{1i}, kcal mol^{-1}	$\Delta H_{1i}/i$, kcal mol^{-1}	$i\,\Delta\nu/\Delta H_{1i}$	$\Delta\nu/\Delta H_{1i}$
Monomer	3600	—	—	—	—	—
Dimer	3300	300	8	4	75	37.5
Trimer	3235	365	13.3	4.45	82	28

superimposition of the bands. Table III shows the validity of $\Delta\nu = f(\Delta H)$ for hydrogen bonds between similar molecules.

With pyrrolidone, which forms dimers only, the position of the band due to hydrogen bonding is independent of concentration up to very high concentrations (see Table IV). Only on reaching pure pyrrolidone does the maximum of the band become displaced toward higher frequencies. One may assume that the bond angles suffer distortion.

TABLE IV. Concentration Dependence of the Maximum (cm^{-1}) of Fundamental Vibrational Bands of Hydrogen Bonds

Bands due to H bonds, C, ml liter^{-1}	Monomer bands						
	Pyrrolidone		N-Ethylacetamide		Laurinlactam		
	ν (3450)	$\Delta\nu$	ν (3455)	$\Delta\nu$	C, g liter^{-1}	ν (3449, 3462)	$\Delta\nu$
0.5	3198	252	3355	100	0.78	3345	104
1	3200	250	3350	105	1.5	3345	104
2	3200	250	3350	105	3.1	3345	104
5	3200	250	3320	135	6.2	3335	114
10	3200	250	3300	155	12.5	3305	144
20	3200	250	3290	165	—	—	—
50	3200	250	3290	165	—	—	—
100	3200	250	3290	165	—	—	—
200	3208	242	—	—	—	—	—
500	3220	230	—	—	—	—	—
Pure	3230	220	—	—	—	—	—

Laurinlactam, having the amine group in the *trans* configuration and thus displaying no preferential dimerization, shows accordingly a concentration dependence of the hydrogen bond band (see Table IV). Similarly, with N-ethylacetamide, Δv (being the difference between the frequency of the H bond band and that of the band due to the monomer) increases with increasing concentration.

These observations about the frequency of the fundamental vibration of hydrogen bonds are in agreement with results from the concentration dependence of the intensity of overtone bands of monomers.

In the case of intramolecular hydrogen bonds, discussed in detail in Section 2.6, the proportionality $\Delta v \propto \Delta H$ is equally well apparent. Figure 28 shows a general summary of our experiments with unsaturated alcohols. Here Δv refers to the fundamental vibration and ΔH has been determined from measurements of the overtone vibrations. One can see in this figure that small differences characterize the alcohols, according to whether they are primary (1–6, 8), secondary (9), or tertiary (7, 10, and 11). Such influences by neighboring groups were also observed by Kuhn.[811]

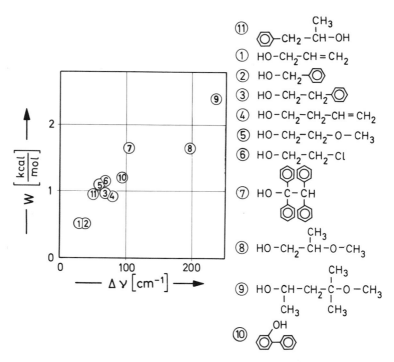

Fig. 28. The H bond energy ΔH of intramolecular H bonds of alcohols as function of the frequency shift Δv.

If one extrapolates the results of Fig. 28 to $\Delta\nu$ corresponding to the intermolecular hydrogen bonds of methanol, one obtains the correct order of magnitude for the hydrogen bond energy of methanol, i.e., 4 kcal mol^{-1}. The strength of the hydrogen bond of an X—H group is proportional to the basicity of the acceptor. Thus $\Delta\nu$ of X—H, where X stands for various acceptors, will be proportional to the heat of mixing of the respective bases in mixtures with other donors. Pimentel and McClellan[1099] have compiled values of $\Delta\nu$ for CH_3OD in various bases. These were plotted against the heat of mixing of one-half mole of $CHCl_3$ with one-half mole of the respective bases. Linear relationships were found for homologous series.

Bellamy and Pace[134] have compared the frequency displacements of various donors interacting with a large number out of 70 different acceptors. A plot of, say, $\Delta\nu$ for methanol as ordinate against $\Delta\nu$ for phenol as abscissa gives a rigorously linear relationship for all the bases. In the same manner, if one compares methanol with pyrrole or with phenyl-acetylene, one obtains different straight lines for N-containing and for O-containing bases. The authors interpret this as due to different proton distances in different donor groups; however, the possibility of charge effects needs to be further examined. Similar results are obtained when comparing values for methanol with those for diphenylamine.

Aromatic systems act as weak acceptors, and a comparison of the results of Bellamy and Pace shows a decrease in the strengths of the hydrogen bonds as follows: phenol > methanol > pyrrole > phenylacetylene > di-phenylamine.[134] Similarly, Bellamy et al.[130] found a linear relationship when plotting $\Delta\nu/\nu$ for the hydrogen bond between H_2O, HBr, $C_6H_5NH_2$, and various acceptors against the corresponding values with pyrrole as the donor. These results have been confirmed when comparing pyrrole with H_2S, CH_3OD, and HI, where H_2O, CH_3OD, and pyrrole have about the same $\Delta\nu/\nu$ values, those for HBr and HI are larger, and those for $C_6H_5NH_2$ and H_2S are smaller. Recently Drago et al.[434] have compared spectroscopically measured frequency displacements $\Delta\nu$ of t-butanol, phe-nols, and 1,1,1,3,3,3-hexafluoro-2-propanol in the presence of various acceptors with calorimetrically determined heats of association ΔH.[434] They find for each donor a linear relationship, $\Delta H = a\,\Delta\nu + b$, for com-plexes with various acceptors. If one now compares the association of an acceptor in the presence of various donors, again a linear relationship is found, of the form $\Delta H = c\,\Delta\nu$.

With the aid of such relationships it is possible to predict the hydrogen bond interaction of other donors with all acceptors which have been studied with given donors simply by measuring only one acceptor system.

2.3. Distance Dependence of Hydrogen Bonds

Some workers discuss the distance dependence of hydrogen bonds with the aid of crystal spectra. Rundle and Parasol[1164] found an asymptotic dependence of the OH frequency on the O—O distance R(O—O) in the crystal. In a smaller distance range Lord and Merifield[869] found a linear relationship between $\Delta\nu$ and R(O—O). In the same way Pimentel and Sederholm[1100] found linear relationships but with different slopes for O—H$\cdots$O, N—H$\cdots$O, and N—H$\cdots$N. They write

$$\Delta\nu(\text{O—H}\cdots\text{O}) = 4.43 \times 10^3(2.84 - R)$$

$$\Delta\nu(\text{N—H}\cdots\text{O}) = 0.548 \times 10^3(3.21 - R)$$

$$\Delta\nu(\text{N—H}\cdots\text{O}) = 1.05 \times 10^3(3.38 - R)$$

where $\Delta\nu$ is in cm^{-1} and R is a limiting distance in Å.

The frequency $\Delta\nu$ of free molecules was estimated by analogy for most of the substances. Further experimental results of a similar kind have been published by other workers.[561,1014,1370]

Lippincott and Schröder[1207] attempted a theoretical interpretation of the $\Delta\nu(R)$ curves by assuming exponential potential functions. They also assumed for all the substances a uniform $\nu_0 = 3700$ cm^{-1}. Such an approximation appears to be of very limited value because a comparison is made between the most diverse substances, such as maleic acid and nickel-dimethylglyoxime. Some fine structure should appear in the frequency position of free OH vibrations and also in the potential curve of the hydrogen bond, arising from the varying charge effects. Bellamy and Owen[131] have recently expressed the function $\Delta\nu = f(R)$ by a Lennard-Jones potential

$$\Delta\nu \ (cm^{-1}) = 50[(d/R)^{12} - (d/R)^6] \tag{27}$$

where d is the collision diameter as given by the gas kinetic theory. As d varies from atom to atom, the authors were able to express the various functions of hydrogen bonding of different groups like OH and NH and also O, N, Cl, F, with the aid of these various d values. The representation of hydrogen bonds by a Lennard-Jones potential is somewhat surprising. One would expect stronger dipole effects with a corresponding lower power of the attraction term. This discrepancy can probably be expressed in the following way: The hydrogen bond length does not depend solely on the interaction of the H bond but also on the potential curve of the O—H valence bond. All the data discussed refer to crystals. The distance R may therefore well be determined by other factors as well. For this reason the comparison of molecules with different electronic structures is not altogether desirable, as the distance is not an independent parameter.

These objections do not arise in the investigation of the temperature dependence of spectra. If in the case of ice spectra one plots the position of the band maxima obtained from the temperature dependence of the spectra[890] against the distance between the molecules derived from the known density, then the result obtained is shown in Fig. 29. From these measurements on the 1.4-μm overtone band it follows that

$$\Delta \nu \propto 1/R^{14} \tag{28}$$

This distance dependence corresponds, in the range accessible to us, to O—O values, as determined by Wall and Hornig[1370] in the spectra of various molecules. From their data it follows for $R > 3$ Å that

$$\Delta \nu \propto 1/R^3 \tag{29}$$

In this range $\Delta \nu = f(R)$ corresponds to a dipole–dipole interaction if we assume the Badger–Bauer rule. Pimentel and McClellan[1099] have calculated

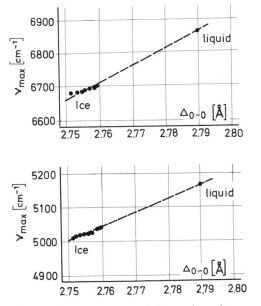

Fig. 29. Frequency ν_{max} of the maxima of two overtone bands of ice as function of the temperature-dependent O—O distance of nearest neighbors. Top: 1.4 μm; bottom: 1.9 μm.

the distance dependence of the interaction energy U_{theor} for dipoles having a dipole length of 1 Å and various charges q. For large distances, relevant to liquids and solutions, there is qualitative agreement with the experimental distance dependence. For $q = 0.3e$ and $R = 2.76$ Å, $\Delta H_{theor} = 3.8$ kcal mol^{-1} for H_2O, and this agrees quantitatively with the experimental results.

From this discussion we should expect a pronounced broadening of the bands due to intermolecular hydrogen bonds, because of the distribution of hydrogen bond length between donor groups and acceptors.

2.4. Angle Dependence of Hydrogen Bonds

Theoretically one would expect that the hydrogen bond energy will depend on the orientation of the molecules. Crystal-structure analyses by neutron diffraction show an approximately linear arrangement of the proton between the two electronegative atoms in the hydrogen bond.[67,68, 680,1428] In hydrates values of the angle β (defined in Fig. 30) of 20–25° are known. Unfortunately, experimental data on the dependence of the hydrogen bond energy on the angle β have so far been lacking. One may attempt to obtain some information about this dependence in solutions from the temperature dependence of the association equilibrium constant. Table V gives the values for the enthalpy change ΔH derived from $\log K = f(1/T)$ found in our investigations.[881,882]

Those angles β that follow from Stuart–Briegleb models are also included. The plus or minus sign in the last column stands for the existence or nonexistence of higher aggregates. Figure 31 gives a survey of the values of $\Delta H_\beta / \Delta H_{\beta=0}$ as a function of the angle β.[881,882] It shows the unequivocal preference for $\beta = 0$. Electron diffraction has shown that acetic acid dimers in the vapor phase have $\beta = 0$.[726]

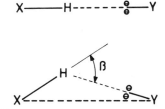

Fig. 30. The H bond interaction between a donor system XH and the free electron pair of an acceptor Y depends on the bond angle β between the axis of the XH group and the axis of the free electron pair orbital.

TABLE V. Enthalpy Change of Various Aggregates

	ΔH_{12}	$\Delta H_{12}/2$	β, deg	ΔH_{13}	$\Delta H_{13}/3$	β, deg
Imidazole	—	—	—	(18)	(6)	—	+
Propionic acid	11.0	5.5	0	15	5	—	—
Pyrrolidone	7.7	3.85	0	—	—	—	—
Caprolactam	7.0	3.5	0	—	—	—	—
Cyclopentanone oxime	8.0	4.0	49	12.8	4.3	0	
Cyclooctanone oxime	8.0	4.0	49	13.3	4.42	0	—
Laurinlactam	4.5	2.25	—	8.9	2.96	—	++
Pyrazole	4.4	2.2	72	15.1	5.0	12	+
Pyrrole	3.7	1.85	(90)	—	—	—	++
Methylaniline	(0.5)	(0.25)	—	—	—	—	+

From the exclusive existence of dimers and trimers in solutions of oximes we infer OH$\cdots$N bonds.[876] X-ray diffraction[622] has indicated that these structures also exist in solid formamide oxime. Similarly, spectroscopic data on cyclic and linear oximes lead to the same result.[547] The free energy ΔG shows a similar angle dependence in the case of oximes.

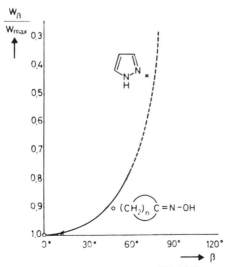

Fig. 31. The H bond energy $\Delta H_\beta / \Delta H_{\beta=0}$ as function of the H bond angle β.[881,882]

This angle dependence explains the results in Figs. 10–12 and is the key to the varying concentration dependence of the α values (Fig. 10). For lactams having the *cis* configuration of the amide groups stable cyclic dimers with $\beta = 0$ are possible,[881] as in the case of carboxylic acids. In both cases the rule holds that hydrogen-bonded associates are particularly stable in cases where the ring has six atoms, not counting the protons.

Lactams with the *trans* configuration of the amide groups show aggregation of a higher order. In this case a molecular model containing six atoms in the ring cannot be constructed. In the case of propionic acid in CCl_4 solution we have found some evidence for trimers at higher concentrations.[881] This is easily explained by addition of a third acid molecule to the freely accessible excess of free electron pairs of the $C=O$ or OH groups.

Wolf *et al.*[1245] also give indications for trimers in the case of the lower carboxylic acids, while in the case of the higher ones this could not be confirmed. In benzoic acid solutions, too, Wolf and Metzger[1427] found only dimers. Presumably trimers are prevented from being formed by steric hindrance in the case of higher carboxylic acids. The trimer of the oximes can also form a six-membered ring with $\beta = 0$.[876,881] The tendency of hydrogen bonds to form angles $\beta = 0$ undoubtedly has great significance for biochemistry. It offers indications for the specificity of biochemical structures. For instance, the pairs of bases in DNA are connected by hydrogen bonds via six-membered rings with $\beta = 0$. In addition, in the α-helix of proteins the individual helix turns are bonded to one another by hydrogen bonds with $\beta = 0$.

Another example for preferential occurrence of $\beta = 0$ is provided by polyethylene oxide derivatives which have been rendered hydrophobic. Here we find dihydrates with $2H_2O$ per ether oxygen to be preferred.[878,879] According to X-ray results, hydrogen bonds with $\beta = 0$ exist in this dihydrate.[743,744,1158] This indication of particularly stable organic hydrates with $\beta = 0$ opens up possibilities of interpreting the action of protective colloids.[878] In the tridymite lattice of ice, too, the H_2O molecules are in six-membered rings with $\beta = 0$.

In the gas hydrate structure[1193,1244,1245] the water molecules are arranged in planar five-membered rings. The angle in these five-membered rings, of $108°$ is only slightly different from the $O-O-O$ angle in ice ($109.5°$). The hydrogen bond energy in this case is only slightly less than in the tridymite arrangement. This difference in energy is easily compensated by additional dispersion interactions involving the enclosed "guest" molecules. No protons or free electron pairs face toward the interior of the cavities formed by the pentagonal dodecahedra, i.e., the water molecules turn their most "hydrophobic" side toward the cavity. The arrangement of planar five-membered rings leads to a decrease in entropy compared to the more mobile six-membered rings in the tridymite structure and may well have an influence on the entropy effects in aqueous solutions of organic substances. This has led to the postulate of "iceberg" structures in solution. No corresponding increase in the number of hydrogen bonds has, however, yet been shown in such solutions.[654]

2.5. Matrix Technique

The IR spectroscopic methods applicable to solutions can be considerably extended by freezing in vapor mixtures of substances to be investigated in inert gases at very low temperatures.[93,1094,1095,1097,1098,1334] The discoverer of this method, Pimentel,[1096,1322] made use of this matrix technique in investigations of hydrogen bonds. Van Thiel et al.[1333] observed various sharp OH bands at the principal vibrational OH valence frequency at a varying concentration ratio of nitrogen to methanol. The authors assigned the individual bands to various cyclic hydrogen bond aggregates (monomers, dimers, etc.) according to the concentration dependence of the individual band intensities, which sometimes display maxima. These maxima appeared when band intensities were plotted against the intensity of the CH band at 2965 cm^{-1}. Similar phenomena could be observed also at various concentrations of H_2O in solid N_2.[1333] Here, owing to the two H_2O bands (symmetric and antisymmetric vibrations) all the bands appear twice. Table VI offers suggested assignments of the individual bands.[884]*

With the aid of the Stuart–Briegleb calotte models it is possible to assign hydrogen bond angle values β to each single species. The angle dependence of hydrogen bond frequency shift $\Delta\nu$ is shown in Fig. 32.[887]

At medium concentrations of H_2O in the N_2 matrix there is a splitting of the free OH vibration, while at low or at high concentrations one or the

---

* In a recent publication Barnes and Hallam[94] have confirmed similar $\Delta\nu$ values. With ethanol in an argon matrix they observed first a splitting of 7 cm^{-1} in the monomer band which they assign to the *trans* and *gauche* form. Such a splitting of about 20 cm^{-1} is also observed in the vibrational overtones of solutions.[887] The band of the dimer is at about 3535 cm^{-1} and is split into three (four?) components. Presumably the configurations *trans–trans*, *trans–gauche*, and *gauche–gauche* of the two components are involved. The trimer band also shows a slight indication of splitting (3453 cm^{-1} and 3437 cm^{-1}).

TABLE VI. Assignment of Bands Observed in the Solid N_2 Matrix

	H_2O						CH_3OH		
	ν_1	$\Delta\nu$	$\Delta\nu/\Delta\nu_{max}$	ν_2	$\Delta\nu$	$\Delta\nu/\Delta\nu_{max}$	ν	$\Delta\nu$	$\Delta\nu/\Delta\nu_{max}$
Free OH in the free molecule	3725	—	—	3640	—	—	3660	—	—
Free OH in the molecule with hydrogen bonds in the other OH group	3700	25	0.06	3620	20	0.05	—	—	—
Hydrogen bond in the dimer	3545	180	0.46	3435	205	0.49	3490	170	0.42
Hydrogen bond in the trimer	3510	215	0.55	—	—	—	3445	215	0.52
Hydrogen bond in the tetramer	3390	335	0.86	3260	380	0.9	3290	370	0.9
Hydrogen bond in the polymer	3335	390	1	3220	420	1	3250	410	1

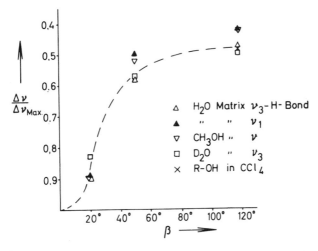

Fig. 32. Plot of $\Delta\nu/\Delta\nu_{\beta=0} = f(\beta)$ of the H bond bands of the fundamental vibrations of H_2O, D_2O, and CH_3OH obtained by the matrix technique. Crosses refer to solutions of alcohol in CCl_4.

other of the two peaks dominates. We have suggested an assignment of the band dominant at high concentration to a free OH vibration, with the second OH group or its free electron pairs forming a hydrogen bond. This hypothesis is confirmed by recent measurements of the matrix bands of methanol in solid N_2 by Tursi and Nixon,[1313] who were able to show a fine structure of the single bands at high resolution. This shows a similar concentration dependence of the relative intensities of the separate components as do the principal bands discovered by Pimentel and collaborators.

We are inclined to assign this fine structure to perturbations of the OH group by various hydrogen bonds in other parts of the molecule. Our assignments of the fine structure of the free OH bands observed by Tursi and Nixon* are shown in Table VII. The assignment of the principal bands of D_2O using the matrix technique is exactly the same as with H_2O. The measurements used for Table VII are those of Tursi and Nixon. Thus a consistent assignment of all the bands is achieved, assuming that aggregates of the hydrogen-bond-containing substance suffer a band displacement $\Delta\nu$, and the free OH vibrations will also display a small effect of this kind owing to interaction with the rest of the molecule. For the latter there thus exists a generalized Badger–Bauer rule.

* I thank Dr. Nixon for sending me the illustrations that made this analysis possible·

TABLE VII. H_2O Stretching Vibration Assignment of the Fine Structure of Free OH[a]

	System I			System II		
	ν_3	$\Delta\nu$	$\Delta\nu/\Delta\nu_{max}$	ν_3	$\Delta\nu$	$\Delta\nu/\Delta\nu_{max}$
Free in the free molecule	3726	—	—	3715	—	—
Free in a dimer	~3703	~23	~0.55	3695	20	0.49
Free in a trimer	3698	28	0.68	3690	25	0.6
Free in a tetramer	3691	35	0.85	(3680)	35	0.85
Free in a polymer	3684.5	41	—	3674	41	—

[a] Splitting in systems I and II is discussed on p. 281–282.

This explanation of the fine structure, which differs from the one proposed by Tursi and Nixon, is also corroborated by the following observations:

1. The appearance of fine structure of the matrix bands of H_2O and their absence in the case of methanol indicates cyclic aggregates which, in the case of methanol, lack free OH bands.

2. In intramolecular hydrogen bonds of the dialcohols there are free OH groups, the free electron pairs of which are bound by the intramolecular hydrogen bond. Because of the β dependence of the hydrogen bond energy, these intramolecular bonds exhibit $\Delta\nu$ which is a function of the alkyl chain length. In Fig. 33 the position of the principal vibrational frequency of the free OH bands is plotted against the frequency of the bands due to the intramolecular hydrogen bonds of the bound OH groups. The free OH vibration is displaced by about 10 cm^{-1} per the 130 cm^{-1} displacement of the band due to the intramolecular hydrogen bond. A surprisingly similar result can be seen from the measurements by Mohr et al.[991] They measured the fundamental vibration of H_2O in CCl_4 solution when different acceptors were added. In these very dilute solutions there is always a free OH vibration ν_{free} and a hydrogen bond vibration due to hydrogen bound to the acceptor $\nu_{H\text{-bond}}$. If one plots ν_{free} against $\nu_{H\text{-bond}}$, one obtains a displacement of ν_{free} by about 10 cm^{-1} for a displacement of $\nu_{H\text{-bond}}$ by 120 $\pm$ 20 cm^{-1} (Fig. 34).

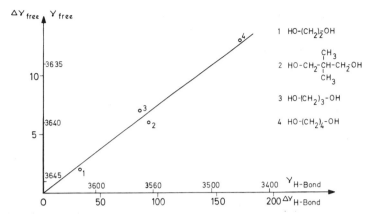

Fig. 33. Frequency ν_{free} and frequency shift $\Delta\nu_{\text{free}}$ of the free OH group of diols during formation of an intramolecular H bond by the second OH group, as function of frequency $\nu_{\text{H-bond}}$ and the frequency shift $\Delta\nu_{\text{H-bond}}$.

3. When the splitting of the free OH or OD band, as obtained by Tursi and Nixon and according to the assignments of Tables VIII and IX, is plotted as in Fig. 35 against the frequency of the corresponding band caused by hydrogen bonding of the aggregate to which the OH or OD group belongs, the result for H_2O is a displacement of the free OH band by 10

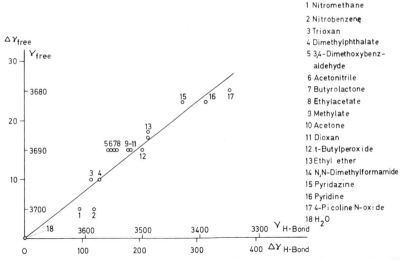

Fig. 34. Correlation between ν_{free} and $\nu_{\text{H-bond}}$ and between $\Delta\nu_{\text{free}}$ and $\Delta\nu_{\text{H-bond}}$ for dilute solutions of H_2O in CCl_4 and different acceptors.

TABLE VIII. D₂O Stretching Vibration Assignment of the Fine Structure of Free OD[a]

	System I			System II		
	ν_3	$\Delta\nu$	$\Delta\nu/\Delta\nu_{max}$	ν_3	$\Delta\nu$	$\Delta\nu/\Delta\nu_{max}$
Free in the free molecule	2764.4	—	—	2756.6	—	—
Free in the dimer	(~2742)	~22	0.55	—	—	—
Free in the trimer	2738	26	0.65	—	—	—
Free in the tetramer	~2730	~34	0.85	—	—	—
Free in the polymer	2724	40	—	—	—	—

[a] The fine structure assignment for D_2O is in good agreement with that for H_2O (see also Ref. 641).

cm⁻¹ per the ~100 cm⁻¹ displacement of the band due to hydrogen bonding of the rest of the molecule. The result for D_2O is 10 cm⁻¹ per 70 ± 10 cm⁻¹.

All the interpretations suggested in Figs. 33–35 give a similar order of magnitude for the displacement of the free OH vibrational band when the rest of the molecule (free electron pairs or the second OH group) is exposed to hydrogen bond interactions. With the exception of matrix bands of high resolution, the fine structure of such small splittings cannot be shown experimentally for normal solutions and liquids. This means that without

TABLE IX. Assignment of the Bands Due to Heavy Hydrogen Bonds of D₂O[1313]

	ν_3	$\Delta\nu$	$\Delta\nu/\Delta\nu_{max}$	ν_1	$\Delta\nu$	$\Delta\nu/\Delta\nu_{max}$
Free I	2764	—	—	2655	—	—
Free II	2756	—	—	—	—	—
Dimer	2618	146	0.5	—	—	—
Trimer	2599	165	0.57	2475	180	0.65
Tetramers	2525	239	0.83	2450	205	0.75
Polymer	2475	289	1	2380	275	—

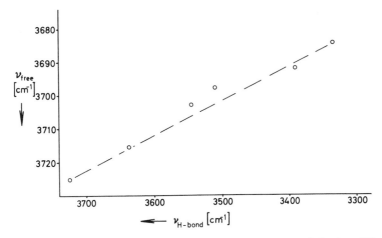

Fig. 35. Frequency ν_{free} of the fine structure components of the free OH vibration of H_2O as function of the frequency $\nu_{H\text{-bond}}$ of the corresponding H bond band, as obtained by the matrix technique.

special precautions only free OH groups can be demonstrated spectroscopically. Cross *et al.*[353] and several subsequent theoretical investigators have discussed nine different stages of aggregation of H_2O. Normally one cannot distinguish spectroscopically the different stages, particularly in liquids.

A further indication of the interpretation of the angle dependence of OH bands as discussed above comes from dilute alcoholic solutions, where one can distinguish a second band due to hydrogen bonding. Its frequency coincides with that of the dimers in the matrix band.[123,133,318,843,934] In solutions of alcohols in CCl_4 one observes a minimum of the dipole moments in the same concentration range where these dimer bands appear.[695,965, 968,969,1208,1329] Assuming the Badger–Bauer rule, which ought to hold for the homologous series shown in Figs. 33–35, one can infer that the energy of the two unfavorable hydrogen bonds in a cyclic dimer corresponds approximately to that of one optimal linear hydrogen bond. The energy of four hydrogen bonds in a cyclic tetramer is higher than that of three linear hydrogen bonds. The appearance of cyclic hydrogen bond aggregates with unfavorable angles is therefore to be expected from energy considerations. As the foreign molecules are situated at the defects in the matrix, the cyclic aggregates will be favored additionally.

We have to give a possible interpretation of the two band systems I and II in Table IX found by Tursi and Nixon in the case of H_2O in N_2 matrices.

Barnes *et al.*[97] have investigated HCl in solid matrices of several frozen gases. While in the noble gases, in CF_4, and in SF_6 there were indications of the HCl rotational structure, this was no longer the case in matrices of N_2, C_2H_4, CO, or CO_2. Instead, together with the principal band of HCl a subsidiary band appeared with a frequency shift of 11.8 cm^{-1} with N_2 and 17 cm^{-1} with CO. The authors interpret this subsidiary band as an interaction effect with various trapping sites[97] of the HCl molecule in the N_2 matrix. This splitting is in good agreement with the one observed by Tursi and Nixon in the case of H_2O in the two systems I and II (Tables VII and IX). The frequency shift $\Delta\nu$ of the individual rotational lines of HCl in noble gas matrices compared to those in HCl vapor, found by Hallam and co-workers,[97] is proportional to the critical temperature of the noble gas, and this can be taken as a convenient measure for the magnitude of the dispersion interactions. The shift for CO_2 corresponds to the order of magnitude of this effect. In distinction, the displacement of the maximum in a matrix of N_2, CO, or C_2H_4 is much larger, so that these molecules clearly show a stronger interaction. This is in agreement with observations of HCl in an argon matrix doped with various other gases. In the case of oxygen there is no change in the band, but a change occurs when N_2 is added.[96] In the case of CO, $\Delta\nu \simeq 79$ cm^{-1} and for C_2H_4 it is as high as 139 cm^{-1}. It is known from other observations that π electrons may act as acceptors for hydrogen bonds.*

Observations with varying concentrations of HCl in argon matrices by Barnes *et al.*[95] have shown different aggregation bands of HCl, attributed to dimers, trimers, tetramers, and various polymers. The corresponding frequency shifts were $\Delta\nu/\Delta\nu_{max} = 0.43$, 0.48, 0.68, and 0.78. Every stage of aggregation corresponds to a band with a fine structure, as in the observations of Tursi and Nixon.

2.6. Intramolecular Hydrogen Bonds

A further method for estimating the angle dependence of hydrogen bonds can be developed in the case of intramolecular bonds. These have been studied in detail, e.g., by Kuhn.[806,808–810,812] In dialcohols OH—$(CH_2)_n$—OH two OH vibrations are observed in highly dilute solutions, one of which can be attributed to a free OH group and the other to an intramolecular hydrogen bond. The frequency displacement between the

---

* Bands due to hydrogen bonding in HCl solutions with various acceptors are discussed in Refs. 587, 806, 808–810, 812, 887, and 1395.

two forms of OH increases with increasing n, the number of CH_2 groups. At the same time the angle of the intramolecular hydrogen bond decreases.[887] If one of the OH groups is replaced by an acceptor X, one still observes intramolecular hydrogen bonds. A series of such observations is summarized in Table X.[887] In this case of intramolecular hydrogen bonds, structures such as those due to intermolecular bonding, like the cyclic dimers, are impossible. It is due to this fact that the results obtained with large β do not agree with those resulting from the matrix technique. For the compound

$$\text{\Large⬡}\!-\!CH_2\!-\!CH\!-\!OH$$
$$|$$
$$CH_3$$

in CCl_4 the different possible kinds of hydrogen bonds are shown in Fig. 36. Detailed investigations have shown that the following bands can be distinguished: 3622 cm^{-1}, free OH vibration; 3600 cm^{-1}, intramolecular hydrogen bond with the phenyl group of the same molecule acting as acceptor; 3555 cm^{-1}, a band of intermolecular hydrogen bonds with the phenyl groups of a neighboring molecule acting as acceptor; 3480 cm^{-1}, a band of energetically (β) unfavorable intermolecular OH$\cdots$OH bonds; 3360 cm^{-1}, a band due to energetically optimum intermolecular OH$\cdots$OH bonds. On lowering the temperature of solutions containing compounds of the type X—(CH$_2$)$_n$—OH, one first observes a lowering of band intensity in the range of the free OH bands and an increase of the intensity in the

TABLE X. Frequency Shifts $\Delta\nu$ (cm^{-1}) in the Bands Due to Intramolecular Hydrogen Bonding X—(CH$_2$)$_n$—OH

X	$n=1$	$n=2$	$n=3$	$n=4$	$n=5$	First overtone		
						$n=1$	$n=1$	$n=3$
—OH	—	30	78	156	153	—	58	150
—OCH$_3$	—	30	86	180	—	—	70	(200)
CH$_2$=CH—	18	40	—	—	—	32	79	—
⬡—	18	28	40	—	—	40	67	—
HC≡C—	—	42	50	—	—	—	—	—

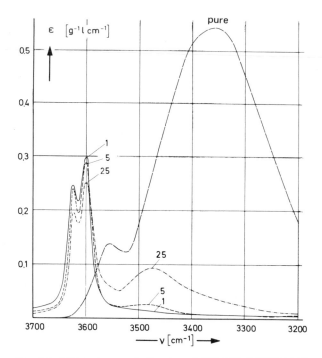

Fig. 36. OH fundamental vibration of $C_6H_5CH_2CH(CH_3)OH$
in CCl_4 (concentration in ml liter^{-1}): 3622 cm^{-1}, free OH;
3600 cm^{-1}, intramolecular H bond; 3555 cm^{-1}, intermolecular
H bond with phenyl group as acceptor; 3480 cm^{-1}, intermolecular OH$\cdots$OH bond with unfavorable β; 3360 cm^{-1}, intermolecular OH$\cdots$OH bond with favorable β.

range of intermolecular hydrogen bonds with $\nu < 3550$ cm^{-1} (fundamental)
or $\nu < 6900$ cm^{-1} (first harmonic). On the other hand, the ratio E_O/E_G
of the two band intensities also changes, where E_O is that due to free OH
vibration and E_G is that due to intramolecular hydrogen bonds.[895] The
decrease of E_O is connected with the formation of new hydrogen bonds at
low temperatures. In addition, the intensity E_G of the band due to intramolecular hydrogen bonds also changes. At equilibrium we get (denoting
the open OH groups by the subscript O and the groups bound to the
intramolecular H bond by the subscript G)

$$C_O \rightleftharpoons C_G \tag{30}$$

$$K = C_O/C_G = E_O\varepsilon_G/\varepsilon_O E_G = kE_O/E_G \tag{31}$$

$$\log K = \log k + \log(E_O/E_G) \tag{32}$$

From plots of $\log(E_0/E_G)$ against $1/T$ one can obtain the enthalpy of formation of intramolecular hydrogen bonds (see also refs. 55 and 56) and this has been experimentally confirmed. In Fig. 28 the enthalpy thus determined is plotted against the frequency difference $\Delta\nu$ between the overtone band of the open OH groups and that due to OH groups bound in intramolecular hydrogen bonds. It is shown that in homologous series, e.g., of primary alcohols, the Badger–Bauer rule holds.

The method of eqns. (30)–(32) applies only where there is no overlapping of the bands. This condition is satisfied better in the overtone region than in the fundamental vibration because $\Delta\nu$ is larger in the overtone bands. Also the intensity of bands due to intramolecular hydrogen bonds is lower in the latter case, so that overlapping effects will also be much less marked.

Occasionally one finds in the literature paradoxical results for the interaction energy of intramolecular bonds determined by means of the fundamental vibrational frequency. Presumably this is due to perturbation effects arising from overlapping. For intramolecular bonds of CH_3—O—CH_2—CH_2—OH and Cl—CH_2—CH_2—OH, $\Delta\nu = 70 \text{ cm}^{-1}$, with enthalpies of 1.1 and 1.2 kcal mol^{-1}, well on the Badger–Bauer curve for primary alcohols of Fig. 28. The structure of the intramolecular bond of Cl—CH_2—CH_2—OH has been determined in the microwave range by Azrak and Wilson[65] (H—Cl distance, 2.609 Å).

We have been able to observe cis and trans configurations of aliphatic alcohols from the details of the free OH overtone bands.[887] From the temperature dependence of the two bands belonging to the two configurations we have estimated the enthalpy of the following transition:

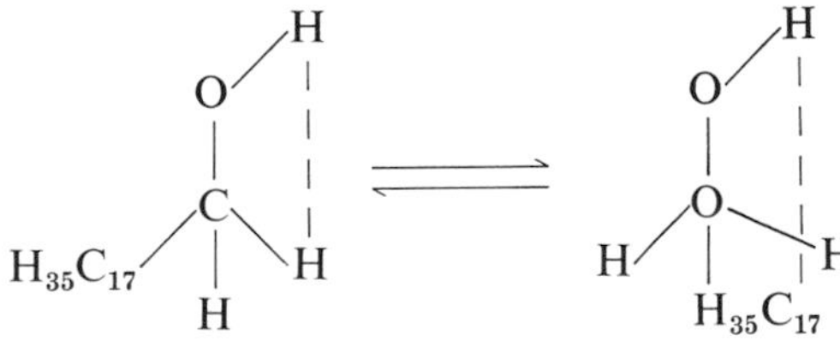

We obtained a frequency shift of $\Delta\nu = 30 \text{ cm}^{-1}$ in the first harmonic and an enthalpy change of 0.4 kcal mol^{-1}. Corresponding to the mechanism of intramolecular bonds E_0/E_G is independent of concentration.[887] This is true also for mixtures of two solvents A and B, even when there are strong acceptors present for hydrogen bonds.[895] It is therefore possible to determine the interaction effect with the solvent independently of the intramolecular bond equilibrium.

For the solute X we may write

$$X + A \rightleftarrows XA, \qquad K_A = [X][A]/[XA] \qquad (33)$$

$$X + B \rightleftarrows XB, \qquad K_B = [X][B]/[XB] \qquad (34)$$

For $[X] \ll [A]$, $[B]$ we have

$$[A] \sim [A]_0 = \varrho_M 1000\gamma_A/\sum \gamma_i M_i, \qquad [B] \sim [B]_0 = \varrho_M 1000\gamma_B/\sum \gamma_i M_i$$

$$[A]/[XA]K_A = [B]/[XB]K_B \qquad (35)$$

$$[XA]/[XB] = ([A]/[B])K_B/K_A = (\gamma_A/\gamma_B)K_B/K_A = (\gamma_A/\gamma_B)k \qquad (36)$$

with

$$[XA] + [XB] \sim [X]_0 \qquad (37)$$

where $[X]_0$ is the concentration of X as weighed,

$$[XA]/([X]_0 - [XA]) = k\gamma_A/\gamma_B$$

$$([X]_0/[XA]) - 1 = (1/\alpha_A) - 1 = (\gamma_B/\gamma_A)(1/k)$$

$$\alpha_A = k\gamma_A/[1 + \gamma_A(k - 1)] \qquad (38)$$

Figure 37 shows experimental values for α_A derived from vibrational overtone results, the solvent system being CCl_4–solvent A. Here γ_A is the mole fraction of benzene, dimethylsulfoxide, or tributylamine. In order to determine α, one may use the extinction at the peak of the solute molecule either in solvent A or in solvent B. The results are the same in both cases. With β-phenylethanol the preference for benzene as partner, as against

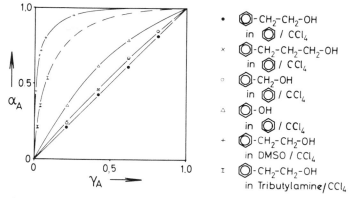

Fig. 37. Fraction $\alpha_A = (X)(A)/(X)_0 = f(\gamma_A)$ of OH groups of X which interact with acceptor A in mixtures A–CCl_4 at 20°C.

CCl_4, is very slight. This is a consequence of the especially strong intra-molecular hydrogen bonds in this alcohol. With benzene the intramolecular bond has the same frequency shift as the intermolecular bond. In the case of γ-phenylpropanol the proportion of intramolecular H bonds is much smaller, owing to the configurational entropy. In this case there is a pref-erential solvation in benzene.

The intramolecular bonds with π electrons become stronger in the presence of neighboring CH_3 groups. Kuhn[806,808-812] gave the following values for intramolecular bands:

$$\Delta\nu = 63 \text{ cm}^{-1} \qquad \Delta\nu = 88 \text{ cm}^{-1} \qquad \Delta\nu = 86 \text{ cm}^{-1}$$

Similar results have been obtained for CH_3 in intramolecular bonds, in the case of $HO{-}(CR_2)_3{-}O{-}CH_3$ in the overtone region. 4-Methoxy-4-methyl-pentan-2-ol has considerably fewer free OH groups than butane-1,3-diol-3-methyl ether.

Certainly there may be steric effects contributing to a reduction in the proportion of free OH groups in the former. The frequency shift is, however, larger for the former than it is for the latter. Thus a stronger intramolecular bond can be assumed to exist in 4-methoxy-4-methyl-pentan-2-ol.[895] Azrak and Wilson[65] have determined the structure of intramolecular hydrogen bondings in the 2-haloethanols by microwave spectra.

2.7. Comparison of Hydrogen Bond Studies in the Fundamental Frequency Range and in the Overtone Regions

Many readers may be surprised at the high accuracy with which hydro-gen bond characteristics can be determined by IR vibrational overtone spectra. In this section, therefore, the comparison of hydrogen bond de-terminations from fundamental and overtone vibrations will be discussed. There are a number of important advantages with regard to equipment for working in the overtone range. The photometric accuracy of the equipment

TABLE XI

	Free OH	H bond	$\Delta\nu$	$\varepsilon_0/\varepsilon_G$
Fundamental vibration ν, cm^{-1}	3623	3342	281	—
ε_{max}, mol^{-1} cm^2	~57	~86	—	~0.65
Overtone vibration ν, cm^{-1}	7072	~6580	~490	—
ε_{max}, mol^{-1} cm^2	2.3[a]	~0.1	—	~23

[a] Accurately determined by graphical extrapolation $\varepsilon_c = f(\log c)$; $\varepsilon_c = f(\varepsilon_c{}^2 C)$; $\varepsilon_c^{1/2} = f(\varepsilon C)$.

is higher and the layer thicknesses which can be used in the absorption cells for investigating solutions lie in a more favorable range, 1–10 mm, resulting in higher accuracy and reproducibility; also background corrections are usually smaller. Table XI shows a comparison of hydrogen bonds in their fundamental and overtone vibrations for solutions of cyclohexanol in CCl$_4$.[890] While the overtone vibration of the H bond is of the same order of magnitude as in the case of free vibration, although accompanied by strong broadening of the band $\int \varepsilon \, d\nu$, the intensity of the fundamental vibration band is strongly increased for ε_{max} as well as for $\int \varepsilon \, d\nu$. While the extinction coefficient of the maximum of the free OH fundamental vibration increases by a factor of 24 compared to the first overtone vibration, for the band due to the hydrogen bond this factor is about 800! The frequency shift between the maxima of the free OH bands and of those due to hydrogen bonds is greater by a factor of 1.75 in the case of the first harmonic ν_1 as compared to that of the fundamental frequency ν_0. This is also true for diols (see Table XII).

The weak band of the hydrogen bond and its larger frequency shift in the overtone range greatly facilitate an accurate quantitative determination of free OH groups. In cases of broader OH bands, such as occur particularly in pure liquids, this advantage may prove quite decisive. Also the band of the cyclic dimers at 3510 cm^{-1} is relatively intensive. The corresponding overtone band of the dimers at 1.455 μm is, however, only weakly indicated. Similar results exist for CH$_3$OH or C$_2$H$_5$OH. Unfortunately, owing to the inaccurate determination of the small thicknesses of the layers (a few microns), the fundamental vibrational spectra will only allow intensities to be expressed within an error of several per cent (Table XIII).

TABLE XII. $\Delta\nu$ of the Intramolecular Hydrogen Bond of Diols of OH Valence Vibration[a]

	$\lambda_{1,\text{free}}$, μm	$\lambda_{1,\text{H-bond}}$, μm	$\Delta\nu_1$, cm^{-1}	$\Delta\nu_0$, cm^{-1}	$\Delta\nu_1/\Delta\nu_0$
OH—(CH$_2$)$_2$—OH	1.4035	1.414	53	30	1.75
OH—(CH$_2$)$_3$—OH	1.406	1.434	139	78	1.78
OH—CH$_2$—C(CH$_3$)$_2$—CH$_2$OH	1.406	1.437	153	86	1.78
OH—(CH$_2$)$_4$—OH	1.406	~1.464	~270	157	~1.77

[a] Subscript zero indicates fundamental, subscript one indicates overtone.

While at the fundamental OH vibrational hydrogen bonding band, $\int \varepsilon \, d\nu$ of CH$_3$OH is about ten times as intense as the band due to free molecules, at the first overtone band of the hydrogen bond, it is only slightly more intense than the band due to free molecules. The bands due to H bonds have been integrated up to the beginning of the wavelength range of the CH bands. The ratio of intensities of overtone to fundamental vibrations—the electrical anharmonicity—is about 0.03 for the free bands and about 0.006 for bands due to H bonds.

We have attributed the maximum of the 1.59-μm (6290-cm^{-1}) band due to hydrogen bonding to the low-intensity combination band system. The intensity of this band is almost impossible to separate from the proper hydrogen bonding band with a maximum at 1.52 μm (6580 cm^{-1}). Through

TABLE XIII. Intensities of Fundamental and Overtone Vibrations of CH$_3$OH Solutions in CCl$_4$ at 30°C[a]

C, ml liter^{-1}	Frequency range, cm^{-1}	Interpretation	$\int \varepsilon \, d\nu$, liters mol^{-1} cm^{-2}
1	3200–3800	Free OH	2,250 (2,320)
500	2800–3700	Bound OH	22,900
1	6800–7500	Free OH	117 (121)
500	6050–7700	Bound OH	139

[a] Figures in parentheses for $\int \varepsilon \, d\nu$ are corrected for $\alpha = 100\%$.

Table XIV. Frequencies for Conformational Isomers of Various Alcohols[a]

Alcohol	v_2, cm^{-1}	v_2', cm^{-1}	$\dfrac{v_2 - v_2'}{v_1 - v_1'}$	v_1, cm^{-1}	v_1', cm^{-1}	$\dfrac{v_1 - v_1'}{v_0 - v_0'}$	v_0, cm^{-1}	v_0', cm^{-1}
Ethanol	10,395	10,373S	(1.5)	7102	7087S	—	3635	—
n-Propanol	10,395	10,363S	(1.3)	7111	7087S	—	—	—
n-Butanol	10,398	10,368S	(1.4)	7105	7083S	—	—	—
iso-Butanol	10,406	10,363S	(1.6)	7109	7082S	—	—	—
Benzyl alcohol	10,336	10,406S	(1.4)	7064	7115S	—	—	—
2-Phenylethan-1-ol	10,395	10,288	(1.65)	7097	7032	—	—	—
3-Phenylpropan-1-ol	10,510	10,465S	(1.8)	7107	7082S	—	3640	3600
sec-Butanol	10,368	10,309S	(1.4)	7088	7047S	—	—	—
1-Phenylpropan-2-ol	10,357	10,277	1.6	7023	7074	2.1	3600	3624
α-Phenylphenol	10,144	10,309	1.8	6945	7037	—	—	—
Tetraphenylglycol	10,320	10,152	1.8	7051	6957	—	—	—
Benzpinacol	—	—	—	6948	7052	2.5	3570	3610

[a] S means the shoulder of a band, its frequency determined with low accuracy. v_i' is the frequency for other conformations. The overtones are thus about twice as sensitive to interaction effects as the fundamental vibration. The second overtone is shifted 1.6–1.8 times as much as the first overtone.

symmetric extrapolation of this 1.52-μm band and comparison with the CD_3OH spectrum, one may estimate the intensity of the 1.59-μm band to be about 9 liters mol^{-1} cm^{-2}.

In addition, from the frequency shifts of different conformations of the OH group of alcohols previously discussed,[887] it was possible to show clearly the increasing frequency shifts with the degree of the overtone. Table XIV summarizes the frequencies of the fundamental vibration v_0 and of the first and second overtones v_1 and v_2 for conformations of various alcohols, as well as the ratios of frequency shifts of two conformations i and i'. In the case of intramolecular hydrogen bonds of the dialcohols the first overtone showed a frequency shift of the bands 1.75 times larger than that of the fundamental frequency band.[887]

The previously discussed[890] overtone association band of HF gas with $\Delta v = 1000$ cm^{-1} has also been observed in fundamental vibration[256] with $\Delta v = 500$ cm^{-1} and also in the second overtone[1367] with $\Delta v = 1400$ cm^{-1}. The relationships between the Δv values of HF are similar to those in Table XIV. For the quantitative determination of the degree of association α it is irrelevant which of the OH or NH bands is used. Figure 38 shows identical results for CH_3OH whether the fundamental or the first overtone vibration is used. Where one finds different values in the literature when comparing the two bands, most probably the reason lies in differing assumptions as to the correction to be applied to allow for the background. This is particularly true for the fundamental vibrational band of the H bond, and this provides one more reason for determining α from overtones. Figure 39 shows the independence of α for various overtones.

The consideration of anharmonicity[892] gives indications about differences in behavior of fundamental and overtone vibrations in hydrogen

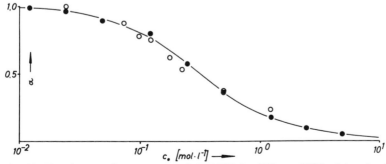

Fig. 38. Fraction α of monomers of CH_3OH in CCl_4 at 30°C, determined by the fundamental (○ at 2.75 μm) and by the first harmonic (● at 1.4 μm).

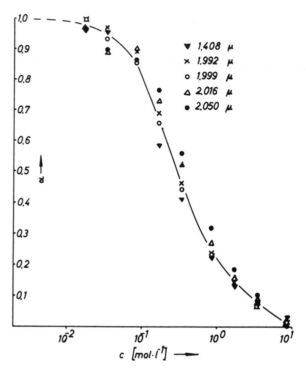

Fig. 39. Fraction α of monomers of C_2H_5OH in CCl_4 at 20°C determined by different overtone or combination bands are shown to be in good agreement.

bond formation. Barthomieu and Sandorfy[107] have shown recently that with different amines the anharmonicity $\omega_e X_e = \frac{1}{2}\nu_{02} - \nu_0$ for bands due to hydrogen bonds is reduced as compared with that for bands of free NH groups. If the anharmonicity was calculated from $\omega_e X_e = -\frac{1}{2}\nu_{03} - \frac{1}{2}\nu_{02}$; they found scarcely any difference between bands of hydrogen bonds and those of free molecules.

For various alcohols Sandorfy et al.[437,494] found an increase of 50–100% in the anharmonicity for hydrogen bond bands as compared to free OH bands if the calculation was carried out on the fundamental and first overtone, while if a similar calculation was performed on the first and second overtones, a large decrease, even to strongly negative values, was found. This result must be modified if one accepts a band analysis of the first overtone of the alcohols according to which the wavelengths of the overtone vibration of hydrogen bonds have to be adjusted to take account of the combination band at 1.53 μm (see Section 2.10).

2.8. Overtone Frequencies of Alcohols

Figures 17, 40, and 41 and Table XV show spectra of the lower alcohols in the overtone region in dilute and concentrated solutions of CCl_4 at 20°C. At 1.4 μm one observes the first overtone of the OH valence band (ν_1 = 3682 cm^{-1}). At ~1.7 μm and ~2.3 μm there are overtones of CH vibrations. Vibrations in the range of 1.9–2.2 μm correspond to combination vibrations (see Table XVIB). The position of the 1.4-μm band is relatively independent of the molecular constitution (see Ref. 887 concerning small differences among primary, secondary, and tertiary alcohols).

The CH vibrations and also the OH combination vibrations at ~2 μm show different fine structures depending on the constitution of each alcohol (see Figs. 17, 40, and 41). In contrast to H_2O, the OH band of the alcohols

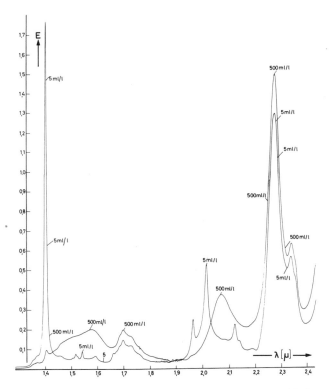

Fig. 40. Overtone spectrum of CH_3OH in CCl_4 at 20°C for two different concentrations (in ml liter^{-1}) and cd = 50 ml liter^{-1} cm = const.

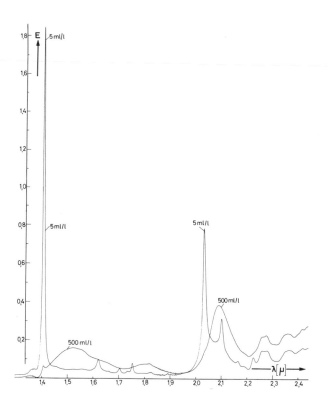

Fig. 41. Overtone spectrum of CD_3OH, same conditions as for Fig. 40. Differences appear in the region of OH–CH combination bands (1.5–1.8 μm; 2.0–2.2 μm) and in the region of CH vibrations (1.65–1.8 μm; 2.2–2.4 μm).

at ∼2 μm consists of several bands even in dilute solutions. For CH_3OH and CD_3OH, where exact information exists about the fundamental vibrational frequencies,[466,661] it is possible to make unambiguous assignments of these bands if it is assumed that one is dealing with a combination of vibrations consisting of the OH valence vibration, OH deformation vibrations, and various CH or CD vibrations (see Table XVI).

The attempted assignment in Table XVIB can also be correlated with the intensity of the bands, indicated in Table XVIA. The order of decreasing intensities in the assignments in Table XVIB is the following, both for CH_3OH and for CD_3OH: $v_1 + v_8$; $v_1 + v_3'$; v_3''; $v_1 + v_4$; $v_1 + v_8'$.

Table XVIB explains why for CD_3OH the band of shortest wavelength of this group of bands is the most intense, and why for CH_3OH line 5 (see Table XVIB and Figs. 40 and 41) is strongest.[890] The strongest line of this group is in the wavelength range 2.01–2.06 µm, as in the case of the other alcohols. One may conjecture that the intense bands are always combinations of OH valence and deformation vibrations.

TABLE XV. First Overtone Vibration of Free OH Valence Vibration in Solutions of CCl_4

R—OH	t, °C	C, liter^{-1}	λ_{max}, µm	ν_{max}, cm^{-1}
H_2O	390	Pure	1.396	7163
H_2O	20	Saturated	1.397	7158
CH_3OH	350	Pure	1.397	7158
CH_3OH	20	3	1.4043	7121
CD_3OH	20	5	1.4041	7122
C_2H_5OH	350	Pure	1.4009	7138
C_2H_5OH	20	3	1.4080	7103
$n\text{-}C_3H_7OH$	20	3	1.4065	7110
$n\text{-}C_4H_9OH$	20	3	1.4075	7105
$\begin{array}{c}CH_3\\ \diagdown\\ \quad CH-CH_2-OH\\ \diagup\\ CH_3\end{array}$	20	5	1.4066	7109
$\begin{array}{c}CH_3-CH-CH_3\\ \vert\\ OH\end{array}$	20	3	1.4114	7085
$\begin{array}{c}CH_3-CH_2-CH-CH_3\\ \vert\\ OH\end{array}$	20	3	1.4108	7088
$\begin{array}{c}CH_3\\ \vert\\ CH_3-C-CH_3\\ \vert\\ OH\end{array}$	20	3	1.4157	7063
naphthalene—OH	20	6 g liter^{-1}	1.4181	7052

TABLE XVI. Attempted Assignment of the OH Combination Vibration of CH$_3$OH and CD$_3$OH[466,661]

A. Attempted assignment of the gas spectra[466,661]

			CH$_3$OH	CD$_3$OH
ν_1	OH	Stretching	3682	3690
ν_3'	CX$_3$	Asymmetric bending	1477	1047
ν_6'	CX$_3$	Symmetric bending	1455	1134
ν_3''	CX$_3$	Asymmetric bending	1415	1075
ν_8	OH	Bending	1346	1297
ν_7''	CX$_3$	Rocking	1233	877
ν_7'	CX$_3$	Rocking	1116	858
ν_4	CO	Stretching	1060; 1034; 1012	988
ν_8'	OH	Out-of-plane bending	~790	~775

B. Attempted assignment of the second combination band in CCl$_4$ solutions

	CH$_3$OH				CD$_3$OH		
	Observed		Calculated		Observed		Calculated
	λ, μm	ν, cm^{-1}	ν, cm^{-1}	Gas spectra	λ, μm	ν, cm^{-1}	ν, cm^{-1}	Gas spectra
1.	2.193	4560	~4472	$\nu_1+\nu_8'$	2.223	4498	4465	$\nu_1+\nu_8'$
1a.	2.144	4660	4699	$\nu_1+\nu_4'$	2.187	4570	4567	$\nu_1+\nu_7''$
							4548	ν_7'
2.	2.123	4710	4721	$\nu_1+\nu_4$	2.163	4623	4678	$\nu_1+\nu_4$
3.	~2.08	~4810	4798	$\nu_1+\nu_7'$	2.098	4766	4765	$\nu_1+\nu_3''$
							4737	$\nu_1+\nu_3'$
4.	~2.059	~4860	4920	$\nu_1+\nu_7''$	—	—	—	—
5.	2.012	4970	5028	$\nu_1+\nu_8$	2.03	4926	4987	$\nu_1+\nu_8$
6.	1.965	5089	5159	$\nu_1+\nu_3'$;	—	—	—	—
			5237	$\nu_1+\nu_6'$;				
			5097	$\nu_1+\nu_3''$				

2.9. Assignment of the Side Bands of the Alcohols

A more detailed investigation of the 1.4-μm OH bands has shown that this band also displays fine structure depending on constitution. The difference, as compared with the 2-μm group of bands, is that all the alcohols have a very strong principal band at 1.4 μm while constitution-dependent satellite bands with much lower intensities appear between 1.45 and 1.65 μm (Fig. 42). The extinction coefficients at the maxima of these side bands are lower by a factor of 50–100 than those of the principal band at 1.4 μm

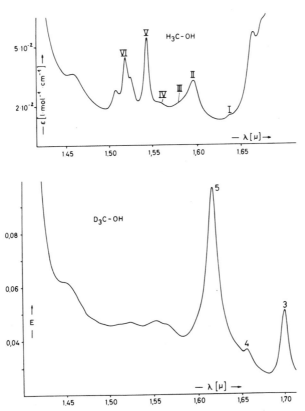

Fig. 42. In dilute solutions of alcohols in CCl$_4$ one can observe very weak side bands of the free OH band. They can be discussed as combination bands of OH and CH, respectively, with CD. The figure shows free OH (and OD) side bands in spectra of dilute solutions of alcohols in CCl$_4$ (see text). Top: H$_3$C—OH, 7.411×10^{-2} mole liter^{-1}; bottom: D$_3$C—OH, 5 ml liter^{-1}, 10 cm.

and for this reason were not considered in most of the publications. We were able to observe them very clearly by using a Cary Photometer Model 14, which possesses an extinction range from 0 to 0.1 over the entire scanning width.

At the third overtone (0.97 μm) of phenol, Kreuzer and Mecke[793] observed side bands in the range of the association bands and these they attributed to the alkyl group. Satellite bands similar to those shown in Fig. 42 can be seen in the published spectrum of pentan-3-ol.[1231]

These side bands for methanol were discussed and interpreted by Fletcher and Heller[491] as a combination of free OH and CH vibrations. The side bands vanish with increasing concentration, as does the principal band at 1.4 μm. Quite obviously they therefore belong to the system of free OH vibrational bands. They can scarcely be caused by impurities or association. In the same frequency range similar groups of bands appear in the vapor. The side bands are dependent on the groups neighboring the OH group. They resemble each other in the primary alcohols and the same is true in the secondary alcohols. This dependence on neighboring groups is very clearly shown by the spectrum of D_3COH (Fig. 42). We conjecture

TABLE XVII. Attempted Assignment of OH Sidebands 1.4–1.75 μm

			CH_3OH				CD_3OH		
			λ, μm	ν_{obs}, cm^{-1}	ν_{calc}, cm^{-1}		λ, μm	ν_{obs}, cm^{-1}	ν_{calc}, cm^{-1}
I.	1.	$+ \nu_x$	1.635	6100	6080	1.	1.747	5724	5758
II.	1a.	$+ \nu_x$	1.615	~6180	6180	—	—	—	—
II.	2.	$+ \nu_x$	1.596	6265	6230	2.	1.732	5780	5780
III.	3.	$+ \nu_x$	~1.58	~6340	~6330	3.	1.70	5880	5883
IV.	4.	$+ \nu_x$	1.56	~6400	~6380	4.	1.6505	6050	6026
V.	5.	$+ \nu_x$	1.542	6485	6490	5.	1.6165	6186	6186
			1.525	6557	—		—	—	—
VI.	6.	$+ \nu_x$	1.516	6600	6609		—	—	—
			1.506	6640	—		—	—	—

$$\nu_x = 1520 \text{ cm}^{-1} \qquad\qquad \nu_x = 1260 \text{ cm}^{-1}$$

that these so called "side bands" correspond to a higher combination vibration of the 2-μm group of bands.

For CH_3OH it is possible to couple the side bands between 1.4 and 1.7 μm with the OH band at ~2 μm if, according to Table XVII, one adds 1520 cm^{-1} to each of the components of the 2-μm band (see also Fig. 42). A similar coupling is possible for CD_3OH by adding 1260 cm^{-1}. The OH deformation vibration ν_8 in the vapor state is given as 1345 cm^{-1} for CH_3OH[661] (1455cm^{-1} in the liquid state) and 1297cm^{-1} for CD_3OH.[466] There is a CH_3 vibration[661] at 1477 cm^{-1} in the vapor state for CH_3OH and one at 1145 cm^{-1} for a CD_3 vibration[466] in CD_3OH [$\nu_2(e)$ for CH_4 is 1526 cm^{-1}; $\nu_2(e)$ for CHD_3 is 1299 cm^{-1}[466]]. This assignment also corresponds to the sharpness of the bands II, V, VI, 3, 4, and 5 (Fig. 42).

It is not impossible that other additives to the frequencies of the group of bands at ~2 μm (see Table XVIA) exist. We merely wanted to show that a certain correlation exists between the 2-μm and the 1.5–1.7-μm groups, without claiming an unambiguous correlation in detail. Also, as far as intensities go, the 1.6-μm group could very well be the next overtone vibration of the 2-μm group.

Such side bands can be shown to exist also in the 1.45-μm NH overtone vibration of pyrrole, where they have been observed in dilute solution at 1.505, 1.555, and 1.585 μm.* There is also at 1.466 μm a shoulder of distinct resolution (called band 2) belonging to the NH main band (called band 1). This shoulder has its equivalent with CH_3OH and CD_3OH at 1.455 μm. Pauling[1078] conjectured that the NH main band is due to an NH bond, with NH in the plane of the molecule. The sideband at 1.466 μm would correspond to a vibration with the NH bond not in the plane of the system of resonance.

The three side bands mentioned all have a slight shoulder corresponding to the main band. It is interesting that in the case of pyrrole the hydrogen bond frequency shift is much smaller than with OH vibrations; thus quite clearly different bands due to hydrogen bonding are responsible for the individual side bands (see Table XVIII). From column 6 one can see that to each side band there corresponds a separate hydrogen bonding band with a frequency shift of 60 cm^{-1} relative to the free vibration. The intense main band, on the other hand, shows a somewhat greater shift between free NH vibration and vibration of NH bound in the hydrogen bond, namely 80 cm^{-1}. Similar results were found for several alcohols.

---

* Wulf and Liddel[1439] interpret these side bands as "interaction effects" between OH and CH groups.

TABLE XVIII. Wavelengths of the First Vibrational Overtone Band of NH in Pyrrole

Band 1						Band 2		
Free NH			Bound NH			$\nu_{free} - \nu_{bound}$		
λ, μm	ν_1, cm^{-1}	$\Delta\nu$	λ, μm	ν, cm^{-1}	cm^{-1}	$\sim\lambda$, μm	$\sim\nu_2$, cm^{-1}	$\sim\nu_1 - \nu_2$
1.4576	6860	—	1.4747	6781	79	1.467	6817	~43
1.5053	6643	217	1.519	6583	60	1.5115	6615	~28
1.554	6435	208	1.569	6373	62	1.559	6414	~21
1.584	6313	122	1.599	6254	59	(1.5925?)	(6279?)	(34?)

2.10. Alcohol Bands Due to Hydrogen Bonds

The foregoing helps to explain the observation that most alcohols have a band due to hydrogen bonding with a band maximum at about 1.58 μm (Fig. 40), while CD_3OH, in concentrated solutions, has such a band with a maximum at about 1.518 μm (Fig. 41).

According to the above discussion it is likely that the band due to hydrogen bonding in the 1.4-μm main band is at 1.5 μm for most alcohols, while, as shown above, the side-band system with its own band due to hydrogen bonding displaces this maximum to 1.58 μm. This explains why with pure methanol such a band appears at 1.58 μm, while the ice band at 1.492 μm is at a much shorter wavelength.[890] Much larger frequency shifts between free OH vibrations and H bond vibrations cannot be expected between CH_3OH and H_2O because from the matrix technique we know that bands due to H bonds have the same frequency shift in both substances and therefore also a similar hydrogen bond energy. This discrepancy between the matrix result for the fundamental frequency and for the first overtone at 1.4 μm may be explained by superimposition of the bands due to hydrogen bonding and the side-band system with CH_3OH. This is directly indicated by the spectrum of CD_3OH.

Table XIX contains a classification of the wavelength positions of different maxima in the range of bands due to hydrogen bonding for highly concentrated solutions. The table also contains a comparison with free OH vibrations and frequency shifts. In some cases the position of the

TABLE XIX. Assignment of the Bands Due to Hydrogen Bonding

Molecule	Free OH, μm	Hydrogen-bond, μm	Δν, cm⁻¹	Free OH, μm	Hydrogen-bond, μm	Δν, cm⁻¹	Free OH, μm	Hydrogen-bond, μm	Δν, cm⁻¹	Free OH, μm	Hydrogen-bond, μm	Δν, cm⁻¹
CH_3OH	1.404	~1.5	450	1.54	1.576	135	2.04 / (1.96)	2.073	140 / (270)	2.123	2.19	150
CD_3OH	1.404	~1.52	530	1.617 / 1.747	1.66 / 1.8	180 / 205	2.02 / —	2.09 / —	207 / —	2.103	2.193 / —	200 / —
C_2H_5OH	1.408	~1.5	440	1.533 / —	1.58 / —	160 / —	2.0 / 2.05	2.08 / ~2.14	200 / ~210	2.12 / 2.14	2.19 / 2.214	180 / 150
$n\text{-}C_3H_7OH$	1.406	~1.5	450	1.53	1.59	230	1.99	2.08	220	2.11; 2.12	2.19	160; 180
$i\text{-}C_3H_7OH$	1.411	~1.495	385	1.535 / —	1.58 / —	180 / —	2.003 / 2.053	2.09 / 2.112	215 / 230	— / 2.12	— / 2.206	— / 180
$n\text{-}C_4H_9OH$	1.405	~1.515	510	1.532 / —	1.59 / —	230 / —	1.995 / —	2.09 / —	250 / —	2.107; 2.12 / 2.06	2.206 / 2.175	200; 180 / 240
$sec\text{-}C_4H_9OH$	1.411	1.5	430	1.538 / —	1.59 / —	220 / —	2.01 / 2.056	2.09 / ~2.175	210 / ~270	2.11	2.2	200
$tert\text{-}C_4H_9OH$	1.4157	1.495	360	1.518	~1.57	220	2.023	2.09	260	2.07; 2.103	2.17; 2.2	2.30; 2.10
CH_3OD	1.888	2.009	330	—	—	—	2.012	2.05	100	—	—	—
CD_3OD	1.888	2.014	340	—	—	—	2.03 / 2.1	2.13 / ~2.19	240 / ~190	—	—	—

maximum of the band due to hydrogen bonding *cannot* be very accurately given. This is particularly true when there is superimposition of two bands due to different hydrogen bonds.

What strikes one here is that the strong main band at ~1.4 μm has a band due to H bonding displaced by about 400–500 cm⁻¹. The second band due to hydrogen bonding at ~1.58 μm may be attributed to a band displaced by about 200 cm⁻¹, the band being a satellite band at 1.5 μm. This shift is much smaller than the one suffered by the 1.4-μm band. This corresponds to results with pyrrole (see Table XVIII). The H bond bands of alcohol near 2.1 μm correspond to a frequency shift of 200 cm⁻¹ as against the free OH vibrations. We must remember that the frequency shift between bands due to free OH and bands due to hydrogen bonds increases with increasing order of the valence overtone vibrations.[880] According to the attempted assignment of the satellite bands, these belong to a combination vibration between simple OH valence and bending vibrations. The 1.4-μm main band belongs, however, to the second overtone of the OH valence vibration. This differential frequency shift of the 1.4-μm main and satellite bands seems understandable.

This attempt at interpretation could also explain the relatively high intensity of the bands due to hydrogen bonds of the satellite system (~1.58 μm) compared to the bands due to hydrogen bonds of the 1.4-μm main band (~1.5 μm). The extinction at the maximum of the 1.4-μm main band is much larger than that of its association band (~1.5 μm). In the case of the 2-μm vibrations the extinctions of the maxima of the hydrogen bond and of the free bands are of the same order of magnitude. If the 1.58-μm bands belong to the free satellite bands, and thus only to simple OH vibration rather than to valence overtone vibration, their extinction maximum should be larger than that of the 1.5-μm bands belonging to the 1.4-μm main band.

Further indications of the complexity of the intensity relationships of H bond bands are observations made on the HOD spectrum (Fig. 43).[886] Apart from a band at ~1.56 μm ($2\nu_3$), there is another very strong band caused by H bonding at 1.71 μm [ν_3(OH) + ν_3(OD)]. With normal ice one also observes $2\nu_3$ (1.493 μm) and $2\nu_1$ (1.56 μm) as a weak shoulder. These observations show that quantitative evaluation of hydrogen bonding bands have to be performed with greater care than the very simple evaluation of the narrow free OH bands. When the extinction at a particular wavelength is taken as a measure of the closed hydrogen bonds it may not seem important whether the intensity of the band at this wavelength is caused by one or several different bands as long as all the bands react in the same way to hydrogen bonds when there is no Fermi resonance to consider and when the hydrogen bonds are of the same kind. The last condition is particularly important.

With some alcohols, particularly with CH_3OH (see Fig. 40), the absorption range of the broad hydrogen bonding band overlaps with the absorption range of the CH vibrations at ~ 1.7 μm. It is not impossible that Fermi

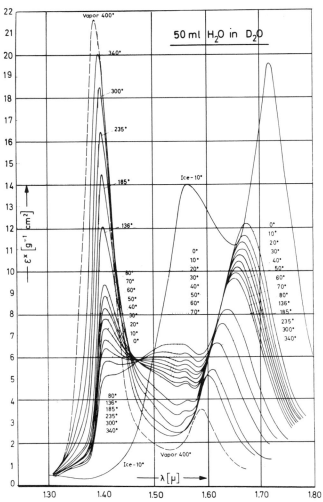

Fig. 43. Temperature dependence of the HOD spectrum (50 ml H_2O/liter D_2O under saturation conditions). 1.4 μm, free $2\nu_3(OH)$ vibration; 1.56 μm, linear bonded $2\nu_3$ vibration; 1.71 μm, $[\nu_3(OH) + \nu_3(OD)]$ H-bonded vibration.

TABLE XX. Vibrational Frequencies for CH₃OH and C₂H₅OH in cm⁻¹

	Monomers			Dimers		Linear hydrogen bond[a]		
	ν_{01}	ν_{02}	ν_{03}	ν'_{01}	ν'_{02}	ν''_{01}	ν''_{02}	ν''_{03}
CH₃OH	3642	7121	10443	3510	~6870	3340	6600	(~9760)
C₂H₅OH	3632	7102	10391	3500	~6825	3330	6580	(~9840)

[a] The analysis of the second overtone vibration ν''_{03} suffers a perturbation by a structure not yet fully elucidated, of the hydrogen bonding band. We are investigating this point.

resonance plays a part in boosting the intensity of the CH band in concentrated solutions.

Table XX shows the values obtained according to our analysis of the first and second overtone bands of the alcohols.

The values with a single prime refer to the energetically unfavorable bonds, and those with a double prime to linear hydrogen bonds. Table XXI shows the resulting values for the anharmonicity. Here $(\omega_e X_e)_{0102} = \frac{1}{2}\nu_{02} - \nu_{01}$ and $(\omega_e X_e)_{0203} = \frac{1}{3}\nu_{03} - \frac{1}{2}\nu_{02}$.

According to Table XX, the anharmonicities $(\omega_e X_e)''_{0102}$ of bands of optimal bonds are about half as large as $(\omega_e X_e)_{0102}$, those of free OH groups. With the aid of our band analysis we thus obtain very similar results for alcohols and amines. It is then easy to calculate the corresponding harmonic frequencies ω_e (see Table XXII):

$$\omega_{e0102} = \nu_{01} + 2(\omega_e X_e)_{0102}, \qquad \omega_{e0203} = \frac{1}{2}\nu_{02} + 3(\omega_e X_e)_{0203}$$

One may summarize by saying that overtone bands are more suitable for the determination of intensities of free OH or NH bands than are the fundamental frequencies. Three factors speak for this conclusion: higher photometric accuracy of the overtone band instruments, greater frequency shift, and lower intensities of the bands due to H bonds in the overtone

TABLE XXI. Anharmonicities in CH₃OH and C₂H₅OH in cm⁻¹

	$(\omega_e X_e)_{0102}$	$(\omega_e X_e)_{0203}$	$(\omega_e X_e)_{0'102}$	$(\omega_e X_e)_{0'1'02}$	$(\omega_e X_e)_{0'2'03}$
CH₃OH	82	79	~75	40	(~45)
C₂H₅OH	81	86	~88	40	(~10)

TABLE XXII. Harmonic Frequencies of CH_3OH and C_2H_5OH in cm^{-1}

	Monomers		Linear hydrogen bonds	
	ω_{e0102}	ω_{e0203}	ω''_{e0102}	ω''_{e0203}
CH_3OH	3806	3798	3420	(3435)
C_2H_5OH	3794	3809	3410	(3320)

range. The last two factors reduce the errors caused by superimposition of free vibration and bond vibration.*

The fundamental frequency is more suitable than the overtone frequency for studies of hydrogen bonding bands.

2.11. Toward a Theory of the Hydrogen Bond

From experimental studies it is known that free electron pairs, particularly in the first row of the periodic system, may act as acceptors for hydrogen bonds. In Section 2.2 a few examples of the connection between this acceptor function and the basicity of free electron pairs were given. Similarly, the donor function of the X—H group runs parallel with the acidity and the free position of the proton. Apart from the well-known OH, NH, or FH groups as hydrogen bond donors, there are weaker hydrogen bonds, such as in chloroform. Even —C≡CH groups can form hydrogen bonds.

When studying these bonds we obtain the total interaction effect by taking the sum of the donor and acceptor effects of the XH group in question. In the case of substances forming strong hydrogen bonds, one may scarcely be able to distinguish whether it is the acidity of the protons or the basicity of the free electron pairs which is decisive for the particularly strong interaction. Mixtures of H-bond-forming substances may give more detailed indications. Figure 44 reproduces the overtone vibration of $0.5\,m$ CH_3OH and $0.5\,m$ phenol in CCl_4. When the concentration of each alcohol is only $0.25\,m$ the mixture absorbs less in this range of "free" OH vibration than the sum of the half-band-intensities would predict (circles in Fig. 44). In the range of the CH_3OH band the total intensity is somewhat increased, but only on the long-wave side of the maximum. Particularly noticeable

---

* For discussion of superimposition of bands, especially in the evaluation of water spectra see Ref. 891.

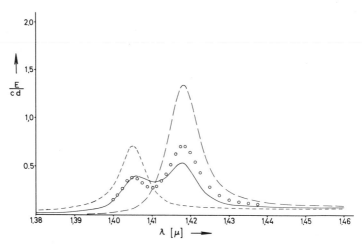

Fig. 44. Nonadditivity of H bond interactions in mixtures (20°C, CCl$_4$).
0.5 m methanol (- - -); 0.5 m phenol (— —); 0.25 m phenol + 0.25 m
methanol: experimental (——), theoretical (○ ○ ○).

in these mixtures is the lower intensity in the range of the phenol band.
Two reasons may be advanced for this: The acceptor effect of the free
electron pair in methanol is stronger than that in phenol; in the case of
phenol the acceptor effect is reduced owing to steric factors or because of
the π electrons of the aromatic ring.

Figure 45 shows that the decrease of the total area under the band
(integrated over the "free" methanol and phenol bands) is larger for highly
concentrated than dilute solutions. Incidentally, the first observations of
this kind were made by Kuhn[807] on the fundamental vibration of the
system phenol–*tert*-butanol. Particularly strong decreases of the band
intensity in mixtures *tert*-butanol–phenol have been observed by us over
the whole concentration range in the case of the overtone vibration. Our
results for the system ethanol–phenol were the same for the overtone
vibration as those for methanol–phenol already described in detail.*

---

* Several authors have found a proportionality between Δv of the band due to H bonding
and the Hammet constant σ. (See, e.g., Refs. 295, 456, and 1020.) Concerning the
connection between Δv and the pK_a values of various acceptors, see Ref. 1261. An
interesting example of the finding that the H bonds of a mixture can be stronger than
the bonds of the single components is provided by the bases of RNA. Küchler and
Derkosch[803] have shown spectroscopically that the heteroassociate adenine–uracil
is more stable in solutions than the homoassociates adenine–adenine or uracil–uracil.
This result has been confirmed with the IR method by Kyogoku et al.[817] and by an
NMR method by Katz.[734]

In the system *tert*-butanol–methanol the intensity of the 1:1 mixture of 0.1 *m* solutions was that predicted by the theory. In contrast, with *tert*-butanol–pyrrolidone the intensity of the butanol band is less than expected, while that of the pyrrolidone band corresponds to expectations. Pyrrolidone is therefore a stronger acceptor than *tert*-butanol.

The simplest interpretation of the hydrogen bond interaction effect is the assumption of a dipole–dipole interaction. Using the simplest dipole–dipole models, Pimentel and McClellan[1089] have been able to represent correctly not only the absolute magnitude of the hydrogen bond energy but also a certain angle dependence. We have estimated the angle dependence of the hydrogen bond energy ΔH and have found it to correspond to simple dipole–dipole models (see Fig. 46). For large angles the cyclic dimer curves have been specially added to the figure,

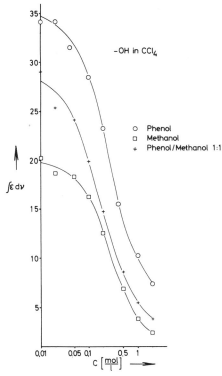

Fig. 45. Plot of $\int \varepsilon \, \partial \nu$ of the band areas of the free OH vibration in experiments of the type of Fig. 44. The nonadditivity can be recognized, especially at high concentrations.

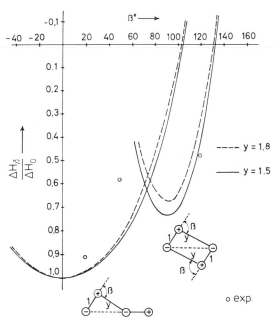

Fig. 46. Theoretical angle dependence of the dipole-dipole interaction for different dipole lengths y and different orientations. Open circles refer to H bond interaction energies determined by the matrix technique.

allowing for the assumptions inherent in the interpretation of the matrix experiments.

The simplest representation of the molecules as dipoles is, of course, too approximate. Lennard-Jones and Pople[838] and Schneider[1195] have suggested somewhat improved point-charge models. The position of the point charges can be selected in conformity with quantum mechanical ideas and the Coulomb interaction of such point charges can then be discussed. Besides this, a series of experiments has been undertaken to calculate hydrogen bonding quantum mechanically, and a recent survey has been provided by Lin.[815]

Particularly remarkable is the suggestion, discussed by Frank and Wen,[505,512] that cooperative mechanisms by charge structures may be involved in the case of large hydrogen bond aggregates. This picture is able to explain, at least qualitatively, many experimental findings. It is particularly in harmony with the observation in alcohols of weaker association of the dimers compared to the higher aggregates. However, this can also be understood as an additive dipole chain. We have not yet found direct spectroscopic indications of the cooperative element of hydrogen bonding. Most of the experiments can be explained on the simpler hypothesis, or so it seems to us, of the angle dependence of hydrogen bonds.

2.12. Hydrogen Bonds in Pure Liquids

Much can be learned about the state of hydrogen bonds in pure liquids by extrapolation from concentrated solutions. For instance, from Fig. 10 it follows that with most hydrogen-bond-forming substances the proportion of hydrogen bonds not formed (but potentially capable of being formed) is less than 10% at room temperature. With methanol there are only about 2% free OH groups present. This statement, which follows from observation on easily accessible solution concentrations, contradicts some theories of liquids. For further statements about the state of pure solvents there are two methods of approximation. First one may neglect changes of the extinction coefficients at the transition to pure liquid and estimate the order of magnitude of the proportion of open hydrogen bonds from the extinction coefficients of the band maxima. On the other hand, one can study the temperature dependence of the spectra of the pure solutes. As an example for the first method, the estimation of hydrogen bonds in polyamides may be mentioned. Figure 47 shows the overtone vibration spectra of N-ethyl-acetamide. At 1.455 μm there is the maximum for free NH vibrations, at 1.53 μm the corresponding band of hydrogen bonding. At 1.55 μm there is a further weak NH band, the hydrogen bond of which absorbs at 1.59 μm. In Fig. 48 are shown a few spectra of pure N-ethylacetamide at different temperatures; also shown are the free NH bands for dilute solutions, the ordinates of which are reduced by a factor of four. With increasing temperature the extinction of the sharp solution bands of the unbound NH vibrations rises. In Fig. 49 an estimate is given of the proportion of free NH groups in these solutions as a function of temperature, from the extinction at the maximum of the free NH vibration. Samples of solid 6-polyamide have a similar spectrum to that of liquid N-ethylacetamide.*

Assuming the extinction coefficients to be similar in both systems, then the course of the curves in Fig. 49 follows. The two upper curves correspond to polyamide samples with cyclohexyl rings adjacent to the NH group. This obviously loosens the structure in a way that leads to the formation

---

* When evaluating the spectra of solid polyamide samples a correction was applied to the concentration of the amide groups, derived from the temperature dependence of the density: $C_t = (\varrho_t/\varrho_{20})C_{20}$; also the temperature dependence of the layer thickness d has been taken into account, $d_t = d_{20}(\varrho_{20}/\varrho_t)^{1/3}$.

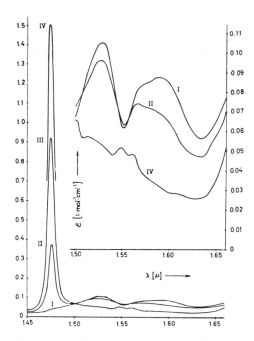

Fig. 47. Extinction coefficients of the NH overtone band of N-ethylacetamide (20°C, CCl₄
solutions). Inset: H bond band on an enlarged
scale. (I) 10.53, (II) 2.106×10^{-1}, (III) 5.265
$\times 10^{-2}$, (IV) 1.053×10^{-2} mol liter⁻¹.

of fewer hydrogen bonds. The temperature dependence of free NH groups
is surprisingly similar in liquid ethylacetamide and in 6-polyamide. This
is true also for the absolute value if one allows for the crystalline regions
in the polyamide samples (known from X-ray results). This means that the
hydrogen bond is in the same state in the amorphous regions of 6-poly-
amides and in N-ethylacetamide.

A comparison of the spectra of dilute solutions and of pure liquids at
moderate temperatures usually shows an increase in intensity of the long-
wave flank of the free XH band for pure liquids. It seems obvious on the
grounds of previously discussed effects of angle and distance dependence
of hydrogen bonds to ascribe this effect to the increased importance of
angle- and distance-dependent hydrogen bonds in pure liquids. As similar
conditions concerning distance dependence* should also exist in solutions,

* Even in CCl₄ solutions of H₂O–dioxane or H₂O–CH₃OH mixtures this asymmetry of
  free OH bands is scarcely observable even at 20°C.[894]

(although the lattice theory of liquids implies a not too broad distribution over various distances), we are inclined to believe that angle dependence and orientation defects are important parameters in pure liquids. In the case of phenol there is a particularly striking broadening of the free OH vibration.

It is also striking that the asymmetry induced by the broadening of the long-wave flank diminishes somewhat with increasing proportion of non-polar groups. In *n*-butanol or cyclohexanol[890] this asymmetry is somewhat less than in ethanol,[505,512] and in ethanol a little less so than in methanol. Greater asymmetries are observed for pyrrolidone[890] and cyclooctanone oxime. One may conjecture, especially in the alcohol series, that the proportion of free OH groups increases in the higher alcohols in nonpolar surroundings, while in methanol the proportion of open OH groups will increase in the vicinity of polar groups or of orientation defects with un-

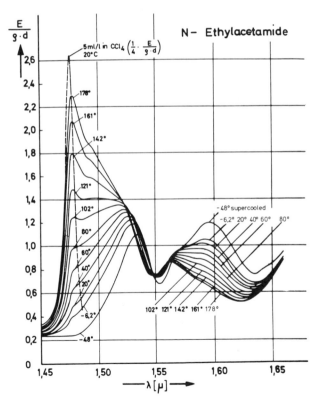

Fig. 48. Temperature dependence of the NH overtone band of N-ethylacetamide. Broken line: free NH band of very dilute solutions in CCl_4 (ordinate scale reduced by a factor of four).

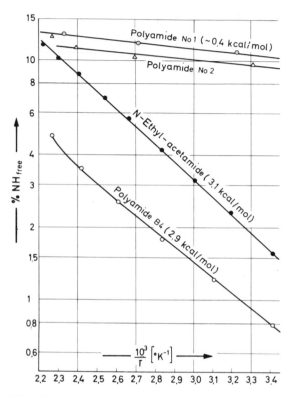

Fig. 49. Estimated percentages α of free NH groups. The temperature dependence of α is the same with N-ethylacetamide and 6-polyamide. By subtraction of the crystalline portions of 6-polyamide, the values for α become identical.

favorable, weak hydrogen bonds, through interaction effects with neighboring OH groups. In H_2O the asymmetry of the "free" OH vibration is correspondingly even greater (see Fig. 8 in ref. 893). If one neglects the asymmetry of the free OH vibrations by taking account only of the short-wave branch of the bands up to the maximum, one finds that this "half-width" and also the position of the maximum are very similar in pure liquids and in their dilute solution.

It can be concluded that in pure solvents one may speak unambiguously of free OH or NH groups. Table XXIII gives some half-widths of the first overtone vibration (measured up to the maximum). One notices a greater band broadening of the OH vibration in pure phenol compared to the solution. This is obviously due to interaction of OH groups which,

TABLE XXIII. Half-widths $\frac{1}{2} \Delta\nu_{1/2}$ (in μm)

	CCl$_4$ solution	Pure liquid
n-Butanol	0.0058	~0.006
Cyclohexanol	0.006	0.006
Cyclooctanone oxime	0.005	0.006
Pyrrolidone	0.005	0.008
Methanol	0.0036	0.015
Ethanol[a]	0.0055	0.008
Phenol	0.0044	0.015
Butan-1,4-diol	0.005	0.01

[a] The increased half-width in ethanol solutions is attributable to the different conformations of the OH groups.[892]

although not forming hydrogen bonds with neighboring OH groups, interact with the π electrons of neighboring molecules. This statement is confirmed by the even greater broadening of the "free" OH bands of butanol, where the band is split in dilute solution because of different conformations and because of the intramolecular hydrogen bond to the double bond.[887] Peaks or maxima appear in dilute solutions at 1.4075, 1.413, and 1.425 μm.[887] From the spectra of pure butanol one recognizes that temperature dependence of the intramolecular hydrogen bond formation contributes to the broadening of the OH band. A similar effect is shown by the spectra of propan-1,3-diol and butan-1,4-diol. It is known that the intramolecular hydrogen bond of propanediol induces a smaller frequency shift with respect to the free OH vibration than does that of butanediol. Correspondingly the asymmetry of the OH band at 1.4 μm in propanediol is greater than in butanediol. In the latter not only does the band due to the intramolecular hydrogen bond suffer a larger shift to longer wavelengths but also, owing to the entropy factor associated with the more numerous conformational possibilities, the band is less intense. The influence of intramolecular hydrogen bonds on the proportion of free OH groups is clearly seen when one compares butan-1,4-diol with butin-1,4-diol; the latter shows larger amounts of free OH.

2.13. Determination of Free OH Groups in Alcohols and in Water

Even though the similarity of half-widths and band maxima in pure liquids and in dilute solutions is a fact, the method described in the previous section for determining free OH groups in the range of the free OH vibration from the extinction coefficients of solutions and the band intensities of the pure liquids is somewhat unsatisfactory, and a different method for the determination of free OH groups might be studied. One way would be to carry out investigations at high temperatures and densities, in the range of the critical density (Fig. 43). In many investigations on water it has been shown that at the critical density and for temperatures above the critical point the band intensity in the range of the free OH vibration remains constant.[877,880,893]

In an approximation process one may choose therefore the super-critical state as the standard state for the liquid, corresponding to completely open hydrogen bonds. On the other hand, the height of the ice band scarcely changes at low temperatures, and therefore ice can be chosen as the standard state for completely closed, linear, optimal hydrogen bonds. Under these two assumptions, one may try to obtain approximate values for the proportion of open OH groups from the band intensity in the range of the maximum for supercritical water; in this way Fig. 50 is obtained.[877,878,880,890,891] Deductions about the state of the several possible hydrogen bonds are much more uncertain than Fig. 50 would indicate. Obviously the situation

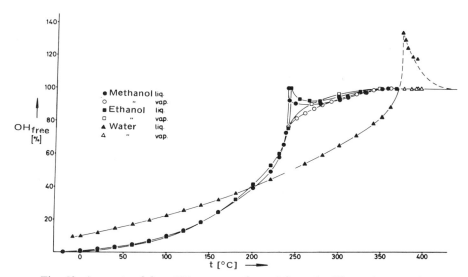

Fig. 50. Amounts of free OH groups estimated from the IR overtone spectra.

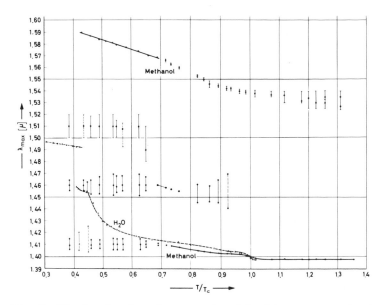

Fig. 51. Wavelengths of maxima or peaks of the 1.4-μm overtone band of H_2O (×) and CH_3OH (●) determined from the temperature dependence of the bands under saturation conditions. The peaks near 1.45 μm are in the region of angle-unfavorable H-bond bands.

is one of distribution over several states, as discussed, e.g., by Falk and Ford[465] or Pople.[1109]

On the other hand, one finds in water spectra, as well as in those of methanol or ethanol, certain band peaks which are indicated in Fig. 51. Besides the clearly distinguishable peaks in the range of linear hydrogen bonds (about 1.5 μm) and maxima (or peaks at low temperature) in the range of the free OH vibration (1.4 μm), further peaks have been observed experimentally in the range 1.45–1.46 μm. This is exactly the range where, according to the matrix theory, the maxima of the cyclic dimers could be expected.*

According to the angle dependence of the hydrogen bond energy discussed in Section 2.4, one should expect, even for the case of a purely statistical distribution over all possible bond angles (Fig. 32), this medium range of angles to be preferred. Since for energetic reasons particular angles are preferred (e.g., the zero bond angle), the distribution function over the

* The band 1.53–1.59 μm belongs (as discussed above) to a different band system. In ethanol, too, there are three maxima or peaks for this region. Obviously this band also marks three such limiting states.

different hydrogen bond states has distinct maxima, and there is thus an additional reason why certain maxima should appear more clearly in the overtone vibrational spectra where overlapping is less pronounced. Studies of fundamental vibrations associated with smaller frequency shifts and very high intensities of the bands due to hydrogen bonding are so uninformative that they cannot be used to confirm or contradict the validity of the results obtained from experiments in the overtone range. Even if one were to take the extreme position of considering a continuum of states in liquid water, it must be conceded that this "continuum" reaches a limit at the concept of the free OH groups. There is thus virtue in considering the state of free OH groups as the state of perfect orientation defects, whether one starts from the continuum theory of liquids or from models based on the existence of various species.

From the point of view of spectroscopy, such an interpretation is more exact because, owing to superimposition of the bands, it is very difficult to analyze the different hydrogen bond states. The question of how far band superimposition in the case of hydrogen bonds interferes with the determination of free OH groups can be answered for the overtone range, where very similar results for the proportion of free OH groups in the case of four different H_2O overtone bands and one HOD band have been obtained all within the limits of experimental error.

Some sceptics may object to this simple method of approximation and argue that insufficient account has been taken of the difference of absorption by free OH groups in different environments (solution, vapor, polar, non-polar, etc.). It must, however, be remembered that $\int \varepsilon \, dv$ is scarcely changed even by large changes in the bands.[892] If one changes the pressure of methanol at supercritical temperatures, relatively large changes are observed in the band, owing to the appearance of a certain vibrational structure; nevertheless $\int \varepsilon \, dv$ remains unchanged. This also holds for the transition to dilute solutions in CCl_4 at room temperature (see Table XXIV).

Similarly, according to Table XXIV, $\int \varepsilon \, dv$ remains constant for saturated vapors at different temperatures; in these measurements hydrogen bond formation may be neglected. It is therefore advisable to check the results in Fig. 50 by a determination of $\int \varepsilon \, dv$. In order to avoid superimposition of bands with those due to hydrogen bonding, it is also advisable to determine $\int \varepsilon \, dv$ only on the short-wave side of the band maximum. By following these procedures, it is found that the results from Fig. 50 remain unchanged.[894] By making such changes in the method of approximation, the reliability of the results is further enhanced.

TABLE XXIV. Band Areas of Methanol Spectra

Density, ϱ/ϱ_c	t, °C	Area	Density, ϱ/ϱ_c	t, °C	Area
0.1	314	1040	0.024	174	1145
0.25	307	1097	0.19	201	1150
0.5	310	1066	0.32	220	1100
1	314	1044	(1)	(243)	(875)
			1	377	1020
Solutions					
2 g liter^{-1} CCl$_4$	20	1066			
2 g liter^{-1} CS$_2$	20	930			

We shall here add a few remarks about the appearance of the curves in Fig. 50. In many experiments a density gradient could be observed just above the critical temperature T_c which extended up to $T > 1.1T_c$.[889] This temperature range is known to be anomalous for all the thermodynamic quantities. One may discuss liquids near their melting point as solids with defects[878,896] and for nonpolar liquids these defects can be regarded as holes, in the hole model of liquids.[896] In the case of liquids with hydrogen bonds, however, the essential feature is the orientation defect of broken hydrogen bonds. The range $T_c < T < 1.1T_c$, with densities near ϱ_c, can be regarded as the range where there is still a residual existence of a liquidlike state. Whereas for water one observes, particularly in the vapor phase, at $T > T_c$ and ϱ_c a constant extinction in the range of free OH vibrations, with alcohols the extinction continues to increase up to about 360°C.[892] This observation implies that in alcohols not all the hydrogen bonds are open at T_c, whereas they must be open at temperatures corresponding to T_c of water. This agrees the result that the hydrogen bonding bands in alcohols display the same frequency shift relative to the free OH vibration as is the case in water. According to the Badger–Bauer rule we have to draw the conclusion that the bond energy for one hydrogen bond is very similar in water and in the alcohols. From observations at the critical temperature we have to infer that T_c for water is essentially determined by the hydrogen bonds. For alcohols, however, T_c is determined both by the hydrogen bonds and by the dispersion forces between the nonpolar groups. In connection with the spectroscopically estimated proportion of free

OH groups, one has to remember that free OH groups in free molecules differ very little from the OH groups of molecules the other parts of which (second OH group or free electron pairs) undergo hydrogen-bond interaction effects. From the calculated free OH groups one cannot therefore infer the proportion of free molecules of monomers; in fact small wavelength changes of band maxima near T_c (see Fig. 51) seem to indicate that below T_c the free OH groups in molecules, the other parts of which are bound, preponderate, while only at T_c do free molecules in the proper sense begin to appear in appreciable proportions.*

Water bands can be constructed at all temperatures in all their details by the addition of three bands, as can be easily demonstrated with the du Pont curve analyzer.[893] This finding supports the idea of accumulation of certain absorption bands, as distinct from a complete continuum—be it in the sense of accumulation of certain frequencies (see Fig. 32) or of preference for certain bond angles.

The spectroscopically determined proportion of free OH groups contradicts some published theories on the structure of water. These theories usually assume a much higher proportion of free OH groups. One may comment that most theories are adjusted to a number of constants and consider only temperatures below 100°C. From the magnitude of the heat of fusion alone it becomes apparent that near the melting point there cannot be such a high proportion of free OH groups as some of these theories have assumed. It is remarkable, on the other hand, that the thermodynamic properties of the alcohols and of water can be quantitatively calculated, without any arbitrary constant, up to the critical temperature purely from the spectroscopically determined proportion of free OH groups.[883] It is even sufficient to take a simplified two-species model for the purpose. The molar specific heat can be expressed, e.g., by the following equation:

$$c_V = c_{V,\mathrm{id}} + (\partial p/\partial T)\,\Delta H_\mathrm{H} + \tfrac{1}{2}Z(T)R$$

where $c_{V,\mathrm{id}}$ is the specific heat of H_2O vapor in the ideal state. No arbitrary constants appear, only the differential derived from the experimentally measured proportion p of free OH groups. Spectroscopic experiments[883] to determine the proportion of different hydrogen bond states have first of all to be brought into harmony with all the observations described in the previous sections, which has not always been the case with statements in the literature, and in the second place they are not unequivocal,

---

* Jakobsen et al.[704] arrive at the same conclusion from spectroscopic studies at higher pressures with heptan-3-ol and pentan-3-ol.

especially if all the extinction coefficients are unknown.[888] A determination of the extinction coefficients of the various hydrogen bond states has not so far proved possible.

As Vand and Senior[1324] have shown theoretically, and the broad absorption bands confirm, we have to reckon with broad bands of different hydrogen bond states. According to the present state of experimental technique, we can determine with reasonable accuracy only the proportion of free OH groups. The appearance of peaks in the absorption bands, or the preference shown for C_{2v} symmetry, determined by Walrafen (see Volume 1) by Raman methods confirms that certain hydrogen bond states are preferentially present. From the spectroscopically determined proportion of free OH groups one may estimate approximately the magnitude of the hydrogen-bonded aggregates, according to the cluster model, by assuming cooperative mechanisms for the orientation defects in hydrogen-bonded liquids.[877,880,893] At room temperatures one finds, using an idealized cluster model, a few hundred molecules for water and 100 molecules for alcohols.[877,880,893] The small proportion of free OH groups at room temperature makes it in any case unlikely that an appreciable proportion of smaller aggregates could exist, as has been assumed by some authors.[459,1404]

The supercritical state was selected in our discussion as the standard state for a liquidlike environment without hydrogen bonds. This approximation should not be mistaken for a vaporlike state. Even at saturation densities below critical temperatures, and depending on the pressure, free or hindered rotation has been shown to occur. Where rotational fine structure is observed, especially if it is within the resolving power of the spectrometer used, a quantitative evaluation becomes very complicated.[874, 875,898] Figure 9 shows that even in dilute solution of water in CCl_4 a band structure is observed. This structure is temperature-independent. We could never observe such a band structure in liquid water up to T_c, and this also argues against the presence of free monomer water molecules in appreciable concentrations. Such an assumption is also contradicted by the much too low vapor pressure of water, considering its molecular weight. According to the hole model of liquids, one may assume that the concentration of vapor molecules can at most be of the same order as the concentration of defects, and in water this cannot be higher than the concentration of rotational defects of free OH groups. One may again conclude that the proportion of monomer molecules in liquid water at room temperature is less than 1 in 10^4.

The effect of salts on the spectrum resembles the effects observed when the temperature of pure water is changed.[897] From this fact it is possible

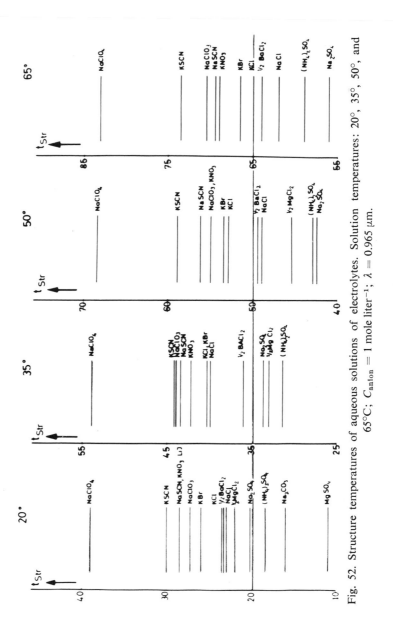

Fig. 52. Structure temperatures of aqueous solutions of electrolytes. Solution temperatures: 20°, 35°, 50°, and 65°C; $C_{\text{anion}} = 1$ mole liter^{-1}; $\lambda = 0.965$ μm.

to derive experimentally a structure temperature for every salt solution (Fig. 52). Thus one obtains an order corresponding to the Hofmeister or the lyotropic ion series. These latter are shown therefore to be coordinated with the structure changes of water. There are salts which have a lower structure temperature than corresponds to the actual solution temperature (structure-promoting salts), and there are also salts with a higher structure temperature than the solution temperature (structure-breaking salts). The latter frequently have large anions which presumably interfere with the structure of water. These salts can have a salting-in effect, particularly for substances with ether groups.[878,897] They increase the hydrophilic nature of water by increasing its proportion of open OH groups. The concept of structure temperature is useful, however, only as a heuristic principle for small temperature intervals. It merely provides a comparative measure of the percentage of free OH groups. It is, however, important whether 25% open OH groups are present at 20°C or at 70°C, because the ratio $\Delta H_{\text{H-bond}}/kT$ is altered. When acids or alkalies are added to water, the half-widths of the bands broaden very considerably,[6,878,897] indicating a pronounced shortening of the life-time of hydrogen bonds, a result which Zundel[725,1382,1458] has discussed in terms of a tunnel effect.

CHAPTER 5

Thermodynamic Properties

F. Franks and D. S. Reid

Unilever Research Laboratory
Colworth House
Sharnbrook, Bedford, England

1. INTRODUCTION

Thermodynamics provides a powerful tool for the study of the interactions between solutes and water in aqueous solutions, and provides many insights into the nature of these interactions. The methods of thermodynamics have, however, one serious limitation in that they give information only on the average properties of the macrosystem, and are unable to give any direct information on the microscopic structure and properties of the system. In the case of dilute aqueous systems, for example, this means that only the bulk properties of the solvent water can be studied directly, although the use of suitable models for the structure of bulk water allows computation of the properties of the postulated hydrated components to be carried out. Such models, however, are not unique, and care must be taken to ensure that the model chosen to represent bulk water is in every way consistent with the known properties of bulk water. An understanding of the behavior of such solutions must therefore be based on information gained from as wide a variety of techniques as possible.

In this chapter it is our intention to consider the range of thermodynamic information which is available relating to the properties of aqueous nonelectrolyte solutions and to discuss possible interpretations of this information, bearing in mind the requirement that any interpretation which is proposed must be consistent with the microscopic behavior of these systems. Since the main subject of this chapter is the thermodynamic properties of aqueous nonelectrolyte solutions, the nonthermodynamic properties of such systems will only be considered when they are required to differentiate between alternative explanations of thermodynamic data.

In the writing of this chapter, no attempt has been made to include all literature references to the thermodynamics of aqueous nonelectrolyte solutions. Rather, a conscious choice has been made of those references which appear to best illustrate the complex nature of these solutions and the present state of knowledge of such solutions. Many of the cited references contain extensive bibliographies of the field and it was felt to be unnecessary to repeat all such citations since they could easily be checked.

Various thermodynamic quantities have been employed when considering the properties of aqueous nonelectrolyte systems and we must first indicate which these are and how they are interrelated.* Three of the most important and most commonly evaluated properties of systems are the changes in enthalpy ΔH, entropy ΔS, and free energy ΔG which occur as the system is modified. These are related by the equation

$$\Delta G = \Delta H - T \Delta S$$

where T is the absolute temperature.

Also frequently measured is the change in heat capacity ΔC_P, where

$$\Delta C_P = [\partial(\Delta H)/\partial T]_P$$

The volume of a system V is also commonly measured, as are the coefficients of expansibility α' and compressibility β, where

$$\alpha' = (1/V)(\partial V/\partial T)_P \quad \text{and} \quad \beta = -(1/V)(\partial V/\partial P)$$

In many cases these quantities are difficult to interpret without further treatment. It is often convenient when dealing with the concentration

---

* The majority of symbols employed in this chapter are defined on their first appearance in the text. It is, however, convenient to define the meanings of certain subscripts and superscripts at this point, since these are employed frequently to qualify the main symbols.

    Superscripts: ($^\circ$) Property refers to the infinitely dilute solution reference state. ($^\ominus$) Property refers to standard state. (^{E}) Denotes an excess (over ideal) property. (^{v}) Vaporization refers to the process liquid → vapor.

    Subscripts: ($_h$) Hydration, refers to the process, pure solute(ideal vapor) → solute (solution). ($_{soln}$) Solution, refers to the process, pure solute(liquid or solid) → solute (solution). ($_m$) Mixing, refers to the process, pure solute + pure solvent → solution. ($_t$) Transfer of solute, under standard conditions, from H_2O to another solvent medium. ($_{1,2}$) Refers to component 1 (i.e., solvent) or component 2 (i.e., solute).

dependence of properties to consider the excess thermodynamic quantities ΔG^E, ΔH^E, ΔS^E, etc. since one is most interested in deviations from ideality. The excess quantity is defined as the difference between the measured quantity and the value one would have expected in an ideal solution:

$$\Delta X^E = \Delta X - \Delta X_{id}$$

where an ideal solution can be defined as one which obeys Raoult's law over the whole range of composition.

An alternative method of treating the results is to evaluate the partial molal thermodynamic quantities, which provide a measure of the effect of one mole of solute in an infinite amount of solution at the stated concentration, and hence gives a clearer insight into solute–solvent interactions. The partial molal quantity $\bar{X}_i$ is evaluated as

$$\bar{X}_i = \left(\frac{\partial X}{\partial n_i}\right)_{T,P,n_1,n_2\cdots}$$

where n_i is the number of moles of component i in the mixture. The apparent molal quantity ϕ_x of a solute is the contribution to the quantity X which one would evaluate as being due to the solute if the solvent component were to have the properties of the pure solvent. It is defined as

$$\phi_x = (X - n_1 X_1^{\ominus})/n_2$$

where $X_1^{\ominus}$ denotes a molar property of the pure solvent. This quantity is also useful when interpreting solution properties. In the case of expansibility and compressibility it is also convenient to consider the molar expansibility α and compressibility K, where

$$\alpha = (\partial V/\partial T)_P \quad \text{and} \quad K = -(\partial V/\partial P)$$

where in this case V is the volume containing one mole. Various other meaningful thermodynamic parameters, e.g., $C_V = (\partial H/\partial T)_V$ and the Joule–Thomson coefficient, are seldom measured in aqueous solutions and so need not be considered in this chapter.

In many cases it is convenient to consider standard-state quantities, $\Delta H^{\ominus}$, etc. This, however, raises the problem of the choice of standard states. While it is clear that in many cases the most useful standard states for the initial system are those of the pure liquid solvent, and the pure solute (either solid, liquid, or gas), the choice of standard state for the final solution is less straightforward. One normally chooses a hypothetical solution which is at unit concentration with, however, many of the properties of the infinitely dilute solution. This choice is dealt with in detail in most thermodynamic textbooks[559,842] and need not be considered further here, except to warn that care must be exercised in the choice of the concentration units to be employed, since in some cases, depending on this choice, the standard-state quantities can be very different. A clear demonstration of this is provided by Arnett and McKelvey,[51] who show that the standard free energy of transfer ($\Delta G_t^{\ominus}$) of propane from D_2O to H_2O changes sign if one changes concentration units from mole fraction or molarity to molality.

Let us now reconsider our choice of standard state for the pure solute. When considering ΔG, ΔH, or ΔS, the choice is, as indicated above, fairly straightforward. One can either choose the pure solid, liquid, or, indeed, gas. If one chooses the gas, and hence evaluates the standard hydration properties ($\Delta H_h^{\ominus}$, etc.), then one evaluates the total solute–solvent interactions in solution (since in the ideal gas there are no interactions). Normally, how-ever, the ideal gas properties are obtained from measurements of the vapor pressure of the pure solute and it is assumed that the vapor is ideal. This can lead to serious errors if the vapor in fact exhibits marked deviations from ideality. Preferably, one should determine vapor imperfections, but in practice this is an extremely arduous task, and so the approximation of ideal vapor is commonly employed. One must therefore be constantly aware of this approximation.

If the liquid standard state is chosen, one obtains from the standard solution parameters an indication of the difference between solute–solute and solute–solvent interactions. For a series of closely similar solute mole-cules it is preferable to employ the ideal vapor as standard state since this eliminates the contribution from solute–solute interactions. The pure solid as standard state for the solute poses many problems, not least the unknown contribution from crystal lattice energies which makes intercomparison of results uncertain. In certain cases a further choice of standard state is possible. This is best illustrated by an example. If one considers the hexose sugars, $C_6H_{12}O_6$, a comparison of their heats of solution from the crystalline state would not be very useful. However, one could employ as standard state the pure elements (by making use of the heats of formation). One would then have a direct comparison of the enthalpies of a series of sugars in solution, since all would have a common standard state.

So far no major problem has come to light in this consideration of the choice of standard state. However, when one considers volumes the situa-tion is much more confused. There is no obvious choice of standard state for a solute which will allow meaningful comparisons to be made. The pure solute standard state is unsatisfactory, as it does not truly reflect molecular dimensions. The true volume of a solute molecule is, however, unmeasurable. Since packing in the pure solute and in solution will be different, it is difficult to see what one should employ as a standard state. Various suggestions have been made (see, e.g., Ref. 518) but as yet no practicable solution has been found, and one must therefore employ the pure solute standard state, despite its imperfections. Alternatively, one may simply compare volumes in the solution standard state and make no use of any information on volume changes during the solution process.

One further aspect of the standard state should be emphasized before leaving this topic. The standard temperature is usually taken as 25°C. While this is a convenient temperature to achieve in a laboratory, there is no particular reason for this choice, and some other temperature could have been chosen. It must be borne in mind that a comparison of solution properties at 25°C may lead to different conclusions from those of a comparison at some other temperature, since at 25°C, as at any other arbitrarily fixed temperature, solutes may be in widely differing regions of their phase diagrams.

In order to obtain a thermodynamic description of a system, one must experimentally determine the thermodynamic parameters of solution which have been mentioned above. Many methods exist by means of which these can be determined and it is our purpose here only to remind the reader of these methods. Free energies can be most readily derived from measurements of vapor pressure, either directly or indirectly (i.e., static or dynamic methods, or isopiestic methods). To produce reliable results is extremely demanding of experimental expertise. Freezing point depression measurements also yield free energies, as do solubility measurements (if the solute is sparingly soluble, this is a very useful technique). Enthalpies can be measured directly in a wide range of calorimeters, or alternatively they can be derived from the temperature dependence of the free energy of the process. Direct measurement is to be preferred where possible, but for sparingly soluble solutes the van't Hoff derivation from the temperature dependence of the solubility is frequently employed. Generally, entropies are derived from the relevant ΔG and ΔH measurements. The heat capacity change ΔC_P can be obtained from the variation of the measured enthalpy with temperature, or alternatively it is possible with a suitably designed calorimeter to evaluate ΔC_P directly.

Volumetric measurements are generally obtained either by dilatometric and pycnometric techniques, which tend to be of limited precision, or by a range of float techniques, by means of which the weight of a float is adjusted until it is of the same density as the solution. Float techniques can be designed to give very high precision. Expansibilities may either be directly measured in a dilatometer or derived from the temperature dependence of the measured volumetric properties of the system. The coefficient of compressibility of the system is conveniently obtained from the velocity of sound in the system by employing the Laplace equation,

$$\beta_s = 1/\varrho c^2$$

where ϱ is the solution density and c is the velocity of sound in the sample. This provides a relatively simple means of obtaining the coefficient of adiabatic compressibility of a system. The coefficient of isothermal compressibility β_T is much more difficult to determine, and therefore is generally not measured. In favorable circumstances β_T can be evaluated from β_s by employing the following relationships:

$$\beta_T - \beta_s = TV(\alpha')^2/C_P \qquad \text{or} \qquad \beta_s^{-1} - \beta_T^{-1} = TV\gamma^2/C_V$$

where γ is the ratio of the heat capacities.

The detailed description of such experimental techniques as have been mentioned is not a subject for this chapter; the reader is referred to textbooks of experimental thermodynamics for this type of information.

1.1. Classification of Solutes

One of the most useful classifications of solutes in aqueous solutions is that due to Rowlinson[1161,1162] and Franks,[514] in which nonelectrolyte solutes are divided into two classes which may be termed "typical aqueous" and "typical nonaqueous." The solutes grouped under the heading "Typical aqueous" are those which exhibit in solution such anomalous thermodynamic properties as are found only in aqueous systems. "typical nonaqueous" solutes have aqueous solution properties which are similar to those of normal nonaqueous solutions. In thermodynamic terms, the two groups of solutes are typified by the relative magnitudes of ΔH^E and $T \Delta S^E$. Those solutes for which $T | \Delta S^E | > | \Delta H^E |$ are classified as "typical aqueous," whereas those for which $T | \Delta S^E | < | \Delta H^E |$ are "typical nonaqueous." In other words, the characteristic of a "typical aqueous" solute is entropy control of the solution thermodynamics. In this chapter typical aqueous solutes have been subdivided into two classes, apolar solutes, and also "mixed" solutes which both have large apolar regions, and a polar group, the apolar regions, however, still providing the dominating influence on the solution properties. Typically nonaqueous solutes are dealt with under the heading of hydrophilic solutes, since solutes belonging to this class have a preponderance of polar groups capable of direct interaction with the solvent water. Two other categories of solute are also considered in this chapter. Tetraalkylammonium ions and other ions with large apolar substituents are considered in terms of the influence of their apolar characteristics on their solution properties, and urea is considered as a solute class on its own, because of the possibly unique characteristics of its interactions with water.

One further point should be made here with regard to solutes in the typically aqueous category. Criteria for the existence of a lower critical solution temperature (LCST) in solution can be summarized briefly as follows. For an LCST to exist, then we must have

$$\frac{\partial^2(\Delta G^{\mathrm{E}})}{\partial x_2{}^2} = -\frac{RT}{x_1 x_2}, \qquad \frac{\partial^2(\Delta H^{\mathrm{E}})}{\partial x_2{}^2} > 0, \qquad \frac{\partial^2(\Delta S^{\mathrm{E}})}{\partial x_2{}^2} \geq \frac{R}{x_1 x_2}$$

and also

$$\Delta G^{\mathrm{E}} > 0, \qquad \Delta H^{\mathrm{E}} \leq 0, \qquad \Delta S^{\mathrm{E}} < 0$$

Thus an LCST results from large negative deviations of the entropy from ideality. Since ΔG^{E} must be positive, it follows that $T \,|\, \Delta S^{\mathrm{E}} \,| > |\, \Delta H^{\mathrm{E}} \,|$, which is the criterion for a typically aqueous solute. Clearly, therefore, for an LCST to exist in aqueous nonelectrolyte solutions requires that the solute is of the typically aqueous category. The requirements for an LCST are quite severe, and such phenomena are rare, but when we consider mixed solutes it will become clear how the delicate balance between polar and apolar groups can, in certain cases, produce an LCST, and also it will be apparent that many aqueous solutions of mixed solutes are close to producing an LCST, though in practice none is observed.

In general, there are insufficient data available on any particular system for $\partial^2(\Delta G^{\mathrm{E}})/\partial x_2{}^2$ to be evaluated, and therefore a more approximate LCST criterion would be useful. Copp and Everett[337] have suggested that the magnitude of ΔG^{E} at $x_2 = 0.5$ is a useful guide. If $\Delta G^{\mathrm{E}}_{x_2=0.5}$ is plotted against T, regions where $\Delta G^{\mathrm{E}}_{x_2=0.5} < \frac{1}{2}RT$ are likely to be a single phase, and those where $\Delta G^{\mathrm{E}}_{x_2=0.5} > \frac{1}{2}RT$ to be two-phase. If $\Delta G^{\mathrm{E}}_{x_2=0.5}(T)$ crosses the $\frac{1}{2}RT$ line, the crossing pattern indicates whether an LCST or upper critical solution temperature (UCST) is likely to occur.

2. BINARY SYSTEMS

2.1. Infinitely Dilute Solutions

2.1.1. *Apolar Solutes*

Experimentally, apolar solutes are found to be almost insoluble in water, and hence their solutions can be considered to be effectively at infinite dilution. Due to their low solubilites, the thermodynamic properties of such solutions have almost invariably been derived by consideration of the temperature dependence of the solubility of the apolar solute. These solutions of apolar solutes in water provide perhaps one of the most striking indications that water is not a "normal" solvent. In this section the thermodynamic properties of binary systems of apolar solutes and water will be considered. The properties of ternary systems with an apolar solute and water as two of the components, which have helped clarify the nature of the interactions between apolar solutes and water, will be considered in Section 3.1.

The alkanes form a series of apolar solutes which have been widely studied. Their aqueous solubilities have been measured by a large number of investigators (for example, Refs. 257, 309, 670, 681, 908, 909, 1001, 1002, 1021). In Table I the solution thermodynamic properties of some hydrocarbons are listed. The results are intriguing. While initially it may not have been surprising that hydrocarbons possess a low solubility in water (and hence have a positive ΔG_{soln}), inspection of Table I shows that for the lower hydrocarbons ΔH_{soln} is negative. The low solubility is therefore a consequence of a negative ΔS_{soln} which overcomes the favorable enthalpy term. The occurrence of "entropy-controlled" solution properties is confined to water, and those solutes for which $|\Delta H^E| < T|\Delta S^E|$ in water have had their solution behavior labeled as "typically aqueous."[514,1161,1162]

The observation that ΔS_h was negative was first made by Butler.[257] In 1937 Butler reviewed his earlier work on the hydration of nonelectrolytes in dilute solution and showed that the ΔS_h was the predominant factor in determining the free energy of hydration of the alcohols. The results of Lannung[827] and Valentiner[1316] led to a similar conclusion for solutions of the rare gases in water. One further observation, by Barclay and Butler,[81] must now be considered. They showed that a plot of ΔH_h versus ΔS_h for the rare gases, etc. in nonaqueous solvents was a straight line. However, a similar plot for these solutes in water is also a straight line, displaced to more negative entropies. This observation must also be considered in attempts to explain the nature of these solutions. Such observations provided the impetus for the development of models describing the nature of aqueous solutions of apolar solutes. Eley[446,447] proposed a two-stage solution process consisting of cavity formation and the introduction of solute, and he ascribed the negative ΔS_h to a loss of translational movement of the solute molecules and to the cavity formation process itself. This model has never gained wide acceptance as a complete explanation, but rather has been incorporated into many other models.

At the present time the most popular explanation is based on the proposal by Frank and Evans[508] that the presence of the apolar solute molecule caused an increase in the order of the water surrounding the solute. This region of increased order was labeled an "iceberg" by these authors, with the warning that the term should not be taken too literally. This postulate of increased structure provides a qualitative explanation of the negative ΔS_h, and if one considers that an increase in temperature would "melt off" some of the extra structure, the positive $\Delta C_{P_2}^\circ$ found for these solutes by D'Orazio and Wood[431] (see Table I) is also explained.

The concept of solvent structuring by apolar molecules introduced by Frank and Evans received further qualitative support from Glew,[564] who showed the existence of close similarities between the thermodynamic properties of methane clathrate hydrate (see Chapter 3) and those of aqueous methane solutions (see Table II). While this does not show that methane in aqueous solution produces extra order in the solvent water akin to a clathrate lattice, it does indicate that the methane probably does have a structural influence on the water.

Once the Frank–Evans concept of structure promotion had been accepted as providing a qualitative explanation of the thermodynamic properties of aqueous solutions of apolar solutes, attempts were made to improve the model and provide a quantitative basis. In one such extension[510] it was assumed that the structure induced in the water was similar to that of the clathrate hydrates. Properties were then assigned to both the monomer water species and the clathrate water species and an attempt was made to account quantitatively for the known solution properties. This resulted in only partial success. One disadvantage of such an approach had already been realized by Claussen and Polglase,[309] who had noted that a crystal hydrate model, while successful in explaining the properties of methane, ethane, and propane in solution, was unable to account for the similar properties of butane in solution, since the n-butane molecule was too large to fit into a clathrate cavity, and also no clathrate hydrate of n-butane was known to exist. Frank and Quist had partially overcome this

TABLE I. Thermodynamic Solution Properties of Apolar Solutes at Infinite Dilution in Water at 25°ᵃ

Solute	$-\Delta H^\circ_{soln}$, cal mol⁻¹	$-\Delta S^\ominus_{soln}$, cal deg⁻¹ mol⁻¹	$\Delta G^\ominus_{soln}$, cal mol⁻¹	$\Delta C^\circ_{P_2}$, cal deg⁻¹ mol⁻¹	$-\Delta H^\circ_h$, cal mol⁻¹	$-\Delta S^\ominus_h$, cal deg⁻¹ mol⁻¹	$\Delta G^\ominus_h$, cal mol⁻¹	Ref.
CH_4	—	—	—	—	3050	31.2	6250	309
	—	—	—	—	3190	31.8	6290	1001, 1002
	2250–2860ᵇ	18.4–16.8ᵇ	2510–3150ᵇ	55ᶜ	—	—	—	1026
C_2H_6	—	—	—	—	3980	33.6	6030	309
	—	—	—	—	4060	34.3	6160	1001, 1002
	2370–1270ᵇ	19.5–16.8ᵇ	3320–3860ᵇ	66ᶜ	—	—	—	1026
C_3H_8	—	—	—	70ᵈ	5700	40.0	6220	309
	—	—	—	—	5060	37.8	6230	1001, 1002
	2090–1450	23.5–21.3	4900	—	—	—	—	1026
C_4H_{10}	—	—	—	72ᵈ	6000	41.4	6340	309
	—	—	—	—	5750	40.6	6380	1001, 1002
	960–720	22.7–21.9	5820–6000	—	—	—	—	1026
C_5H_{12}	300±1600	24	6840	—	6700	58	10520	1021
C_6H_{14}	−600±1700	23	7440	—	7000	58	10440	1021
	−112±30	—	—	—	—	—	—	1141
Cyclohexane	−62±20	—	—	—	—	—	—	1141

Benzene	−191±30	—	—	—	—	—	—	1141
	−580	13.5	4610	97[d]	7550	36.4	3310	207, 1026
Toluene	−640	15.7	5330	—	9380	39.8	2490	207, 1026
Ethylbenzene	−390	19.0	6070	—	9710	44.3	3510	207, 1026
m-Xylene	−410	19.1	6110	—	9790	44.2	3390	207, 1026
p-Xylene	−460	19.0	6120	—	9650	44.1	3490	207, 1026
Ne	—	—	—	21[d]	1400	27.3	—	13
	—	—	—		1830	28.8	6750	1316
Ar	—	—	—	58[d]	2880	30.7	—	13
	—	—	—		2740	30.2	6260	1316
Kr	—	—	—	40[d]	3780	32.6	—	13
	—	—	—		3690	32.3	5935	1316
Xe	—	—	—	44[d]	4120	33.0	—	13
	—	—	—		4290	33.6	5720	1316

[a] Only data whose accuracy is known with a high degree of confidence are included.
[b] Hypothetical values.
[c] D'Orazio and Wood. [431]
[d] Alexander et al. [15]
Solution standard state: hypothetical solution at unit mole fraction.

TABLE II. Reaction Scheme Employed by Glew for 0°C

I	$CH_4(g) \rightarrow CH_4(l)$	$\Delta H_I = -4.621$ kcal mol^{-1}
II	$CH_4(g) + nH_2O(l) \rightarrow CH_4(H_2O)_n(s)$	$\Delta H_{II} = -12.830$ kcal mol^{-1}
III	$CH_4(l) + nH_2O(l) \rightarrow CH_4(H_2O)_n(s)$	$\Delta H_{III} = -8.228$ kcal mol^{-1}
IV	$CH_4(g) + nH_2O(s) \rightarrow CH_4(H_2O)_n(s)$	$\Delta H_{IV} = -4.553$ kcal mol^{-1}
V	$H_2O(l) \rightarrow H_2O(s)$	$\Delta H_V = -1.436$ kcal mol^{-1}

I	Transfer of one mole of methane from gas phase to liquid solution.
II	Formation of solid gas hydrate from gaseous methane and liquid water.
III	Formation of solid gas hydrate from liquid water and dissolved methane.
IV	Formation of solid gas hydrate from gaseous methane and solid ice.[a]
V	Formation of ice from liquid water.

[a] $\Delta H_{II} - \Delta H_{IV} \approx \Delta H_{III}$, i.e., $nH_2O(l) \rightarrow nH_2O(s) \equiv nH_2O(l) \rightarrow (H_2O)_n(s)$. Also, $\Delta H_I \approx \Delta H_{IV}$.

criticism by assuming that any formation of clathrate cages in solution was not necessarily complete. However, the disadvantage remains that any clathrate model is of necessity based on the assumption of cavities of definite radius.

Another attempt to account for the structural effects was made by Nemethy and Scheraga.[1026] They accepted the Frank and Wen[512] model for water of the "flickering cluster" and tried to put it on a statistical mechanical basis. They then proceeded to try to account for the properties of hydrocarbon solutions by considering what the likely effect of a hydrocarbon molecule would be on the equilibrium between water clusters and water monomers. This calculation was based on the assumption that a water molecule possessing four hydrogen-bonded water neighbors (i.e., in a cluster) could, in addition, accommodate a solute molecule neighbor, thus lowering the energy of the water molecule, while nonbonded water molecules (monomers) could only accept a solute neighbor by replacement of a water neighbor. This would lead to a raising of the energy of the nonbonded water molecule. Hence the inert solute would be expected to increase the proportion of clusters in the water. Despite the doubts which have been expressed as to the lack of internal consistency of the statistical treatment employed by Nemethy and Scheraga, the qualitative explanation of the nature of the structuring effect in a solution containing apolar solute would seem to be eminently satisfactory.

A further attempt to explain the thermodynamic properties of these solutions was made by Ben-Naim.[143] He again accepted a two-structure model for liquid water, and considered that the role played by the solute was one of shifting the solvent structure equilibrium in the direction of enhanced structure. While in detail his approach is different than those of Frank and Evans, Frank and Quist, and Nemethy and Scheraga, the essential conclusions are little altered and the differences might be considered to be solely in the postulated mechanism of the structural effect.

The concept that apolar solutes cause structuring in water by somehow reinforcing the stability of water clusters is supported indirectly by the following evidence: If, indeed, cluster stabilization occurs, it is in no way unreasonable to think of such stabilization in solution to be in some manner analogous to clathrate formation, since it is known that many apolar solutes do, indeed, form crystalline clathrate hydrates. One might therefore expect that the stabilizing effect would depend on the radius of solute molecule, since it would somehow have to fit into a cavity, or partial cavity in the water cluster. In this, it would be analogous to the clathrate hydrates, with the solute molecules fitting into the cavities produced by the clathrate lattice. If one considers only those clathrate-formers that enter a single, simple cavity (i.e., one excludes such clathrate-formers as the tetraalkyl-ammonium salts), there is an optimum range of sizes of molecule which may fit into such a solvent cage to give a stable clathrate. Evidence for the validity of the analogy would be provided if it could be shown that the solution properties of apolar solutes likewise exhibited a dependence of size, with an optimum size range similar to that found in the clathrate hydrates. It can be shown that such a size dependence does exist, as illustrated, for example, by the dependence of solubility of apolar solutes in water on the molecular diameter of the solute (Fig. 1). The optimum size range for maximum solubility is similar to the optimum size range for maximum clathrate stability, indicating that the concept of "apolar structure-making" involving the stabilization of some cluster framework by a solute molecule occupying a cavity, or partial cavity, in the cluster does, to a degree, fit the known properties of these solutions. This is not intended to imply that long-lived clathrate structures necessarily exist in solution, but only that the stabilization of water structure by apolar solutes in solution resembles, in many respects, the stabilization of water in a clathrate lattice. What is almost certain is that the nearest-neighbor correlations in liquid water are increased by apolar solutes.

The views discussed so far are probably the most widely held at the present time, although an alternative explanation for the loss in entropy

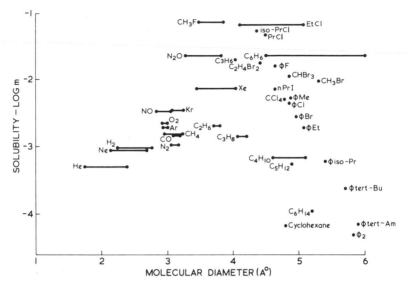

Fig. 1. Aqueous solubility of simple substances as function of solute size. Horizontal lines indicate ranges of molecular size as obtained by different methods (gas viscosity, van der Waals *b* coefficient, second virial coefficient, etc.).

in solution was suggested by Aranow and Witten.[42] According to their model, when a hydrocarbon molecule or the hydrocarbon chain of an alcohol or amine passes from the vapor state into aqueous solution the chain mobility is restricted. As a result, the internal entropy is greatly reduced. They suggest that it is this loss in entropy rather than any increase in water structure which provides the major contribution to the total entropy. While this effect is operative, however, the available thermodynamic evidence would suggest that it is not the major reason for the negative solution entropies. The entropies of solution of the inert gases, for example, are negative, although loss of rotational and vibrational freedom cannot be invoked as explanation. In addition, relaxation measurements have shown that the tetrahydrofuran molecule rotates as freely in aqueous solution as it does in the pure liquid (see Chapter 10). The primary cause of the entropy change would therefore appear to be structural changes in the solvent. Support for this view is provided by nonthermodynamic evidence, e.g., NMR data indicate increases in the reorientation times of water molecules in the neighborhood of predominantly nonpolar substances.[659]

Miller and Hildebrand[977] have also proposed an alternative explanation for the behavior of apolar solutes in water, based on the Pople[1109]

"bent hydrogen bond" description of liquid water. Their treatment, however, appears rather superficial. They dismiss the mixture models for water,
mainly on the basis of the conclusion by Wall and Hornig,[1370] from a
Raman study of the O—D stretching vibration in HDO, that "a mixture
model with well-structured lattice regions *must* be discarded," and they
ignore the almost overwhelming weight of evidence (see Volume 1, Chapter
14) including the more extensive Raman studies of Walrafen (see Volume 1,
Chapter 5 and references therein), indicating that water is best represented
as a mixture of species. It is stated that "the extraordinarily large heat
capacity of water results from the capacity of its hydrogen bonds to absorb
thermal energy. This necessarily becomes less with water molecules which
are in contact with a non-wetting surface of Teflon or paraffin, and likewise,
we propose, with molecules of an inert gas."* Since the heat capacities
of solutions of apolar solutes are much higher than one would expect from
a simple summation of the heat capacities of the water and the solute,
it is clear that the explanation given by Miller and Hildebrand, involving
a reduction in hydrogen bonding and a consequent reduction in heat capacities, is unable to account for the experimental facts and so must be discarded.

So far we have been considering the solution thermodynamic properties
of the hydrocarbons, as derived from solubilities. Similar results are obtained for the rare gases by the same type of technique. Results are shown
in Table I. A general disadvantage of such studies is that values obtained
for ΔH_{soln} by the van't Hoff technique can be greatly in error.† It is therefore
useful to be able to compare such derived heats with heats obtained by
direct calorimetric measurement. In these systems, however, the low solubility of the solute renders calorimetric study difficult. Nevertheless, two such
studies have been carried out. Alexander[13] has measured the calorimetric

---

* In support of this statement, they quote Rowlinson,[1161] "The addition to water of
 any molecules containing inert groups must reduce the total number of hydrogen
 bonds... ." However, in the second edition of his well-known book, Rowlinson[1162]
 adds the comment, "Extremely dilute solutions in water are exceptions."
† The van't Hoff treatment of equilibrium constants is subject to a number of problems:
 (a) Since the evaluation of ΔH involves a differentiation step, it makes severe demands
 on the accuracy of the experimental solubility data; also ΔH depends on the interpolation equation (if any) employed to express solubility as function of temperature.[15]
 (b) It is assumed that the solution process is adequately described by the equation:
 solute(pure) → solute(aq), i.e., that the solute activity is equal to its concentration and
 that the solvent activity is sensibly unchanged. If this is not true, the van't Hoff heat will
 not agree with the calorimetric heat. (c) It is assumed that no cooperative effects are
 involved in the expression for $\ln K$. (d) Whereas calorimetry yields integral heats,
 the van't Hoff method gives rise to the differential heat.

$\Delta H_h{}^\circ$ for some of the rare gases. His results are given in Table I. Reid et al.[1141] have measured the calorimetric ΔH_{soln}° for benzene, n-hexane, and cyclohexane, with the results given in Table I. The agreement between the van't Hoff and the calorimetric heats indicates that the use of van't Hoff heats is in order when comparing the properties of apolar solutes.

One such comparison has been carried out by Krishnan and Friedman,[797] who have evaluated the thermodynamic transfer properties of hydrocarbons from water to dimethylsulfoxide (DMSO) and to propylene carbonate (PC), solvents where there can be no structuring as envisaged by Frank and Evans. In this analysis it was assumed that the partial molal enthalpy of an infinitely dilute solute species in a solvent was the sum of the following contributions: (1) the enthalpy of the solute at infinite dilution in the gas phase; (2) the enthalpy increase in the process of making a cavity in the solvent to accommodate the solute molecule; (3) the enthalpy due to van der Waals interactions between solute and solvent, and polarization–dipole and dipole–dipole interactions; (4) the enthalpy of formation of solute–solvent hydrogen bonds.

In addition, for aqueous solutions only, a further contribution to the enthalpy had to be included, that due to structural changes produced in the solvent by the solute (or by the cavity). The results of this analysis provide an indication of the magnitude of the structural effect of apolar solutes in water, and show it not to depend linearly on the number of carbon atoms in the molecule. This approach is discussed further in Section 2.1.2.

Up to this point we have been considering only the thermodynamic properties of solutions of apolar solutes in H_2O. Some investigations have also been carried out into the properties of some apolar solutes in D_2O, enabling the thermodynamic transfer parameters for such solutes from H_2O to D_2O to be evaluated.[142,791] Since it is normally assumed that D_2O is a more structured liquid than is H_2O, the transfer parameters for apolar solutes should help throw some light on the nature of the solute–solvent interaction. It can be seen (Table III) that the results are consistent with the concept of "structure promotion" by apolar solutes, the effect being more pronounced in D_2O than in H_2O.

It is interesting that the statistical model for water and for hydrocarbon solutions developed by Nemethy and Scheraga was applied both to H_2O and D_2O. While, as discussed previously, there are inconsistencies in the mathematical derivation of the model, it has nevertheless achieved a fair degree of success in giving a semiquantitative description of aqueous solutions of apolar solutes, based on the Frank and Evans concept, and it accounts for the properties of solutions both in H_2O and D_2O.

TABLE III. Thermodynamic Properties for Transfer of Apolar Solutes from H$_2$O to D$_2$O at 25$^{\circ a}$

Solute	$\Delta G_t{}^{\ominus}$, cal mol^{-1}	$\Delta H_t{}^{\ominus}$, cal mol^{-1}	$\Delta S_t{}^{\ominus}$, cal deg^{-1} mol^{-1}
Argon	−48.3	−237	−0.63
Propane	−14.4	−222	−0.70
Butane	−17.4	−142	−0.42

^{a} Standard states: hypothetical solution at unit mole fraction.

2.1.2. "Mixed" Solutes

As a class, mixed solutes have been more widely studied in aqueous solution than any other nonelectrolytes. Since at infinite dilution one can study the role of solute–solvent effects in the absence of solute–solute effects, a great deal of effort has been devoted to the study of such solutions. In particular, the infinite-dilution thermodynamic properties of the alcohols are of great interest, as they constitute the only series of mixed solutes which have been studied widely by a large range of techniques. Franks and Ives[516] have given a comprehensive review of the thermodynamic properties of alcohol–water mixtures up to 1965 and it would therefore be inappropriate to repeat this. We will therefore concentrate more on recent developments in the understanding of the properties of such mixtures.

In recent years new results have become available (Table IV) for $\Delta H_{\text{soln}}^{\circ}$ and $\Delta C_{P_2}^{\circ}$ of the alcohols which seem to have eliminated the uncertainties present in earlier determinations. In particular, Hill[14,667] has carried out a definitive series of measurements at various temperatures of the heats of solution of alcohols ranging from methanol to the butanols. His results have been confirmed by Krishnan and Friedman[797] and Arnett et al.[48] An interesting feature of the results is that they demonstrate that earlier indications[63,257] of a constant increment in $\Delta H_h{}^{\circ}$ for each addition of a —CH$_2$— group to the alcohol were fortuitous. It was found that, as in the case of the hydrocarbons, the alcohols had negative entropies of solution. Hill made use of the vapor pressure measurements of Butler et al.[258] together with estimates of the fugacities of the pure alcohols to calculate the free energies of solution of the alcohols. These, in combination with the measured heats of solution, were used to evaluate the entropies of solution. In addition, data for the heats of vaporization of the alcohols

TABLE IV. Thermodynamic Properties of Alcohols at Infinite Dilution in Water at 25°

Alcohol	$-\Delta H^\circ_{soln}$, cal mol⁻¹	$-\Delta S^\ominus_{soln}$, cal deg⁻¹ mol⁻¹	$\Delta G^\ominus_{soln}$, cal mol⁻¹	$\Delta C^\circ_{P_2}$, cal deg⁻¹ mol⁻¹	$-\Delta H^\circ_h$, cal mol⁻¹	$-\Delta S^\ominus_h$, cal deg⁻¹ mol⁻¹	ΔG°_h, cal mol⁻¹	Ref.
MeOH	1733	6.6	250.4	18.0	10,797	33.4	840.5	667
	1730	—	—	—	10,800	—	—	797
	1754	—	—	23.1	—	—	—	48
EtOH	2415	10.7	780.4	33.9	12,598	39.8	740	667
	2420	—	—	—	12,600	—	—	797
	2433	—	—	39.2	—	—	—	48
n-PrOH	2422	13.4	1586	49.9	13,790	44.3	569	667
	2430	—	—	—	13,790	—	—	797
	2419	—	—	56.4	—	—	—	48
i-PrOH	3102	14.5	1221	49.2	.14,016	45.4	478	667
	3080	—	—	—	13,970	—	—	797
	3124	—	—	55.4	—	—	—	48
BuOH	2217	15.3	2352	62.2	14,495	47.1	444	667
	2200	—	—	—	14,820	—	—	797
	2249	—	—	71.6	—	—	—	48
t-BuOH	4137	18.8	1466	63.3	15,309	50.5	244	667
	4110	—	—	—	15,310	—	—	797
	4172	—	—	66.2	—	—	—	48

Solution standard state: hypothetical solution at unit mole fraction.

have enabled the enthalpy, free energy, and entropy of hydration to be calculated. It is apparent that the apolar region of the alcohol molecule is affecting the solvent water in a manner similar to an apolar solute. In this case, however, the presence of the polar functional group provides sufficient extra interaction to solubilize the whole molecule. The heat capacities of solution also show the effect of the apolar group, being large and positive, and the behavior of the alcohols can, indeed, be adequately described as that of "solubilized hydrocarbons."

Arnett et al.[48] have determined the heat capacities of solution of a large number of low-molecular-weight alcohols. As indicated from Hill's results, the ΔC_{P2}° values were found to be large and positive. Where comparisons could be made with Hill's results, the ΔC_{P2}° values determined by Arnett were found to be rather higher, though the $\Delta H_{\text{soln}}^\circ$ values at 25° were in good agreement. This probably indicates that the absolute accuracy of ΔC_{P2}° determined by Arnett is lower than the accuracy obtained by Hill, but it is reasonable to assume that the relative magnitudes of Arnett's ΔC_{P2}° values are not seriously in error. Bearing this assumption in mind, the correlation of ΔC_{P2}° with various molecular parameters could be investigated. It was found that for n-alcohols some degree of correlation existed with carbon number and molar volume. This observation is consistent with the postulate that the apolar regions control the solution behavior, since the molar volume is a crude indicator of the size of the hydrocarbon chain. Chain branching was found to lower ΔC_{P2}°.

If, instead of using the pure liquid alcohol, the vapor was used as the standard state, then chain branching was found to increase $(\bar{C}_{P2}^\circ - C_P^{\text{v}})$. This is a consequence of the increase in C_P^{v} which occurs on chain branching, so that any interpretation of changes in ΔC_{P2}° ascribed to branching would be hazardous. The heat capacity of solution from the vapor phase, $(\bar{C}_{P2}^\circ - C_P^{\text{v}})$, was compared to ΔS_h°, yielding a correlation which, if we accept the postulate that the negative ΔS_h° is an indication of solvent structuring by the solute, supports the idea that the heat capacity of solution will reflect solvent structuring.

Another interesting approach to investigate the "solvent-structuring" concept was that of Krishnan and Friedman,[797] who measured ΔH_{soln} of alcohols in water, DMSO, and PC at high dilution, and hence obtained the enthalpies of transfer of the alcohols from one solvent to another. These transfer enthalpies were interpreted in terms of a model which postulated additive group contributions to the solvation enthalpies in all three solvents. For water, an additional, nonadditive, "structural" contribution was also found to be necessary. It proved possible to evaluate both the group con-

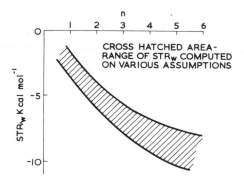

Fig. 2. Structural contribution to $\Delta H_h{}^\circ$ of alcohols as function of carbon number, according to Friedman and Krishnan (Volume 3, Chapter 1).

tributions and the magnitude of the "structural" term in water. The structural contribution to the enthalpy of solution in water was found to range from about -2.5 kcal mol^{-1} for methanol to about -8.5 kcal mol^{-1} for amyl alcohol, and it was seen to be leveling off at about this chain length as shown in Fig. 2. This suggested that if the structural effect was due to the cavity required to accommodate a coiled hydrocarbon chain, it was an effect characteristic of small cavities, since the cutoff was much sharper than would be expected if the effect were proportional to the surface of the cavity produced by a coiled chain. This in turn would tend to suggest that the surface polarization of water can develop on the cavity surface only after a certain flatness is reached, and that once this polarization develops it interferes with the buildup of water structure around the cavity.

Before leaving the subject of the thermal properties of the alcohols at infinite dilution two other points should be emphasized. In the discussion of apolar solutes mention was made of the Barclay–Butler[81] plot and of the interpretation of this plot proposed by Frank and Evans.[508] Using Hill's data, an improved Barclay–Butler plot can be constructed for the alcohols (Fig. 3) and it is found that this line is parallel to the line for the hydrocarbons but displaced from it. It would therefore appear that the postulated structuring effects which produce the anomalous slope of the Barclay–Butler plot are still present, and that the polar group provides an effectively constant increment in both enthalpy and entropy, thus displacing the line. A similar conclusion can be reached by a procedure in which an alcohol is formally considered as consisting of a functional group C—OH together with a series of alkyl chains.[522] By this formalism, any alcohol is equivalent to methanol plus the relevant alkanes, less hydrogens. It can be shown that ΔH_h° and ΔS_h° of alcohols, up to C_5, can be satisfactorily accounted for by this formalism, thus indicating that the apolar group hydration is unaffected by the presence of the C—OH group, and the heat and entropy of hydration of the —C—OH group can be estimated.

Before moving on to discuss the volumetric properties of alcohols in infinitely dilute solution some thermal data available for other mixed solutes will be considered. In this case few systematic studies exist from which enthalpy, entropy, and free energy can be derived, so that interpretation of the results is more hazardous. Franks and Watson[526] have measured the heats of solution of a series of dialkylamines in water and calculated the heats of hydration, the heats of vaporization being obtained from the temperature dependence of the vapor pressure. Where possible, free energies of hydration were computed from literature data and the entropies of hydration derived. In general, the results (Table V) produce a picture similar to that obtained by Hill, with negative entropies of solution and large, positive heat capacities, both presumably reflecting the properties of the apolar regions of the solute. In addition, the Barclay–Butler plot is found again to be parallel to the line for aqueous solutions of apolar solutes. Measurements of the heats of solution have also been made for a series of cyclic ethers.[518] In this case lack of data prevented evaluation of the

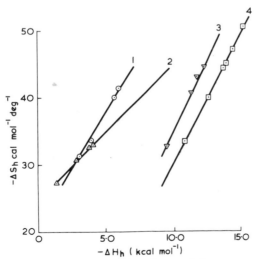

Fig. 3. Barclay–Butler plots (see text) for aqueous solutions of (1) alkanes, (2) rare gases, (3) cyclic ethers, and (4) simple alcohols at 25°.

TABLE V. Thermodynamic Properties of Some "Mixed" Solutes at Infinite Dilution in Water at 25°^{ca}

Solute	$-\Delta H^\circ_{soln}$, cal mol⁻¹	$-\Delta S^\ominus_{soln}$, cal deg⁻¹ mol⁻¹	$\Delta G^\ominus_{soln}$, cal mol⁻¹	ΔC°_{P2}, cal deg⁻¹ mol⁻¹	$-\Delta H^\ominus_h$, cal mol⁻¹	$-\Delta S^\ominus_h$, cal deg⁻¹ mol⁻¹	$\Delta G^\ominus_h$, cal mol⁻¹	Ref.
THF	3574 3568	17.7 —	1715 —	— 38.2	11,296 11,180	40.6 —	805 —	260, 261 518
THP	3484 3609^b	20.2 —	2551 —	— 43.2	11,682 10,875^b	43.0 —	1151 —	260, 261 518
1,4-Dioxan	2295 2347 2390	11.1 — —	1017 — —	— 10.0 —	11,466 11,178 —	35.8 — —	−780 — —	260, 261 518 995
2-Me-THF	4240	21.6	2195	—	12,280	44.5	973	260, 261
1,3-Dioxolane	1103	8.3	1370	—	9,540	32.6	180	260, 261
Pyrrolidine	6247 6280	24.6 —	267 —	— —	15,193 14,520	46.9 —	−1206 —	260, 261 1170
Piperidine	6248 6365	24.6 —	1079 —	— —	15,634 15,290	49.6 —	−833 —	260, 261 1170
Hexamethylene-imine	5764	26.2	2052	—	16,314	52.6	−634	260, 261
N-Methyl-pyrrolidine	7219	29.2	1490	—	15,158	51.8	297	260 261
N-Methyl-piperidine	7170	31.3	2172	—	15,720	54.0	382	260, 261

Diethylamine	7756	28.3	671	30.9	15,563	52.1	−40	526
Di-n-propylamine	7965	34.1	2186	38.2	17,560	60.1	370	526
Di-i-propylamine	8542	—	—	67.4	16,629	—	—	526
Di-n-butylamine	7222	40.2	3440	72.4	18,151	63.2	701	526
Di-i-butylamine	5263	—	—	83.5	15,564	—	—	526
Di-sec-butylamine	7973	—	—	92.0	17,871	—	—	526
Formic acid	162	—	—	—	11,233	—	—	778
Acetic acid	281	—	—	10	12,619	—	—	778
	284	—	—	—	—	—	—	(c)
n-Propylamine	5884	—	—	39.9	13,324	—	—	778
i-Propylamine	6560	—	—	42.5	13,157	—	—	778
n-Butylamine	5576	—	—	55.9	14,111	—	—	778
i-Butylamine	5585	—	—	53.0	13,654	—	—	778
t-Butylamine	7012	—	—	50.9	14,096	—	—	778

[a] Only data whose accuracy is known with a high degree of confidence are included.
[b] This sample may have contained a volatile impurity.
[c] E. M. Arnett and J. V. Carter (unpublished, quoted in Ref. 778). Solution standard state: hypothetical solution at unit mole fraction.

free energies and entropies of solution. Again, large, positive heat capacities of solution were obtained, except in the case of dioxan, which had a smaller positive heat capacity of solution. Also of note was the slope of $\partial(\Delta C_{P_2}^\circ)/\partial T$, which for dioxan was near zero, while for the other ethers in the study it was negative. These observations would tend to indicate that the behavior of dioxan is closer to that of a hydrophilic solute, or on Rowlinson's classification, giving rise to typically nonaqueous solution behavior.

Another study which has thrown some light on the calorimetric behavior of mixed solutes in solution was carried out by Konicek and Wadsö,[778] who measured the limiting enthalpies of solution of some amines, carboxylic acids, and amides in water. These compounds were considered as being fairly simple model systems for proteins in aqueous solution. The dependence of $\Delta H_{\text{soln}}^\circ$, ΔH_h°, $\Delta C_{P_2}^\circ$, and $\bar{C}_{P_2}^\circ$ on the chain length of the solute molecule was considered for the different homologous series. $\Delta H_{\text{soln}}^\circ$ and $\Delta C_{P_2}^\circ$ exhibited few regularities, as might be expected from the complex nature of the process they describe. Regularities were apparent in the increment in ΔH_h° per —CH$_2$ for the different series of straight-chain solutes; chain branching produced no clear pattern. The most interesting conclusion from this study arose from the consideration of $\bar{C}_{P_2}^\circ$. Figure 4 shows plots

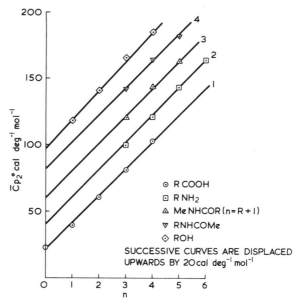

Fig. 4. Limiting partial molar heat capacities (25°) of aqueous solutions of homologous series as function of carbon number; according to Konicek and Wadsö.[778]

of $\bar{C}_{P_2}^{\circ}$ versus chain length for the different solute series considered. From the constancy of the increment in $\bar{C}_{P_2}^{\circ}$ per —CH$_2$, independent of the polar group, it would appear that the water structure around the apolar group is only slightly affected by the polar group, a conclusion which is in accord with our earlier deduction from the additive nature of the enthalpies of solvation of alcohols.

In general, then, the thermal properties of mixed solutes at infinite dilution are characterized by the relative magnitudes of the excess heats and entropies. The heat capacities of solution and the partial molal heat capacities also clearly demonstrate the importance of the apolar character of these solutes. Recent work which confirms this view involved a thermodynamic study of dilute aqueous solutions of cyclic amines[260] and cyclic ethers.[261] Free energies, enthalpies, and entropies of hydration have been evaluated. The effects of the balance between polar and apolar groups can be clearly seen in the results shown in Table V.

The volumetric, or *PVT*, properties of mixed solutes at infinite dilution are in many ways fruitful sources of information on such solutions, though the problem of choosing a suitable standard state for the purposes of intercomparison of data does tend to reduce the value of such results. As before we will first of all consider the *PVT* behavior of the alcohols as typical mixed solutes and then move on to consider what additional information has become available from measurements on dilute solutions of other mixed solutes.

If we consider first the partial molar volumes, these have been reported for the alcohols by Friedman and Scheraga[534] and Alexander[12] and others. In general, $\bar{V}_2^{\circ}$ has been obtained by linear extrapolation of ϕ_v against concentration. Franks and Smith,[524] who used more sensitive experimental and extrapolation techniques, evaluated $\bar{V}_2^{\circ}$ for the butanols and also showed that the common practice of linear extrapolation was questionable. It was suggested that in the absence of solute–solute interactions one might expect ϕ_v to have no concentration dependence (i.e., zero slope) in the limiting region, and it was shown that such behavior would be consistent with the requirements of the Gibbs–Duhem relationship. The results for the butanols obtained by Franks and Smith did, indeed, suggest that the limiting slope of ϕ_v against concentration was zero. This zero limiting slope is taken to be an indication of the absence of any long-range interaction between solute molecules. The finite limiting slopes observed for ϕ_v versus $\sqrt{c}$ in electrolytes (and predicted by Debye–Hückel theory) are a consequence of long-range (electrostatic) interactions between solute particles (ions). Bearing in mind the possibility of errors in $\bar{V}_2^{\circ}$ arising from invalid extrapolation, it is interesting to compare $\bar{V}_2^\circ$ data for different alcohols. Since the alcohols vary greatly in molecular size and shape, some effort must be made to relate all the $\bar{V}_2^\circ$ values to a common standard state. Since at present no fully satisfactory standard state has suggested itself, the normal practice is to subtract from $\bar{V}_2^\circ$ the volume $V_2^\ominus$ of the pure liquid, and thus to compare $\bar{V}_2^{\circ E}$ for the different alcohols.

The results of Friedman and Scheraga for a series of alcohols showed $(-\bar{V}_2^{\circ E})$ to increase with temperature and also showed it to increase with molecular size. They attempted, with limited success, to correlate their results with the predictions of Nemethy and Scheraga's[1025,1026] theory, assuming that the polar and apolar portions of the molecule produced additive volume effects. Since the results of Franks and Smith for the butanols raise doubt as to the reliability of the earlier data of Friedman and Scheraga, it would appear pointless to discuss them in more detail. Suffice it to say that the increasingly negative values of $\bar{V}_2^{\circ E}$ with increasing solute size probably reflect stabilization of aqueous structure by the solute, and the negative $\bar{\alpha}_2^{\circ E}$ $(= \partial \bar{V}_2^{\circ E} / \partial T)$ values would seem to signify that structure promotion persists over a considerable temperature range. The results of Franks and Smith for the butanols lead to similar conclusions. While it is not expedient to attach any great significance to the values of $\bar{V}_2^{\circ E}$, because of the unsatisfactory nature of the standard state, the fact that $\bar{V}_2^{\circ E}$ is large and negative can be taken to reflect solvent-structure stabilization and comparison of $\bar{\alpha}_2^{\circ E}$ data shows that, while n- and iso-butanol will tend to behave as normal solutes above 40°, t-butanol at this temperature is still a powerful structure promoter.

In addition to the *PVT* properties of the alcohols, data are available on the volumetric properties of some other interesting systems containing a mixed solute. Nakanishi et al.[1016] have shown that, while $\bar{V}_2^{\circ E}$ may always be negative (reflecting the large void volume of water), its magnitude is not simply related to the molar volume of the solute, but rather to the ratio of polar to nonpolar groups in the solute molecule. This was demonstrated by a study which encompassed related molecules with different polar/apolar group ratios. Support for this observation has been provided by a study of the cyclic ethers,[518] where it has been shown that the molar volume of the pure solute is not even a reliable indication of solute size, since the molar volumes of pure cyclic ethers of similar size, e.g., tetrahydropyran and dioxan, are very different. In addition to showing that the addition of a polar group to the ether ring reduces $| \bar{V}_2^{\circ E} |$, Franks et al. also observed a startling effect on $\bar{\alpha}_2^{\circ E}$, which even more clearly illustrates the major importance of the ratio of polar to nonpolar groups. Comparing tetrahydrofuran (THF) to tetrahydropyran (THP), $\bar{\alpha}_2^{\circ E}$ becomes more negative as the nonpolar group increases in size. From THF to tetrahydrofurfuryl alcohol (THFA), and from THP to tetrahydropyran-2-carbinol (THPA), $\bar{\alpha}_2^{\circ E}$ becomes less negative as the polar groups are increased. From THP to dioxan $\bar{\alpha}_2^{\circ E}$ changes sign, and the polar character is increased even more. Indeed, it is questionable whether dioxan should be considered a mixed solute or a hydrophilic solute. The concentration dependence of the volumetric properties of these ethers is also interesting and is discussed in a later section.

A further aspect of the *PVT* properties of infinitely dilute alcohol solutions which remains to be considered is the compressibility. It has recently been shown[520] that the apparent molar adiabatic compressibilities ϕ_K of hydrophobic solutes are characteristically different from those of hydrophilic solutes. This difference can be partially understood on the basis that the order promotion by hydrophobic solutes produces an extra amount of "structure" in the solvent, some of which can be destroyed by the application of pressure. Thus, compared to a hydrophilic or an electrostrictive solute, the ϕ_K of a hydrophobic solute will be more positive (or less negative). It is possible that in the future use may be made of this observation to diagnose the degree of hydrophobicity of a solute. The alcohols, as "mixed" solutes, would be expected to exhibit hydrophobic type behavior, as indeed is the case.

The transfer properties of mixed solutes from H_2O to D_2O can also help throw light on the nature of these solutions. As is discussed more fully by Arnett and McKelvey,[51] the enthalpies of transfer of alcohols from H_2O to D_2O[51,791] indicate the important role of the structure promotion by apolar regions.

In this section, while we have discussed the solution thermodynamics of a range of mixed solutes at infinite dilution, we have not considered the effects of apolar groups on the ionization thermodynamics of carboxyl or amino groups. Rather we have considered only the solution properties of the neutral molecules. Ionization phenomena are an important related topic and are fully discussed in Volume 3, Chapter 3.

2.1.3. Hydrophilic Solutes

It was indicated earlier that, due to their limited solubilities, and consequent experimental difficulties, the apolar solutes have been studied infrequently. The hydrophilic solutes, such as the sugars, which have available a considerable array of functional groups capable of hydrogen bonding with the solvent water have also been little studied. Here, however, experimental difficulties are unlikely to be the reason for this neglect. It is, we fear, possibly a consequence of the widely held belief that sugars and other hydrophilic solutes form ideal solutions in water and therefore are of little intrinsic interest and do not merit very close study. In the investigations which have included hydrophilic solutes most effort has gone into isopiestic measurements, which are not suitable for the evaluation of properties at infinite dilution; very few studies have in fact yielded information about the properties of such solutes at or near infinite dilution.

What data are available on such solutes at infinite dilution have shown that subtle effects exist, especially in solutions of monosaccharides. Measurements of $\phi_v{}^\circ$ and $\phi_K{}^\circ$ on sugar solutions[520] have shown that there is a dependence of the partial volumetric properties of the solute on the orientation (steric position) of hydroxyl groups. Such discrimination can be understood in terms of an observation by Warner[1379] that the oxygen spacings in many hydrophilic solutes are such that they match the second nearest-neighbor distance in water as obtained from the X-ray radial distribution function (see Volume 1, Chapter 8). Further, it has been shown that the mutarotation equilibria of sugars are consistent with such structural compatibility.[719,720] It might therefore be expected that the degree and type of hydration of the solute molecules should depend on the steric compatibility with bulk water structure, and this has been demonstrated by Tait et al.[1275] No long-range effects need therefore be invoked and at infinite dilution one would expect the properties of the solute to be dependent on the number of groups with favorable oxygen spacings, or in practice, on the relative numbers of equatorial and axial groups.

The studies by Franks et al. do, indeed, show the effect of increasing equatorial substitution, in particular, $\phi_K{}^\circ$ for myo-inositol is seen to be almost as negative as $\phi_K{}^\circ$ for NaCl, where strong electrostrictive hydration plays the dominant role.

The thermal studies on record for hydrophilic solutes are limited to measurements by Taylor and Rowlinson[1280] and Sturtevant[1265] on the heats of solution of glucose in water, some calorimetric studies which have yielded information on the mutarotation equilibria of sugars[26,1265,1277] a few investigations of heats of solution of glycerol[436,733,1142] which have yielded infinite dilution values, and also some studies of dimethylsulfoxide solutions[340,750,1127] and hydrogen peroxide solutions.[552]

In those cases where heats of solution have been measured their exothermic nature indicates that extensive hydrogen bond formation is taking place (Table VI). However, in contrast to the case of apolar solutes, the

TABLE VI. Thermodynamic Properties of Some Hydrophilic Solutes at Infinite Dilution in Water at 25°

Solute	$-\Delta H^\circ_{soln}$, cal mol^{-1}	$-\Delta S^\ominus_{soln}$, cal deg^{-1} mol^{-1}	$\Delta G^\ominus_{soln}$, cal mol^{-1}	$\Delta C^\ominus_{P_2}$, cal deg^{-1} mol^{-1}	$-\Delta H^\circ_h$, cal mol^{-1}	Ref.
Glycerol	1253	2.7	−458	18.3	—	[a]
	1280	—	—	1.0	—	1142
α-Glucose	−2630	−4.0	1438	—	—	1280, 1407
	−2570	—	—	—	—	1265
Glucose (glass)	~1110	—	—	—	—	1280
Sucrose	−1400	−2.4	680	—	—	1407
DMSO	4328	—	—	—	16,980	1127
H$_2$O$_2$	819	—	—	−8.9	13,160	552

[a] R. C. Wilhoit, personal communication.
Solution standard state: hypothetical solution at unit mole fraction.

heat capacities of solution provide no indication of extensive structuring of the solvent, values near zero being common. This would tend to indicate that ΔC_P reflects hydrogen bond formation between solute and solvent, as might be expected, rather than an increase in solvent–solvent hydrogen bonds. This is consistent with the concept of specific site hydration, proposed by Franks and co-workers, which must be compatible with water structure. The mutarotation heats[719] also support the specific site hydration theory, the negative ΔH of mutarotation $(\alpha \rightarrow \beta)$ being in accord with a greater hydration potential of equatorial hydroxyl groups as compared with axial ones.

An interesting result reported by Ono and Takahashi[1046] illustrates clearly the role of the hydroxyl orientation. They estimated the relative enthalpies of glucose and mannose in solution from measurements carried out on the glucose $\rightarrow$ fructose and mannose $\rightarrow$ fructose conversions. Their results (Fig. 5) show that glucose in solution has a lower enthalpy than mannose. This is understandable in terms of a combination of two effects: (a) Steric repulsions of hydroxyl groups on the molecule cause a raising of the energy, and (b) there are different interactions of axial and equatorial hydroxyl groups with water.

The lower enthalpy in solution of β-glucose as compared to α-glucose can be ascribed to the more favorable interactions which may be assumed to exist between the hydroxyl groups of the β-glucose (all of which are equatorial) and the solvent water, in comparison to the interactions between the α-glucose hydroxyl groups and water. In contrast to this, the lower enthalpy of α-mannose in solution as compared to β-mannose can be ascribed mainly to the existence in β-mannose of a steric repulsion between the equatorial hydroxyl on C(1) and the axial hydroxyl on C(2). In α-mannose both these hydroxyl groups are in the axial position, with a consequent reduction in the steric repulsion. It is not possible at present to assess the relative importance of these two effects when one is considering

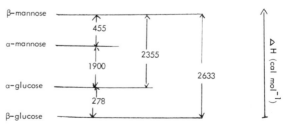

Fig. 5. Relative enthalpies of glucose and mannose anomers in aqueous solution at 25°.

the factors which result in mannose having a higher enthalpy in solution than glucose. More data are required on the relative enthalpies of a series of monosaccharide solutes in water before such an assessment could be attempted.

2.1.4. Tetraalkylammonium Halides

In considering the properties of the tetraalkylammonium salts, it is not our purpose to discuss any behavior which can be explained by conventional theories of electrolyte solutions. Rather, it is those aspects of their behavior that on the basis of electrolyte theory would be termed anomalous which we wish to discuss, as it is fairly certain that the anomalies are due, in general, to the large apolar moieties on these ions. In general terms, therefore, we will examine those solution properties of tetraalkylammonium ions (and other similar ions) that are a reflection of their partial nonelectrolyte character.

Considering such ions at infinite dilution in water, the contribution made by the apolar groups to the overall observed behavior can be seen clearly in the thermal properties. In particular, the heat capacities of solution are found to be large and positive,[512,1168,1180,1266] an observation which is typical of the effect of apolar groups on water. That it is, indeed, the apolar regions of the ion which are responsible for the large positive heat capacity of solution is indicated by the dependence of $\Delta C_{P_2}^\circ$ on the nature and size of the substituent groups on the ion,[1180] the Bu_4N^+ ion having a more positive $\Delta C_{P_2}^\circ$ than the Pr_4N^+ ion, for example. Clearly, therefore, the apolar groups govern the interactions of the ions with water, and in place of the electrostrictive structure-making and -breaking effects observed for simple ions, the tetraalkylammonium ions exhibit apolar solution effects. That there does, indeed, appear to be a large "structural" contribution to the thermal properties from the apolar groups is made clear by a series of studies by Krishnan and Friedman,[797–799] who, employing an approach similar to that described in Section 2.1.2 for alcohols, have evaluated the "structural contribution" to the enthalpy of solution of these salts by means of a comparison of their transfer enthalpies between the solvents H_2O, D_2O, PC, and DMSO. This study included the alkali metal ions, so that a comparison could be made between the behavior of "conventional" ions and the tetraalkylammonium ions. While for transfer from DMSO to PC there is a fairly smooth variation in ΔH_t, decreasing with increasing ionic radius, for both alkali metal and tetraalkylammonium ions, for the transfer from water to PC, ΔH_t is found to decrease smoothly with radius for the alkali metal ions, but it increases with radius in the tetraalkylammonium ion series up to Am_4N^+. This is clearly a function of the size of apolar groups on these ions, and, indeed, when the difference in structural effect in H_2O and D_2O is evaluated (as was done for the alcohols) the resulting "isotope effect" is in agreement with the structural contribution computed for alcohols in H_2O and D_2O. This would tend to confirm the importance of apolar groups in governing the interactions between R_4N^+ ions and water.

The solvation enthalpies computed by Krishnan and Friedman would suggest that Me_4N^+ is a net structure breaker, while Pr_4N^+, Bu_4N^+, and Am_4N^+ ions are net structure makers. Et_4N^+ would appear to be neither. Similar conclusions have been drawn by other investigators[739-741] employing different criteria of structure-making and structure-breaking.

The volumetric properties of solutions of the tetraalkylammonium salts are also of interest. Whereas for most simple ions there is a volume decrease on solution, the magnitude of which decreases with increasing ion size, Franks and Smith[523] have shown that if the tetraalkylammonium ion size is estimated from molecular models, the volume decrease on solution increases in magnitude with increasing ion size. This type of behavior parallels the behavior of homologous series of other alkyl derivatives, again indicating the important role played by the apolar groups on these ions. Millero and Drost Hansen[983] evaluated ϕ_v for a series of tetraalkylammonium salt solutions between 20 and 40°C at 1 deg intervals, and hence obtained apparent expansibilities, which they related to $\bar{a}_2°$, the partial molal expansibility at infinite dilution. $\bar{a}_2°$ was small for ammonium chloride but increased with increase in the size of the alkyl substituent groups. This was interpreted as resulting from expansibility changes in the water produced by these ions. This structural effect increases with increasing ion size and decreases with increasing temperature. A further difference between simple ions and the tetraalkylammonium ions becomes apparent in the temperature dependence of the apparent expansibility. For the more common electrolytes $\partial^2\phi_v°/\partial T^2$ is negative, whereas for the tetraalkylammonium salts it is positive.[585,983]

In an interesting paper Hepler[650] has considered the relationship between the concept of "structure-making" and "structure-breaking" solutes, and the thermal expansion of aqueous solutions. Arguing that, since increasing pressure would break up the bulky "structured" regions in water, $(\partial \bar{C}_{P2}/\partial P)_T$ would be expected to be negative, he employs the relationship

$$(\partial \bar{C}_{P2}/\partial P)_T = -T(\partial^2 \bar{V}_2/\partial T^2)_P$$

and concludes that "structure-breaking" solutes should have negative $\partial\bar{\alpha}_2^\circ/\partial T$ and "structure-making" solutes positive $\partial\bar{\alpha}_2^\circ/\partial T$. On this basis, the tetraalkylammonium ions would be classed as "structure-making" ions.

The apparent molal adiabatic compressibilities $\phi_{K_s}^\circ$ of a series of tetra-alkylammonium salts have been derived by Conway and Verrall[334] from ultrasonic velocity measurements. For these salts $\phi_{K_s}^\circ$ is negative and decreases as the number of carbon atoms in the solute molecule increases. The results are interpreted in terms of changes in the local compressibility of the solvent near the ions. It is assumed that three limiting types of local water structure near ions can be distinguished: (a) an "icelike" (tetra-hedrally coordinated) configuration, less compressible than bulk water, (b) an "unbonded" configuration, with a higher compressibility, and (c) electrostricted water, with a compressibility lower than (a). The dependence of $\phi_{K_s}^\circ$ on anion, increasing with increasing ion size in the order $I^- >$ Br$^-$ $>$ Cl$^-$, was consistent with a corresponding decreasing electrostriction. Moving from Me$_4$N$^+$X to Bu$_4$N$^+$X, $\phi_{K_s}^\circ$ indicates that the amount of water of type (a) is increasing at the expense of electrostrictive water. These results point once more to the role of the apolar regions of these ions in determining their solution properties.

2.1.5. Urea

The thermodynamic properties of urea in infinitely dilute solution are fairly well characterized, due largely to the importance of this solute as a denaturing agent in protein chemistry. A useful compilation of what may be regarded as the "best quality" data available on urea solutions is provided by Stokes.[1252] In this paper he deals mainly with the properties of urea solutions at finite concentrations, but he also includes some extrapolated values at infinite dilution. Of particular interest is $\phi_{C_P}^\circ$, the apparent molar heat capacity, and also ϕ_v°. $\phi_{C_P}^\circ$ is obtained from the data of Gucker and Ayres,[605] while the volumetric measurements are due to Stokes. As they refer to infinite dilution, $\phi_{C_P}^\circ = \bar{C}_{P_2}^\circ$, and $\phi_v = \bar{V}_2^\circ$. At 25°C, $\bar{C}_{P_2}^\circ$ is close to $C_{P_2}^\ominus$ for pure solid urea, $\bar{C}_{P_2}^\circ$ decreases rapidly with temperature in comparison to $C_{P_2}^\ominus$, which has only a small temperature dependence. This results in $\Delta C_{P_2}^\circ$ becoming negative at low temperatures. Since nonelectrolyte solutes which are classed as structure makers have characteristically large $\Delta C_{P_2}^\circ$, this would indicate that urea is not a structure-making solute in this sense. It does not, however, indicate that urea is necessarily a structure-breaking solute, though it does suggest that it may be so. The volumetric behavior of urea is also interesting in that ϕ_v° decreases rapidly with temperature. This can be explained by considering urea as a structure breaker in its overall effect, since at low temperatures, where water is more ordered, the effect would be correspondingly larger, resulting therefore in the observed decrease in $\phi_v{}^\circ$ as the temperature decreases.*

2.2. Aqueous Solutions at Finite Concentration

2.2.1. *Mixed Solutes*

The behavior of mixed solutes in water at finite concentrations provides possibly one of the most fascinating examples of the interplay of opposing effects that can be found in solution thermodynamics. At one and the same time the solute contains groups which can hydrogen-bond with the water, and so tend to hold it in solution, and also apolar groups which, by virtue of their large negative entropy of hydration, tend to force the solute out of solution. The solution properties thus vary from solutes which are miscible with water in all proportions to solutes which have only a limited solubility, depending on the relative importance of the polar and apolar regions. In between these extremes is a region where the effects are finely balanced, and in some cases so fine is the balance that an LCST results. In other cases micelle formation occurs, and this could be looked on as a compromise which, in a sense, allows the polar groups to remain in solution while removing the apolar regions from solution. It is not surprising that the major effort in the study of aqueous nonelectrolyte systems has been directed to this particular area.

Again the alcohols constitute the most thoroughly studied groups of solutes of this class, and our discussion of the properties of such solutions will center mainly on what has been learned from studies of water–alcohol

---

* At this point it is timely to mention the work of Abu-Hamdiyyah,[5] if only to warn the reader against taking too seriously the arguments put forward. This work has been the subject of detailed criticism by Holtzer and Emerson,[678] who emphasized the many flaws in its logic. In addition, they expressed strong opinions on the utility of the concept of water structure in rationalizing the properties of aqueous solutions. This aspect of their paper is considered in some detail in Chapter 1. Returning to the paper by Abu-Hamdiyyah, he asserts that a urea molecule can enter a hydrogen-bonded water cluster without causing appreciable distortion and is thus a structure-making solute. In support, he cites the high solubility of urea, the small heat capacity of solution, the dielectric increment, the viscosity increase on solution, and the "near ideality" of such solutions. The superficiality of these arguments is exposed by Holtzer and Emerson,[678] who point out that we do not know the mechanism of solution, so that the solubility provides little information. Also, the effect of structural changes on the heat capacity cannot be assessed. Similarly, they show that the viscosity, dielectric increment, and near-ideality of the solutions provide no evidence for the conclusions reached by Abu-Hamdiyyah.

mixtures. The properties which characterize such solutions can be summarized as follows:

1. At low temperatures and concentrations negative deviations from
Raoult's law are observed, but at higher temperatures ($>10°$) the mixtures
show positive deviations.

2. The excess heats of mixing ΔH^E show a complex dependence on
concentration. Thus, it has been found that in mixtures of low alcohol
content $\Delta H^E < 0$, but as the mole fraction x_2 of alcohol increases, ΔH^E
goes through a minimum, and indeed in some cases at high x_2, ΔH^E also
goes through an endothermic maximum. The shape of the $\Delta H^E(x_2)$ curve
appears to be governed by the number of carbon atoms and the configuration of the alkyl group (Fig. 6). $\Delta C_P{}^E$ is always positive, but here, too, one
finds a complex concentration dependence.

3. The excess volumes ΔV^E are negative, and the $\Delta V^E(x_2)$ curve has
an inflection corresponding to a minimum in $\bar{V}_2{}^E$. No such inflections have
been reported in the thermal properties.

4. There is a minimum in the coefficient of adiabatic compressibility
$\bar{\beta}_2(x_2)$ curve which corresponds to the minimum in $\bar{V}_2{}^E(x_2)$.

These results indicate that there are three concentration regions of
interest in dilute alcohol solutions. It is clear that at a concentration $x_2{}'$

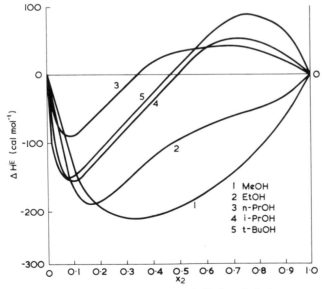

Fig. 6. $\Delta H^E(x_2)$ isotherms ($25°$) for alcohols.

corresponding to the minima in $\bar{V}_2{}^E$ and $\bar{\beta}_2$ there is a change in the nature of the interactions in solution. This concentration is fairly well defined and appears to depend mainly on the number of carbon atoms, their steric configuration, and the number and position of polar groups. At a higher concentration x_2' which corresponds to the minimum in ΔH^E, many of the physical properties give indications of an LCST, reflecting the precarious balance of the polar and apolar regions of the solute molecule.

Having thus briefly summarized the most interesting features of alcohol–water mixtures, it is necessary to consider the observations in more detail and also to amplify them by drawing on data available for other systems containing solutes of the same "mixed" classification. Let us first consider the thermal properties of such solutions.

The concentration dependence of ΔH^E of mixed solutes in water is of great interest. As has already been indicated, these solutes typically produce an S-shaped $\Delta H^E(x_2)$ curve (see, for example, Refs. 140, 212, 214, 972, 973, 995, 1017). The interpretation of such curves is difficult, since they are almost certainly the result of opposed contributions from a variety of complex interactions. If, for example, on mixing water with the solute, hydrogen bonds are broken endothermically and new ones are formed exothermically, ΔH^E will be the difference between two large effects. In general, however, it can be said that regions of exothermic mixing will be due largely to the enhancement of water–water interactions, with water–solute hydrogen bonding producing secondary effects. If we consider the dilute solution region, below the concentration x_2', it would appear reasonable to suppose that due to the structural influence of the solute, the water–water hydrogen bonding is enhanced, thus producing an increasingly exothermic ΔH^E. Beyond x_2', where $-\Delta H^E$ is a maximum, there may be a progressive breakdown in structure, resulting in a more endothermic ΔH^E, since now the ratio of hydrogen bonds broken to hydrogen bonds formed is increasing.

The general shape of the $\Delta H^E(x_2)$ and $\Delta V^E(x_2)$ curves for these systems has been accounted for by Mikhailov and Ponomarova,[972,973] who employed a model based on the following equilibria:

$$H_2O(\text{void}) \rightleftharpoons H_2O(\text{network}) \tag{a}$$

$$H_2O(\text{void}) \rightleftharpoons H_2O(\text{structureless region}) \tag{b}$$

$$\text{nonelectrolyte}(\text{void}) \rightleftharpoons \text{nonelectrolyte}(\text{structureless region}) \tag{c}$$

and derived expressions for the thermodynamic parameters (see Chapter 1). They distinguish three concentration regions. At low concentrations the solute is located preferentially in the voids and the displaced water molecules are incorporated in an "icelike" network. At intermediate concentrations addition of solute is accompanied by a transfer of H_2O(network) to H_2O (void), and at high concentrations the water structure has completely disintegrated. The calculated $\Delta H^E(x_2)$ and $\Delta V^E(x_2)$ curves are found to be of the same general shape as those observed by experiment. In the high concentration region, where the water structure has been completely destroyed, the system behaves generally in accordance with regular solution theory, confirming to some extent the notion that the anomalous solvent properties of water are a consequence of its three-dimensional structured nature.

Also of interest are the relative magnitudes of ΔH^E and $T\,\Delta S^E$. It has been shown that the positive deviations from Raoult's law exhibited by these solutions (i.e., ΔG^E positive) are a consequence of $T\,|\,\Delta S^E\,| > |\,\Delta H^E\,|$, in other words, the solution properties are under entropy control. In this respect the positive deviations from Raoult's law exhibited by these solutions arise from different causes than the positive deviations which may be observed in certain other nonaqueous mixtures, since in these other mixtures such deviations arise from $\Delta H^E > 0$. The relative magnitudes of $T\,|\,\Delta S^E\,|$ and $|\,\Delta H^E\,|$ in aqueous solutions are in themselves an indication of the precarious balance which exists, since $T\,|\,\Delta S^E\,| > |\,\Delta H^E\,|$ is *one* of the criteria for an LCST, i.e., only in systems whose mixing properties are under entropy control can an LCST occur. Careful examination of these systems does, indeed, show how precarious the balance is in the lower alcohols and ethers. For example, Malcolm and Rowlinson[941] have found that for dioxan–water mixtures over a wide temperature range the conditions for an LCST just fail to be met, and indeed addition of a small amount of a third component is sufficient to cause phase separation. The properties of *t*-BuOH–water and THF–water mixtures are similarly close to producing an LCST,* and indeed for THF[954] the existence of an LCST has recently been established at 72°C, i.e., just above the normal boiling point at atmospheric pressure. With the higher alcohols the apolar regions, and hence $T\,|\,\Delta S^E\,|$, become all-important, and one finds that the system separates into two phases at all temperatures, with no LCST being apparent, since now the positive deviations from Raoult's law have increased to such an extent that a one-phase system is unstable in certain concentration ranges.†

---

* This is clearly demonstrated by small-angle X-ray scattering.[78]

† Many of these systems exhibit negative temperature coefficients of solubility (as can be seen from their phase diagrams), indicating that, if freezing did not occur, a LCST would be observed at some temperature below the freezing point of the system.

One further consequence of $T \,|\, \Delta S^E \,| > |\, \Delta H^E \,|$ is that at low temperatures negative deviations from Raoult's law should be observed. This has indeed been found to be the case[769,786] in alcohol solutions at high dilution and low temperatures ($<10°$). As the concentration, or the temperature, is raised, the sign of ΔG^E changes. While mathematically it is easy to see the change in sign of ΔG^E with temperature as being due to the increasing importance of the $T \Delta S^E$ term, it is difficult to visualize the physical significance of such an observation. It is also difficult to account satisfactorily for the concentration dependence, so that it is best to leave this behavior unexplained, bearing in mind the possibility that for these systems the ranges of temperature and concentration studied may be, in structural terms, very wide indeed.

The study in which the existence of negative deviations was established is worthy of close examination.[786] Freezing point depression and other methods were employed to evaluate the water activity in a range of dilute nonelectrolyte solutions, and, expressing the activity coefficient γ_1 by the expansion

$$\ln \gamma_1 = Bx_2{}^2 + Cx_2{}^3 + \cdots$$

B and C were evaluated. These were then related, through McMillan–Mayer theory, to pair and triplet interactions. The results for water–alcohol mixtures, summarized briefly, show alcohol molecules in solution to attract each other, and that this tendency increases with rising temperature. Pairwise interactions are more important for methanol and ethanol, while triplet interactions are more important for the higher alcohols. In more concentrated solutions ($x_2 > 0.1$) of alcohols, $\gamma_1 > 1$ at all temperatures, requiring higher terms in the virial expansion, and implying that associations of large numbers of solute molecules are important. Similar measurements for water–dioxan indicated that pair interactions are relatively unimportant, tending to support the evidence from other techniques which suggests that dioxan has properties which would lead one to classify it as somewhere between a typical "mixed solute" and a hydrophilic solute. The increase in the pairwise interaction with temperature appears to parallel the expected behavior which would result from "hydrophobic bonding," so that this result might, at first sight, be taken as convincing evidence for the occurrence of "hydrophobic bonding" between the apolar regions of solutes in solution. However, as will be seen later, this same investigation produced similar results for certain of the sugars, the pairwise interactions of the same order of magnitude also increasing with temperature, and since it is difficult to visualize "hydrophobic bonding" in a hydrophilic solute, it would appear that the McMillan–Mayer approach may not be as useful, and readily interpretable, as might appear at first sight to be the case. A fuller discussion of the results for hydrophilic solutes will be found in the appropriate section. Suffice it to say here that the results of Kozak et al.[786] seem only to indicate that pairwise interactions increase with increasing size of the solute molecule.

So far we have not considered $\Delta C_{P_2}^{E}$ for these solutions. As might be expected, $\Delta C_{P_2}^{E}$ is found to be large and positive, indicative of solute-induced changes in the intermolecular structure of the water. A typical example of this is provided by the temperature dependence of ΔH^{E} for acetone–water mixtures (Fig. 7). Thus, at 20° mixing is exothermic over most of the composition range, while at 90° it is endothermic over most of the composition range.[1028] Again, it is not possible at present to analyze such observations more thoroughly, as the necessary theoretical background has yet to be developed.

Up to this point we have mentioned only briefly the phase behavior of these systems, indicating that in certain cases conditions are right for the occurrence of an LCST. Several studies have been carried out which show the change in the phase behavior of homologous series of solutes,

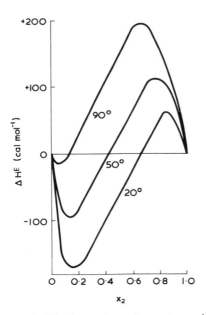

Fig. 7. ΔH^{E} isotherms for water–acetone mixtures as function of mole fraction of acetone. These curves are typical of aqueous solutions of "mixed solutes."

for which the increasing chain length results in $|\Delta S^{\mathrm{E}}|$ increasing faster than $|\Delta H^{\mathrm{E}}|$, since ΔH^{E} is to some extent at least governed by the polar group, whereas ΔS^{E} is largely determined by the apolar group. In suitable systems a gradation of phase behavior is found, from solutes which are completely miscible at all temperatures, through solutes which exhibit an LCST, finally reaching solutes which exhibit a miscibility gap at all temperatures. The conditions for an LCST were indicated in the introductory section, the discussion being based on a study by Copp and Everett[337] on triethylamine–water mixtures. The most extensive information derives from mixtures of water and substituted pyridines. The classic studies of these systems are those by Andon et al.,[30] who derived ΔG^{E} from vapor pressure measurements. Both the work of Copp and Everett and of Andon et al. illustrate the usefulness of the approximate working criteria for an LCST which can be derived from the rigorous thermodynamic requirements, since the approximate relationships can readily be employed to predict the likelihood of phase separation, whereas the rigorous relationships require data which are not generally available. As mentioned previously, THF–water mixtures also exhibit an LCST. In all cases the solutes fall into the "mixed" classification, since only with this class of solute does one find a negative entropy of solution (typical of apolar groups) coupled with a sufficiently exothermic ΔH^{E} to produce appreciable solubility.

One further aspect of certain mixed solutes in this concentration region is of interest—micelle formation. In a way, micelle formation could be considered as a phase separation, but of a special type, for here the apolar residues of solute molecules (or ions) aggregate to form a "nonaqueous" environment, while the polar groups remain on the cluster surface and in contact with the water. It is not the purpose of this chapter to give a detailed account of the properties of micellar solutions, since this is a vast subject in its own right. Rather, it is to indicate how the special properties of mixed solutes in aqueous solution give rise to this type of behavior for many systems.

In connection with micelles, it is appropriate to discuss the concept of "hydrophobic bonding," which has already been mentioned in connection with the work of Kozak et al.[786] The term "hydrophobic bond" is in a way rather unfortunate, since the concept as first defined[735] does not really imply a bond, but rather a *non*-bonded interaction. The idea first arose from considerations by Kauzmann[735] of the factors affecting the conformational stability of proteins in aqueous solution. Since the "solution" of an apolar group in water is accompanied by an unfavorable entropy change, Kauzmann considered that there would be a tendency for the apolar groups of the protein to come in contact with one another in solution, lessening their area of contact with the water. This tendency of apolar groups to come together in aqueous solution and thus minimize energetically unfavorable interactions with the water is termed "hydrophobic bonding." As such, it is one of the important factors in micelle formation. Nemethy and Scheraga[1027] demonstrated in a semiquantitative manner the validity of this concept, and from their discussion it was clear that, at least at fairly low temperatures, the strength of the hydrophobic bond would increase as the temperature increased. This is in marked contrast to the behavior of normal chemical bonds. Ben-Naim[149] has also provided a theoretical treatment showing that there will be a tendency for apolar groups in water to come together and thus minimize their interaction with the water.

As indicated above, the "hydrophobic bond" plays an important role in protein conformational stability in aqueous solution, and also in the stabilization of micelles. Further, the tendency of mixed solutes to aggregate into dimers, trimers, etc. in aqueous solution, as evidenced by the work of Kozak, Knight, and Kauzmann, would, if it does indeed occur, appear to be a reflection of the importance of hydrophobic bonding. Later, when considering the tetraalkylammonium ions, it will be seen that these, too, have a tendency to aggregate in solution, forming cation–cation pairs. This, also, would appear to be a manifestation of a hydrophobic interaction.

Let us now move on to consider the PVT properties of dilute solutions of mixed solutes. There is available a fairly large body of data on this subject, and we propose to consider only those investigations where the results are relevant to the understanding of the nature of such solutions. Experimentally, it has been found that even at high dilutions $\bar{V}_2^E(x_2)$ plots for many mixed solutes have a finite, nonzero slope. Typical examples of this are the alcohols[517,524] and the cyclic ethers.[518,995] It is our opinion that such slopes are of necessity an indication of solute–solute interactions. Since it is unlikely that there exists any direct long-range interaction between these solute molecules comparable to the electrostatic interactions between ions, it would appear reasonable to assume that the long-range interaction is transmitted specifically by the solvent, and so one could envisage this interaction as some form of either reinforcement or interference occurring between the structured regions of solvent affected by individual solute molecules. The minimum which is observed in many $\bar{V}_2^E(x_2)$ curves could be considered as being the point at which reinforcement of solvent sheaths begins to be replaced by interference between solvent sheaths, since there is no longer sufficient solvent to support the full structuring ability of the solute molecules. Except in this region of concentration, it is difficult to interpret volume effects, since at higher concentrations both mixing volumes ΔV_m and excess volumes $\bar{V}_2^E$ include contributions from solute–solute, solute–solvent, and solvent–solvent effects. However, those investigations which have been carried out at sufficiently low concentrations enable us to comment on the role of both the polar and the apolar regions of the mixed solute molecule. Experimentally, it has been found that the slope of the $\bar{V}_2(x_2)$ curve is dependent on the ratio of polar to apolar groups in the solute molecule.[1016] As the proportion of polar groups increases, the negative slope of $\bar{V}_2(x_2)$ decreases in magnitude, and in the case of very polar (hydrophilic) solutes the slope is near zero, indicating that as the apolar nature of the solute decreases, its capacity to influence water structure at long range decreases. The positive slopes which are observed beyond the characteristic minimum are taken to be indicative of structure breaking, and in the case of a highly polar solute such as H_2O_2,[986] it is thought that this effect predominates at all concentrations, since $\bar{V}_2(x_2)$ has at all times a positive slope. If one considers the alcohol data, it can be seen (Fig. 8) that as the alkyl group increases in size, $\partial \bar{V}_2^E / \partial x_2$ becomes more negative, and the minimum in $\bar{V}_2^E(x_2)$ moves to lower x_2. This effect appears to be a maximum for t-BuOH. Since it appears that the t-butyl group is of a size most favorable to clathrate-type structure stabilization, it would not appear unreasonable to expect that this group would produce the most extensive structural effect in water, and the volumetric behavior of the alcohols would certainly appear to conform with this notion.

The structuring effects of some mixed solutes can be seen clearly when one considers another aspect of their volumetric behavior, namely their ability to shift the temperature of maximum density, θ, of water. It has been shown[525] that a solute which mixes ideally with water gives rise to a negative $\Delta\theta$. This ideal contribution $\Delta\theta_{id}$, when subtracted from the measured $\Delta\theta$, leaves a structural contribution, which can only be positive if the excess expansibility $\bar{\alpha}_2^E \ (= \partial \bar{V}_2^E / \partial T)$ is negative. In several cases such negative excess expansibilities, or positive $\Delta\theta_{str}$, have been demonstrated, providing convincing evidence for the "structure-making" effects of mixed solutes. Wada and Umeda[1363] demonstrated a positive $\Delta\theta$ for lower alcohols, ethers, and ketones, and later it was also shown that $\bar{\alpha}_2^{\circ E}$ of aqueous butanols was negative,[524] giving rise to positive $\Delta\theta_{str}$. Certain amines also produce a positive $\Delta\theta_{str}$,[525] as do certain cyclic ethers,[518] but not dioxan. The case for assuming that a positive $\Delta\theta_{str}$ corresponds to a stabilization of the intermolecular structure of water receives support from the observation that for D_2O, $\theta = 11°$, which is believed to reflect the stronger hydrogen bonds in this liquid.

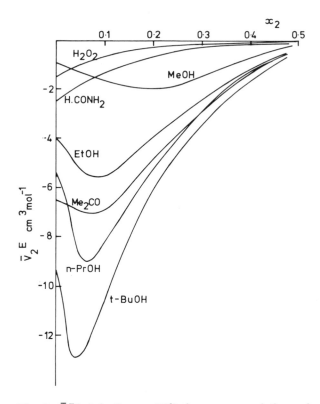

Fig. 8. $\bar{V}_2^E(x_2)$ isotherms (25°) for aqueous solutions of mixed (MeOH, EtOH, PrOH, t-BuOH, Me₂CO) and hydrophilic (H₂O₂, H.CONH₂) solutes.

Let us now further consider the expansibilities of dilute solutions of mixed solutes. As has been indicated, it is expected that a solute with a negative $\partial \bar{V}_2^E/\partial T$ will be structure making, and if $\partial \bar{V}_2^E/\partial T$ is positive, the solute will be a structure breaker. Since at sufficiently high concentrations all solutes would be expected to exert a structure-breaking effect, a structure-making solute (at low concentrations) will exhibit a negative $\partial \bar{V}_2^E/\partial T$ which changes sign at higher concentrations. Two typical examples of such behavior are provided by THF and propylene oxide (PO).[995] A further use to which expansibilities have been put is described by Neal and Goring,[1019] who propose a method of distinguishing between hydrophobic and hydrophilic types of solution behavior.

A further property of aqueous alcohol solutions which appears to reflect the occurrence of structural interactions in solution is the thermal pressure coefficient γ $[= (\partial P/\partial T)_V]$. Preliminary measurements[913] of γ

as a function of concentration have shown that $\gamma(x_2)$ goes through a pronounced maximum in the concentration range in which other volumetric anomalies occur. The most interesting feature of this particular anomaly is its magnitude, γ^E being of the same order as γ_{id}.

2.2.2. Hydrophilic Solutes

The concentration dependence of the thermodynamic properties of hydrophilic solutes has been studied more widely than have the infinite dilution properties. As indicated in Section 1.1, such solutes are characterized by $|\Delta H^E| > T|\Delta S^E|$, a behavior typified by the polyhydroxy solutes (e.g., glucose[1280]) and DMSO.[750,1127] In the past two widely differing approaches have been taken in order to explain the properties of such solutions. In the first approach, that of the semiideal solution,[1253] it is proposed that the solute forms a series of hydrogen-bonded complexes with the water, and that each individual complex then behaves ideally in its solution in water. Thus, in brief, the solution properties are explained in terms of a series of well-defined solute–solvent equilibria, and no solute–solute interactions need be invoked. The second approach[786] is based on the McMillan–Mayer theory of solutions, and essentially describes the solution in terms of solute–solute interactions. That both approaches achieve some degree of success is a reminder of the extreme complexity of aqueous solutions and of the inadequacy of thermodynamics to discriminate between two realistic models.

The semiideal solution theory has been applied mainly to isopiestic activity data which have been obtained for hydrophilic solutes in aqueous solutions.[1152,1186] It is assumed that the water can form a series of hydrogen-bonded complexes with the solute, each functional group of the solute molecule having a characteristic binding constant. For polyhydroxy solutes all the hydroxyl groups are assumed to be capable of hydrogen bonding with a water molecule, and it is also assumed that the stepwise binding constants K are equal. These assumptions in effect lead to a one-parameter fit (K) of the isopiestic solvent activity data, and in the case of glucose, sucrose, glycerol, and mannitol this fit is found to be surprisingly good when one considers the crudeness of the basic assumptions. It would therefore seem to be a reasonable first approximation to describe aqueous solutions of hydrophilic solutes by this approach. Due to the oversimplifications employed, it would not appear to be worthwhile to attempt an interpretation of the variations of the K values with different solutes, so that the theory does not hold any promise when one wishes to study the subtle differences which may exist between the solution properties of closely related solutes.

Kozak et al.,[786] using the McMillan–Mayer theory approach, reached entirely different conclusions on the nature of solutions of hydrophilic solutes. As indicated in Section 2.2.1, they employed activity data, obtained both from freezing point depression and from isopiestic measurements, together with heats of solution to evaluate the second and third osmotic virial coefficients of a range of aqueous mixtures (see Chapter 1), including some solutions of hydrophilic solutes. From the virial coefficients the extent of pairwise and triplet solute–solute interactions could be estimated. Their results led them to postulate that pairwise interactions occurred, the extent of these interactions increasing with increasing number of solute functional groups capable of participating in hydrogen bond formation. They found raffinose (with 11 hydrogen-bonding groups) to have stronger interactions than sucrose (with eight), mannitol (with six), and glycerol (with three). A similar pattern was indicated for triplet interactions. Once again, however, this is an approach which holds little promise when it comes to interpreting the subtle differences between closely related solutes, since solutes with the same number of functional groups capable of hydrogen bonding gave similar results for the extent of pairwise interactions.

Recently it has been shown that a specific hydration model provides a better description of such solutions. Tait et al.[1275] arrived at this model from consideration on the NMR and dielectric relaxation data obtained for monosaccharide solutions. To explain their results, they had to postulate a model of monosaccharide hydration in which the hydration water was assumed to be in equilibrium with a lattice (or bulky) structured component of the solvent water. This would therefore account for the absence of long-range interactions in such solutions, and hence the small concentration dependence exhibited by the bulk properties of the solution. The distribution of equatorial and axial hydroxyl substitution would be expected to affect the primary hydration to some extent, producing subtle variations in solution properties from one solute to another. Franks et al.[520] have considered the volumetric and other thermodynamic properties of such solutions in terms of the three approaches outlined above, and have shown that the specific hydration model does, indeed, provide a better description of these solutions. It was, however, pointed out that the Stokes and Robinson[1253] treatment provided a useful first approximation when one was considering the simpler thermodynamic functions, such as ΔG^{E}, and it was only on studying the derivatives of ΔG^{E}, such as volume, that the inadequacies of the theory became important. Schönert[1204] has also shown that in the case of sucrose $\partial(\ln \gamma)/\partial T$ and $\partial(\ln \gamma)/\partial P$ cannot be fitted by the simple mean hydration number model of Stokes and Robinson.

It is concluded, therefore, that at the present time the best explanation of the thermodynamic properties of aqueous solutions of hydrophilic solutes is in terms of the specific site hydration model. The absence of long-range interactions in solution is a consequence of the postulated equilibrium between hydration water and the lattice component of bulk water, as the solute will not perturb the solvent water at a distance beyond the hydration sheath. This is in marked contrast to the behavior of solutes with apolar groups, where these are believed to produce an altered structural arrangement, giving rise to long-range interactions in solution.

2.2.3. *Tetraalkylammonium Halides*

Even at fairly low concentrations the tetraalkylammonium salts exhibit some remarkable deviations from Debye–Hückel theory, and these have highlighted the role played by the apolar groups in the solution properties of these ions. The osmotic and activity coefficients for solutions of such salts have been determined by many workers. The most systematic study has been that of Lindenbaum and Boyd,[851] who studied a series of tetra-alkylammonium halides by an isopiestic technique. For concentrations below 1 m the osmotic coefficients for the chlorides increase with the size of the cation $Bu_4N^+ > Pr_4N^+ > Et_4N^+ > Me_4N^+$, while bromides and iodides show the reverse order. At higher concentrations the osmotic coefficients for the salts of the larger cations decrease sharply. The dependence of osmotic coefficients and activity coefficients on cation size and on concentration was explained as the result of three effects: The chlorides showed the "structure-making" ability of the cations, while for the bromides and iodides, "water-structure enforced ion pairing" was assumed to be the dominant process. At higher concentrations micelle formation was believed to be important for the larger cations. Similar explanations were found necessary to explain the osmotic coefficients of a series of tri-*n*-alkylsulfonium halides.[848]

In an extension of this work Lindenbaum *et al.*[852] have studied the variation of osmotic coefficients with temperature in solutions of tetra-alkylammonium halides. In combination with data available for 25°, they were able to estimate the partial molal entropy of the solvent in these solutions. The results could be understood in terms of the effect of anion size and of the increasing temperature on the local structure of the solvent water, and the relationship between these effects and the "structuring" ability of the apolar regions of the R_4N^+ ions.

Further evidence for the importance of the apolar groups was provided by a study of the entropies and heats of dilution of tetra-*n*-alkylammonium halides in aqueous solution.[847] The heats of dilution for a given halide decreased in the order $Bu_4N^+ > Pr_4N^+ > Et_4N^+ > Me_4N^+$, which is in accord with the expectation that the hydration of these ions would follow this order. The apparent molal excess entropies for Bu_4N^+ and Pr_4N^+ reflected the structure-making ability of these large ions. Rather surprisingly, the anion did not appear to have much effect on the apparent molal excess entropies. This suggested that the structural contributions to the enthalpy and entropy were so great that they masked the differences due to the relative sizes of the anions, and any ion pairing. However, typically, structural contributions to enthalpy and entropy tend to cancel out in the free energy, and so ion pairing effects can be observed in the free energy. This entropy/enthalpy compensation is a common phenomenon in aqueous solution, and is discussed in detail by Ives and Marsden.[701] (See Volume 3, Chapter 3.)

The concentration dependence of the volumetric properties of the tetraalkylammonium salts in solutions is also different from that of normal 1:1 electrolytes, initial experiments having suggested that plots of ϕ_v versus $\sqrt{c}$ had negative limiting slopes.[1392] However, Franks and Smith[523] have shown that at very low concentrations the tetraalkylammonium bromides tend to conform to the behavior expected of univalent electrolytes. The negative deviations from expected Debye–Hückel behavior at higher concentrations have been explained in terms of structural stabilization[523,1392] due to the proximity of the hydration shells of two ions. Overlap of the shells is presumed to result in structural reinforcement, reducing the volume requirements of the ion. Thus the deviations are a direct consequence of the apolar nature of the ions. This concept has been extended by Friedman,[532] who has attempted to account for the properties of such solutions in the following manner (see Volume 3, Chapter 1): The pair potential function $U(r)$ is considered to consist of several terms: a core repulsion term, an electrostatic term, a term which takes into account the overlapping hydration cospheres of the ions, and a term describing the behavior of a cavity in a dielectric medium:

$$U(r) = U_{core} + U_{es} + U_{hyd} + U_{cav}$$

By considering the situation when hydration cospheres overlap, he shows that the pair correlation functions $g(r)$ for tetraalkylammonium salts indicate the existence of both cation–anion and cation–cation ion pairs. The cation–cation pairs are the result of "structure-enforced" ion pairing. Further experimental evidence for cation–cation ion pairs has been obtained from volumetric studies in ternary systems containing potassium halides and tetraalkylammonium halides in H_2O and D_2O.[1390,1391]

The H_2O–D_2O isotope effect on $\bar{V}_2$ of tetraalkylammonium ions is found to be consistent with the concept of these ions as "apolar structure makers," the volume change on transfer from H_2O to D_2O being positive.[332]

Finally, it is worth noting that Desnoyers et al.[410] have employed an interesting approach to the volumetric behavior of alkali halides, in which they consider $\phi_v = \phi_v{}^\circ + 1.86\sqrt{c} + hc$. The h parameter is related to a structural interaction model, two solutes being considered to attract one another if their structural influences are compatible, and to repel if incompatible. They have claimed that this model is capable of accounting for most of the excess thermodynamic functions of tetraalkylammonium salts in solution.

2.2.4. Urea

As in the case of urea solutions at infinite dilution, a useful review of reliable data is given by Stokes.[1252] Another compilation is that due to Kresheck and Scheraga.[790] In an attempt to explain the concentration dependence of the thermodynamic properties of urea, Stokes employed an association model of the form

$$U + U \rightleftharpoons U_2$$
$$U_2 + U \rightleftharpoons U_3, \text{ etc.}$$

and employed Flory–Huggins (volume-fraction) statistics to derive the thermodynamic parameters governing these equilibria. He included in his treatment a reasoned criticism of earlier attempts to treat urea solutions in terms of a series of association equilibria and pointed out that Schellman's[1189] implicit assumption that the molal osmotic coefficient of an ideal solution is unity was incorrect and hence his calculations, and the more extensive calculations of Kresheck and Scheraga, based on Schellmann's, were in error. By employing volume fraction rather than mole fraction statistics, Stokes obtained a sensibly constant value for the heat of association (i.e., independent of the degree of polymerization), and he also computed satisfactorily constant values for the equilibrium constants for the various association steps under a range of conditions. All this would tend to indicate that urea–urea interactions alone can satisfactorily explain the activity variations in urea solutions, as well as the heats of dilution of such solutions. However, one must remember that this is not a unique explanation. Frank and Franks[509] have also considered the thermodynamic properties of urea solutions, employing in this case a radically different model, according to which urea is assumed to mix ideally with a postulated "dense" (non-hydrogen-bonded) species of water but, for geometric reasons, is considered to be barred from tetrahedrally ordered water clusters (the "bulky" species) (see Chapter 1). After setting up the relevant partition functions to describe this situation, calculations show that on the basis of this formalism, urea would cause structure-breaking by lowering the chemical potential of the dense species and therefore shifting the equilibrium between dense and bulky water in favor of dense water. By using what appeared to be reasonable values for the adjustable parameters of their model, as reasonable a fit was obtained for the activity coefficients and relative enthalpies of urea in aqueous solution as is provided by the association model of Stokes. The decision whether either model does in fact describe the real situation is not an easy one. In strictly thermodynamic terms, either model is satisfactory, so that any decision as to the validity of the models must be based on nonthermodynamic observations. Fortunately, spectroscopic methods should·be capable of discriminating between a formalism which requires the existence of dimers, trimers, etc. of urea, and one in which urea is effectively an "inert" diluent.

Recently, such a spectroscopic study has been performed[485] employing the NMR technique in a study of aqueous urea solutions (Chapter 10). From a detailed consideration of their results the authors conclude that there is no evidence for the existence of long-lived dimers or higher aggregates of urea in solution, and hence the Stokes model[1252] is not supported, whereas the statistical structure-breaking model[509] may well be closer to reality. This result provides a further illustration of the point made in the discussion of the paper by Holtzer and Emerson[678] (see p. 5), that thermodynamics does not provide the primary evidence for any model, but rather that thermodynamics can only discriminate between a model which is wrong and one which *might* be realistic.

2.3. Dilute Solutions of Water in Nonelectrolytes

One further concentration region of interest in binary aqueous nonelectrolyte systems remains to be considered, that of dilute solutions of water in the nonelectrolyte. This region is the subject of a comprehensive review by Christian *et al.*[298] and will therefore be considered only briefly in this chapter. In such solutions the long-range structure of water which is so important to the properties of dilute aqueous solutions is no longer in existence, and the only aqueous species of any importance are monomers, dimers, etc. The properties of such solutions, therefore, are those of typical nonaqueous solutions, which indeed they are. It is generally believed that water dissolves in aliphatic and aromatic hydrocarbons and other apolar solutes, primarily as the monomer. Thermodynamic evidence from cryoscopic measurements,[1086] vapor pressure measurements,[297] and partial molal volumes[951] support this conclusion, though at concentrations close to saturation there is some indication of associated species. In slightly polar solvents vapor pressure[715,950] and other nonthermodynamic evidence indicates that the water is associated to some extent (see Chapter 4). Spectral evidence indicates that trimeric species are of greatest importance. For solutions of water in polar organic solutes the main polymeric species appear to be dimers and trimers.

In addition to the monomer–dimer–trimer self-association of the water molecules, there also exists the possibility of specific solute–solvent interactions, the water-forming molecular complexes with the organic solvent. This is a complex subject, a wide range of hydrates existing in solution. The reader is referred to the review by Christian *et al.*[298] which treats this topic in some detail.

2.4. Summary of Binary Systems

The preceding discussion of the thermodynamic properties of binary aqueous solutions of nonelectrolytes clearly illustrates the difference in the character of the interactions between apolar groups and polar groups with water. The interaction of an apolar group with water is one which has no parallel in other solvent systems, and is in many respects a manifestation of the structured nature of liquid water. Most of the interest in aqueous nonelectrolyte solutions arises from the unusual behavior which is a consequence of this interaction. While the interaction between polar groups and water would appear to be more "normal" when viewed in the light of other solvent systems, water itself is such an abnormal liquid that even "normality" must be viewed as unusual. Though an increasing effort is being devoted to studies of aqueous nonelectrolyte systems, we are still far from understanding their full complexity, and much work is still required before we reach a proper understanding of the many facets of the behavior of such solutions. In Table VII we have attempted to summarize the present state of knowledge of the solution properties of certain binary aqueous non-electrolyte systems, in an attempt to illustrate the large gaps which exist in our knowledge of the thermodynamic properties of such systems. The number of $+$ signs is an indication of the quantity and quality of available data.

TABLE VII. Extent of Present Knowledge of the Thermodynamic Properties of Some Typical Binary Aqueous Nonelectrolyte Solutions

Solute type	Infinite dilution		Finite concentrations	
	Property	Ref.	Property	Ref.
Hydrocarbons	ΔH_h° (+++)	207, 309, 515, 908, 909, 1001, 1002, 1021, 1141	—	—
	$\Delta G_h^{\ominus}$ (+++)	207, 309, 515, 908, 1001, 1002, 1021	—	—
	$\Delta S_h^{\ominus}$ (++)	207, 309, 515, 908, 1001, 1002, 1021	—	—
	$\Delta C_{P_2}^{\circ}$ (+)	431	—	—
	$\bar{V}_2^{\circ}$ (+)	949	—	—
Alcohols	ΔH_h° (++++++)	14, 48, 667, 797	ΔH_m (++++)	135, 136, 140, 214, 751, 769, 823
	$\Delta G_h^{\ominus}$ (++++++)	14, 258, 667	ΔG_m (+++)	259, 751, 986, 1121
	$\Delta S_h^{\ominus}$ (+++++)	14, 667	ΔS_m (+++)	751, 986, 1121
	$\Delta C_{P_2}^{\circ}$ (+++++)	14, 48, 667	ΔC_{p_m} (+)	769
	$\bar{V}_2^{\circ}$ (+++)	12, 524, 534	ΔV_m (+++)	12, 517, 524, 534, 1015, 1016
Amines	ΔH_h° (++++)	260, 526, 778, 1436	ΔH_m (+++)	169, 176, 337, 1435, 1436
	$\Delta G_h^{\ominus}$ (++++)	260, 299, 383, 526	ΔG_m (+++)	169, 337, 772-774
	$\Delta S_h^{\ominus}$ (++++)	260, 526	ΔS_m (++++)	169, 337
	$\Delta C_{P_2}^{\circ}$ (+++)	526, 778	ϕ_v (+++)	334
	$\bar{V}_2^{\circ}$ (+++)	334, 1336	—	—

TABLE VII. (*Continued*)

Solute type	Infinite dilution		Finite concentrations	
	Property	Ref.	Property	Ref.
Rare gases	ΔH_h° (+++)	13, 1316	—	—
	$\Delta G_h^\ominus$ (+++)	1316	—	—
	$\Delta S_h^\ominus$ (+++)	1316	—	—
Cyclic ethers	ΔH_h° (++++)	261, 518	ΔH_m (++++)	579, 995, 1017, 1339
	$\Delta G_h^\ominus$ (+++++)	261	$\bar G$ (+++)	66, 579
	$\Delta S_h^\ominus$ (+++++)	261	$\Delta C_{P,m}$ (+++)	995
	ΔC_P° (+++)	518	ΔV_m (+++)	995
	$\bar V^\circ$ (+++)	518	—	
Amides	ΔH_h° (+++)	51, 778	ΔH_m (+++)	53
	ΔC_P° (+++)	778	$\bar G$ (+++)	53
	$\bar V_2^\circ$ (+++)	54, 53	ΔV_m (+++)	54, 53
Amino acids	ΔH_{soln}° (+++)	690, 789	—	—
	ΔG_{soln}° (+++)	690	$\bar G$ (++++)	449, 691, 1235
	ΔS_{soln}° (+++)	690	$\bar S$ (+++)	1149
	$\bar C_P^\circ$ (+++)	788	—	—
	$\bar V_2^\circ$ (+++)	449	—	
Monosaccharides	ΔH_{soln}° (++++)	1265, 1280	$\bar H$ (++++)	1280
	$\bar V_2^\circ$ (+++)	520	$\bar G$ (+++++)	1152, 1186, 1253, 1280
	Mutarotation properties	1046	$\bar S$ (++++)	1280
			$\bar V$ (+++)	1280

3. TERNARY SOLUTIONS

3.1. Ternary Solutions Containing Both H_2O and an Apolar Solute

In Section 2.1.1 it was made clear that, especially in recent years, the study of ternary aqueous systems with an apolar solute as one component has greatly helped to clarify the nature of the interactions between solute and solvent. Some of the most striking examples of the utility of this approach are provided by the work described in this section. One fruitful field of study has been that of ternary systems containing water, urea, and an apolar solute. One of the most interesting thermodynamic studies on such systems is that of Wetlaufer et al.[1398] who determined, by means of solubility measurements over a range of temperatures, the thermodynamic transfer functions of eight hydrocarbons from water to 7 M solutions of urea and 4.9 M solutions of guanidine hydrochloride, both of which substances are important as protein denaturants. While Wetlaufer et al. were mainly interested in obtaining data for application to studies of protein denaturation and made little use of the data to consider the nature of the ternary solution systems per se, Frank and Franks[509] later made use of these data in order to test their "statistical structure-breaker" theory of urea solutions (see Section 2.2.4), and at the same time employed their solution model to give further insights into the nature of the interactions in aqueous hydrocarbon solutions. With regard to hydrocarbons, the model suggests that the hypothesis by Frank and Evans[508] that the "extra" negativeness of $\bar{S}_2$ for hydrocarbon gases in water arises from a structure equilibrium shift in the water, induced by the apolar surface, may not be unique. An alternative mechanism which might also produce this effect could arise from the fact that water received some of the hydrocarbon solute into a different type of environment from that which it would encounter in a "normal" solvent (see Chapter 1). Consideration of the two binary aqueous mixtures, namely water–alkane and water–urea, has also made possible an explanation of the observed hydrocarbon solubility changes which are produced by the addition of urea.[509,1398]

Another study of ternary aqueous/apolar mixtures which has yielded results of interest was by Wen and Hung[1389] on systems containing water, hydrocarbon, and tetraalkylammonium salt. They measured the solubilities of methane, ethane, propane, and butane in water and in 0.1 to 1.0 m solutions of ammonium bromide, $(HOC_2H_4)_4NBr$, and R_4NBr at four temperatures and derived the thermodynamic parameters for transfer. The results show that ammonium bromide salted out all the gases, while the large tetraalkylammonium ions had a salting-in effect. Both $(CH_3)_4NBr$ and $(HOC_2H_4)_4NBr$ salted out the smaller gas molecules and salted in the larger gas molecules. Their explanation for this behavior considers two effects: (a) indirect interactions between R_4NBr and RH through changes in the structure of the aqueous medium, and (b) direct, hydrophobic interaction between the two solutes. Since, as Eley[446] suggested, it is likely that gas solubility is dependent on the number of cavities existing in water (see Fig. 1 and discussion) and the cavities presumably occur in the water clusters, then the number of cavities (or voids) will be greater at low temperatures, and so the gas solubility will fall off with increasing temperature. The "hydrophobic structure-making" due to the tetraalkylammonium salts should also decrease the availability of voids, and hence reduce the solubility. Alternatively, if one accepts that "hydrophobic hydration" falls off with increasing temperature, as evidenced by the decrease in solubility of hydrocarbons, then whatever mechanism is employed to explain this decrease in hydration, one would expect competition between the hydrophobic hydration of the tetraalkylammonium salts and the hydration requirements of the alkanes, resulting in a reduced solubility of the alkane which would be most marked at low temperatures, where the hydration is more extensive. At low temperatures, where the hydration of Bu_4N^+ is most marked in comparison to the smaller R_4N^+ ions, one would then expect a much lower solubility of hydrocarbons in solution of this ion. At higher temperatures, where the hydration is much reduced, the reversal of salting-out efficiency is readily rationalized in terms of the reduced importance of the hydration, allowing mechanism (2) to become the governing factor. If this factor did assume greater importance, one would expect its efficiency to increase with increasing size of the apolar groups on the RN_4^+ ion, i.e., the solubility of hydrocarbons would be greatest in Bu_4NBr solutions. The observed change in salting out efficiency between 5° and 25°, namely

$$Bu_4N^+ > Pr_4N^+ > Et_4N^+ \qquad (5°)$$

$$Et_4N^+ > Pr_4N^+ > Bu_4N^+ \qquad (35°)$$

would appear to be satisfactorily explained in terms of these mechanisms, with competition for hydration being important at low temperatures, where the hydration is most stable, and hydrophobic interactions becoming more important at higher temperatures, where the long-range hydration is significantly reduced. That the range and magnitude of the salt effect in $(EtOH)_4NBr$ are similar to those of Me_4NBr though in size $(EtOH)_4NBr$ is similar to Pr_4NBr can be understood on the basis of the "subunit" approach explained in Section 2.1.2. EtOH would be equivalent to Me—COH, and so the hydrophobic hydration would be that of a Me group. The hydration of a —COH group does not appear to be extensive, and does not interfere with the hydration of the hydrophobic section of the molecule, so that one would, on this basis, expect the effect of $(EtOH_4)NBr$ to be similar to that of Me_4NBr, as observed.

An interesting series of studies on ternary aqueous systems, including one apolar solute, is that of Ben-Naim and co-workers[143,144,146,153,154] dealing with the thermodynamics of dissolution of argon in mixtures of water with a second solute, normally an organic compound. By employing a two-structure model in which static and relaxational contributions to the thermodynamic functions are defined, it is possible to obtain an indication of the thermodynamic functions for transfer of argon from pure water to the solution. Depending on the sign of the static contribution to the transfer entropy, one can classify a solution as being more or less structured than water. It is not entirely clear, however, whether this is a true indication of structural effects or whether it is an indication of cooperation or antagonism between the solution structures which may be induced both by argon and by the other solute, analogous to similar effects in electrolyte solutions, discussed by Steigman and Dobrow.[1247]

Some results obtained by Ben-Naim for the solubility of argon in various mixed solvents are summarized in Table VIII. On the basis of his model, he concludes that the following solutes increase the structural integrity of the water: methanol, ethanol, n-propanol, and n-butanol, which are all mixed solutes in our classification. Glycerol, sucrose, and glucose are thought to decrease the structural integrity of the water. These are all hydrophilic solutes, and it may be that they produce a structural effect which is incompatible with the apolar hydration of argon. Simple electrolytes also decrease the solubility of argon. The salting out observed as resulting from addition of tetraalkylammonium ions is difficult to account for on Ben-Naim's model, but the discussion of Wen and Hung[1389] on the effect of tetraalkylammonium salts on the solubility of alkanes may be relevant here. Clearly, one must exercise extreme caution when attempting to interpret interactions taking place in ternary systems.

3.2. Ternary Solutions Containing an Electrolyte

Studies of the behavior of electrolytes in aqueous organic solvents may also help to throw light on the nature of the interactions in binary aqueous nonelectrolyte mixtures. Most studies of the thermodynamics of

TABLE VIII. Solubility Coefficients of Argon in Binary Aqueous Solvent Mixtures[a]

		Argon solubility $\geq$ solubility in pure water				Argon solubility $<$ solubility in pure water		
			Ostwald absorption coefficient				Ostwald absorption coefficient	
Second component	Concentration		5°	25°	Second component	Concentration	5°	25°
MeOH	$x_2 = 0.015$		49.60	35.96	Glycerol	$x_2 = 0.015$	43.80	31.93
EtOH	$x_2 = 0.015$		49.80	35.85	p-Dioxan	$x_2 = 0.015$	46.24	35.00
n-PrOH	$x_2 = 0.015$		48.45	35.45	Glucose	0.5m	41.55	30.40
n-BuOH	$x_2 = 0.015$		48.00	34.90	Sucrose	0.5m	38.80	28.85
					LiCl	1m	37.01	27.47
					NH_4Cl	1m	38.20	28.27
					KI	1m	33.41	25.51
					Me_4NCl	0.5m	44.07	32.14
					Et_4NI	0.5m	44.88	33.00
					Pr_4NI	0.5m	43.45	32.20

[a] Ostwald absorption coefficients of argon in pure water are 48.07 (5°) and 34.08 (25°).

transfer of ions from pure water into a mixed aqueous solvent have concentrated on $\Delta G_t^{\ominus}$, which does not appear to depend markedly on the structure of the solvent. However, $\Delta H_t^{\ominus}$ and $\Delta S_t^{\ominus}$ appear to contain appreciable "structural contributions." Determinations of $\Delta H_t^{\ominus}$ for electrolytes in some aqueous binary mixtures[46,1227] showed this to be positive, passing through a maximum in very dilute solution. As the concentration of the organic solvent increases further, $\Delta H_t^{\ominus}$ becomes negative. Feakins et al.[474] showed this behavior to be consistent with the idea that such organic solutes at low concentrations cause an increase in solvent structure. The structural effect can also be clearly seen in plots of the partial ionic entropies of a series of ions in a mixed solvent against the partial ionic entropies of the same ions in water.[521] For methanol–water mixtures a series of straight lines are obtained. From the relative positions of these lines it can be deduced that mixtures containing 10 and 20% methanol have a greater structural integrity than pure water. A similar conclusion has been reached on considering the variation of $\bar{S}_i^{\circ}$ with solvent composition.

Further studies in ternary aqueous electrolyte systems have helped clarify the anomalous nature of solutions of tetraalkylammonium salts. The mixing properties of such salts with solutions of normal 1:1 electrolytes have been studied by a variety of techniques. Two examples should suffice to indicate the usefulness of such approaches. Studies of the heats of mixing of ternary salt solutions have shown clearly[27,1429] that the sign of ΔH_m is governed by the effects of the ions on the solvent structure, and that the tetraalkylammonium ions in general promote structure, but that the structure so produced is different from that produced by electrostrictive hydration. Volumes of mixing[1390,1391] in ternary systems have also shown the differences in the solvent effects produced by tetraalkylammonium ions compared to the simple alkali metal ions.

The phase behavior of ternary systems is an extremely complex subject which is in general beyond the scope of this chapter. Certain observations are, however, relevant to the discussion of LCST's in Section 2.2.1. It was indicated that in some cases solutes satisfied, or came close to satisfying the approximate criteria for an LCST without actually giving rise to the phenomenon. An indication in certain cases of the closeness of phase separation can be obtained by adding a small amount of a third component to the solutions, with resultant phase separation. Such indications of the closeness of phase separation have been observed in water–dioxan, where the addition of a small amount of salt causes phase separation, though water and dioxan by themselves are miscible in all proportions and do not exhibit an LCST. This observation confirms the view of Malcolm and Rowlinson[941] that dioxan in water is very close to satisfying the conditions for an LCST. Addition of salt to *t*-butanol–water mixtures produces a similar effect. The effect of a third component on the position of the LCST in the triethylamine–water system has been thoroughly studied,[621] the addition of a third component which was appreciably soluble in only water or triethylamine lowering the temperature of phase separation.

4. RELEVANCE TO BIOLOGICAL SYSTEMS

The work described in this chapter is of interest not only for itself but also because of the insights it provides as to the nature and extent of the interactions taking place in biological systems, and especially for the clear indication it provides as to the important role that water and solvation must play in such systems. The important constituents of biological systems, apart from small molecules, are the protein, nucleic acid, and carbohydrate polymers. Proteins contain both polar and apolar groups, and hence hydrophilic and hydrophobic interactions play a major role in the determination of protein conformation in solution. Our understanding of protein solutions, although still quite rudimentary, has been greatly aided by studies of solutions of apolar and especially mixed solutes. Water is also important to the conformational stability of nucleic acids. Understanding the behavior of carbohydrate polymers, being primarily hydrophilic in constitution, requires more investigations of small-molecule hydrophilic solutes. It is clear, however, that once again water plays a major role in determining the biological behavior of such molecules. An understanding of the solution behavior of aqueous nonelectrolytes can thus be seen as an essential prerequisite to a full understanding of the complex processes occurring in biological systems.

CHAPTER 6

Phase Behavior of Aqueous Solutions at High Pressures

G. M. Schneider

Ruhr-Universität Bochum
Abteilung für Chemie
Germany

1. INTRODUCTION

Three different types of two-phase equilibria have to be considered in fluid mixtures: liquid–gas, liquid–liquid, and the so-called "gas–gas" equilibria. They all have the common feature of existing between two fluid phases of different densities separated by a meniscus.*

The aim of this chapter is to give a review of these three kinds of phase equilibria at high pressures (say above 300 bar) for the special case of aqueous systems that show particularly complicated and interesting effects (Section 2–4). In Section 5 the experimental results are briefly discussed from a phase-theoretical, and in Section 6 from a thermodynamic, point of view. For a detailed discussion of the equilibria (especially liquid–liquid immiscibility, azeotropy, appearance of solid phases) and of the thermodynamics of aqueous mixtures at normal and low pressures see this volume, Chapter 5 and Volume 3, Chapter 1. More detailed reviews of the phase behavior of fluid mixtures at high pressures have recently been published.[1199–1201]

---

* Under certain conditions, however, the phases can have the same density (isopycnic systems, appearance of barotropic phenomena) or the same refractive index (isooptic systems) where the meniscus can be invisible.

2. LIQUID–GAS EQUILIBRIA

The different types of phase behavior at high pressures are most easily understood and classified by the pressure dependence of the critical phenomena. At a critical point the intensive properties of two phases in equilibrium become identical. Whereas pure substances are characterized by a critical point for the gas–liquid equilibrium, binary systems exhibit a critical line in the three-dimensional $T–P–x$ space (where $x = $ concentration), and systems with n components exhibit an $(n − 1)$-dimensional critical surface in the $(n + 1)$-dimensional $T–P–x_1–x_2– \cdots –x_{n-1}$ space for all kinds of fluid–fluid equilibria.

In binary systems with which this chapter is mainly concerned the critical points of the binary mixtures are situated at the extreme values of all isobaric $T(x)$ cuts or all isothermal $P(x)$ cuts through the two-phase region in the $T–P–x$ space. The line that connects the critical points of all binary mixtures is the critical curve. A very simple example for the liquid–gas equilibrium is schematically given in Fig. 1.*

For practical purposes it is convenient to discuss the critical phenomena of binary systems by means of the $P(T)$ projections of the critical lines. For many systems these $P(T)$ projections are continuous lines between the critical points (CP) of the pure components I and II and may exhibit a pressure maximum, a temperature minimum or maximum, or may run monotonically or nearly linearly between CP I and CP II. For each case examples have been given elsewhere.[742,1146,1161,1199–1201,1456]

Because of the great experimental difficulties only few binary aqueous mixtures have been studied up to the liquid–gas critical region despite the considerable theoretical and practical interest in such measurements. Therefore only some of the types of critical curves cited above have been established for binary aqueous solutions. The most important types are given schematically in Fig. 2. Figure 2(a) corresponds to the system $NH_3(I)$–$H_2O(II)$.[1309] The fact that the critical $P(T)$ curve is nearly a straight line indicates that this systems is not far from ideal; Fig. 1 is a schematic three-dimensional representation of this case. Figure 2(b) corresponds to binary mixtures of H_2O (component II) with ethanol,[764,1058,1060] 1-propanol,[99,804] or acetone[604] (component I) that form positive azeotropes up to the critical region; for ethanol–H_2O and acetone–H_2O, however, the azeotrope is

---

* In all figures the critical lines are represented by solid lines, the phase diagram of the pure components by dashed lines, and the three-phase lines by dot-dash lines; CP is the critical point of a pure component; C, E, K are critical end points; and Q_1 is a quadruple point.

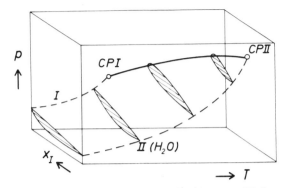

Fig. 1. The T–P–x surface for the liquid–gas equilibrium of a simple binary mixture, schematic; CP = critical point of pure component, dashed lines give vapor pressure curve of pure component.

limited at a minimum temperature of 32 and 85°C, respectively. For type 2(b) several $P(x)$ isotherms are schematically represented in Fig. 2(c), the cusplike isotherm for $T =$ const $= T_3$ being especially interesting; important thermodynamic conclusions can be derived from such a phase behavior (see Rowlinson[1161]). For phenol–H_2O the azeotrope seems to disappear with increasing temperatures and pressures even before the critical range is reached.[43] The system SO_3–H_2O shows a negative azeotrope and seems to have a maximum temperature on the critical line.[1262]

In Fig. 3 the $P(T)$ projections are given for some important cases where the liquid–gas equilibria and the crystallization diagram interfere in the T–P–x space. For type 3(a) the three-phase line LGS_I runs through a pressure maximum between the triple point of pure component I and the

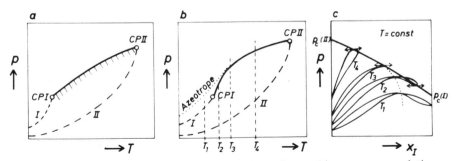

Fig. 2. The $P(T)$ projections of the T–P–x surface of some binary aqueous solutions, schematic; see text. (a) Without azeotrope, found for NH_3–H_2O; (b) with a positive azeotrope, found for 1-propanol–H_2O; (c) $P(x)$ isotherms for type (b).

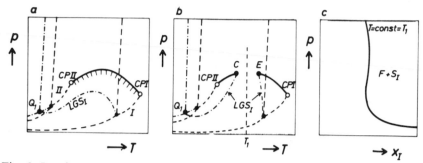

Fig. 3. Interference of crystallization and gas–liquid critical phenomena for aqueous binary mixtures (schematic). (a) Type found for NaCl–H_2O; (b) type found for SiO_2–H_2O; (c) $P(x)$ isotherm for type (b).

quadruple point Q_1 of the system where a liquid and a gaseous phase coexist with the pure solid components I and II. This curve represents the "vapor pressure curve" above a solution saturated with solid component I; it can be determined quite easily by experiment in a closed autoclave filled with component II and an excess of solid component I.[501,502,1043] This type is important for systems with components differing strongly in their triple-point temperatures and has been found for NaCl–H_2O and aqueous solutions of boric acid, borates, borax, etc.; for NaCl–H_2O the maximum of the LGS_I curve is situated at approximately 600°C, 400 bar with NaCl concentrations of 69 wt% in the saturated liquid and 0.04 wt% in the coexisting gas phase.[1241]

For type 3(b) the pressure maximum of the LGS_I curve is so pronounced that the critical curve LG is cut twice at the critical end points C and E; between C and E pure solid component I is in equilibrium with a homogeneous, highly compressed fluid phase [see $P(x)$ isotherm in Fig. 3(c)]. This type of behavior has been detected in the SiO_2–H_2O system, the critical end point E being at 1080°C, 9700 bar and 75 wt% SiO_2.[748,749]

3. LIQUID–LIQUID EQUILIBRIA

3.1. Pressure Dependence of Critical Solution Temperatures in Binary Systems

For the pressure dependence of liquid–liquid equilibria only the classical work of Kohnstamm, Timmermans, and their co-workers[1111,1294,1295] existed up to 1963; with only a few exceptions their measurements were limited to the pressure range below 200 bar. New measurements were started in 1963; pressures up to 7 kbar were used and some new types of pressure dependence of immiscibility phenomena were observed.[1199–1201,1250] Especially interesting phase separation effects were found for aqueous mixtures. In Fig. 4 all types presently known for aqueous binary solutions are represented schematically. Some examples are given in Figs. 5–9, where solution temperatures are plotted against pressure for mixtures with $x = $ const $\approx x_c^{\mathrm{UCST}}$ at 1 bar for Fig. 5, or $x = $ const $\approx x_c^{\mathrm{LCST}}$ at 1 bar for Figs. 6–8 (for references see Ref. 1199). Since critical concentrations depend only slightly on pressure for liquid–liquid equilibria, these $T(P)$ cuts for $x = $ const are also characteristic for the critical $T_c(P)$ curves themselves.

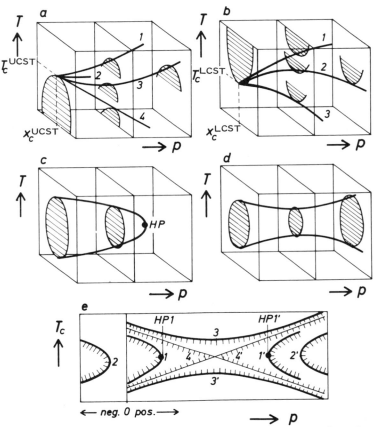

Fig. 4. Pressure dependence of critical solution temperatures, schematic, see text; $T_c = $ critical solution temperature, HP = hypercritical solution point. (a) Upper and (b) lower critical solution temperatures, (c)–(e) closed solubility loops.

Upper critical solutions temperatures (UCST) may either rise with increasing pressure [type 1 in Fig. 4(a); e.g., phenol–H_2O (Fig. 5)], may remain constant within a limited pressure range [type 2 in Fig. 4(a); approximately realized for phenol–H_2O (Fig. 5) and 3-methylpyridine-D_2O (Fig. 9) at low pressures], run through a temperature minimum (type 3 in Fig. 4(a); e.g., 3-methylpiperidine–H_2O (Fig. 8)], or decline [type 4 in Fig. 4(a); e.g., isobutyric acid–H_2O, acetonitrile–H_2O (Fig. 5)].

Lower critical solution temperatures (LCST) may either rise with increasing pressure [type 1 in Fig. 4(b); e.g., triethylamine–H_2O (Fig. 6)], run through a temperature maximum [type 2 in Fig. 4(b); e.g., 4-methyl-piperidine–H_2O, nicotine–H_2O, 2,6-dimethylpyridine–H_2O, and many other systems (Fig. 6)], or decline with increasing pressure [type 3 in Fig. 4(b); approximately realized for 2,4,6-trimethylpyridine–H_2O in Fig. 6].

For *closed solubility loops* in the isobaric $T(x)$ cuts an especially interesting pressure dependence was found. With increasing pressure the loops may become smaller and finally disappear completely at a so-called hyper-critical solution point (HP) in the $T–P–x$ space where $dT_c/dP = \pm\infty$ (Fig. 4c); at higher pressures the constituents are miscible in all proportions. Examples of this phase behavior are methylethylketone–H_2O, 2-butanol–H_2O, 2-buto-xyethanol–H_2O, 2-methylpyridine–D_2O; some results are given in Fig. 7.

Figure 4(d) shows another type of pressure dependence of closed loops. With increasing pressure the loop first shrinks but does not disappear

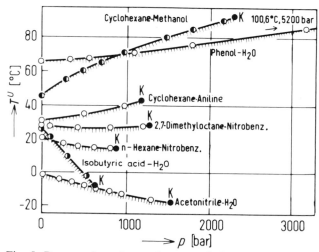

Fig. 5. Pressure dependence of upper solution temperatures for $x = $ const $\approx x_c^{UCST}$ at 1 bar. For references see text; $K = $ appearance of a solid phase.

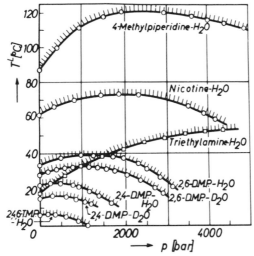

Fig. 6. Pressure dependence of lower solution temperatures for $x = $ const $\approx x_c^{\mathrm{LCST}}$ at 1 bar. For references see text; DMP = dimethylpyridine, TMP = trimethylpyridine.

completely as in Fig. 4(c) but instead becomes greater again. 3-Methyl-piperidine–H_2O (Fig. 8) and 3-methylpyridine–D_2O (Fig. 9) are examples of this phase behavior.

The $T_c(P)$ curves of the types presented schematically in Fig. 4(e) have been observed in methylpyridine–water systems (Fig. 9).[1199] Type 1–1′ in Fig. 4(e) has been found for 2-methylpyridine–D_2O, the points HP1 and HP2 being situated at ~200 and 2000 bar, respectively. Binary mixtures of 2-, 3-, and 4-methylpyridine with H_2O and of 4-methylpyridine with D_2O (Fig. 9) exhibit $T_c(P)$ curves of type 2′ in Fig. 4(e), the curves 2 being displaced to negative pressures. The phrase "high-pressure immiscibility" has been proposed for phase separation effects of type 2′ in Fig. 4(e).[1199–1201] Type 3–3′ in Fig. 4(e) is equivalent to type 4(d) and has been found for 3-methylpyridine–D_2O (Fig. 9), whereas the curves 4 and 4′ correspond to a transition type between 1–1′ (or 2–2′) and 3–3′. The phrase "hypercritical solution point" has been proposed for points such as HP in Fig. 4(e).[1199]

Similar systems have recently been studied by Steiner and Schadow[1250] at pressures up to 1240 bar; immiscibility phenomena hitherto unknown were observed in the system methylisopropylketone–H_2O, where $P(x)$ isotherms with two pressure maxima were found.*

* This result has to be checked.

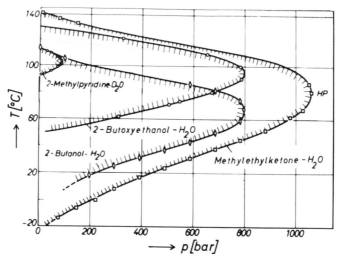

Fig. 7. Pressure influence on the solution temperatures of closed solubility loops for $x = \text{const} \approx x_c^{LCST}$ at 1 bar. For references see text; HP = hypercritical solution point.

For a review of liquid–liquid equilibria at high pressures and for further examples on aqueous and nonaqueous solutions the reader is referred to articles by Schneider[1199–1201] where a whole pattern of demixing phenomena in binary systems under pressure is derived and a review of the shapes of the immiscibility surfaces in the $T–P–x$ space is given.

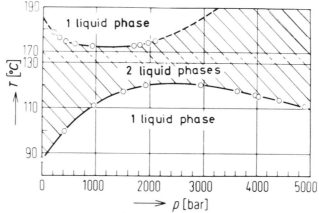

Fig. 8. Pressure influence on liquid–liquid immiscibility in the system 4-methylpiperdine–H_2O for $x = \text{const} \approx x_c^{LSCT}$ at 1 bar.

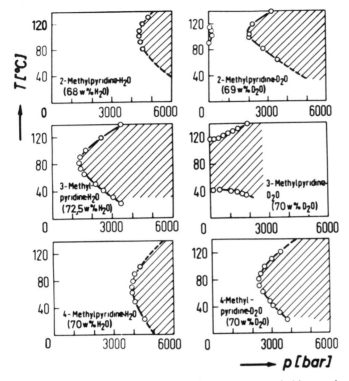

Fig. 9. Pressure dependence of solution temperatures in binary mixtures of methylpyridines with H_2O and D_2O [see Fig. 4(e); the $T(P)$ curves for $x = $ const given are characteristic for the $T_c(P)$ curves].

3.2. Extension to Ternary Systems and Influence of Added Salts

The solution temperatures of binary mixtures may be strongly influenced by the addition of a third component.

Up to now only few measurements exist for ternary nonelectrolyte systems. Schneider[1197] studied the system 3-methylpyridine–H_2O–D_2O up to 2000 bar and showed that depending on the H/D ratio of the water all transition types between type 2–2' (corresponding to 3-methylpyridine–H_2O) and type 3–3' (corresponding to 3-methylpyridine–D_2O) in Fig. 4(e) could be observed. Hirose et al.[673] studied the system isopropanol–H_2O–benzene and found that with increasing pressure the upper solution temperatures of the ternary mixtures rise for mixtures rich in isopropanol, decline for benzene-rich solutions, and run through a temperature minimum for solutions rich in H_2O. The liquid–liquid equilibria in the system benzene–n-heptane–H_2O have been studied up to 300 bar by O'Grady.[1043]

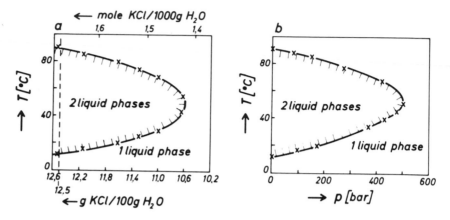

Fig. 10. Liquid–liquid immiscibility in the system 1-propanol–H$_2$O–KCl (wt% H$_2$O/wt% 1-propanol = const = 1.5). (a) Salt influence for P = const = 1 bar. (b) Pressure influence for constant concentration of KCl (12.5 g of KCl per 100 g of H$_2$O).

The solution temperatures of aqueous solutions of nonelectrolytes are strongly influenced by the addition of salts or other electrolytes. Besides their practical applications, structural information may be obtained from such solubility studies, since the salting-in or salting-out effects can be explained, at least qualitatively, by the ordering or disordering effect of the ions on the water structure[1202] (see Volume 3, Chapter 1).

Experiments at high pressures[457,1036,1199,1202]* have shown that there are evident analogies between the influence of pressure and dissolved salts on the immiscibility phenomena in liquid systems. An example is given in Fig. 10, where upper and lower solution temperatures are plotted for the system 1-propanol–H$_2$O–KCl, the mass ratio H$_2$O/1-propanol being kept constant at a value of 1.5. Figure 10(a) shows the solution temperatures as a function of the amount of KCl added at normal pressure and Fig. 10(b) shows them as a function of pressure for a constant KCl content of 12.5 g of KCl per 100 g of H$_2$O.[1202] There is a perfect analogy between the curves, increasing pressures showing the same effect as decreasing salt content.

In Fig. 11 the influence of dissolved salts on the immiscibility phenomena is shown for the system 3-methylpyridine–H$_2$O at high pressures and a constant mass ratio 3-methylpyridine/H$_2$O of 3/7.[457,1199] The salt-free system corresponds to the type 2–2′ in Fig. 4(e). The high-pressure

---

* J. Baldenhofer, P. Engels, and G. M. Schneider, University of Karlsruhe, unpublished results.

immiscibility region of the salt-free system (curve 2' in Fig. 4e) is displaced to higher pressures by the addition of an in-salting salt (e.g., NaI) and to lower pressures by an out-salting salt (e.g., Na_2SO_4). At a salt concentration of 0.007 M the hypothetical immiscibility region at negative pressures (curve 2 in Fig. 4e) has already reached positive pressures corresponding now to the type 1–1' in Fig. 4(e). With further increasing salt content there is a continuous transition from type 1–1' to type 3–3'. At a Na_2SO_4 concentration of 0.032 M approximately the same solution temperatures are obtained as for the salt-free system 3-methylpyridine–D_2O. Figure 11 gives evidence not only for the high degree of similarity between the different types of liquid–liquid immiscibility in Fig. 4(e) but also for the existence of immiscibility regions at negative pressures.

The phase behavior of aqueous solutions with added salts under high pressures can be extremely complicated, as is demonstrated by Fig. 12 for the system ethanol–H_2O–$(NH_4)_2SO_4$: For a mixture of ethanol and H_2O with a mass ratio of 3/7 with 16.1 wt% $(NH_4)_2SO_4$ four different solution temperatures (two lower and two upper) exist in some pressure ranges, e.g., at 240 bar.[1036] For further details see Refs. 457, 1036, 1199, 1202.

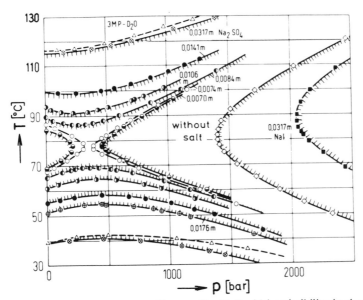

Fig. 11. Salt and pressure effects on liquid–liquid immiscibility in the system 3-methylpyridine–H_2O (wt% water/wt% 3-methylpyridine = const = 7/3; see text).

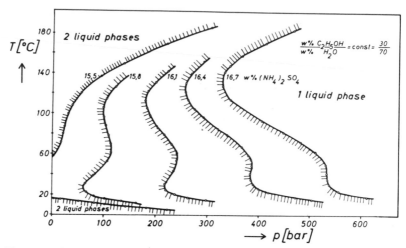

Fig. 12. Salt and pressure effects on liquid–liquid immiscibility in the system ethanol–H_2O–$(NH_4)_2SO_4$ (wt% H_2O/wt% ethanol = const = 7/3; see text).

3.3. Effect of Solid Phases

The crystallization behavior of aqueous mixtures is very complicated, e.g., due to the formation of gas hydrates (see this volume, Chapter 3). Here only the solidification of binary aqueous solutions that show liquid–liquid immiscibility will be discussed for some important and interesting cases.[1199,1160,1198]

The phase diagram for the normal case is well known and is represented schematically in Fig. 13(a) or 13(c): What type occurs depends on the relative values of the freezing temperatures of the pure components I and II. The pressure dependence of the two types is quite different. For Fig. 13(a),

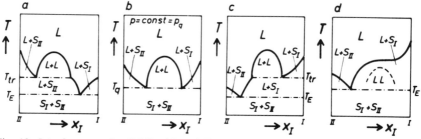

Fig. 13. Interference of solid–liquid and liquid–liquid equilibria in binary mixtures; schematic, see text; T_{tr} = triple point temperature LLS, T_q = quadruple point temperature LLSS).

where at the triple point temperature T_{tr} solid component II ($= H_2O$) coexists with two liquid phases, T_{tr} will decline with increasing pressure as the freezing temperature of pure H_2O does up to 2100 bar, whereas the freezing temperature of the pure component I will normally increase (e.g., acetonitrile–H_2O).[1198,1199] For Fig. 13(c), where pure component I coexists with two liquid phases, T_{tr} will normally increase with rising pressure (e.g., phenol–H_2O).[1160] The temperature T_{tr} has been determined as a function of pressure for several systems up to the kilobar region. In special cases the liquid–liquid immiscibility range disappears below the crystallization surface in the T–P–x space, e.g., at the critical end points K in Fig. 5; at higher pressures liquid–liquid equilibria can only be realized in supercooled solutions (see Fig. 13d).[1160,1198,1199]

The interesting transition type shown in Fig. 13(b), lying between types 13(a) and 13(c), is likely for aqueous mixtures of organic compounds having a rather low freezing temperature. With increasing pressures the phase diagram may change from type 13(a) to type 13(c), showing at a pressure $P = P_q$ the phase behavior of Fig. 13(b) with a quadruple point temperature T_q where four phases (two liquid phases, pure solid components I and II) are in equilibrium. Such a phase behavior has recently been found for the first time in the systems acetonitrile–H_2O ($P_q = 1240$ bar, $T_q = -24.3°C$), aniline–H_2O, o-xylene–H_2O, and others.[1198]

4. GAS–GAS EQUILIBRIA

The term "gas–gas equilibrium" or "immiscibility of gases" has been applied to some phase separation effects in fluid systems at high pressures and temperatures, especially *above* the critical temperature of the less volatile substance.

The different types for binary mixtures are shown schematically in Fig. 14. The critical curve is interrupted and consists of two branches. The branch starting from the critical point of the more volatile component I ends at a so-called critical end point C on the three-phase line LLG, where two liquid and one gaseous phases are in equilibrium, whereas the branch beginning at the critical point of the less volatile component II either immediately tends to higher temperatures and pressures (curve 1 in Fig. 14, so-called "gas–gas equilibrium of the first type") or goes through a temperature minimum first and then runs steeply to increasing pressures and temperatures even above the critical temperature of pure component II (curve 2 in Fig. 14, so-called "gas–gas equilibrium of the second type").

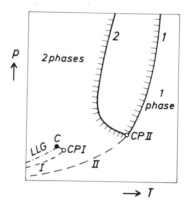

Fig. 14. $P(T)$ projections of the T–P–x surface
for binary systems showing gas–gas equilibria
of the first (curve 1) and of the second (curve 2)
types.

These phase separation effects were predicted by van der Waals in 1894 and
were discussed in detail by Kamerlingh Onnes and Keesom in 1907. Type 2
was found for the first time in 1940 by Krichevskii and co-workers for the
N_2–NH_3 system.[794,795] Type 1 was reported for the first time in 1952 by
Tsiklis for the system He–NH_3.[1307] Gas–gas equilibria have recently been
reviewed by several authors.[1161,1199–1201,1296,1297,1308,1312]

Since 1940 similar phase separation effects have been found in about
40 systems, including nearly 20 aqueous mixtures, which are cited in Table I;
the temperature range, maximum pressure of the investigations, and the
references are also given. According to Table I gas–gas equilibria of the
second type have been found in binary mixtures of H_2O with N_2,[1310,1311]
CO_2,[1298] Xe,* and many hydrocarbons,[22–24,229,331,360,1135,1136] whereas
gas–gas equilibria of the first kind have only been observed up to now in
the Ar–H_2O system.[839,840]

The critical $P(T)$ curves of many of the systems cited in Table I are
plotted in Figs. 15 and 16. In Fig. 15 only the vapor pressure curves of
H_2O and NH_3 and the critical curves of binary mixtures of NH_3 with
N_2,[794,795] Ar,[1308,1312] and He[1307] and of H_2O with NH_3,[1309] CO_2,[1298]
Ar,[839,840] C_2H_6,[360] C_6H_6,[23] and NaCl[1241] are shown. In Fig. 16 only the
vapor pressure curve of pure H_2O and the branches of the critical curves
that start from CP H_2O are plotted for binary aqueous solutions of eth-

* H. Welsch, University of Karlsruhe, Germany, unpublished results.

TABLE I. Binary Aqueous Systems in Which Gas–Gas Equilibria Have Been Found Experimentally[1199–1201,1308,1312]

System	T, °C	P_{max}, bar	Ref.
A. Systems with gas–gas equilibria of the first type			
Argon	270–440	3000	839, 840
B. Systems with gas–gas equilibria of the second type			
Nitrogen	300–350	6000	1310, 1311
Xenon	300–500	3000	[a]
CO_2	250–350	3500	1278, 1298
Methane	300–450	3000	[a]
Ethane	200–400	3700	360
n-Butane	355–364	1100	360, 1310
n-Pentane	340–352	620	331
2-Methylpentane	330–355	500	331
n-Heptane	350–355	550	331
	350–420	1250	229
Propene	325–350	2500	839, 840
Cyclohexane	130–363	200	1135
	270–410	1800	229
Benzene	200–357	200	1136
	287–300	735	331
	260–370	3700	22–24
(with D_2O)	300–340	3000	229
Toluene	305–310	455	331
	310–360	2000	22, 24
o-Xylene	310–380	2000	22, 24
Ethylbenzene	310–380	2000	22, 24
n-Propylbenzene	320–400	2000	22, 24
1,3,5-Trimethylbenzene	320–400	2000	22, 24

[a] H. Welsch, University of Karlsruhe, Germany, unpublished results.

ane,[360] *n*-butane,[360] cyclohexane,[229] benzene,[23] and methyl-substituted benzenes;[22,24] additionally, the branches of the critical curves that correspond to liquid–liquid equilibria are given for the naphthalene–H_2O[22,24] and the biphenyl–H_2O system.[229]

Both NH_3–H_2O[1309] and NaCl–H_2O[1241] systems (Fig. 15) have continuous critical curves between the critical points of the pure components; no phase separation in the liquid phase occurs in these systems. For Ar–H_2O[839,840] and NH_3–He[1307] the critical curve tends directly to higher temperatures and pressures and corresponds to a gas–gas equilibrium of the first type. For all other systems in Figs. 15 and 16, except for naphthalene–H_2O[22,24] and biphenyl–H_2O,[229] the critical curve has qualitatively the same shape, with a temperature minimum and a steep ascent to higher temperatures and pressures according to a gas–gas equilibrium of the second type. The curves for naphthalene–H_2O[22,24] and biphenyl–H_2O[229] that belong definitely to liquid–liquid equilibria and end on the three-phase line *LLG* fit to the other curves remarkably well (see Section 5).

Since high pressures are reached in these experiments, the molar volumes of the two phases are of the same order, and it depends on the average molecular weight of the separate phases if *barotropic* phenomena will be observed. Thus a barotropic behavior was frequently found in

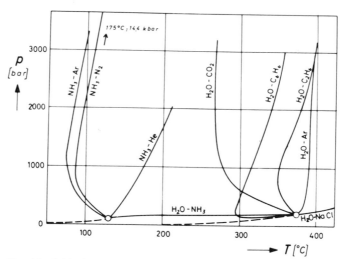

Fig. 15. Critical $P(T)$ curves of binary H_2O and NH_3 systems showing gas–gas equilibria; for references see text; dashed lines are vapor pressure curves of H_2O and NH_3.

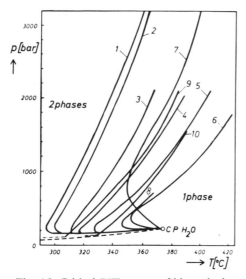

Fig. 16. Critical $P(T)$ curves of binary hydrocarbon–water mixtures showing gas–gas equilibria (curves 1–8) and liquid–liquid equilibria
(curves 9 and 10). For references see text; only
the branches starting from CP H_2O are represented for the systems 1–8; dashed line is
vapor pressure curve of H_2O; dot-dash curve
is three-phase line LLG. (1) Benzene–H_2O;
(2) benzene–D_2O; (3) toluene–H_2O; (4) o-
xylene–H_2O; (5) 1,3,5-trimethylbenzene–H_2O;
(6) cyclohexane–H_2O; (7) ethane–H_2O; (8)
n-butane–H_2O; (9) naphthalene–H_2O (LL);
(10) biphenyl–H_2O (LL).

systems showing gas–gas immiscibility, especially if component I consists
of large and/or weakly interacting particles and component II of comparatively small and/or strongly interacting molecules (e.g., H_2O). Thus
barotropic phenomena have already been detected in the systems Ar–
H_2O[839,840] and CO_2–H_2O.[1298] But barotropism is not necessarily combined
with gas–gas immiscibility.

The number of systems with gas–gas immiscibility will probably increase continuously in the future when progress in experimental techniques
will allow the investigation of mixtures of compounds that differ progressively in size, shape, and polarity, and it seems that the occurrence of
gas–gas equilibria will be the rule for binary mixtures of water with apolar
or weakly polar nonelectrolytes.

5. PHASE-THEORETICAL ASPECTS

In Sections 2–4 the three possible types of two-phase equilibria in fluid systems, namely liquid–gas, liquid–liquid, and gas–gas, were briefly discussed. For years the limits between these different kinds of phase behavior appeared to have been well established. However, new measurements during recent years have shown that there are continuous transitions between these three types of phase equilibrium and that there is a real continuity between liquid–gas, liquid–liquid, and gas–gas equilibria.[1199–1201] In the present section this continuity will be demonstrated for binary hydrocarbon–H_2O mixtures.[22–24,229,1199–1201]

Because of their very low mutual miscibility, mixtures of hydrocarbons with water have not in the past been investigated, and for practical applications (e.g., steam distillation, extraction) complete immiscibility is still commonly assumed. During recent years some measurements of the solu-

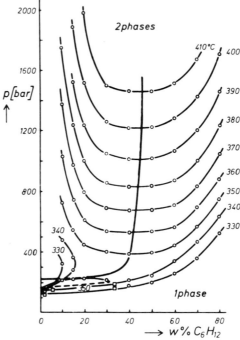

Fig. 17. $P(x)$ isotherms for the system cyclohexane–H_2O. For $T < 350°C$ the parts at low pressures are omitted; full line is the $P(x)$ projection of the critical curve (see also Fig. 16).

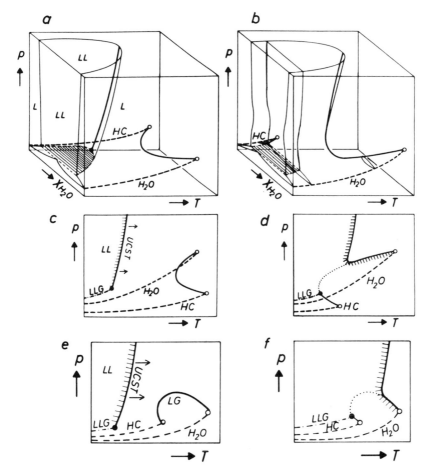

Fig. 18. T–P–x surfaces and $P(T)$ projections for binary mixtures of H_2O with hydro-carbons (HC); schematic, see text. (a, c) Type found for naphthalene–H_2O, biphenyl–H_2O. (b, d) Type found for benzene–H_2O and aqueous solutions of methyl-substituted benzenes. (e) No aqueous system known. (f) Type found for cyclohexane–H_2O, n-butane–H_2O.

bility of hydrocarbons in water were extended to ranges of temperature and pressure where hydrocarbons and water are miscible in all proportions and where parts of the critical curves could be determined (see Table I and Figs. 15–17).

In Fig. 18 the phase behavior of hydrocarbon–water mixtures is deduced from a superposition of liquid–liquid and liquid–gas equilibria.[1200] Figures 18(a) and 18(c) correspond to a binary system with a critical curve LG showing a temperature minimum and a phase separation into two liquid phases at lower temperatures, with the UCST rising with increasing pressure. This type has recently been found for the naphthalene–H_2O[22,24] and biphenyl–H_2O[229] systems. They are the first hydrocarbon–H_2O systems where complete miscibility in all proportions has been found in ranges of temperature and pressure that belong definitely to the liquid state; however, the critical curve LG with a temperature minimum shown in Figs. 18(a) and 18(c) has not yet been determined experimentally.

The lower the mutual miscibility, the more the critical curve LL in Figs. 18(a) and 18(c) is displaced to higher temperatures, until finally the type of Figs. 18(b) and 18(d) is observed, which corresponds qualitatively to the phase behavior of aqueous solutions of benzene and alkyl-substituted benzenes (see Fig. 16). For benzene–H_2O[23] and 1,3,5-trimethylbenzene–H_2O[22,24] the origin of the steeply ascending branches from liquid–liquid equilibria seems to be evident. The curves for naphthalene–H_2O[22,24] and biphenyl–H_2O[229] that belong definitely to liquid–liquid equilibria fit remarkably well to these curves. In Figs. 18(e) and 18(f) it is shown how the critical curves of systems such as cyclohexane–H_2O,[229] ethane–H_2O,[360] n-heptane–H_2O[229] etc. that show gas–gas equilibria of the second type can be explained by the same hypothesis.

The qualitatively similar shape of all curves above 1000 bar shows that there are continuous transitions between all types of critical curves as shown in Fig. 16. Additionally, a marked extractive effect of H_2O on different hydrocarbons is demonstrated. This concept of a continuity between all kinds of phase equilibria in fluid mixtures has also helped with our understanding of the great variety of phase relationships and critical phenomena in other systems (e.g., mixtures of hydrocarbons with CO_2 and CH_4, mixtures of rare and inert gases, etc.[1199–1201]

6. THERMODYNAMIC DESCRIPTION

In Section 5 the continuity of all kinds of phase equilibria in fluid mixtures has been phenomenologically shown for some recent results on binary hydrocarbon–H_2O mixtures. This continuity is equally demonstrated by the fact that there is no real distinction in the thermodynamic description between a gas–liquid, a liquid–liquid, and a gas–gas critical point.[1161]

A classical description of the phase equilibria in fluid mixtures derives from the assumption that the molar Gibbs free energy is an analytic function of the mole fractions x and the temperature T for a fixed pressure at and near the critical point of the mixture. This classical treatment has been extensively described by several authors.[614,1161]

Therefore only the most important results for binary mixtures need to be cited. For the pressure dependence of a critical temperature we have

$$\frac{dT_c}{dP} = \frac{(\partial^2 V/\partial x^2)_c}{(\partial^2 S/\partial x^2)_c} = \frac{T_c(\partial^2 V/\partial x^2)_c}{(\partial^2 H/\partial x^2)_c} \tag{1}$$

which can be simplified for the case of critical solution temperatures of liquid–liquid equilibria:

$$\frac{dT_c}{dP} = \frac{T_c(\partial^2 V^E/\partial x^2)_c}{(\partial^2 H^E/\partial x^2)_c} \tag{2}$$

or

$$\frac{dT_c}{dP} \approx \frac{T_c V_c^E}{H_c^E} \tag{3}$$

In eqns. (1)–(3), S, V, H, V^E, and H^E are the entropy, the volume, the enthalpy, the excess volume, and the excess molar enthalpy, respectively. Equation (3) only holds if V^E and H^E have the same functional form of the type $V^E = af(T, x)$ and $H^E = bf(T, x)$, e.g., for "regular solutions," where the excess molar Gibbs free energy G^E is given by $G^E = A(T, p)x_1x_2$. [614,756, 1011,1161,1199–1201]

According to eqns. (2) and (3) some knowledge about the thermodynamic mixing functions can be deduced from the pressure dependence of critical solution temperatures at temperatures and pressures where hardly any direct measurements have been performed. Some of these thermodynamic conditions, based on a detailed discussion of eqns. (2) and (3), are schematically represented in Fig. 19. [1199,1200] The most interesting conclusions are as follows:

(a) For $dT_c/dP = 0$ it follows that $(\partial^2 V^E/\partial x^2)_c = 0$ or that $V_c^E \approx 0$. With the simplifications mentioned V^E would have to change its sign from negative to positive with increasing pressure for type 3 in Fig. 4(a), type 2 in Fig. 4(b), the type of Fig. 4(d), and all the types of Fig. 4(e).

(b) For $dT_c/dP = \pm\infty$ the result is that $(\partial^2 H^E/\partial x^2)_c = 0$ or that $H_c^E \approx 0$ (see type of Fig. 4c and types 1–1' and 2–2' in Fig. 4e).

(c) For closed solubility loops the types represented in Figs. 4(e) and 9 and with the simplifications mentioned above, H^E has to change from negative to positive with increasing temperature at constant pressure, and V^E has to do the same with increasing pressure at constant temperature. The change of sign of H^E has been confirmed at normal pressures for aqueous solutions but no direct determinations of H^E have yet been performed

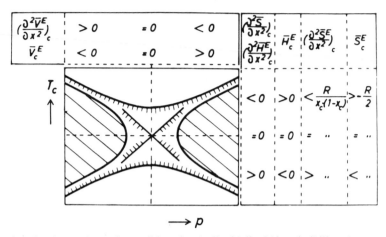

Fig. 19. Thermodynamic conditions for the liquid–liquid immiscibility phenomena in methylpyridine–water mixtures (according to Figs. 4e and 9; see text).

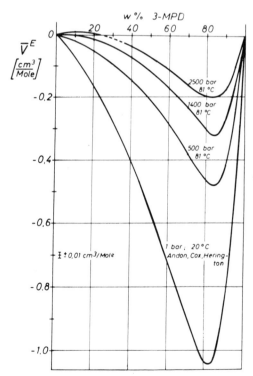

Fig. 20. Excess volume $\bar{V}^E$ in the system 3-methyl pyridine–H_2O as function of temperature, pressure, and composition (see text).

at high pressures, such experiments being difficult because of the large heat capacities of the high-pressure vessels. The change of sign of V^E with increasing pressure has recently been verified for an aqueous binary system of type 2–2′ in Fig. 4(e) (3-methylpyridine–H_2O) by a direct method; some results are given in Fig. 20.[457,1199] The change of sign of V^E with increasing pressure can be understood by the fact that in aqueous solutions V^E is normally negative and that the compressibility of the mixtures is often lower than the compressibility of the pure components; thus V^E is likely to be displaced in the direction of positive values with rising pressure. These effects can be deduced from a simple cluster model of H_2O and aqueous solutions.[457]

(d) For gas–gas equilibria eqn. (1) holds. It has been shown by several authors that for upper critical temperatures $(\partial^2 H/\partial x^2)_c < 0$. Thus the sign of the slope of the $T_c(P)$ curve is determined by the sign of $(\partial^2 V/\partial x^2)_c$. The PVT behavior of aqueous solutions in the gas–gas critical region has only been studied for few systems, e.g., for benzene–H_2O,[22,24] CO_2–H_2O,[1298] Ar–H_2O.[839,840]

It has been shown that the agreement between the classical theory and the experimental results is qualitatively good. However there is evidence that the classical theory seems to be quantitatively inadequate because of the nonvalidity of the assumption that G is analytic at or near critical points.[1011,1161,1200] More experimental and theoretical work seems to be necessary in order to attain further progress on these problems. A recent review of the theoretical descriptions of gas–gas equilibria[1200] has highlighted these approaches as being especially interesting for predictions.

7. CONCLUSIONS

Whereas phase equilibria in fluid systems have been intensively studied over a large temperature range for many years, a systematic investigation of the pressure range above several hundred bars has become possible only during recent years through the rapid development of experimental techniques. Up to now the results from such experiments are still rather sparse. Nevertheless some interesting phenomena could be found, especially for aqueous mixtures, e.g., high-pressure immiscibility and gas–gas equilibria. There is no doubt that the so-called gas–gas equilibria, only rarely found previously, will become more and more important, and it is quite probable that the exhibition of gas–gas equilibria will be the rule rather than the exception for systems with relatively weak interactions between unlike molecules, such as mixtures of H_2O with apolar or weakly polar substances.[1200]

The experimental techniques[1201] for the investigation of phase equilibria in fluid systems at high temperatures, high pressures, and possibly under strong corrosion conditions will allow the study of mixtures with components which differ more and more in volatility, structure, size, and polarity up to supercritical conditions, e.g., water with gases, organic compounds, molten salts, solids, etc. All these experiments may additionally present a certain practical or technological interest for mineralogy, geochemistry, and the earth sciences (especially for problems connected with hydrothermal syntheses and with the formation and migration of oil and natural gases), for separation processes (particularly extraction*), reactor research, high-pressure chemistry (especially for the application of these mixtures as reaction media with variable density, solubility, and dielectric constant), wet air oxidation,[862,1285] etc. Additionally, some thermodynamic information may result from such measurements in a range of temperature and pressure where nearly no direct measurements exist (see Section 6). Thus the study of the phase behavior of mixtures, and especially of aqueous solutions under high pressures, will remain an interesting and important field of experimental and theoretical research.

---

* Recently a patent has been granted for the demineralization of sea water with cumene by an extractive method using gas–gas equilibria.[1288]

Dielectric Properties

J. B. Hasted

Birkbeck College
University of London
England

The study of the dielectric properties of aqueous solutions of nonelectrolytes includes both the static or low-frequency dielectric constant and the relaxation spectrum or variation of the dielectric constant as the frequency is varied. Since the most important nonelectrolytes are organic liquids, the dielectric properties of these pure liquids must also be studied and understood, as well as those of pure water, which were discussed in Volume 1 Chapter 7 of this treatise. The dielectric properties are to some extent structure-sensitive, that is, they depend upon the correlation or interaction between the electric dipoles, both permanent and induced, principally the permanent dipole moments associated with OH and similar groups. The strongest electric fields acting on the water dipoles are those produced in the neighborhood of electrolytic ions. These will be considered in Volume 3, Chapter 8, except insofar as it will be necessary to consider their relevance to nonelectrolyte solutions.

1. THE STATIC DIELECTRIC CONSTANTS OF POLAR LIQUIDS

The static dielectric constant ε_s of a material has been related to the electric polarization first of all by the Clausius–Mosotti equation, and subsequently to the molecular electric dipole moments by the theories of Debye[387,388] and Onsager.[1047] Onsager realized that the essential difficulty of the Clausius–Mosotti equation lies in the choice of the Lorentz local electric field, the derivation of which is an artificial procedure; the use of this local field does not lead to agreement with the observed static dielectric constants for polar liquids. Onsager determined the local electric field acting on a molecule in a liquid by regarding the molecule as a real cavity embedded in a medium with homogeneous dielectric constant extending right up to the surface of the molecule. The polarization per unit volume is then equal to $(\varepsilon_s - 1)(2\varepsilon_s + 1)/9\varepsilon_s$. The Onsager theory predicts correct static dielectric constants for so-called "unassociated" or "normal" liquids but fails to give even the approximate values for "associated" liquids such as water, alcohols etc.

Kirkwood[762] generalized the Onsager theory to take into account the dielectric inhomogeneity of the medium in the vicinity of the molecule due to the hindering of rotation of its neighbors by the molecule; he calculated an expression for the polarization P_0:

$$P_0 = V(\varepsilon_s - 1)(2\varepsilon_s + 1)/9\varepsilon_s = \tfrac{4}{3}\pi N_0[(\boldsymbol{\mu} \cdot \boldsymbol{\mu}^*/3kT) + \alpha] \qquad (1)$$

where $\boldsymbol{\mu}$ is the molecular dipole moment in the surroundings of the condensed medium; it is not identical with the free molecule dipole moment $\boldsymbol{\mu}_0$, and must be calculated from a model; n is the refractive index, V the molar volume, N_0 the Avogadro constant, and α the polarizability of the molecule. $\boldsymbol{\mu}^*$ is the average dipole moment of a sphere containing at its center a dipole held aligned in a fixed direction.

The calculation of the static dielectric constant of water from eqn. (1) was discussed in Volume 1, Chapter 7. It is carried out by setting up a structural model of the medium, from which the calculation of $\boldsymbol{\mu}^*$ is made. $\boldsymbol{\mu}^*$ is often symbolized in terms of a dipole correlation factor $g = \boldsymbol{\mu}^*/\boldsymbol{\mu}$. The sensitivity of the correlation factor g to structural models or to a statistical assembly of models may be assessed from the pure-water calculations.

Mixtures of polar liquids with water can be investigated along the same lines,[1055] but first of all it is necessary to understand the static dielectric constants of the pure liquids. Most of the liquids completely miscible with water are themselves polar, and possess correlation parameters g which are a measure of the extent of mutual hindering of rotation of neighboring dipoles. Table I lists values of g calculated[1056] for a number of polar liquids from the observed static dielectric constants, which are also listed.

TABLE I. Correlation Factors g Calculated from Static Dielectric Constants ε_s

	$T, °C$	μ_0, D	ε_s	g
n-Amyl alcohol	20	—	15.8	3.43
n-Butyl alcohol	20	—	18.0	3.21
n-Propyl alcohol	20	—	19.5	3.07
Ethyl alcohol	20	—	24.6	3.04
Methyl alcohol	20	—	32.8	2.94
sec-Butyl alcohol	20	—	—	2.83
tert-Butyl alcohol	—	—	—	2.38
Isobutyl alcohol	—	—	—	3.36
Benzyl alcohol	—	—	—	2.08
Hydrogen cyanide	20	2.80	116	4.1
Hydrogen fluoride	0	~1.8	83.6	3.1
Hydrogen peroxide	0	2.13	91	2.8
Ammonia	15	1.48	17.8	1.3
Diethyl ether	20	1.15	4.4	1.7
Acetone	20	2.85	21.5	1.1
Nitrobenzene	20	3.90	36.1	1.1
Ethyl bromide	20	1.80	9.4	1.1
Pyridine	20	2.20	12.5	0.9
Benzonitrile	25	3.90	25.2	0.8

Calculations have been made by Oster and Kirkwood[1056] of the correlation parameters for aliphatic alcohols, on the basis of an arrangement in chains by means of hydrogen bonds:

The chains are taken to be of infinite length, each molecule being correlated in orientation only with members of its own chain; the average configura-

tion of a specified chain, due to infinite curling up, is approximately spherical. Free rotation around the hydrogen bonds is allowed, but bending is
excluded. For chainwise coordination we have

$$g = 1 + 2 \sum_{i=1}^{\infty} (\cos \gamma_i)_{av} \tag{2}$$

where γ_i is the angle between the dipole moments of a molecule and a
chain neighbor separated from it by n bonds. Further, we have

$$(\cos \gamma_i)_{av} = f(\cos^2 \tfrac{1}{2}\theta)^i \tag{3}$$

$$f = (\mu_H - \mu_R)(\mu_H + \mu_R \cos \theta)/(\mu_H{}^2 + \mu_R{}^2 + 2\mu_R\mu_H \cos \theta) \tag{4}$$

where μ_H and μ_R are the components of the dipole moment of the alcohol
molecule ROH along the OH and OR bonds, respectively, and $\theta = 105°$
is the ROH bond angle. By summation of the series obtained by substitution
of eqn. (3) into eqn. (2), one obtains

$$g = 1 + 2f \cot^2(\theta/2) \tag{4a}$$

The calculation of liquid-phase dipole moments is performed using
Onsager's sphere formula:

$$\mu = [(2\varepsilon_s + 1)/(2\varepsilon_s + n^2)][(n^2 + 2)/3]\mu_0 \tag{5}$$

where n is the optical refractive index. The calculation from eqns. (2)–(4a)
yields $g = 2.57$ for the alcohols listed in Table I, whose correlation parameters calculated from experiment vary from 2.94 to 3.43. The fact that the
experimental values of g exceed the theoretical value suggests that there is
a residual correlation in the orientations of molecules in different chains.

For the normal alcohols g increases with paraffin chain length; for
tert-butyl alcohol g is low, since steric interference with hydrogen bond
formation is likely.

Orientational correlation, of which the parameter g is a measure, is
governed not by the molecular dipole alone, but also by the location of the
permanent charge distribution in the interior of a polar molecule. Neighboring molecules of hydrogen-bonded liquids, possessing exposed charge
distribution, can assume their relative configurations by strong interactions.
But when the charge distribution is shielded, as in nitrobenzene, such
configurations are excluded by intermolecular repulsive forces, and the
orientational correlation is not great.

Some modification of the Kirkwood theory was introduced by Fröh-lich,[535] which could result in minor, but not qualitative, changes in the foregoing description. In Volume 1, Chapter 7 this modification was dis-cussed in relation to water. Fröhlich derived the theory independently, and included in it the Onsager reaction field calculated with the aid of the static dielectric constant of the continuous medium in which the Kirkwood correlation sphere is imbedded. This leads to eqn. (B141) in the appendix to his book[535]:

$$(\varepsilon_s - \varepsilon_\infty)(2\varepsilon_s + \varepsilon_\infty)/\varepsilon_s(\varepsilon_\infty + 2)^2 = (4\pi N_0/9kTV)g\mu_0^2 \qquad (6)$$

Here ε_∞ is the dielectric constant at a frequency sufficiently high for the intermolecular contributions to have no time to contribute but not suffi-ciently high for the intramolecular contributions (induced polarization and vibrations of the molecule) to be damped out. The free molecule dipole moment μ_0 replaces the medium-surrounded dipole moment μ. Only for $\varepsilon_\infty = 1$ would this equation reduce to eqn. (1). However, the use of eqn. (5) to calculate μ from μ_0 would make the two equations equivalent with a factor of $(2\varepsilon_s + 1)/(2\varepsilon_s + n^2)$, provided that ε_∞ is taken as n^2. For pure water there is a dispersion process at submillimeter wavelengths[278,432] which in Volume 1, Chapter 7 was interpreted as intermolecular, so that it was not included in ε_∞, which was taken as equal to n^2. For alcohols the calculations of Oster and Kirkwood included n^2 rather than ε_∞. Only for $\varepsilon_\infty \gg n^2$ would serious discrepancies develop; the opening up of the submillimeter band and the interpretation of dispersions as intramolecular resonance effects could influence the situation as far as alcohols are con-cerned.

The temperature variation of the correlation parameter g may be interpreted in terms of the energy of the hydrogen bonds which are respon-sible for the association of the molecules in the liquid. The technique was worked out by Cole and his colleagues, and was applied to alcohols,[362] to N-substituted amides,[361] and to nitriles which associate as linear chains.[327,363,844] A weighted average correlation factor for all the individual species can be calculated. For a linear chain containing n dipoles it is simply $g = n$. No calculations for aqueous solutions have yet appeared, but it is possible that the technique could be applied.

For alcohols it is assumed that there is a distribution of chains of varying lengths governed by the association equilibria

$$(\mathrm{ROH})_n + \mathrm{ROH} \rightleftharpoons (\mathrm{ROH})_{n+1} \qquad (7)$$

i.e., with breakage of the chains possible only at the ends. The equilibrium constant K is given in terms of the free energy ΔG by

$$K = C_{n+1}/C_n C_1 = (1/C_0)e^{\Delta G/RT} \tag{8}$$

where C represent polymer concentrations and C_0 the gross concentration as a monomer. The ratio α is defined as

$$\alpha = C_1/C_0 = (1 - KC_0\alpha)^2 \tag{9}$$

$$C_n/C_0 = \alpha^n(KC_0)^{n-1} \tag{10}$$

The mean correlation parameter $\bar{g}_n$ for an n-membered chain is defined as

$$n\bar{g}_n = \sum_i^{i=n} g_{in} \tag{11}$$

where g_{in} is the correlation parameter of the ith member. It can be expressed as

$$
\begin{aligned}
n\bar{g}_n &= n + 2f(n-1)\cos^2\theta + 2f(n-2)\cos^4\theta + \cdots + \\
&\quad + 2[n - (n-1)]\cos^{2(n-1)}\theta \\
&= 1 + 2f\cot^2\theta - [(2f\cos^2\theta)/n][(1 - \cos^{2n}\theta)/\sin^4\theta] \tag{12}
\end{aligned}
$$

where

$$f = (\mu_H + \mu_R)(\mu_H + \mu_R\cos 2\theta)/(\mu_H^2 + \mu_R^2 + 2\mu_H\mu_R\cos 2\theta) \tag{13}$$

μ_H and μ_R are the group moments along OH and OR, and 2θ is the ROH bond angle.

An approximate expression can be obtained for $\bar{g}_n$, namely,

$$\bar{g}_n = 2.40 - (2.00/n), \quad n > 1; \qquad g_1 = 1 \tag{14}$$

The overall mean correlation parameter is then

$$\bar{g} = (1/C_0)\sum_{n=1}^{\infty} n\bar{g}_n C_n = 2.39 - 2.00\sqrt{\alpha} + 0.61\alpha \tag{15}$$

For a linear chain containing n members the calculation is simpler, the weighted average correlation factor being

$$\bar{g} = \left(\sum_{n=1}^{\infty} n^2 C_n\right)\Big/C_0 = (1 + KC_1)/(1 - KC_1) \tag{16}$$

Therefore

$$KC_0 = (g^2 - 1)/4 \qquad (17)$$

so that a plot of $\log(g^2 - 1)$ against T^{-1} should be linear with gradient $\Delta H/2.30R$.

The heats of association ΔH of molecules deduced in this way are listed in Table II.

While the temperature variation of g can provide the heat of association, its absolute value can be related to the bond angle or angles involved in the chains. A good example is N-methyl acetamide,[844] which forms chains of configuration

for which the correlation factor for infinite chains is

$$\bar{g}_\infty = 1 + 2(\cos\theta_1)\{[\cos(2\pi - \theta + \theta_2)]/\cos(2\pi - \theta + \theta_1 + \theta_2)\} \qquad (18)$$

where θ_1 is the angle between the NH moment and the molecular moment and θ_2 is the angle between the CO moment and the molecular moment. The average chain formation angle θ can be calculated from the absolute value of g at absolute zero, obtained by extrapolation. We find $\theta = 160°$.

TABLE II. Heats of Association Calculated from Temperature Variation of Static Dielectric Constants

	ΔH, kcal mol^{-1}
Hydrogen cyanide	-4.6
Cyanoacetylene	-2.8
N-Methylacetamide	-4.48
N-n-Butylacetamide	-3.80
Methyl alcohol	-6.50
Ethyl alcohol	-8.36
n-Butyl alcohol	-7.70

2. THE STATIC DIELECTRIC CONSTANTS OF NONELECTROLYTIC AQUEOUS SOLUTIONS

There exist some established data for the static dielectric constants ε_{ss} of nonelectrolytic aqueous solutions.[330,439,1445] Usually the range of concentration covered is incomplete, and in much of the work only the dielectric increment, defined in eqn. (23) below, was determined. This rests upon the assumption, which is only very approximately true, that ε_{ss} is a linear function of concentration of the solute. A list[1445] of dielectric increments δ [eqn. (23) below] is given in Table III.

The Kirkwood theory (in its original form) was first applied to the magnitudes of these increments by Oster.[1055] The polarization P_m of the mixture is

$$P_m = x_1 P_1 + x_2 P_2 \tag{19}$$

where the subscript 1 represents water and 2 represents the other polar liquid; x represents molar fraction. According to the original formulation of the Kirkwood equation,

$$\frac{(\varepsilon_{ss} - 1)(2\varepsilon_{ss} + 1)}{9\varepsilon_{ss}} V_m = \frac{(\varepsilon_{s1} - 1)(2\varepsilon_{s1} + 1)x_1 V_1}{9\varepsilon_{s1}}$$
$$+ \frac{(\varepsilon_{s2} - 1)(2\varepsilon_{s2} + 1)x_2 V_2}{9\varepsilon_{s2}} \tag{20}$$

where V_m is the mean molar volume. Additivity of volumes is assumed:

$$V_m = x_1 V_1 + x_2 V_2 \tag{21}$$

so that, approximately,

$$\varepsilon_{ss} \simeq \varepsilon_{s1} + \left[\frac{(\varepsilon_{s2} - 1)(2\varepsilon_{s2} + 1)}{2\varepsilon_{s2}} - (\varepsilon_{s1} - 1) \right] \frac{x_2 V_2}{V_m} \tag{22}$$

In the limit of low concentration the static dielectric constant of the mixture is taken to be a linear function of concentration,

$$\varepsilon_{ss} = \varepsilon_{s1} + \delta c_2 \tag{23}$$

where c_2 is the molar concentration of solute. Therefore we can write for the dielectric increment

$$\delta = \{[(\varepsilon_{s2} - 1)(2\varepsilon_{s2} + 1)/2\varepsilon_{s2}] - (\varepsilon_{s1} - 1)\}V_2/1000 \tag{24}$$

TABLE III. Dielectric Increments of Dipolar Ions and Related Substances in Water at 25°C

Substance	Dielectric increment	Ref.
Glycine	22.6	1446
	23.0	649
	26.4	413
	30	646
	22.65 ± 0.03	1053
α-Alanine	23.2	1446
	23.6	649
	27.7	413
	23.20 ± 0.05	1053
α-Aminobutyric acid	23.2	1446
α-Aminovaleric acid	22.6	1446
dl-α-Valine	25	415
l-α-Leucine	25	415
β-Alanine	34.6	1446
	35	416
	42.3	649
β-Aminobutyric acid	32.4	1446
	36	420
γ-Aminobutyric acid	51	416
γ-Aminovaleric acid	54.8	1446
δ-Aminovaleric acid	63	416
ε-Aminocaproic acid	77.5	1446
	73	416
ζ-Aminoheptylic acid	87	416
l-Asparagine	28.4	413
	20.4	594
l-Glutamine	20.8	594
d-Glutamic acid	26	415
l-Aspartic acid	27.8	413
dl-Proline	21	415
N-Phenylglycine	∼30	421
Ornithine	51	420
Sarcosine	24.5	415

TABLE III. (*Continued*)

Substance	Dielectric increment	Ref.
d-Arginine	62	419
Taurine	41	415
Creatine	32.2	594
Glycocyamine	30	594
Acetylhistidine	62	594
Glycine dipeptide	70.6	1446
	70	415
	70.5	273
	80	646
Betaine *m*-aminophenol	32	421
Betaine *p*-aminophenol	45	421
Betaine of *p*-amino-*trans*-cinnamic acid	100	421
Betaine of *p*-sulfanilic acid	73	421
Betaine of *m*-sulfanilic acid	60	421
4,4'-Diphenylbetaine	155	594
Dimethylphenylglycine	17	421
o-Aminobenzoic acid	Low	418
m-Aminobenzoic acid	41	274
p-Aminobenzoic acid	~0	418
$m\text{-}(CH_3)_3N^+C_6H_4CH{=}CHCOO^-$	71	422
$p\text{-}(CH_3)_3N^+C_6H_4CH{=}CHCOO^-$ (*trans*)	100	422
$p\text{-}(CH_3)_3N^+C_6H_4CH{=}CBrCOO^-$ (*trans*)	102	422
$m\text{-}(CH_3)_3N^+C_6H_4CH{=}C(C_6H_5)COO^-$ (*cis*)	25	422
$m\text{-}(CH_3)_3N^+C_6H_4CH{=}C(C_6H_5)COO^-$ (*trans*)	90	422
Phenol	−6.6	413
Benzoic acid	−67	413
Aniline	−7.6	413
Urea	3.4	413
	3.15	417
	2.72	1443
Thiourea	4	413
Methylurea	3.7	415
Ethylurea	1	416

TABLE III. (*Continued*)

Substance	Dielectric increment	Ref.
Propylurea	1	416
Urethan	−4.3	415
Biuret	−6.3	415
Semicarbazide	0	415
Thiosemicarbazide	∼0	415
Dimethylurea (asymmetric)	∼0	415
Dimethylurea (symmetric)	3	415
Malonamide	4.3	415
Succinamide	∼−1	415
Acetamide	∼−0.8	415
dl-Malamide	2	420
Benzamide	−4.1	413
Sulfamide	7	414
Nitromethane	−2	420
Hydantoin	−6.4	416
Glycine tripeptide	113	1446
	128	273
Glycine tetrapeptide	159	1446
Glycine pentapeptide	215	1446
Glycine hexapeptide	234	1446
Glycine heptapeptide	290	1444
Glycylalanine	71.8	594
Alanyglycine	71	594
Leucylglycine	62	273
	68.4	594
	65.7±0.3	1053
Glycylleucine	54	416
	74.6	594
	70	273
N-Methylleucylglycine	67	594
Glycylphenylalanine	70.4	594
	110.5±0.7	1053
Phenylalanylglycine	56.7	594

TABLE III. (*Continued*)

Substance	Dielectric increment	Ref.
d-Leucylglycylglycine	120.4	594
	54	416
	112	273
	120.4±0.3	1053
ε,ε'-Diguanidodi(α-thio-n-caproic acid)	151	594
ε,ε'-Diaminodi(α-thio-n-caproic acid)	131	595
Lysylglutamic acid	345	595
Glycine betaine	24–27	420
	18.2	440
α-Aminovalerianic acid betaine	60	646
ξ-Aminopentadecyclic acid betaine	220 (70°C)	646
π-Aminoheptadecylic acid betaine	190 (80°C)	646
Pyridine betaine	18.5	440
	20.5	646
o-Benzbetaine	18.7	440
	20	421
m-Benzbetaine	48.4	440
	58	421
p-Benzbetaine	72.4	440
	68	421
	62	646
Thiobetaine	23	420
N-Dimethylanthranilic acid	12	440
	16.7	421
Betaine o-aminophenol	5.6	421
Pyrrolidine	−1.0	416
Pyridine	−4.2	416
	−3.8	860
Dimethylpyrone	~−3	420
2,5-Dioxopiperazine	−10	415
Glycine anhydride	−10	415
Succinimide	−10	415
Acetoacetic ether	−5.9	420
Hydroxylamine	−0.8	420

TABLE III. (*Continued*)

Substance	Dielectric increment	Ref.
Acetanilide	−4	413
Acetonitrile	−1.74	413
m-Dihydroxybenzene	−6	420
o-Dihydroxybenzene	∼--6	420
p-Dihydroxybenzene	−6.4	420
Methyl acetate	−5	420
1,2-*p*-Aminophenylarsenic acid	1.2	420
Diglycine	70.4±0.4	1503
Triglycine	114.5±0.3	1505
Tetraglycine	165.8±1.0	1503
Pentaglycine	202.2±2.3	1503
Hexaglycine	240±25	1503
Leucylalanine	57.7±0.5	1503
Alanylglycylglycine	117.4±0.2	1503
Alanylleucylglycine	124.5±1.2	1503
Glucose	−4.27	522a
Mannose	−4.25	522a
Ribose	−2.72	522a
Galactose	−3.28	522a
Tetrahydrofuran	−4.85	522a
Tetrahydrofurfuryl alcohol	−4.74	522a
Myoinositol	−2.17	522a
Mannitol	−2.48	522a
Sorbitol	−2.75	522a
α-Methyl-D-glucoside	−6.35	522a
Pyrazine	−6.4	1112
2-Methylpyrazine	−7.3	1112
2-6-Dimethylpyrazine	−8.5	1112
Quinoxaline	−9.3	1112
2-Methylquinoxaline	−11.1	1112
1-4-Diazabicyclo[2,2,2]octane	−7.4	1112

A comparison of experimental and calculated dielectric increments is given in Table IV.

The agreement is good, considering the crudeness of the assumptions; better agreement is achieved by using the measured partial molar volume of the solution, but the values have not all been reported in the literature. However, the calculated dielectric increment is nearly always slightly less negative than the measured value. A unique case for which it is considerably greater (-8.3 compared with -6.15) is that of dioxan. It is likely that there are hydrogen bonds between the water and the oxygens of the dioxan which partially destroy the highly coordinated structure of water, with consequent decrease in the correlation parameter.

Although the derivation of eqn. (24) was made on a molecular basis, it is of interest to consider the possible applicability of dielectric mixture theory.[1318] Such theory has been applied to heterogeneous mixtures and has been tested by experiments with artificial systems. The theory applies to a continuous dielectric medium, particles of which are imbedded in a second continuous dielectric medium. No attempt is made to take into account individual molecular dipoles or the interaction between them. But although liquid mixtures are generally believed to interpenetrate on a

TABLE IV. Comparison of Observed and Calculated Dielectric Increments at 25°C

	ε_{s2}	v_2	$-\delta_{calc}$, eqn. (24)	$-\delta_{obs}$	$-\delta_{calc}$, mixture theory
Methyl alcohol	33.7	40.5	1.79	1.4	1.61
Ethyl alcohol	25.1	58.7	3.11	2.6	3.33
n-Propyl alcohol	21.8	75.1	4.22	4.0	4.94
t-Butyl alcohol	11.4	94.2	6.26	6.3	8.80
Acetone	21.5	73.4	4.13	3.2	4.85
Diethyl ether	4.5	103.0	7.57	7.1	12.61
Glycol	41.2	55.0	2.04	1.8	1.87
Aniline	7.2	90.1	6.40	7.6	7.88
Methyl acetate	7.3	80.0	5.65	5	8.36
Pyridine	12.5	80.6	5.27	4.2	7.07
Acetonitrile	38.8	52.5	2.08	1.7	1.92
Nitromethane	39.4	54.2	2.11	2.0	1.95

molecular scale, it is of interest to consider the contrary hypothesis, namely that larger particles of one liquid are imbedded in the other. This is of particular interest for alcohol–water mixtures, since the pure alcohol has been interpreted as consisting of very long molecular chains folded into spheres, while water is more of a three-dimensional network. Thus it is conceivable that the alcohol–water mixture might respond to mathematical treatment as a heterogeneous mixture.

For mixtures containing spherical particles there is some disagreement between different mixture formulas, and experimental tests have been made using polystyrene latex spheres[450] and glass spheres[1145] in liquids. However, it is not difficult to show that for the aqueous solutions considered above all the mixture formulas lead to decrements in excess of those observed in experiments. On the other hand, the molecular treatment discussed above [eqn. (24)] is not unrealistic, which suggests that inter-molecular structure formation is a better hypothesis than the heterogeneous mixture of continuous dielectrics.

The original Maxwell–Lewin formula[841,1108] for spheres in a polar medium leads to

$$\delta = \tfrac{3}{2} V_2(\varepsilon_1 - \varepsilon_2)/1000 \tag{25}$$

For methyl alcohol the calculated $-\delta = 2.7$, whereas the experimental value (Table IV) is 1.4. If nonspherical shapes of species 2 are assumed, the calculated $-\delta$ values are still larger.

Certain more modern mixture formulas have been successful for polar spheres in nonpolar media and deserve consideration. They are due to Bruggeman,[238] Böttcher,[213] and Looyenga.[867,868] The volume fraction ϕ_2 is related to the mixture dielectric constant and to ε_1 and ε_2 as follows:

$$\phi_2 = 1 - [(\varepsilon_2 - \varepsilon_m)/(\varepsilon_2 - \varepsilon_1)](\varepsilon_1/\varepsilon_m)^{1/3} \quad \text{(Bruggeman)} \tag{26}$$

$$\phi_2 = (\varepsilon_m - \varepsilon_1)(2\varepsilon_m + \varepsilon_2)/3\varepsilon_m(\varepsilon_2 - \varepsilon_1) \quad \text{(Böttcher)} \tag{27}$$

$$\varepsilon_m = [(\varepsilon_2^{1/3} - \varepsilon_1^{1/3})\phi_2 + \varepsilon_1^{1/3}]^3 \quad \text{(Looyenga)} \tag{28}$$

There is not very much difference between mixture dielectric constants calculated with these formulas; we therefore calculate values of $-\delta$ only from the last. It will be seen from Table IV that these values are not by any means in such good agreement with the observed values as are those calculated using the molecular model. At laboratory temperatures the mixing of two polar liquids is sufficiently intimate, and the dipolar and hydrogen bonding interactions so anisotropic, that the heterogeneous mixture theories become inapplicable.

Pottel and Kaatze[1112] have treated their data for nitrogen-containing heterocyclic compounds and substituted tetraalkylammonium bromides by means of a mixture formula due to Brown[236]

$$\phi_2 = (\varepsilon_m - \varepsilon_1)(2\varepsilon_1 + \varepsilon_2)/(\varepsilon_m + 2\varepsilon_1)(\varepsilon_2 - \varepsilon_1) \tag{28a}$$

For $\varepsilon_1 \sim \varepsilon_m$, $\varepsilon_1 \gg \varepsilon_2$, this formula does not differ greatly from that of Böttcher, eqn. (27); however, the authors substitute the square of the optical refractive index for ε_2. For the tetraalkylammonium bromides the formula reproduces the increments well, suggesting that the solute–water structure is minimal; but for the nitrogen-containing heterocyclics (Table VIII) the calculated increments are, as usual, too large.

Data exist for static dielectric constants of mixtures of a number of organic liquids with water over the entire range of concentration (completely miscible liquids). The most extensive set of data is due to Akerlöf and his colleagues.[9,10] Akerlöf's technique was unsatisfactory for loss angles greater than a certain value, and for this reason some of his measurements are inaccurate for high concentrations of solutes of low dielectric constant. Some more recent measurements for t-butyl alcohol solutions are to be preferred[232] and some unpublished measurements are also available for pyridine solutions.[231] These data are listed in Table V.

A new analysis of these data in terms of the correlation factors can be made as follows. The polarization P_s of the solution is calculated from the static dielectric constant ε_{ss} using the Kirkwood–Fröhlich equation

$$P_s = \{[(\varepsilon_{ss} - n^2)(2\varepsilon_{ss} + n^2)/9\varepsilon_{ss}]\}V \tag{29}$$

It is considered sufficiently accurate to interpolate the refractive indices n linearly between those of the two liquids:

$$n = n_1 + (n_1 - n_2)x_2 \tag{30}$$

It is assumed that the polarization is the sum of the contributions from each set of dipoles:

$$P_s = (4\pi N_0/3kT)(x_1 g_1 \mu_1{}^2 + x_2 g_2 \mu_2{}^2) \tag{31}$$

The dipole moments μ_1 and μ_2 are calculated from those of the free molecules μ_{01} and μ_{02} using the Onsager relation in the form

$$\mu_1 = \tfrac{1}{3}(n^2 + 2)\mu_{01} \tag{32}$$

$$\mu_2 = \tfrac{1}{3}(n^2 + 2)\mu_{02} \tag{33}$$

Now it is impossible to separate two correlation factors g_1 and g_2 from a

TABLE V. Static Dielectric Constants of Aqueous Mixtures[a]

Methyl alcohol, wt%	20°C	30°C	40°C	50°C	60°C
0	80.37	76.73	73.12	69.85	66.62
10	75.84	72.37	68.90	65.66	62.77
20	71.02	67.48	64.13	61.06	58.24
30	66.01	62.71	59.53	56.59	53.94
40	61.24	58.06	54.82	52.17	49.52
50	56.53	53.47	50.40	47.82	45.28
60	51.53	48.58	45.64	43.22	41.22
70	46.46	43.63	41.04	38.81	36.68
80	41.46	38.98	36.66	34.62	32.74
90	36.80	34.62	32.56	30.67	28.91
100	32.35	30.68	29.03	27.44	25.97
	(32.8 accepted value)				

Ethyl alcohol, wt%	20°C	40°C	50°C	60°C	80°C
0	80.37	73.12	69.85	66.62	60.58
10	74.60	67.86	64.53	61.49	55.70
20	68.66	62.41	59.22	56.40	50.81
30	62.63	56.73	53.79	51.04	45.88
40	56.49	51.08	48.36	45.80	40.93
50	50.38	45.30	42.92	40.66	36.51
60	44.67	40.02	37.72	35.66	31.82
70	39.14	34.88	32.86	30.87	27.30
80	33.89	29.83	28.10	26.31	23.20
90	29.03	25.64	24.08	22.51	19.80
100	25.00	22.20	20.87	19.55	—
	(24.6 accepted value)				

n-Propyl alcohol, wt%	20°C	40°C	50°C	60°C	80°C
0	80.37	73.12	69.85	66.62	60.58
10	73.52	66.81	63.66	60.65	54.77
20	66.54	60.24	57.23	54.49	49.01
30	59.21	53.46	50.72	48.19	43.00
40	51.68	46.55	44.08	41.76	37.53
50	44.29	39.70	37.38	35.39	31.42
60	37.51	33.54	31.49	29.71	26.22
70	31.56	28.20	26.42	24.92	21.84
80	26.83	23.89	22.39	20.95	18.28
90	23.34	20.67	19.37	18.07	15.81
100	20.81	18.25	17.11	15.88	13.86
	(19.5 best value)				

TABLE V. (*Continued*)

Isopropyl alcohol, wt%	20°C	40°C	50°C	60°C	80°C
0	80.37	73.12	69.85	66.62	60.58
10	73.11	66.33	63.12	60.24	54.83
20	65.72	59.56	56.61	53.87	49.01
30	58.40	52.71	50.18	47.58	43.13
40	51.07	45.86	43.54	41.35	37.31
50	43.68	39.16	37.03	35.05	31.49
60	˙36.28	32.45	30.67	28.90	25.67
70	29.57	26.30	24.85	23.34	20.67
80	24.44	21.63	20.26	19.03	16.70
90	20.95	18.48	17.11	16.02	13.83
100	18.62	16.23	15.06	14.03	11.91

Glycerol, wt%	20°C	40°C	60°C	80°C	100°C
0	80.37	73.12	66.62	60.58	55.10
10	77.55	70.41	63.98	58.31	—
20	74.72	67.70	61.56	56.01	—
30	71.77	64.87	58.97	53.65	—
40	68.76	62.03	56.24	51.71	—
50	65.63	59.55	53.36	48.52	—
60	62.03	55.48	50.17	45.39	41.08
70	57.06	51.41	46.33	41.90	38.07
80	52.27	46.92	42.32	38.30	34.70
90	46.98	42.26	38.19	34.47	31.34
100	41.14	37.30	33.82	30.63	27.88

Acetone, wt%	20°C	25°C	30°C	40°C	50°C
0	80.37	78.54	76.73	73.12	69.85
10	74.84	73.02	71.37	68.07	65.01
20	68.58	66.98	65.34	62.28	59.45
30	62.48	61.04	59.47	56.77	54.17
40	56.00	54.60	53.23	50.82	48.52
50	49.52	48.22	46.99	44.81	42.81
60	42.93	41.80	40.75	38.86	37.04
70	36.51	35.70	34.63	33.03	31.44
80	30.33	29.62	28.74	27.50	26.20
90	24.61	23.96	23.38	22.32	21.26
100	19.56	19.10	18.67	17.80	16.98

TABLE V. (*Continued*)

1,4-Dioxan, wt%	0°C	20°C	40°C	60°C	80°C
0	88.31	80.37	73.12	66.62	60.58
10	78.86	71.43	64.70	58.60	53.07
20	69.16	62.38	56.26	50.75	45.77
30	59.34	53.30	47.88	43.01	38.63
40	49.37	44.19	39.54	35.39	31.67
50	39.50	35.25	31.46	28.08	25.05
60	29.84	26.60	23.72	21.15	18.86
70	20.37	18.20	16.26	14.52	12.97
80	12.19	10.99	9.91	8.93	8.05
90	6.16	5.71	5.30	4.91	4.56
100	2.109	2.102	2.098	2.094	2.090

Ethylene glycol, wt%	20°C	40°C	60°C	80°C	100°C
0	80.37	73.12	66.62	60.58	55.10
10	77.49	70.29	63.92	58.02	—
20	74.60	67.52	61.20	55.36	—
30	71.59	64.51	58.37	52.59	—
40	68.40	61.56	55.48	49.81	—
50	64.92	58.25	52.30	46.75	—
60	61.08	54.53	48.75	43.68	39.13
70	56.30	50.17	44.98	40.19	35.94
80	50.64	45.45	40.72	36.36	32.52
90	44.91	40.43	36.35	32.58	29.27
100	38.66	34.94	31.58	28.45	25.61

tert-Butanol, wt%	20°C	40°C	50°C	60°C	80°C
0	80.37	73.12	69.85	66.62	60.58
10	71.75	64.91	61.84	58.83	52.87
20	62.93	56.77	53.82	50.87	45.14
30	54.17	48.33	45.44	42.96	38.47
40	45.38	40.01	37.35	34.93	30.59
50	36.59	32.16	29.80	27.80	24.24
60	28.91	25.02	23.01	21.30	18.41
70	22.30	19.00	17.41	16.29	13.94
80	17.23	14.60	13.44	12.51	10.45
90	12.97	10.93	10.00	9.36	7.66
100	—	8.44	7.67	6.96	5.90

TABLE V. (*Continued*)

tert-Butanol mole fraction	25°C	30°C
0	78.48	—
0.0546	64.80	—
0.1131	50.65	—
0.1871	36.14	—
0.4170	20.01	—
0.7352	12.76	—
0.8038	11.92	11.27
0.8543	—	11.14
0.8705	11.79	—
0.9229	11.84	11.09
0.9797	12.30	11.21
1.0000	12.52[b]	11.30[b]

iso-Butanol mole fraction	25°C
0.5803	19.51
0.7882	17.59
0.8688	17.47
0.8948	17.43
0.9207	17.48
0.9922	17.67
1.0000	17.50[b]

Pyridine, mole fraction	25°C	Pyridine, mole fraction	25°C
0	78.5	0.186	49.7
0.008	77.35	0.276	42.3
0.011	76.3	0.303	40.2
0.016	74.75	0.328	38.85
0.023	73.9	0.481	30.42
0.025	73.7	0.497	29.84
0.030	72.25	0.549	27.32
0.039	71.15	0.581	26.03
0.056	66.85	0.591	25.61
0.077	63.85	0.661	22.80
0.094	60.85	0.666	22.70
0.098	60.65	0.754	19.67
0.113	57.55	0.874	16.11
0.164	52.05	0.921	14.78
0.173	51.65	1.00	13.06

[a] The headings in the first column give the liquid in mixture with water.
[b] Extrapolated values.

single value of P_s without some assumption. One possible assumption is that the correlation factor for one molecule remains unchanged with varying concentration while that for the other molecule varies continuously. This assumption is physically unrealistic, but nevertheless some conclusions reached are qualitatively similar to those discussed below, which are reached with the aid of the assumption that the two factors g_1 and g_2 are identical. In this case

$$g = (3P_s kT/4\pi N_0)(x_1\mu_1^2 + x_2\mu_2^2) \tag{34}$$

The assumption is most plausible in dilute solutions, when P_s is usually dominated by the most common dipole; the single parameter g then approximates to a quantitative measure of the effect of the addition of the other liquid upon the dipole correlation of the dominant solvent.

The use of the Onsager value of μ_1 for pure water and the resulting single value of g is open to criticism; the correlation in pure water has been discussed in Volume 1, Chapter 7. However, the Onsager technique has been chosen so as to achieve internal consistency for all the studies made. Unfortunately, the temperature variation of the refractive indices of many of the liquids is unreported, but the error introduced by neglecting this is probably not large.

A limitation of the application of Kirkwood theory to media containing two species of molecule is that there will be cross-correlation terms which are difficult to take into account. The correlation of dipoles of species 2 with dipoles of species 1 contributes a term $\mu_{12} \cdot \mu_1$ in addition to the 1–1 correlation term $\mu_{11} \cdot \mu_1$ already considered. It is possible that the separation of the four terms

$$\mu_{12} \cdot \mu_1 + \mu_{11} \cdot \mu_1 + \mu_{21} \cdot \mu_2 + \mu_{22} \cdot \mu_2$$

by measurements over a wide range of frequency could be achieved. From the static dielectric constant, however, nothing more satisfactory than a single correlation factor can be obtained.

Figure 1 shows the variation of the calculated correlation factors with x_2. Free molecule dipole moments and refractive indices of liquids are taken from the tables in Ref. 635; the dipole moment of the glycerol molecule has of necessity been estimated. Calculations are made only at the extreme values of temperature studied experimentally.

The simplest aqueous mixtures are those containing liquids with small correlation factors (acetone) and those which are (apparently) nonpolar (1,4-dioxan). For these there is a steady monotonic fall of correlation factor as x_2 increases. The solutes are all efficient at "breaking the structure" of

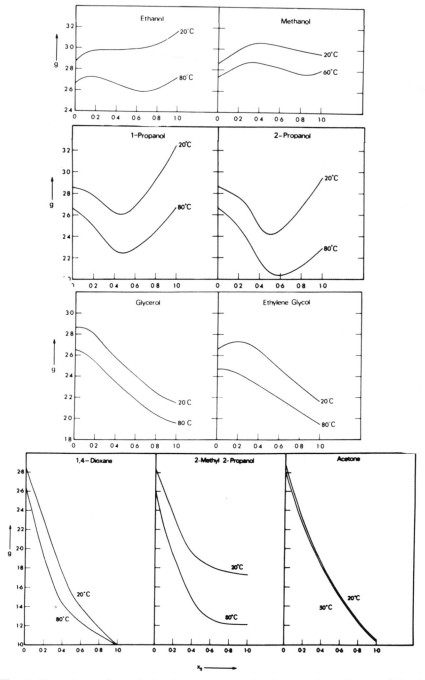

Fig. 1. Dependence of correlation factors upon molar fractions for mixtures of simple organic liquids with water.

water. On the interpretation of Haggis et al.[618] (Volume 1, Chapter 7), this behavior corresponds to a breaking of the hydrogen bonds.

The temperature dependence of the structure-breaking effect is of some interest in the t-butyl alcohol–water mixture. It is apparent that the temperature variation of correlation factor is more marked in mixtures of high t-butyl alcohol content than it is in water. This reflects the structural interpretations discussed earlier. The temperature variation for t-butyl alcohol given by Dannhauser and Cole[362] is actually larger still, partly because of differences arising from the use of ε_∞ instead of n^2 in the Kirkwood formula. Whereas the temperature variation of g for t-butyl alcohol is supposed to arise from variation in chain length ($8 > \Delta H^\ddagger > 6$ kcal mol^{-1}), that for water is supposed to arise from a change in the relative proportions of highly bonded and less highly bonded water molecules; because the structure is a three-dimensional network rather than a coiled chain the statistics of hydrogen bond breakage result in a weaker temperature variation of correlation factor.

The polyhydric alcohols, ethylene glycol and glycerol, also possess correlation factors much smaller than that of water; and these mixtures also show an almost monotonic decrease of correlation factor with decreasing molar fraction of water. However, at lower temperatures there is a region of low alcohol concentration where the correlation factor is slightly higher than that of pure water. This effect is also present in ethyl alcohol–water mixtures. The correlation factors of the two liquids are not very unequal but there is a tendency for small proportions of ethyl alcohol to increase the hydrogen bonding of water; there is also a tendency for small proportions of water to decrease the hydrogen bonding in alcohol.

In methyl alcohol–water mixtures there is a structure-making effect, maximizing the correlation factor at $x_2 = 0.4$. By contrast, both propyl and isopropyl alcohols, at $x_2 \simeq 0.5$, have a marked structure-breaking effect, the correlation factors of these mixtures being well below those of both the alcohols and water. The temperature variation suggests that the structure of the 50% mixture is rather more alcohol-like than water-like.

The addition of water to pure propanol, isopropanol, or t-butyl alcohol results in a lowering of the correlation factor; the comparatively rare water molecules are apparently able to achieve breakage of the alcohol chains.

3. CALCULATION OF DIPOLE MOMENTS OF SOLUTES

It is sometimes possible by application of dielectric theory to calculate the solute dipole moment from measurements of the static dielectric constant of the solution.

The Onsager model is the starting point for this calculation. The molecular dipole is considered to be at the center of a spherical cavity of molecular dimensions. The field acting on the dipole arises partly from the applied field acting on the dielectric continuum outside the cavity and partly from the polarization induced in the dielectric continuum by the dipole itself. In a pure liquid the Onsager expression relating the dielectric constant and dipole moment is

$$\frac{M}{d}\left(\frac{\varepsilon_s - 1}{\varepsilon_s - 2} - \frac{n^2 - 1}{n^2 + 2}\right) = \frac{3\varepsilon_s(n^2 + 2)}{(2\varepsilon_s + n^2)(\varepsilon_s + 2)} \frac{4\pi N\mu^2}{9kT} \qquad (35)$$

where M is the molecular weight and d is the density. For polar liquids in which there is no valence-bond correlation between neighboring dipoles the theory is successful. Therefore it might be supposed that it could be applied to calculate dipole moments of solute molecules which do not correlate by hydrogen bonding with the neighboring water molecules.

The theory was first applied in this form[796,988] but an error arose because the molecular cavity was taken to be spherical, which is often a poor approximation. Buckingham[243,244] applied the solution of the Onsager theory with ellipsoidal cavity to calculating the dipole moment of solutes in polar liquids. Molecules which dissolve in water are in many cases difficult to vaporize or to dissolve in nonpolar liquids, so that some importance is attached to the ability to calculate accurate values of the dipole moments from analysis of the dielectric properties of the aqueous solution.

The Buckingham expression for the static dielectric constant of a mixture of i components of molar fractions x_i and free molecule dipole moment μ_{0i} is

$$\frac{(2\varepsilon_s + 1)(\varepsilon_s - n^2)}{3(2\varepsilon_s + n^2)} \frac{\sum_i x_i M_i}{d} = \frac{4\pi N}{9kT} \sum_i x_i \mu_{0i}^2 f_i g_i$$
$$= \sum_i x_i P_{0i} f_i g_i \qquad (36)$$

where P_{0i} is the orientation polarization of the ith component and M_i its molecular weight, d is the density of the mixture, and N is the total number of molecules per unit volume. The square of the refractive index of the solution, n^2, is sufficiently well approximated by linear interpolation between the squares of the component refractive indices.

In the absence of an applied field the moment of a spherical specimen of the solvent containing a solute molecule of moment μ_{0i} is $f_i \mu_{0i}$. The factor

g_i is related to the energy w of a molecule of moment μ_{0i} by the expression

$$w = -g_i\mu_{0i}\mathbf{E}\cos\theta \tag{37}$$

when a uniform field $\mathbf{E}$ exists in the solution. Here θ is the angle between $\mathbf{E}$ and the dipole axis. Employing the assumption that the molecules are ellipsoids of internal dielectric constant n_i^2, possessing permanent dipoles along one of the axes, surrounded by continuous medium of dielectric constant ε_s, it can be shown that

$$f_i = (2\varepsilon_s + 1)[1 + (n_i^2 - 1)A_i]/3[\varepsilon_s + (n_i^2 - \varepsilon_s)A_i] \tag{38}$$

$$g_i = \varepsilon_s[1 + (n_i^2 - 1)A_i]/[\varepsilon_s + (n_i^2 - \varepsilon_s)A_i] \tag{39}$$

The shape factors A_i, which depend upon the semiaxial ratios, have been calculated.[1053]

For dilute solutions the molecular polarization of the solvent, water, may be assumed to vary linearly with concentration, and can be calculated from the experimental data for a well-behaved solution of a liquid of known orientation polarization, which does not "associate" with water. Acetone has been chosen for this purpose.

On the basis of Buckingham's theory, dipole moments of a number of polar molecules have been derived from static dielectric constant data.

The extensive list of increments in Table III provides data from which further dipole moment calculations of this type might be made.

Table VI lists dipole moments calculated from eqns. (36)–(39). In a few cases comparisons with dipole moments obtained by other means can be made. It is likely that the dipole moments of amino acids (≥ 10 D) will usually be well approximated by the values derived in this way. However, the monosaccharides, which may have smaller dipole moments, may not be so well approximated. The conditions which must be satisfied for the model to be applicable are: (a) the molecule must be sufficiently rigid for the variation between individual molecules to be small (it should be understood that the theory predicts the root mean square dipole rather than the mean value of the dipole); (b) the interaction of the molecule with water molecules must be sufficiently weak for the dielectric behavior of the water to be unaffected and for the solute molecule dipole to be unaffected by hydration.

It would seem from an examination of the four molecules pyridine, acetone, tetrahydrofuran, and tetrahydrofurfuryl alcohol that dipole moments of order 2–4 D are likely to be up to 50% high when calculated from the aqueous solution dielectric constant. However, some confidence

TABLE VI. Dipole Moments Derived from Static Dielectric Constants of Aqueous Solutions

Molecule	μ_{0i}, D		Ref.
	Derived	Other source	
Acetone	3.7	2.85 (gas)	409
Pyridine	3.1	2.5 (gas)	244
Glycine	13.3	14.2 (calc)	244
β-Alanine	17.5	17.4 (calc)	244
Tetrahydrofuran	2.4	1.63 (gas)	409
Tetrahydrofurfuryl alcohol	3.5	2.3 (gas)	409
D-Galactose	5.3	—	409
L-Arabinose	4.3	—	1041
D-Glucose	4.7	—	409
D-Mannose	4.8	—	409
D-Ribose	5.1	—	409
Myoinositol	5.0	—	409

in the internal consistency of the calculations is to be derived from the fact that although the dielectric increments are temperature-dependent, the calculated dipole moments are not.[409] Thus the correlation between solute molecule and water dipoles does not change with temperature. Since the correlation between dipoles in pure liquids and pure alcohols probably changes quite markedly with temperature, the implication is that when there is correlation it must be temperature-dependent; therefore the correlation between solute molecule and water might be presumed to be small. The temperature variation of overall correlations, shown in Fig. 1, is relevant to this discussion.

It is clear that where there are grounds for supposing that there is strong correlation between the two types of molecule (e.g., monohydroxylic alcohols) the model would not be of great value. For example, its application to dioxan–water mixtures yields a negative dipole moment for dioxan. But if the calculation is carried out assuming a dioxan hydrate, a positive dipole moment can be derived. Although the frequency-dependent dielectric data discussed below show no real evidence of the presence of the large energy of activation of relaxation which would be required by the hydrogen bond breakage of a stable hydrate, nevertheless considerable hydrogen bonding between the dioxan and water molecules must occur.

4. SOME DISCUSSION OF EXPERIMENTAL TECHNIQUES

A discussion of the entire technique of dielectric constant and loss measurement would be out of place in this chapter. Nevertheless the present situation in this field makes it important to raise several questions. It will be apparent from a study of this chapter that the state of knowledge of the static, or low-frequency limiting dielectric constants, of solutions is reasonably good; but our knowledge of the frequency variation of the dielectric constant and loss is very incomplete; therefore it is necessary to consider briefly the best means by which this deficiency can be made good.

For kHz frequencies up to almost 1 MHz, impedance bridge techniques have been the traditional method of measurement. But in the MHz band, and up to almost 1 GHz, the most suitable traditional technique for lossy liquids such as aqueous solutions (sometimes exhibiting electrolytic conductivity) is the resonance technique. The electrical side of these techniques is well developed, and can be referred to in textbooks of electrical measurement and dielectrics.[359,668,1258]

For bridge measurements on aqueous solutions the most important feature is the cell in which the liquid is contained. Such a capacitance cell consists of two conducting surfaces between which the liquid is contained; a guard ring is also included in the modern "three-terminal cell." Of the many designs reported, a convenient modern version, due to Vidulich and Kay,[1338] is illustrated in Fig. 2.

For resonance measurements[234,1442] on aqueous solutions the characteristic frequency and selectivity of a resonator of fixed geometry are determined both when it is empty and when it is immersed in the liquid. Typical designs of resonator, concentric rings with half-turn coil, and plane metal spiral, are shown in Fig. 3.

At still higher frequencies (≥ 0.5 GHz) transmission-line techniques are of value. When the sample dimensions extend over an appreciable part of a wavelength it is necessary to use either transmission-line or waveguide techniques. An outline discussion of microwave techniques was given in Volume 1, Chapter 7 and reference should be made to the appropriate textbooks.[91,555,668,1284]

The great drawback of microwave techniques is that an individual apparatus is suitable for taking measurements only over a narrow frequency band, essentially at a single frequency, when considered in terms of a dielectric relaxation process covering two decades of frequency. The expense and sophistication of each single microwave dielectric apparatus have had the effect of limiting the range of frequency coverage that is normally available. Many workers have been content to work at three different frequencies, reasonably widely spaced over the water relaxation band. For aqueous solutions the complexities of multiple relaxation have not hitherto been adequately unfolded by existing microwave studies.

It is for this reason that "time-domain" as opposed to "frequency-domain" techniques have an important role to play.

One form of time-domain technique[693] consists in applying a voltage step function to a capacitor containing the dielectric. The time variation of the transmitted charge $q(t)$ is recorded over as wide a range of times as possible. The current response $i(t)$ of a filled capacitor having unit vacuum capacitance to unit voltage step is related to the complex permittivity by the transform

$$\varepsilon'(\omega) - j\varepsilon''(\omega) = \int_0^\infty i(t)e^{-j\omega t}\, dt \qquad (40)$$

Modern techniques of data sampling and recording, together with fast Fourier transform (FFT) computing programs, have enabled this type of experiment to be successful. Since the charge q varies approximately linearly with the logarithm of time, equal intervals of $\log t$ are the most advantageous choice of sampling times.[1177]

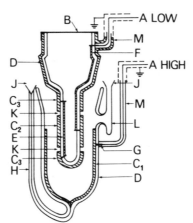

Fig. 2. Schematic diagram of dielectric cell.[1338] (A) Leads to bridge, (B) aluminium foil cap, (C) platinum metal films, (D) Pyrex glass, (E, F) strips of fired silver paste, (G) platinum microprobes, (H) Pyrex capillary tube, (J) filling tubes, 10/30 taper glass joint, (K) guard gaps in platinum films, (L) overflow tube, (M) glass tubing around cell leads.

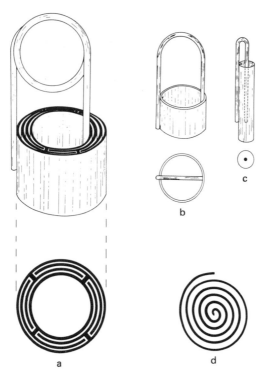

Fig. 3. Resonators for dielectric constant measurements.[234,1442] (a) Cylinder type: natural
period in air, 8×10^{-8} sec; size 13×10 cm. (b)
Cylinder type: natural period, 10^{-8} sec; size
5×7 cm. (c) Cylinder type; natural period,
2×10^{-9} sec; size, 8.6 cm. (d) spiral type: natural
period, 10^{-9} sec; diameter, 1.7 cm.

The capacitance time-domain technique, which depends upon current
measurement, is unsuitable in the ultra-high-frequency and GHz range,
in which much of the important aqueous solution information is to be
found. A concentric line reflection technique ("time-domain reflectometry"
or TDR) is more appropriate.[477,1267,1403] A cell is constructed from commercially available $Z_0 = 50$ ohm impedance precision concentric line of
7 mm diameter. It is partly filled with liquid and connected as in Fig. 4(a)
to a 35-psec risetime voltage step generator and sampling oscilloscope.

From the oscilloscope trace (Fig. 4b) of the first reflection from the
air–liquid interface the frequency variation of the complex permittivity
can be calculated. The potential falls from a value proportional to ϱ_∞ to a
value proportional to ϱ_0; these are the limiting values of the reflection coeffi-

cient. A voltage step of magnitude V_0 is in reflection of magnitude

$$V = V_0\varrho = V_0(Z - Z_0)/(Z + Z_0) \tag{41}$$

where

$$Z = Z_0/\sqrt{\hat{\varepsilon}} \tag{42}$$

Therefore the complex permittivity is related to the complex reflection

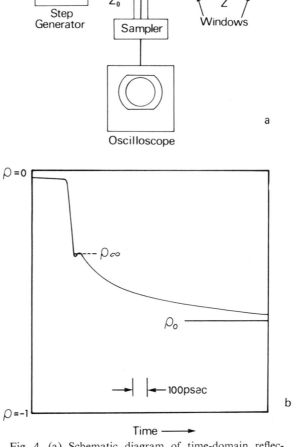

Fig. 4. (a) Schematic diagram of time-domain reflectometry apparatus. (b) Time variation of reflection coefficient ϱ from the interface air–amyl alcohol at $T = 25°C$.

coefficient by

$$\hat{\varepsilon} = [(1 - \hat{\varrho})/(1 + \hat{\varrho})]^2 \tag{43}$$

The complex permittivity $\hat{\varepsilon}(\omega)$ must be transformed into the time domain by means of the Laplace transform:

$$\varepsilon(t) = (1/2\pi) \int_{-\infty}^{+\infty} \hat{\varepsilon}(\omega)e^{jt}\,d\omega \tag{44}$$

The most satisfactory method of analysis of the TDR data is as follows.[1267] The reflected pulse and the incident pulse (which can be approximated by a pulse reflected from a precision short circuit) are transformed into the frequency domain by making use of the following modification of the Shannon sampling theorem due to Samulon[1177]:

$$F(\omega) = \frac{e^{i\omega T/2}}{2i\,\sin(\omega T/2)} \sum_{n=-\infty}^{+\infty} f(nT) - f(nT - T)e^{-i\omega nT} \tag{45}$$

provided that $F(\omega) = 0$ for $\omega > \pi/T$. The pulses to be compared are individually sampled (n samples) at a regular sampling interval T such that the pulse under analysis has no effective frequency contribution above a cutoff frequency $1/2T$. The resulting Fourier analysis yields the complex reflection coefficient

$$\varrho = re^{i\theta} \tag{46}$$

which is then converted to $\hat{\varepsilon}$ using eqn. (43). A time marker system must be employed, so that the pulses to be compared may be accurately referred to a standard point in time.

The front air–liquid interface is bounded by a dielectric window of $Z = 50$ ohms in order to minimize meniscus effects. For aqueous solutions the conductivity is often such that the pulse is sufficiently attenuated for the reflection from the rear interface, whether short-circuited or open-circuit, to be negligible. If this is not the case, the filled concentric line must be lengthened; mechanical instability introduces electrical instability, so that compact lines of long delay are sometimes necessary. Helical lines have been produced by fitting a wire and spacers inside a spiral groove cut into a cylinder.

The TDR technique has already been shown to yield satisfactory data for water over a frequency range from 10 MHz to 12 GHz.[860] Equipment is now available commercially which is claimed to extend the high-frequency limit to 18 GHz.

5. DIELECTRIC RELAXATION OF POLAR LIQUIDS

The partial alignment of the dipoles, which is effected by the application to the liquid of a static electric field, does not remain after the removal
of the field. The dipoles relax to their equilibrium state of random orientation but they take a finite time to do this. Hence the application of a sufficiently rapidly time-varying electric field will not cause the dipoles to align
with it, so that the observed dielectric constant will be much smaller,
containing only contributions from the atomic and electronic polarization.
If it is assumed that the relaxation to equilibrium is exponential with time,

$$P(t) = P_0 e^{-t/\tau} \tag{47}$$

then the frequency dependence of the dielectric constant can be calculated.
The lag between the phase of the field and the alternating motion of the
dipoles introduces a dielectric loss, the field contributing energy which
is dissipated as heat in the medium.

For the application of a harmonically time-dependent field of angular
frequency $\omega = 2\pi\nu$, described by

$$E(t) = E_0 e^{j\omega t} \tag{48}$$

the frequency-dependent complex dielectric constant $\hat{\varepsilon}$, which is related
to the refractive index n and absorption coefficient $\varkappa$ by the equation

$$\hat{\varepsilon} = \varepsilon' - j\varepsilon'' = (n^2 - \varkappa^2) - j(2n\varkappa) \tag{49}$$

is given by

$$\hat{\varepsilon} = \varepsilon_\infty + (\varepsilon_s - \varepsilon_\infty)/(1 + j\omega\tau) \tag{50}$$

where ε_s is the static dielectric constant for time-invariant field and ε_∞
the residual dielectric constant appropriate to frequencies much higher than
that of the dipole orientation relaxation process.

Equation (50) was first obtained by Debye[387] for spherical polar
molecules imbedded in a continuous viscous fluid; but this specialized
model is not necessary, since the expression can be derived more generally.
Rationalization yields

$$\varepsilon' = \varepsilon_\infty + [(\varepsilon_s - \varepsilon_\infty)/(1 + \omega^2\tau^2)] \tag{51}$$

$$\varepsilon'' = (\varepsilon_s - \varepsilon_\infty)\omega\tau/(1 + \omega^2\tau^2) \tag{52}$$

where τ is the relaxation time. These are known as the Debye or Debye–

Drude equations; they are expressed graphically in Fig. 16, Volume 1, Chapter 7. For angular frequency equal to τ^{-1}, $\varepsilon' - \varepsilon_\infty = \varepsilon'' = \frac{1}{2}(\varepsilon_s - \varepsilon_\infty)$. This equality is a good test that the polarization has a simple exponential decay function with a single time of relaxation.

The relation between the observed "macroscopic" relaxation time of eqns. (50)–(52) and the actual "microscopic" relaxation time of a *single* dipole is discussed in Volume 1, Chapter 7 [eqns. (53)–(57)].

An important graphical expression of the Debye equations is a plot of ε'' versus ε' for corresponding values of ω. It was shown by Cole and Cole[326] that this plot is a semicircle with center on the ε' axis at $\varepsilon' = \frac{1}{2}(\varepsilon_s - \varepsilon_\infty)$. But for the Debye equations to be shown to be satisfied for a particular set of data, it is necessary not only that the graph be a semicircle, but also that the points be distributed around the semicircular Cole–Cole plot in the manner prescribed by eqns. (51) and (52).

When there are two widely differing relaxation times the Cole–Cole diagram takes the form of two round hills with a hollow valley between. As the times approach each other the valley disappears and the two round hills merge into a single semicircle.

Unless the times are very different in magnitude, the separation of more than one relaxation time from a set of data is not simple. When two times differ by less than a factor of five, one cannot distinguish from data taken at a few frequencies whether there are two relaxation times or more than two. In the case of mixtures of two polar liquids, where two and only two times might at first sight be expected, it would be possible to determine these times, but it would be difficult to determine whether or not further characteristic relaxations were in fact present.

A distribution of relaxation times of full width at half-height $\Delta\tau/\tau = 2\pi\alpha$ gives rise to a Cole–Cole plot which is a shallow arc rather than a semicircle. An empirical expression[326] can be based on the following construction. If the center of the arc lies on a line below the ε'' axis, making an angle α with the point $\varepsilon' = \varepsilon_\infty$, $\varepsilon'' = 0$, then it can be shown that

$$\hat{\varepsilon} = \varepsilon_\infty + \{(\varepsilon_s - \varepsilon_\infty)/[1 + (j\omega\tau)^{1-\alpha}]\} \tag{53}$$

The parameter α describes the extent of the distribution of relaxation times. The deduction of its value for water has been considered in Volume 1, Chapter 7, Section 7; it is found that $\alpha \simeq 0.012$. But a possible interpretation was made in which there were equal proportions of two species of relaxation times $\tau_1 \simeq 0.8\tau_0$ and $\tau_2 \simeq 1.2\tau_0$. The resulting $\hat{\varepsilon}(\omega)$ remains indistinguishable from that of eqn. (53) until a full statistical analysis is made. Even the demonstration that the data do not conform to a single relaxation time requires a least squares analysis such as was reported in Volume 1, Chapter 7. This is an illustration of the difficulty of separating relaxation times owing to the extremely large "linewidth," probably larger than any other linewidth in physics. A similar or more complicated situation exists for aqueous solutions.

A second relation is sometimes used in the description of a spread of relaxation times:

$$\hat{\varepsilon} = \varepsilon_\infty + \{(\varepsilon_s - \varepsilon_\infty)/[1 + i^{(1-\alpha)}(\omega\tau)^{1-\beta}]\} \tag{54}$$

This relation is discussed in Volume 3, Chapter 8, in connection with aqueous electrolytic solutions; it has not been applied to the mixtures considered in this chapter.

Application of the theory of rate processes to dielectric relaxation leads to an equation

$$1/\tau = (kT/h) \exp(\Delta S^\ddagger/R) \exp(-\Delta E^\ddagger/RT) \tag{55}$$

where $\Delta E^\ddagger$ is the activation energy of the relaxation process and $\Delta S^\ddagger$ its entropy change, and R, T, and h have their usual significance. Thus a plot of $\log \tau$ against T^{-1} should be linear, with slope determined by the activation energy.

The relaxation processes for water were discussed in Volume 1, Chapter 4, and the frequency dependence of the complex dielectric constant was illustrated in Fig. 15 in that chapter. In addition to the principal relaxation time, which is slightly "distributed" with $\alpha \simeq 0.012$, there is a further process, probably to be interpreted as a relaxation, at a frequency ~ 100 times higher but contributing only $\sim 3\%$ of the dielectric constant. For the principal relaxation process, whose small spread has been discussed above, $\Delta E^\ddagger \simeq 5$ kcal mol^{-1}, which is believed to correspond to the breakage of a single hydrogen bond. Some curvature of the semilogarithmic temperature plot illustrates the limitations of a purely bond-breading model, but a statistical interpretation of this has been made.[618] The activation energy of the higher frequency process is 1.5 kcal/mole.

The relaxation processes for the pure liquids which are miscible with water have not all been measured or understood, but for alcohols the investigations of Cole and his colleagues have also been interpreted in terms of breakage of hydrogen bonds.

The frequency dependence of the permittivities of the monohydroxyl alcohols can be analyzed as a single major relaxation, together with a

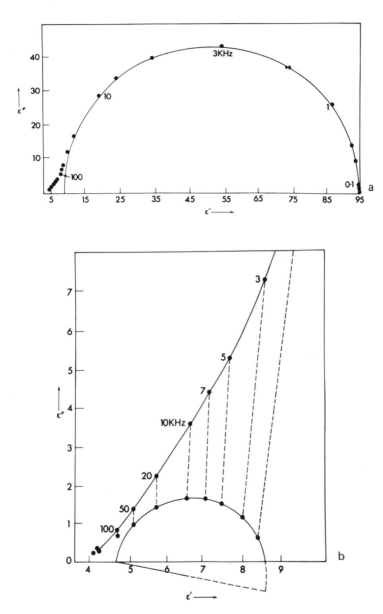

Fig. 5. Cole–Cole diagram of frequency dependence of dielectric properties of ethanol + 5% water at $T = 128°K$. The high-frequency relaxation ($T = 115°K$) is analyzed by subtraction of component of low-frequency relaxation. A possible third relaxation process can be discerned.

further subsidiary relaxation at a higher frequency. The analysis of the data for ethanol ($+1\%$ water) in terms of a sum of two semicircular arcs is shown in Fig. 5.

The activation energy for the principal relaxation process in monohydroxylic alcohols is of the order of the hydrogen bond energy (e.g., 4.9 kcal mol^{-1} for ethanol between 130 and 160°K). It has been attributed to reorientation ·of a molecular OH moment by breaking of its intermolecular hydrogen bond and subsequent forming of a bond to the oxygen of another molecule. The high-frequency relaxation process has been attributed to the movement of a hydrogen bond from one bonding center on an oxygen atom to an unoccupied one on the same atom (illustrated in Fig. 20 in Volume 1, Chapter 7). This is difficult to reconcile with the high activation energy for the process ($\simeq$10 kcal mol^{-1}), which would imply breakage of two hydrogen bonds. However, the necessary rearrangement of one or more chains might well require this high activation energy.

Polyhydroxylic alcohols such as glycerol can be accurately fitted[370] to a skewed-arc locus which was first proposed by Cole and Davidson.[328] This is illustrated in Fig. 6 and is represented by the equation

$$\hat{\varepsilon} - \varepsilon_\infty = (\varepsilon_s - \varepsilon_\infty)/(1 + j\omega\tau_0)^\alpha \tag{56}$$

Most, but not all, of the organic liquid data which have been fitted to this equation are taken at low temperatures. Davidson and Cole showed that their skewed-arc equation corresponded to a distribution function of relaxation times given by

$$
\begin{aligned}
F(\tau) &= [(\sin \alpha\pi)/\pi][\pi/(\tau_m - \tau)]^\alpha &&\quad\text{for}\quad \tau \le \tau_m \\
&= 0 &&\quad\text{for}\quad \tau > \tau_m
\end{aligned}
\tag{57}
$$

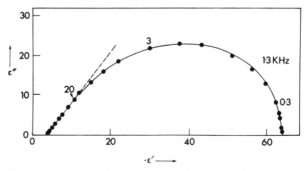

Fig. 6. Analysis of frequency dependence of dielectric properties of glycerol ($t = -50$°C) on skewed-arc locus.

This is a continuous distribution of relaxation times up to a maximum value τ_m, corresponding to conditions of maximum restraint or maximum rigidity in the immediate environment of the reorienting molecule. A closely similar dispersion function has been deduced by Glarum, in the form

$$\frac{\hat{\varepsilon}(j\omega) - \varepsilon'}{\varepsilon_s - \varepsilon'} = \frac{1}{y} + \frac{(y-1)}{y[1 + (a_0 y)^{1/2}]} \tag{58}$$

where

$$y = 1 + j\omega\tau \quad \text{and} \quad a_0 = l_0/D\tau$$

The relaxation is supposed to reflect not single-molecule behavior, but a cooperative process in the entire medium. The polarization varies with time as

$$P(t) = P_0 e^{-t/\tau}[1 - x(t)] \tag{59}$$

where the exponential term is the variation in the absence of local defects and $x(t)$ is the molar fraction of molecules which have already relaxed as a result of the diffusion of defects which have a diffusion coefficient D and length l_0. When $a_0 = 1$, $\alpha = \frac{1}{2}$ and the locus of the data is a skewed arc. But when $a_0 \gg 1$, that is, when the diffusion of defects is extremely slow compared with the relaxation time, the locus becomes a semicircle. For the diffusion-dominated process a high-frequency asymptotic approach to the ε' axis at $45°$ angle is predicted. This is slightly different from the behavior of eqn. (43), and it appears that for glycerol, 1,2-propanediol, and certain other molecules the data lie closer to the Glarum equation.

6. DIELECTRIC RELAXATION OF AQUEOUS MIXTURES

We have seen that there is often difficulty in analyzing the frequency dependence of the permittivity of pure polar liquids; a very wide range of frequency coverage is necessary. Neither in water nor in alcohols are the relaxation processes free from complications. It would seem likely that further complications are to be expected in the behavior of mixtures of polar liquids, and even in mixtures of nonpolar liquids with water. It is discouraging to find that the existing data do not cover such a wide frequency range as would be necessary for full analysis.

The central question can be formulated as follows: Are the relaxation processes of pure polar liquids merely modified quantitatively by the mixing, or are they superseded by some composite relaxation process or processes?

Since a relaxation frequency difference of at least a factor of five is needed for two Debye absorptions of equal intensities to show two separate centers when superposed, an apparently single absorption of a mixture must not be accepted as such until a careful analysis of the contour has been made. Some of the early relaxation study, which was made partially by variation of temperature as opposed to frequency, was unable to resolve closely spaced relaxation times.

The water–ethyl alcohol and water–acetone systems were studied by Nukasawa[1037] over the temperature range 5–35°C at a single frequency (2.87 GHz). It was assumed that the static dielectric constants were temperature-invariant over this temperature range and the data were analyzed by means of a Cole–Cole diagram. The relaxation times and spread parameters ($\alpha \simeq 0.05$) were deduced, although the precise values of ε_s and ε_∞ used are not known. The temperature variations of the relaxation times were used to deduce activation energies and enthalpies of the relaxation process; they vary with molar fractions as shown in Fig. 7. The maxima occurring at about $x_2 = 0.2$ are of interest in that a similar (smaller) maximum occurs in the correlation parameter for ethyl alcohol–water (Fig. 1a), although not for acetone–water. However, the inadequacy of single-frequency data, coupled with the small temperature range, makes it necessary to regard these conclusions as enticing rather than significant. Much more extensive studies are required.

Complete analyses of the entire concentration range of liquids completely miscible with water have not often been achieved, but studies have often been made of the effect on the relaxation time of water of the addition of low concentrations of other liquids, polar and nonpolar. By analogy with eqn. (23), which treats the static dielectric constant, one can write for the relaxation time of the solution

$$\tau_s = \tau(H_2O) + c\,\delta\tau \tag{60}$$

where c is the molar concentration of solute. The linear dependence of τ upon concentration is only an approximation; its adequacy can be judged by studying the small selection of data shown in Fig. 8. The dependence of spread parameter α on concentration, shown in Fig. 9, is rather more nonlinear.

Studies of comparatively dilute solutions of 12 organic molecules in water at three frequencies ($\sim$3, 10, 25 GHz) were made over the temperature range 5–60°C by Hasted and co-workers;[618,643 644] the existence of improved values of ε_∞(water) (Volume 1, Chapter 7), as well as good

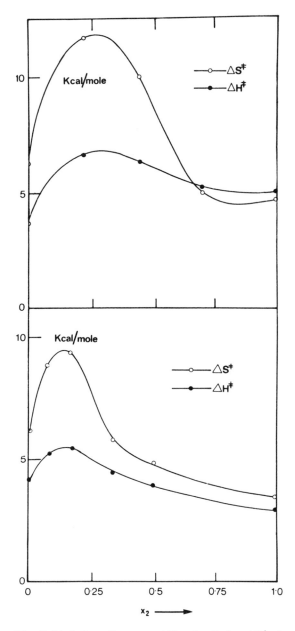

Fig. 7. Variation of entropy $\Delta S^{\ddagger}$ and enthalpy $\Delta H^{\ddagger}$ of activation of relaxation of water–alcohol (upper) and water–acetone (lower) mixtures with molar fraction of the organic component.

static dielectric constants for a number of the solutions,[9] make possible a new analysis of these data, including the determination of spread parameters α. The results are shown in Table VII. The method of analysis, due to Cole and Cole,[326] in terms of eqn. (41) makes use of the relation

$$\omega\tau = (u/v)^{1/(1-\alpha)} \qquad (61)$$

u and v are the chords respectively from $(\varepsilon', \varepsilon'')$ to $(\varepsilon_s, 0)$ and from $(\varepsilon', \varepsilon'')$ to $(\varepsilon_\infty, 0)$ on the Cole–Cole diagram. A straight-line double-logarithmic plot of $(u/v)(\omega)$ for each data point yields both spread parameter α and principal relaxation time τ. The ε_∞ values appropriate to pure water were used for this analysis.

Data for aqueous solutions of 1-propanol[643] and dioxan[644] over a range of temperature are tabulated in Tables VIII and IX. A semilogarithmic plot (Fig. 10) shows that the energies of the principal relaxation processes are similar to that of pure water, except that at the higher temperatures those for the 1-propanol solutions are distinctly smaller; their

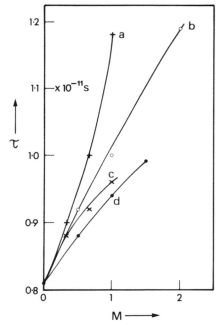

Fig. 8. Dependence of principal relaxation time upon molarity for aqueous solutions of (a) *t*-butanol, (b) 2-methyl pyrazine, (c) *n*-propylamine, (d) pyrazine.

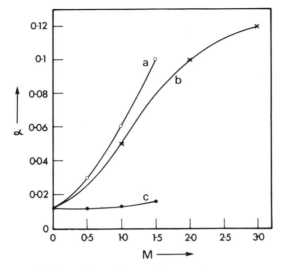

Fig. 9. Dependence of spread parameter α upon molarity for aqueous solutions of (a) propionic acid, (b) 1,4-diazabicyclo-[2-2-2]-octane, (c) pyrazine.

activation enthalpy is more temperature-sensitive than that of pure water. The variance in pure water was interpreted in Volume 1, Chapter 7 in terms of a statistical theory of the temperature variation of the number of water molecules hydrogen-bonded to three neighbors; this governs the rate of formation of the two-bonded molecules which undergo relaxation by hydrogen bond breakage. Thus the 1-propanol molecules are able to maintain a higher proportion of three-bonded water molecules in the higher temperature range.

The data in Table IX and Fig. 10 show that the relaxation times of dioxan–water mixtures are much longer than that of water, although the temperature variation is almost identical. There is an appreciable spread of relaxation times, but for data at two frequencies only, the accuracy in spread parameter α will not be better than ± 0.1. It is possible that there is another relaxation process in addition to the principal one. A tentative suggestion was made in Volume 1, Chapter 7 that the small relaxation time spread in pure water arose from there being approximately equal contributions from the relaxation of asymmetrically and symmetrically two-bonded water molecules, whose activation energies might be expected to be different. Relaxation of an asymmetrically two-bonded molecule involves the transfer of a hydrogen atom, belonging to one water molecule, between two sp orbitals on the oxygen atom of a neighboring water molecule. It is

TABLE VII. Analysis of Microwave Permittivity Data[618] for Aqueous Solutions

Solute (25°C)	Molarity	2.044 GHz		5.937 GHz		14.91 GHz		ε_s	τ, 10^{-11} sec	α
		ε'	ε''	ε'	ε''	ε'	ε''			
Water	—	—	—	—	—	—	—	78.2	0.81	0.012
n-Propanol	0.33	72.6	12.5	61.3	29.3	31.6	33.3	76.9	0.90	0.01
	0.66	70.7	13.0	58.2	30.2	29.5	32.0	75.6	0.945	0.035
	1.0	70.0	14.6	54.9	30.3	27.1	31.3	74.4	1.04	0.055
t-Butanol	0.33	73.7	12.9	59.9	29.6	31.6	33.3	76.6	0.90	0.01
	0.66	71.6	14.0	55.1	30.1	—	—	74.9	1.00	0
	1.0	69.9	15.3	50.4	30.2	—	—	73.3	1.18	0
Diethylketone	0.17	75.2	10.6	62.2	29.0	33.2	35.0	77.0	0.85	0
	0.33	74.6	10.8	61.1	29.3	31.6	34.1	76.0	0.89	0
Phenol	0.25	74.6	11.5	61.3	28.2	32.1	33.6	76.6	0.90	0
	0.5	71.3	12.1	58.4	27.9	29.1	31.2	73.9	0.94	0
Aniline	0.125	75.5	10.3	62.9	28.4	33.6	35.2	77.0	0.825	0
	0.25	74.1	10.4	61.9	28.3	32.5	34.2	75.9	0.855	0
Propionic acid	0.5	73.3	12.7	59.1	28.5	31.4	33.3	75.7	0.885	0.025
	1	69.4	13.4	55.4	28.5	28.9	31.4	72.5	0.95	0.06
	1.5	65.7	13.0	51.4	28.7	26.4	29.6	69.8	0.97	0.10
Ethylamine	0.6	72.4	12.9	57.6	30.0	30.1	33.1	75.5	0.93	0.04
	1.16	—	—	55.2	30.8	27.9	31.7	74.5	1.03	0.06
Pentane dioic acid	0.33	73.6	13.1	60.6	28.4	32.5	33.1	77.0	0.84	0.08
	1.0	65.6	14.8	53.5	26.2	28.6	28.8	70.1	0.96	0.11

Ethylene diamine	0.525	74.3	12.0	59.7	29.2	31.3	33.0	77.0	0.88	0
	1.57	68.5	13.5	51.5	30.2	24.7	29.5	71.9	1.11	0
α-Alanine (*pH* 7)	0.5	70.9	11.7	—	—	31.5	32.9	72.8	0.87	0
	1.0	64.4	10.7	—	—	29.0	30.1	66.3	0.89	0
	1.5	57.7	8.1	—	—	26.7	27.4	59.0	0.84	0
β-Alanine (*pH* 7)	0.5	71.3	10.9	—	—	31.9	33.3	73.1	0.89	0
	1.0	64.8	10.0	—	—	29.9	30.5	66.3	0.84	0
Na propionate	0.5	71.6	14.1	55.7	27.5	30.0	31.3	75.5	0.87	0.11
	1.0	65.2	13.3	49.6	26.4	26.4	27.7	68.5	0.92	0.11
Na butyrate	0.567	68.0	13.4	—	—	29.9	31.3	71.4	0.91	0.08
Ethylamine HCl	0.5	70.6	14.3	56.7	28.0	31.2	32.6	74.1	0.90	0.08
	1.04	62.3	12.8	53.0	28.4	28.8	30.6	68.5	0.94	0.13
n-Propylamine HCl	0.33	72.4	13.9	59.1	27.0	31.9	32.9	76.2	0.88	0.05
	0.66	67.9	12.5	54.2	25.1	29.2	30.7	71.5	0.92	0.05
	1.0	64.2	13.0	—	—	27.1	28.6	68.0	0.96	0.01
Diethylamine HCl	0.33	70.5	11.0	—	—	30.6	32.0	72.7	0.92	0
	0.66	66.0	10.0	—	—	27.3	29.0	67.9	0.90	0
Tetraethylammonium	0.2	71.9	11.7	—	—	31.5	32.5	74.0	0.87	0.02
	0.4	67.8	13.0	—	—	28.4	30.4	70.9	0.94	0.04
	0.6	—	—	50.3	26.3	25.7	28.1	67.5	0.99	0
Tetramethylammonium iodide	0.125	74.7	12.3	62.0	27.9	33.5	33.9	77.2	0.87	0
	0.25	71.9	13.8	60.4	28.0	32.8	32.9	75.3	0.87	0.04
Guanidine chloride	0.36	71.6	10.8	60.0	26.8	33.4	32.5	74.3	0.85	0.02
	0.71	—	—	56.5	24.8	32.4	30.3	69.9	0.83	0.03
	1.07	—	—	52.4	21.9	31.2	28.5	64.0	0.77	0.06

TABLE VIII. Analysis of Microwave Permittivity Data[643] for 1-Propanol–water Mixtures

t, °C	3.26 GHz		9.12 GHz		23.8 GHz		ε_s (Åkerlöf)	ε_∞	τ, 10^{-11} sec	α
	ε' (±1%)	ε'' (±2%)	ε' (±1%)	ε'' (±2%)	ε' (±1%)	ε'' (±2%)				
0.41 M 1-Propanol										
10	75.5	20.0	53.7	36.0	20.5	30.0	82.2	4.3	1.36	0.01
20	74.5	14.0	58.5	31.3	28.0	33.0	78.7	4.2	1.02	0.01
30	73.0	10.6	62.2	26.5	34.0	34.0	75.1	4.2	0.81	0.00
40	71.3	8.5	63.0	22.0	38.5	33.5	71.5	4.1	0.67	0.00
50	68.8	6.2	62.5	18.0	42.5	32.0	68.3	4.1	0.575	0.00
60	65.8	4.5	62.0	15.0	45.0	29.0	65.1	4.2	0.52	0.00
0.82 M 1-Propanol										
10	72.3	21.4	49.0	36.0	18.5	28.0	80.4	4.3	1.49	0.04
20	71.7	14.9	55.0	32.0	25.0	31.5	76.9	4.2	1.12	0.02
30	70.7	10.4	58.8	27.2	31.0	33.0	73.5	4.2	0.86	0.00
40	69.2	7.6	60.3	22.5	36.0	32.5	70.0	4.1	0.69	0.00
50	67.4	6.0	60.5	19.0	39.5	31.7	66.8	4.1	0.58	0.00
60	65.6	4.5	60.0	16.0	43.0	28.7	63.6	4.2	0.52	0.00

TABLE IX. Analysis of Microwave Permittivity Data[644] for Dioxan–Water Mixtures

t, °C	Proportion of dioxan by weight, %	3.253 GHz		23.77 GHz		ε_s (Akerlöf)	ε_∞	τ, 10^{-11} sec	α
		ε' ($\pm 2\%$)	ε'' ($\pm 2\%$)	ε' ($\pm 2\%$)	ε'' ($\pm 2\%$)				
25	14.2	67.2	12.5	24.3	27.7	70.0	3.9	1.03	0
25	24.9	56.5	12.4	18.6	21.9	61.1	3.65	1.22	0
25	33.2	45.3	11.3	15.0	18.3	48.8	3.5	1.24	0.14
25	45.2	36.0	10.8	11.3	13.9	43.6	3.2	1.67	0.13
35	33	44.8	8.93	18.3	19.4	46.4	3.5	0.964	0.045
45	33	42.2	7.07	21.7	20.5	43.9	3.5	0.760	0.09
55	33	40.4	5.62	—	—	41.6	3.5	0.610	0.09

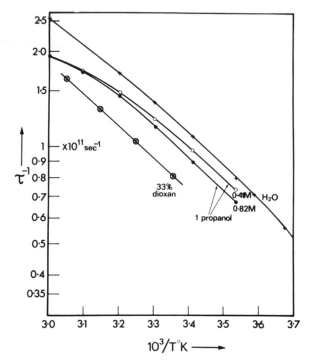

Fig. 10. Semilogarithmic plot of principal relaxation frequencies of aqueous solutions of 0.41 M and 0.82 M 1-propanol and 33% dioxan compared with that of pure water.

possible that in dioxan–water mixtures a similar process can take place, involving the orbitals on the oxygen atom of dioxan; if so, its activation energy is seen from Fig. 5 to be indistinguishable from that of pure water. The suggestion has been made that the dioxan molecule forms a hydrate in aqueous solution. This is equivalent to the statement that the energy of the dioxan–water hydrogen bond is not lower than that of the water–water hydrogen bond. The present discussion suggests that they are of equal magnitude.

Some studies have been made of solutions of small quantities of water in various alcohols. The low-temperature data of Hassion and Cole[642] are displayed in Fig. 11 as semilogarithmic plots of relaxation frequency against inverse temperature, the slopes of which are approximately proportional to activation enthalpy. It is seen that both the ethanol relaxation times are affected by the addition of water. The alcohol chain length is apparently increased by the addition of water; possibly some cross-linking is involved.

More recent studies at higher temperatures,[1181] displayed in Fig. 12, confirm the increase in relaxation time which small proportions of water molecules produce. The effect is hardly detectable for methanol.

An attempt has been made by Pottel and Kaatze[1112] to analyze their data for aqueous solutions of nitrogen-containing heterocyclic compounds (Table X) and tetramethylammonium bromide[1275] in terms of a hydration sheath surrounding the solute molecule. In this table the penultimate column shows values of α deduced from eqn. (61), while the previous column shows values deduced from the location of the center of the Cole–Cole arc.

The simplest model assumes that N water molecules in each hydration sheath relax at a different time $\tau_{sh} = p\tau_w$ from that of the pure water in the remaining volume of the solution, τ_w. The hydration sheath water contributes a certain fraction q of the entire static permittivity of the solution. The frequency variation of the complex permittivity is taken to be a sum of two Debye terms (subscripts w and sh, respectively, for pure

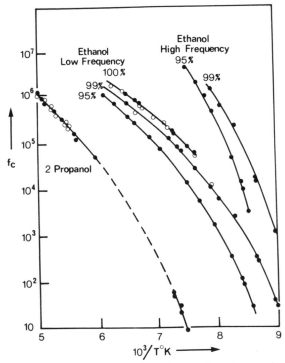

Fig. 11. Semilogarithmic plot of relaxation frequencies of ethanol and 2-propanol containing small quantities of water.

water and sheath water):

$$\hat{\varepsilon} - \varepsilon_\infty = \frac{(\varepsilon_{s,w} - \varepsilon_\infty)(1 - q)}{1 + j\omega\tau_w} + \frac{(\varepsilon_{s,\text{sh}} - \varepsilon_\infty)q}{1 + j\omega\tau_{\text{sh}}} \qquad (62)$$

where $q = Nc/c'$, c and c' being, respectively, the solute and solvent molarities. No attempt is made to apply any of the mixture relations mentioned above [eqns. (25)–(28a)], although some wide frequency band calculations of this type have been made elsewhere.[406] Equation (62) can only be a rough

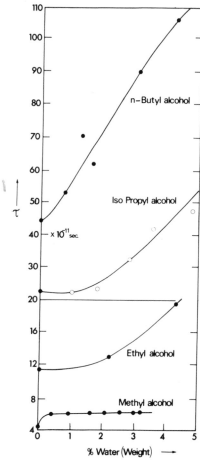

Fig. 12. Variation of principal relaxation times of *n*-butyl alcohol, isopropyl alcohol, ethyl alcohol, and methyl alcohol with concentration of admixed water ($T = ?\ 25°C$).

TABLE X. Analysis of Microwave Permittivity Data[1112] **for Aqueous Solutions**

Solute ($T = 25°C$)	c, mol liter^{-1}, $\pm 2\%$	ε_s	ε_∞	$\alpha \pm 0.015$	$\alpha \pm 20\%$	τ, 10^{-11} sec, $\pm 1\%$
Pyrazine	0.5	75.0	5.8	0	0.012	0.88
	1.0	71.8	6.4	0	0.013	0.94
	1.5	68.5	5.4	0.015	0.016	0.99
2-Methylpyrazine	0.5	74.8	6.9	0	0.02	0.92
	1.0	70.8	5.7	0.012	0.021	1.00
	2.0	63.4	4.6	0.043	0.041	1.19
	4.0	48.4	4.8	0.063	0.06	1.83
2,6-Dimethylpyrazine	1.0	70.0	4.7	0.042	0.045	1.05
	1.5	65.1	4.8	0.053	0.06	1.22
Quinoxaline	0.5	73.6	5.4	0.019	0.022	0.91
	1.0	68.9	4.3	0.043	0.039	0.97
	2.0	58.7	4.3	0.071	0.07	1.22
	4.0	40.5	4.9	0.10	0.11	2.11
2-Methylquinoxaline	1.0	67.2	4.1	0.057	0.07	1.05
	1.5	61.6	4.6	0.078	0.09	1.26
1,4-Diazabicyclo-[2,2,2]-octane	1.0	70.7	4.6	0.056	0.05	1.21
	2.0	63.6	4.3	0.095	0.10	1.86
	3.0	56.0	4.7	0.115	0.12	3.24
Pyridine	0.5	76.9	4.4	0.023	0.022	0.90
	1.0	74.3	5.1	0.022	0.031	0.99
	2.0	70.6	4.2	0.058	0.06	1.20
	4.0	62.3	4.2	0.102	0.09	1.77

approximation to the measured $\hat{\varepsilon}(\omega)$. The observed linear variation of principal relaxation time τ with solute concentration is not consistent with it. Nevertheless some interesting deductions concerning the nature of the postulated hydration sheath can be made with its aid.

Using the fact that the observed principal relaxation time τ_s is determined by the condition $(d\varepsilon''/d\omega)_{\omega\tau=1} = 0$, one obtains

$$\frac{q}{1-q} = \frac{(y^2 - 1)(p^2 + y^2)^2}{p(y^2 + 1)^2(p^2 - y^2)} \tag{63}$$

where $y = \tau/\tau_w$.

From the Cole–Cole equation (53), we get

$$\alpha = 1 + \left(\frac{1}{\varepsilon''} \frac{d\varepsilon'}{d(\omega\tau)}\right)_{\omega\tau=1} \cos\frac{\alpha\pi}{2} \tag{64}$$

Combining eqns. (63) and (64), we get

$$\frac{p}{y} = \frac{y - [(1-\alpha)/\cos(\alpha\pi/2)]}{[y(1-\alpha)/\cos(\alpha\pi/2)] - 1} \tag{65}$$

Therefore the observed τ and α lead to unique values of τ_{sh} and N. The values of $p - 1$ for the data of Table X (with n-propanol from Table VII and dioxan from Table IX) are presented in Fig. 13. They demonstrate

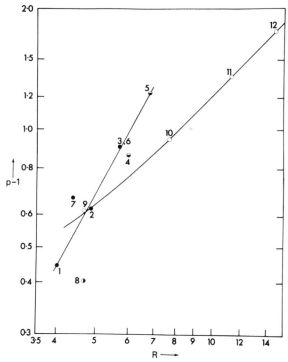

Fig. 13. Variation of hydration sheath relaxation time enhancement ratio p with molar volume ratio R. Key: (1) Pyrazine. (2) 2-Methylpyrazine. (3) 2,6-Dimethylpyrazine. (4) Quinoxaline. (5) 2-Methylquinoxaline. (6) 1,4-Diazabicyclo-[2,2,2]-octane. (7) Pyridine. (8) Dioxan. (9) Tetramethylammonium bromide. (10) Tetraethylammonium bromide. (11) Tetrapropylammonium bromide. (12) Tetrabutylammonium bromide. (13) n-Propanol.

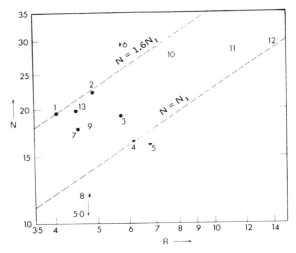

Fig. 14. Variation of hydration sheath population N with molar volume ratio R. See legend to Fig. 13 for key.

dependence upon the ratio $R = V_{solute}/V_{solvent}$ of solute to solvent molar volumes. The calculated N are displayed in Fig. 14. Two guide lines are shown in this figure, representing sheath thickness in terms of the number N_1 of water molecules in the inner hydration sheath, i.e., with surface directly adjacent to the solute molecule. This is calculated from the model of spherical solute molecules and cubic water molecules:

$$N_1 = (36\pi)^{1/3}(V_{solute}/V_{solvent})^{2/3} \tag{66}$$

The conclusion to be drawn is that the effect on the relaxation time does not penetrate much beyond the inner sheath. However, on this model the contribution made by the solute molecule to the dielectric constant is neglected. The approximation is not always satisfactory but it may be of value when the permittivity of the pure solute is $\gtrsim 0.1$ of that of pure water. For pyridine $\varepsilon_s = 12.3$.

6.1. Hydration Sheaths of Long Lifetime

Until recently measurements of relaxation times of aqueous solutions had not extended to sufficiently low frequencies to allow the separation of any but the principal relaxation time of the bulk water. We have seen that this time is not identical with that of pure water, but relaxation times widely different from that of pure water had not been reported.

TABLE XI

	$10^{11}\tau\eta_0$	$\Delta H\ddagger$, kcal mol^{-1}
Process 1	1.7 ± 0.2	(4.5)
Process 2	5.2 ± 0.9	5.5 ± 1
Process 3	11.0 ± 4.5	9 ± 2

Measurements on monosaccharides by Tait et al.[1275] using time-domain technique have extended over the wide frequency range 10^7–10^{10} Hz. Three separate relaxation times were found, using both the original direct time conversion of the data[477] and the full Fourier transform analysis.[860] Their values at 5°C and activation energies are given in Table XI. Here η_0 is the viscosity of water. The fastest relaxation time is that of the bulk of the liquid, but its activation energy cannot be deduced quantitatively from these measurements. It is concluded, on the basis of the contribution to the total static dielectric constant and other evidence, that the *second* process corresponds to the reorientation of the glucose molecule. The *third* process therefore corresponds to the relaxation of the immediate hydration sheath of the molecule. On the other hand, it might be regarded as a measure of the rate of exchange of molecules from the hydration sheath to the bulk liquid state. The high energy of activation presumably corresponds to the "breakage" of more than one hydrogen bond.

This type of hydration represents a different situation from what has hitherto been proposed for nonelectrolyte solutions. But it may also be present in other solutions; for these there are insufficiently wide ranges of data available to decide the question.

7. CONCLUSIONS

The static dielectric constants of aqueous solutions have been reasonably well studied, and information is to be obtained from them both as to the dipole moment of the solute molecules and as to the structural interactions or correlations between the water and solute molecules. There is scope for further investigations of the temperature dependence of the static dielectric constants of solutions, and there is a paucity of information about the dielectric properties of small proportions of water in other solvents.

However, the situation as regards the frequency dependence of the dielectric properties of nonelectrolytic aqueous solutions is not nearly so satisfactory. The number of frequencies at which measurements have been made is sometimes inadequate, although some deductions can of course be made from what is available. It is fortunate that the "spot frequency" microwave techniques are now to be reinforced by the extended-range "time-domain reflectometry" techniques discussed in Section 4.

The paradox of the dielectric study of solutes dissolved in water is that they lower the correlation parameter which governs the static dielectric constant and yet they raise the relaxation time or times. It might be argued that the solute cannot at the same time break up the water structure and enhance it.

It is probable that the correlation parameter g is a reliable measure of the extent or density of hydrogen-bonding structure in liquid water. Most of the structural models assumed in Volume 1 will allow the calculation of correlation parameters g, and two attempts to do so were outlined in Chapter 7. In one of these[618] the statistical breakage of four-coordinated tetrahedral icelike linkages was considered; in the other the bending of the linkages was invoked. A lowering of g by the addition of solute implies that the highly coordinated molecules are interfered with, since these make the largest contribution to g; an interference with the three-dimensional network, whether by bending or breakage, is implied. Since the addition of solute to the water might in any case be expected to interfere with the network by a purely volume effect, one could possibly expect a lowering of g, varying linearly with molar fraction x_2. Strictly, it is only a falling of g below the linear $g(x_2 = 0)$ to $g(x_2 = 1)$ function which could with some certainty be attributed to "structure breaking." Dioxan–water mixtures (Fig. 1d) illustrate this effect.

On the other hand, the relaxation spectrum provides a rather different kind of information. The principal relaxation time of pure water is often interpreted as involving the breakage of a hydrogen bond, because the activation energy of the process is of the correct order. The water molecule that is only bonded by one bond to one neighbor can freely rotate without involving an activation energy greater than 1 kcal mol^{-1}, but one that is bonded to two neighbors requires for reorientation an activation energy $\simeq 5$ kcal mol^{-1}.

The semilogarithmic plots of relaxation frequency of aqueous solutions indicate that the principal relaxation process is still of the order necessary to break one hydrogen bond. The curvature of these plots, together with analysis of the relaxation time spreads, and correlation with the nuclear magnetic resonance relaxation times, must continue to provide data to which structural models can be fitted. The existing data can be interpreted on the supposition that the solute liquids increase the water relaxation time by the contribution of the orientation process of the water molecules adjacent to the molecules. At the same time the solute molecules diminish the proportion of highly bonded water molecules, thereby diminishing the correlation parameter. Some water dipoles, instead of being correlated to neighboring water dipoles and so contributing powerfully to a high static permittivity, are now correlated to a relatively feeble solute molecule and contribute less powerfully.

CHAPTER 8

Spectroscopic Properties

M. J. Blandamer
Department of Chemistry
The University
Leicester, England

and

M. F. Fox
School of Chemistry
City of Leicester Polytechnic
Leicester, England

1. INTRODUCTION

Application of spectrophotometric techniques to the study of water–water and water–solute interactions in aqueous solutions has followed the development of recording spectrophotometers. In principle, spectroscopic studies of aqueous solutions can be divided into three closely related classes. In the first class absorption spectra of water in aqueous solutions are compared to those of pure water at the same temperature and pressure. The majority of investigations in this class have used aqueous salt solutions and the differences observed are often dramatic. In the second class absorption spectra of solutes (cosolvents) in water are compared with the spectra of solutes in either the gas phase or in a nonpolar, nonaqueous solvent. In the third class changes in absorption spectra of one solute in water are monitored as another solute (cosolvent) is added. The first solute is present at very low concentration and acts as a probe because its characteristic (and usually intense) absorption spectra are very sensitive to environmental change. Ideally, detailed knowledge of the spectra of such a probe is necessary but it often happens that the absorption spectra of the probe in aqueous solution aid in the understanding of the spectroscopic characteristics of such species.

Various parameters characterize absorption spectra. Most interest is directed toward the positions of absorption band maxima, given as either the wavenumber, $\bar{\nu}_{max}$, or its reciprocal, the wavelength λ_{max}. The energy E_{max} ($= hc\bar{\nu}_{max}$, where h is Planck's constant and c is the velocity of light *in vacuo*) is the difference between the energies of ground and excited states and increases on going from the infrared through the visible to the ultraviolet regions. In the visible and ultraviolet regions E_{max} is the difference between electronic energy states. Interpretation of absorption spectra would be difficult if it were not for the Franck–Condon principle (see Refs. 105 and 703). This principle is usually extended to include the solvent molecules around the absorbing species; the solvent configuration around the excited state is then taken to be identical to that around the ground state.

Absorption spectra provide information concerning the instantaneous arrangement of solvent and solute molecules[444] and in this sense should be very informative. However, it is not usually clear how such information may be extracted from the broad bands spanning a wide range of frequencies which make up the spectra. In this respect nuclear magnetic and electron spin resonance studies (see below) are at an advantage because time averaging over a period long compared to molecular reorientation times gives narrow lines and the spectra are simplified.

Changes in $\bar{\nu}_{max}$ following a change in an intensive variable such as temperature, pressure, or mole fraction are sometimes characterized by the ratios $d\bar{\nu}_{max}/dT$, $d\bar{\nu}_{max}/dP$, and $d\bar{\nu}_{max}/dx_2$. We adopt the convention that species 1 is water and thus species 2 is, say, the cosolvent in an aqueous binary mixture.

Absorption bands are also characterized by intensity and band shape parameters; ε_{max} measures the intensity of an absorption band at $\bar{\nu}_{max}$. If the band is Gaussian, a quantity $\Delta\bar{\nu}_{1/2}$ measures the bandwidth at half height. The integrated intensity over the complete absorption band is related to the oscillator strength, which is a measure of the "allowedness" of the transition (see Refs. 105 and 703). Both intensity and band shape may change as T, P, and x_2 are varied.

So far it has been assumed, at least implicitly, that absorption spectra comprise single, well-resolved absorption bands, the ideal experimental situation. In practice, complications often occur and an absorption band may show one or more shoulders on the experimental spectral trace. Sometimes it is clear that the overall band envelope is a convolution of two or more overlapping bands. The individual bands may be obtained from the observed spectra using either an analog curve resolver[1377] or digital computer analysis.[1254] These analyses become more difficult, and the outcome less satisfactory, as the extent of band overlap increases. Sometimes each band may characterize one of two species in equilibrium and, as the point of equilibrium is changed, then the absorption bands gain and lose intensity with the various absorption spectra passing through isosbestic points. The subject of isosbestic behavior is complex and caution must be used when drawing conclusions from the presence or absence of such features in a set of absorption spectra.[241,323,467,999,1212] On the other hand, failure to observe isosbestic points can also lead to erroneous conclusions. Suppose that the absorption spectra of a probe solute A are measured in a series of water plus cosolvent mixtures. The data can be summarized in a plot of $\bar{\nu}_{max}(A)$ against x_2 and discussed in terms of a band shift. In fact, the changes may be more correctly interpreted in terms of an equilibrium between A and A', where $\bar{\nu}_{max}(A')$ and $\bar{\nu}_{max}(A)$ are not very different and where $\varepsilon_{max}(A')$ and $\varepsilon_{max}(A)$ are nearly equal. The measured $\bar{\nu}_{max}$ changes slowly at first, when small amounts of A' are present, then rapidly and then finally slowly, when the concentration of A is small. The solvent dependence of the measured $\bar{\nu}_{max}$ is more correctly assigned to a band change.[193]

Closely related to the subject of electronic absorption spectra is the analysis of the deactivation of an electronic excited state by fluorescence.[992] Fluorescence of solutes in aqueous mixtures has been reported but detailed analyses have been concerned with kinetics of deactivations in the context of compensation effects in aqueous systems.[899,1369]

This chapter also includes reference to recent ESR studies (for introduction see Refs. 59, 64, 269). ESR spectra are usually presented in derivative form and are characterized by a number of parameters including g-factors (the magnetogyric ratio for the electron), hyperfine splitting constants a, linewidths ΔH, and line intensities A. Usually the most informative features in ESR spectroscopy are the patterns of hyperfine lines. If the radicals giving rise to the observed spectrum are tumbling rapidly in solution, then all anisotropic dipolar interactions average to zero. Hyperfine coupling constants (and g-factors) may be sensitive to changes in electron distribution and molecular shape which result from changes in the solvent environment (cf. Ref. 611). Linewidth parameters can be analyzed to probe dynamic processes in aqueous systems; thus the linewidth of an ESR signal[686] is governed by the efficiency of two relaxation processes, spin–lattice T_1 and spin–spin T_2. In the present context T_2 is the more interesting quantity and several different mechanisms can contribute to the overall

T_2 relaxation for a system. If very low concentrations, $c \leq 10^{-5}$ mol dm^{-3}, of a paramagnetic probe are used, then direct spin–spin relaxation mechanisms between paramagnetic species are avoided and such contributions to the linewidth can be ignored.

All dipolar interactions are averaged out in the limit that the radicals are tumbling sufficiently rapidly such that enough orientations of the radical are sampled, on the ESR time scale, to make $\sum (1 - 3\cos^2\theta) \to 0$. An indication of the time over which a given orientation persists in a system, the correlation time, can be estimated from the Debye equation, $\tau_c = 4\pi\eta a^3/kT$, where η is the bulk (shear) viscosity of the medium and a is the radius of the spherical paramagnetic solute. With increase in τ_c the linewidth increases and, to a first approximation, the dependence of linewidth on solvent can be correlated with the viscosity of that solvent. On the ESR time scale a τ value of 5×10^{-11} sec is fast and one of approximately 5×10^{-8} sec is slow.[601,1366]

2. EXPERIMENTAL INVESTIGATIONS

It is not possible to review in detail every reported investigation in this branch of the subject. However, to illustrate both the range of application and analytical arguments employed, several examples are considered in this section.

2.1. Spectra of Water in Aqueous Solutions

2.1.1. *The Low-Frequency Region*

Raman spectra of water have a band in the 175-cm^{-1} region which has been assigned[1374,1375] to a hydrogen bond stretch vibration of a group of intermolecularly hydrogen-bonded water molecules. When the temperature is raised the intensity of the band decreases because the equilibrium between water in clusters and nonhydrogen-bonded water changes to favor the latter, using a two-state model for water (see Volume 1, Chapter 5). At a fixed temperature the intensity of this band decreases when urea is added to water.[1376] A quantity $I_2 c_2/I_1 c_1$ is obtained, where I_1 and I_2 are the integrated intensities of the bands for water (put equal to 100) and aqueous solutions containing c_1 and c_2 moles of water, respectively, in 1 dm^3 of solution. For urea in water this ratio equals 0.8, indicating that the concentration of intermolecularly hydrogen-bonded water decreases when urea is added to water. In other words, urea is a structure breaker. In contrast, for sucrose in water the ratio $I_2c_2/I_1c_1 > 1$, and so in these terms sucrose is a structure former.[1376] The shape of the Raman band for sucrose solutions is indistinguishable from that for pure water. It seems likely that water–sucrose hydrogen bonds increase this quantity above unity and that the increase does not necessarily arise from an increase in the fraction of intermolecularly hydrogen-bonded water molecules. Specific solute–solvent interactions are also thought to be responsible for the subtle changes which occur in the Raman spectra of water when dimethylsulfoxide (DMSO), $0.5 < c < 3.5$ mol dm^{-3}, is added to 6.2 m D$_2$O in H$_2$O. A broad, intense component in the 2515-cm^{-1} region is assigned to interactions between S=O and D—O.[1378]

The form of analysis outlined above is widely used to study solute–solvent interactions. Thus, changes in a ratio of parameters obtained from spectra summarize the changes in spectra, the reference being the spectra of either water or a probe in water. Consequently, changes can often be compared with the effect of temperature on the ratio of parameters. This method of analysis is clearly illustrated in the following example.

2.1.2. The Near-Infrared Region

The high intensities and large bandwidths of the pure water absorption bands in the fundamental and overtone regions generally preclude investigations of the effect of added solutes on these spectra. These difficulties can be overcome by measuring the spectra of O—H (D) in D$_2$O (H$_2$O). The near-infrared absorption spectra of 6.0 m HOD in D$_2$O recorded at fixed temperatures over the range $280 < T < 334°$K show an isosbestic point at 1.468 μm.[1433] With increase in temperature a band (a shoulder at low temperatures) at 1.416 μm increases in intensity and is assigned to free O—H groups in the H$_2$O/D$_2$O system (using a mixture model for water; see Volume 1). At the same time the intensity of absorption in the 1.5–1.6-μm region decreases and is assigned to hydrogen-bonded O—H groups.[1433] The assignment of the latter band agrees with the observation of an intense band at 6427 cm^{-1} for HOD/D$_2$O ices at 263°K.[891,894] The equilibrium between the free and bonded states is characterized by a ratio R, equal to the ratio of the absorbances at 1.556 μm ($\bar{\nu} = 6427$ cm^{-1}) and 1.416 μm ($\bar{\nu} = 7062$ cm^{-1}). Therefore R measures the ratio of the bonded to unbonded O—H groups in "water." The ratio R changes when a solute is added to the system at fixed temperatures. The value of R is usually normalized such that $R = 1.00$ for the HOD/D$_2$O system alone at a fixed temperature. A decrease in R resulting from addition of a salt, e.g., potassium chloride, at a fixed temperature is analogous to the effect on R of increasing the temperature and so this solute is called a structure breaker. Conversely, if R increases, then the solute is a structure former. In the original report[1433] the polymer polyvinylpyrrolidone was found to have such an effect. The technique can be used to study the effect of simple solutes,[44,1271] such as t-butyl alcohol (Fig. 1), although in this case t-butyl alcohol-d_1 must be used, otherwise the concentration of protons changes as more alcohol is added. The ratio R increases when t-butyl alcohol is added to this system and such a change is seen to be the more dramatic when set against the prediction that a simple dilution of the system by a noninteracting solute will simply decrease R. This technique, despite some uncertainties in the assignment of the bands, provides a useful method of characterizing the action of a solute on water–water interactions. In addition, the effective temperature of water in the solution can be calculated in the sense originally proposed by Bernal and Fowler.[161] The "effective temperature" is equal to that where the R value of water (HOD in D_2O) is equal to that for the solution. Indeed, by comparing the effects of added solute on R at fixed temperatures, the temperature sensitivity of structure-breaking and structure-making effects may be examined. Finally, it is important to note that a given set of spectra pass through the isosbestic point. Without this condition, analysis of changes in R in terms of the effect of solutes on water structure is invalid.

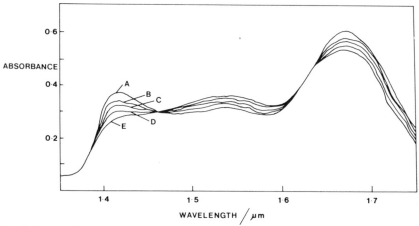

Fig. 1. Near-infrared absorption spectra of 6 M HOD in deuterium oxide containing various mole fractions of t-butyl alcohol-d_1 at 298°K. (A) $x_2 = 0$, (B) $x_2 = 0.01$, (C) $x_2 = 0.02$, (D) $x_2 = 0.03$, and (E) $x_2 = 0.04$.[1132]

2.2. Spectra of Solutes in Aqueous Solutions

In the context of infrared spectroscopy, numerous attempts have been
made to relate the shift of a particular vibrational absorption band for a
solute with a change in a property of the solvent. Such correlations either
emphasize the importance of solvent permittivity, e.g., KBM plots[109,1395]
or compare the shifts of absorption bands for two chemically similar solutes
in a range of solvents, e.g., BHW plots.[130] In these experiments water
cannot normally be used because its own intense spectra mask that of the
solute. Infrared spectra of various aqueous solutions containing biochemical
polymers have been reported[1074] using the window which extends from
6.5 to 10 μm. In general, the effect of water on absorption spectra of a
solute can only be conveniently examined in the visible and near-ultraviolet
regions where water is transparent. Considerable efforts have been made to
link shifts in $\bar{\nu}_{max}$ for a solute absorbing in this region with a change in some
macroscopic property of the solvent. Some success has been achieved. For
example the dependence of $\bar{\nu}_{max}$ on solvent for various 2-2′ cyanine dyes can
be summarized in a linear plot of $\bar{\nu}_{max}$ against $(n^2 - 1)/(2n^2 + 1)$, where n
is the refractive index measured at the Na_D line.[119,932,1396] The solvents
include methyl alcohol, water, and diiodomethane. The band shifts to lower
energy with increase in n, a bathochromic or red shift. Sometimes the position of an absorption band maximum, $\bar{\nu}_{max}$ or E_{max}, for a particular solute
may be taken as a measure of solvent polarity.[1140] For example, E_{max}
for 1-ethyl-4-methoxycarbonylpyridinium iodide is quite sensitive to solvent.
The act of light absorption can be represented as $Py^+I^- \xrightarrow{h\nu} Py^\bullet I^\bullet$, the
transition destroying the polar "ion-pair" ground state. On going to a more
polar solvent, the ground state is stabilized so that the low-energy charge-
transfer band maximum moves to higher energies. Kosower suggested that
E_{max} (expressed in kcal mol^{-1}) defined the Z value for the solvent.[784] The E_T
value for a solvent, another measure of solvent polarity, is obtained from E_{max}
(in kcal mol^{-1}) for pyridinium N-phenobetaine.[428] Here E_{max} is very sensitive to solvent as can often be appreciated from the colors of the various
solutions, red in methyl alcohol and green in acetone. Values of χ_B are similarly determined from E_{max} (again in kcal mol^{-1}) for the merocyanine dye,
4-[5-(5-ethyl-3-methyl-1,2,3,5-tetrahydropyrido[1,2-a]benzimidazolyl)-2,4-
pentadienylidene]-2,2-dimethyl-1,3-cyclobutanedione.[230] The values of Z,
E_T, and χ_B for water are 94.6 (extrapolated), 63.1, and 68.9, respectively.[586]
Rarely does water stand out in these analyses of solvent polarity, although a
change in $\bar{\nu}_{max}$ of at least one dye[201] on going from a nonaqueous solvent
to water is a result of a dye dimer formation rather than a simple band shift.

2.2.1. *Spectra of Dyes in Water*

The absorption spectra of many dyes in water show a new feature not found in the spectra of dyes in other polar solvents, where Beer's law is obeyed. This anomalous behavior of dyes in water is well illustrated by the spectra of 3:3'-diethylthiocarbocyanine *p*-toluenesulfonate in water.[1397] With increase in dye concentration the intensity of the main band (*M* band) at 553 nm decreases and a new band appears at higher energies, the *H* band, at 510 nm.[1397] Over a small range of low concentrations isosbestic points are observed when the spectra are superimposed. This behavior is observed for a wide range of dye solutes in water.[577,578,816,1222,1223] and this phenomenon has been studied using fluorescence[645] and temperature jump[1190] techniques as well as by absorption spectroscopy. The most detailed studies have concerned methylene blue in water.[79,180,216,918,919,920,1004,1005,1062]

The growth of an *H* band with increase in concentration is called metachromism. Various interpretations have been put forward but it now seems agreed that the *H* band can be assigned to dye dimers formed in equilibrium with dye monomers, $D^+ + D^+ \rightleftharpoons (D^+)_2$. Electrochemical experiments[1062] have shown that the activity of bromide in aqueous solutions of 3:3'-dimethyl-9-ethylthiocarbocyanine bromide is independent of the extent of dye–dye association, while ESR experiments[180] show that metachromism in radical cations (e.g., tetracyanoquinodimethane) is accompanied by spin pairing. The equilibrium between monomers and dimer is responsible for the isosbestic points in the absorption spectra. Actually, these are only apparent isosbestic points; with further increase in concentration the spectra do not show isosbestic points. Thus, for methylene blue in water, equilibria involving monomers, dimers, and trimers are involved.[216,1005] The stability of the dye dimer is assigned in part to hydrophobic interaction between two dye cations. For example, the standard entropy $\Delta S^\ominus$ for dimerization is $+14.7$ J mol^{-1} °K^{-1} (mole fraction scale; 298°K) despite an estimated negative electrical contribution to $\Delta S^\ominus$. The entropy increase can be attributed in part to release of enhanced water structure around the hydrophobic dye monomers when these molecules associate.

2.3. Spectra of Probes in Aqueous Solutions

A solute can act as suitable spectroscopic probe if its absorption spectra are very sensitive to environment. In practice, this sensitivity can be traced to a number of factors. For some systems the absorption spectra of a solute in water change when another solute (or cosolvent) is added because an equilibrium between two states of the probe is perturbed and because the two states of the probe have quite different absorption spectra. In other situations the added cosolvent produces a change in the electronic energy states of the probe and $\bar{\nu}_{max}$ changes. Here both ground and excited states are defined to a major extent by the probe, the solvent perturbing these energy states. Quite the most dramatic changes occur where the solvent plays an integral part in defining the electronic excited state. This occurs in a class of spectra called charge-transfer-to-solvent (ctts).[187]

2.3.1. Dyes in Solution

As noted above, absorption spectra of certain ionic dyes in water show both H and M bands, the spectra of dilute solutions showing isosbestic points. Further, for a solution of methylene blue in water an increase in temperature results in an increase in the intensity of the M band and a loss in intensity of the H band, the superimposed spectra again showing iso-bestic points,[180] Fig. 2. These changes can be summarized in plots of the

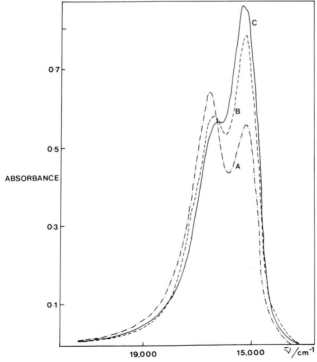

Fig. 2. Absorption spectra of methylene blue ($c = 2.69 \times 10^{-4}$ mol dm^{-3}) in water at three temperatures: (A) 295, (B) 335, and (C) 357°K.

ratio M/H for the intensities of the M and H bands against changes in intensive variable, e.g., temperature or concentration of added solute. It is found, for example, that this ratio is very sensitive to mole fraction of added alcohol.[180]

2.3.2. Iodide in Aqueous Solution

The ultraviolet absorption spectra of a dilute aqueous solution of potassium iodide ($c \simeq 10^{-5}$ mol dm^{-3} in 1-cm-pathlength cells) show an intense absorption band, $\bar{\nu}_{max} = 44{,}200$ cm^{-1} and $\varepsilon_{max} = 1.3 \times 10^4$ at 298°K. The energy E_{max} is significantly larger than the ionization potential (IP) of iodide; the energy of the excited-state orbital is defined to a large extent by the surrounding medium, a ctts transition. This picture of the excited state is certainly consistent with the marked solvent,[187] temperature,[187] and pressure,[187,190] sensitivity of E_{max}. The excited state is described by a centrosymmetric s orbital centered on the iodine atom and the transition is represented as $[I^-]_s \rightarrow [I^\bullet + e]_s$. The extent of penetration of this excited-state orbital into the solvent is a matter for debate[187] and two models, the diffuse[1248] and the confined,[1233] have been proposed. The confined model pictures the excited-state orbital as confined within a potential energy well formed by neighboring solvent molecules. Thus, for iodide in water the potential energy well is formed by the lone pairs of electrostricted water molecules. If the radius of the potential energy well is r, characterizing the excited state, the confined model produces the following simple expression for E_{max} : $E_{max} = IP + (h^2/8mr^2)$, where m is the mass of the electron. With increase in temperature the radius increases, E_{max} decreases, and the band moves to lower energies. For most solvents $E_{max}(I^-)$ is a linear function of temperature and the slope dE_{max}/dT is characteristic of the solvent. When a second solvent is added the band shifts but generally E_{max} is not a linear function of mole fraction. Indeed, E_{max} can be quite sensitive to added cosolvent, as is illustrated in Fig. 3. Often these changes are linked to the effect which the added cosolvent has on water–water interactions which, in turn, modify ion–solvent interactions.

Caution must be used in interpreting spectroscopic data. It is recalled that analyses of thermodynamic data are aided by definition of an ideal liquid mixture such that excess quantities describe the differences between real and ideal situations. A similar analytical procedure would be extremely helpful in the case of spectroscopic data but there is often no clear indication of an ideal behavior. For example, in the case of iodide in aqueous t-butyl alcohol mixtures E_{max} values are known for iodide in the pure solvents. (Actually the spectra of iodide in t-butyl alcohol is complicated by

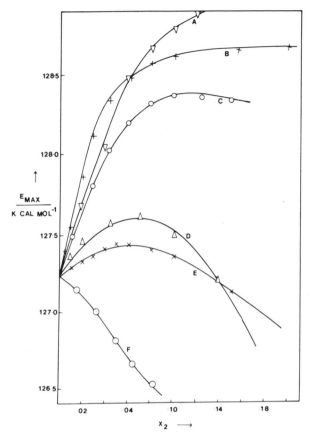

Fig. 3. Effect of added cosolvent on E_{max} for iodide in water at 273.4°K; (A) ethyl alcohol, (B) t-butyl alcohol, (C) 1,3-dioxolane, (D) acetone, (E) acetonitrile, and (F) ammonia.[1432]

ion-pair formation.[187] However, no information is available to define how E_{max} would depend on x_2 if the mixing were ideal. Overall, E_{max} should shift to higher energies (Fig. 3) because E_{max}(I⁻ in t-butyl alcohol) $> E_{max}$ (I⁻ in water), whereas for water plus acetone, E_{max} should shift to lower energies because E_{max}(I⁻ in acetone) $< E_{max}$(I⁻ in water). In practice, a nonideal shift is usually calculated by comparing the observed shift with that predicted if E_{max} were a linear function of x_2.

The above comments can be applied with only slight modification to most spectroscopic studies using probes in binary mixtures. They apply, for example, to the spectra of metal complexes in aqueous solution.

2.3.3. *Metal Complexes in Aqueous Solution*

The colors of certain inorganic complexes are sensitive to the nature of the solvent. Examples of such complexes include Lifshitz's salts, salts of $[Ni(stilbenediamine)_2]^{2+}$, which can exist in blue (octahedral) and yellow (square-planar or distorted tetragonal) forms. A change in the color of a given salt solution can be achieved by adding a second solvent, the color change being attributed to changes in the species coordinated along the z axis of the complex. A series of studies[554] has been reported where the complex salt is the meso-stilbenediamine complex of Ni(II) acetate. A solution of this salt is an organic solvent is blue but addition of water changes the color through green to yellow. If β is the ratio of yellow to blue forms as determined from the absorbance at 440 nm and the total salt concentration and the data are summarized in plots of $\log \beta$ against solvent composition, then the plots are not normally linear. Thus, for t-butyl alcohol plus water the data points in a plot of $\log \beta$ against $\log S$ (S is the total mixed molarity of solvent at 298°K) fall on two straight lines, indicating a change in the nature of the solvent dependence at a particular mole fraction, $x_2 \simeq 0.1$.

2.3.4. *Thermal Perturbation Spectra of Solutes in Water*

In the experiments reviewed above the absorption spectra of a probe in solution are measured using the solvent, water plus cosolvent, as a reference, both sample and reference being held at the same constant temperature and pressure. In another series of experiments two identical solutions are used as reference and sample but these are kept at slightly different temperatures. The measured trace (spectrum) is a function of the temperature sensitivities of both $\bar{\nu}_{max}$ and ε_{max}. Thus, as the spectrum is scanned the sample will, say, transmit more, and then less light, than the reference. Such spectra have been reported when the probe is N-acetyl-L-tyrosine (ATA) and the solvent is water plus t-butyl alcohol[1103] ($T = 277°K$ for sample and $T = 288°K$ for reference); (see also Ref. 210). The trace shows a sharp minimum near 286 nm in a plot of absorbance against wavelength. This minimum reaches an extremum when $x_2 = 0.08$. The magnitude of the extremum in water plus ATA containing various cosolvents has the order t-BuOH > PrOH > EtOH > MeOH.

This technique may have wider application, although the interpretation is not straightforward because changes in absorbance and $\bar{\nu}_{max}$ contribute toward the observed "spectrum."

2.4. Electron Spin Resonance Spectra of Radicals in Aqueous Solutions

This technique has an assured future in this field. Several recent applications of this technique to the study of aqueous solutions are now examined.

2.4.1. ClO_2 and SO_2^-

These molecules are isoelectronic (19 electrons). The ESR spectra of SO_2^- in water show an intense single line for ^{32}S ($I = 0$) together with four weak hyperfine components produced by ^{33}S ($I = \frac{3}{2}$) in natural abundance, 0.74%. The ESR spectra of ClO_2 in water comprise a quartet of lines, the spectra from radicals containing ^{37}Cl ($I = \frac{3}{2}$) and ^{35}Cl ($I = \frac{3}{2}$) not being resolved.[57] The linewidths for these two radicals are very different;[425] for ClO_2, $\Delta H_{ms} = 5$ G and for SO_2^-, $\Delta H_{ms} = 0.73$ G at 295°K. The marked difference in linewidths cannot be accounted for in terms of either g or hyperfine anisotropy but in terms of differences in spin–rotational relaxation times.[58,59,425,1418] The more rapid the relaxation, the broader is the line, and, to a first approximation, the line broadens as the viscosity of the solvent decreases. In water the relaxation time is longer for SO_2^- than ClO_2 because the former interacts directly with the water and so rotation is restrained. In contrast, ClO_2 is accommodated by water in, say, cavities, in the water lattice. In consequence, solute–solvent interactions are weak, rotation is rapid, the relaxation time is short, and broad lines are thus observed.

2.4.2. Nitroxide Radicals

A wide range of stable nitroxide radicals can be prepared; moreover, these paramagnetic molecules are reasonably soluble in water and aqueous solutions.[601,1366] These radicals have been extensively used as spin labels in ESR investigations of enzyme,[1256] membrane, and micellar systems (see also Ref. 282). The ESR spectra of di-t-butyl nitroxide in water comprise three lines (for ^{14}N, $I = 1$). The linewidths are not equal, the central line being the narrower of the three. The difference in linewidths is explained as follows. Three spectra of this radical can be identified following analysis of the ESR spectra of single crystals containing nitroxide. Each spectrum is characteristic for the mutually perpendicular axes x, y, and z. The x axis is parallel to the N—O bond and the z axis is parallel to the p orbital on nitrogen. The three spectra are shown diagrammatically in Fig. 4, which emphasizes the hyperfine anisotropy. There is a small g anisotropy and so the central line of each group is not in the same position. When the mole-

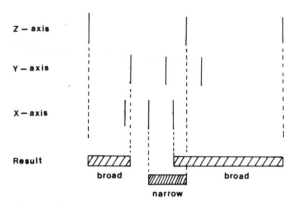

Fig. 4. Diagrammatic reconstruction of ESR spectra of a nitroxide in solution showing linewidth pattern in observed spectra.

cules tumble rapidly the correlation time is short and three narrow lines are observed* (Fig. 5). With decrease in tumbling rate the correlation time increases and the spectra therefore record the effect of many radical orientations and thus the linewidth increases. However, as illustrated in Fig. 4, the wing lines are broadened to a greater extent than the central line, where a smaller anisotropy is involved. The ratio of the measured intensities of the middle- A_m and high-field A_h lines is more sensitive to change in solvent then the measured linewidths. It follows, therefore, that the ratio A_m/A_h is a measure of rotational freedom for the radical in solution. If an added cosolvent enhances this freedom and so increases the reorientation rate, than this ratio decreases. In this manner the ESR technique provides a useful method of measuring the effects of cosolvents on the rate of molecular tumbling in solution. Certain precautions must be observed; dilute solutions must be used, otherwise the lines are broadened by direct spin–spin relaxation mechanisms. Further, the added cosolvent must not promote nitroxide association (and thus spin pairing), otherwise a serious loss of signal intensity will result.

2.4.3. Asymmetric Solvation

The ESR spectra of m-dinitrobenzene radical anions in, for example, acetonitrile show that the two nitrogens are equivalent.[191] However, the spectra of such anions in nonpolar solvents often show a nonequivalence between the two nitrogens together with a marked variation in linewidth

* D. Jones and M. C. R. Symons, unpublished data.

across the complete spectrum, a linewidth alternation. This nonequivalence stems from ion-pair formation,[253,1270] where the cation is located close to one nitrogen atom and the alternating linewidth stems from a momentary nonequivalence between the two nitrogens, the cation migrating (relatively[310]) between the two nitro groups. A similar nonequivalence between nitrogens and associated alternating linewidth phenomena can be produced by asymmetric solvation, i.e., in the absence of ion-pair formation. Here a momentary intense radical–solvent interaction involves one of the two nitrogroups in the *m*-dinitrobenzene.[603] The way this asymmetry produces the linewidth phenomena can be illustrated by reference to Fig. 6, where it is assumed that the splitting from one nitrogen is much larger than from the other in a nonequivalent situation. The spectra are in fact more complex (Fig. 7), and in Fig. 6 hyperfine splittings from the ring hydrogens are not considered. The top half of Fig. 6 shows the three hyperfine components, ($m_N = 0, \pm 1$) from one nitrogen (N1) with a large splitting constant, each line being split by a second nitrogen (N2) having a smaller splitting constant to produce a total of nine lines. The lower half of the figure summarizes the situation, say a moment later, where the splittings are reversed, a large splitting arising from N2 and a smaller one from N1. The figure shows how the two wing lines and the central line are unaffected; however, the remaining lines are broadened such that the nine-line spectrum has the

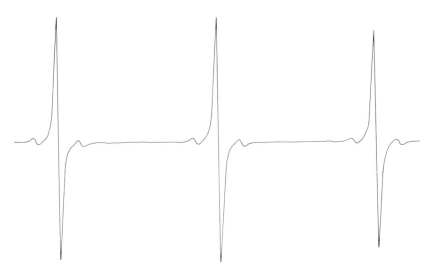

Fig. 5. ESR spectra of di-*t*-butyl nitroxide ($c \simeq 10^{-2}$ mol dm^{-3}) in water at room temperature and at *X*-band frequencies: satellite lines due to ^{13}C (D. Jones and M. C. R. Symons, unpublished data).

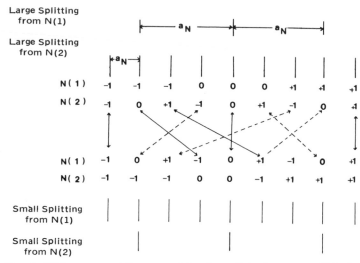

Fig. 6. ESR spectra (diagrammatic) of *m*-dinitrobenzene which is asymmetrically solvated, thereby producing two nonequivalent nitrogen atoms; see text for description.

following pattern of widths—narrow, broad, broad, broad, narrow, broad, broad, broad, narrow. The mean lifetime of the asymmetric units can be estimated from the linewidths, e.g., $\sim 2 \times 10^{-9}$ sec in dimethylformamide and ethanol[610] and 2×10^{-6} sec in water.[717] This phenomenon, asymmetric solvation, can be exploited by following the changes when a cosolvent is added. Thus, when *t*-butyl alcohol is added the lifetime of asymmetrically hydrated *m*-dinitrobenzene in water increases to reach an extremum when $x_2 \simeq 0.02$ and then falls to immeasurably short times, $< 10^{-7}$ sec when $x_2 \simeq 0.06$.

Asymmetric solvation is also responsible for the marked solvent dependence of the ESR spectra of 2,6-dimethyl-*p*-benzosemiquinone radical anion. Changes in differential solvation of the two oxygen atoms resulting from changes in solvent composition alter the spin density within the ring and change the difference Δa between hyperfine coupling constants to ring and methyl protons[181,1038,1271] (but for a different view see Ref. 311).

3. INTERPRETATION OF EXPERIMENTAL DATA

Despite an apparent wealth of spectroscopic information, interpretation of the data is often not straightforward. To a large extent, this branch of the subject is in its formative stage. Consequently, spectroscopic data

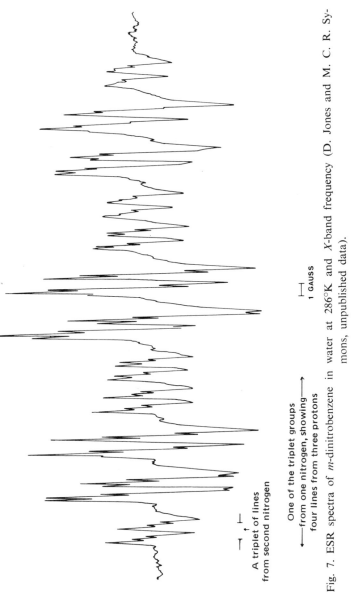

A triplet of lines
from second nitrogen

One of the triplet groups
—from one nitrogen, showing→
four lines from three protons

1 GAUSS

Fig. 7. ESR spectra of *m*-dinitrobenzene in water at 286°K and *X*-band frequency (D. Jones and M. C. R. Symons, unpublished data).

are interpreted in a rather general fashion using such terms as "structure making" and "structure breaking" but with relatively little regard to further subtleties of behavior. Again, many studies have been concerned with the effects of added t-butyl alcohol on water because the changes in observed parameters are often more marked than when other solutes and cosolvents are added. Nevertheless, sufficient information is available to identify patterns of behavior so as to form a basis for extension to other aqueous systems. For the purposes of this review the following account subdivides binary aqueous systems into those where the mixing is typically aqueous and those where it is typically nonaqueous.[514]

If G^E is positive and if $T|\Delta S^E| > |H_m|$, the system is said to be typically aqueous-dominant entropy of mixing with tendency to a lower critical solution temperature (LCST). If $|\Delta H_m| > T|\Delta S^E|$, the system is typically nonaqueous-dominant enthalpy of mixing. Here G^E may be (a) positive, e.g., water plus acetonitrile, with a tendency to phase separation at a UCST, (b) negative, e.g., water plus hydrogen peroxide, but in both cases the sign and magnitude of G^E are controlled by ΔH_m.

3.1. Typically Aqueous Mixtures

Most spectroscopic investigations of the effects of nonpolar solutes on water have concentrated their attention on this type of system with particular reference to alcohol plus water mixtures. The changes in spectroscopic properties for these systems can usually be satisfactorily interpreted in terms of water structure-breaking/water structure-making actions by the solute.[161,508] In some cases the structure-making action can be compared to an effective cooling of the water, a slowing down of molecular reorientations.[519] In other cases the structure-making action of a solute is described in terms of the formation of a liquid clathrate arrangement of water molecules,[516,564] such that the solutes occupy guest sites in the hydrate lattice[709] (Chapter 10). For most part these models have developed from analyses of thermodynamic data. Spectroscopic data generally support these models because the data are not markedly in conflict with this approach. Proponents of the liquid clathrate model argue that when an organic component such as t-butyl alcohol is added to water the system remains homogeneous as a result of alcohol–water hydrogen bond formation. In addition, water–water interactions around the hydrophobic alkyl groups are enhanced, the water structure resembling that in a water clathrate. As more alcohol is added the cospheres of enhanced water structure overlap, mutually reinforcing water–water interactions. This trend continues until there is insufficient water to maintain the integrity of the water structure. At this point, e.g., $x_2 = 0.04$ for t-butyl alcohol plus water at 298°K, a new set of phenomena come into play although description using a simple model is not straightforward.[203] In one approach the mixture is described by stressing the role of alcohol–alcohol association where such associated species[645] occupy a water cavity and where an equilibrium is established between alcohol molecules in singly and doubly occupied cavities. A second approach accounts for the properties of the mixture in terms of marked local concentration fluctuations which become more marked as x_2 increases to reach a maximum at a particular composition,[78,203] e.g., $x_2 = 0.1$ for aqueous t-butyl alcohol at 298°K. In an extreme case these fluctuations may become so intense that phase separation results. For the present these concepts can be said to identify two regions: (a) a liquid clathrate region, and (b) a region where the structure has been broken down. Spectroscopic studies identify two such regions[1271] but in some cases the spectra are sensitive to liquid clathrate formation and in other cases to phenomena in the "broken-down" region. One of two trends in spectroscopic data is usually observed: (a) a rapid change indicating sensitivity to changes in the liquid followed by a region where the measured quantity is insensitive, (b) no change, indicating insensitivity to changes in x_2 followed by a rapid change. Thus, E_{max} for iodide in aqueous t-butyl alcohol mixtures and the ratio R (Section 2.1.2) follow pattern A in Figs. 3 and 8. In contrast, Δa for 2,6-dimethyl-p-benzoquinone[181,1271] follows pattern B in Fig. 9.

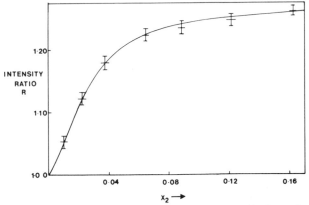

Fig. 8. Ratio of the intensities of the near-infrared absorption bands at 1.556 and 1.416 μm for 6 mol dm^{-3} HOD in D_2O as a function of mole fraction x_2 of $(CH_3)_3COD$ at 298°K.

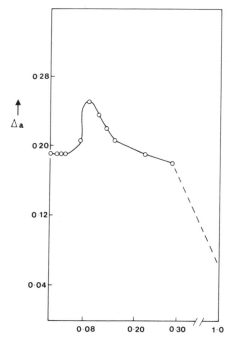

Fig. 9. Difference between methyl and ring
proton hyperfine coupling constants Δa for
2,6-dimethyl-p-benzosemiquinone radical anions in water plus t-butyl alcohol at room temperature. (The cation is tetra-n-butylammo-
nium; ion-pair formation does not modify the
spectra in pure t-butyl alcohol to any appreciable extent.)

3.1.1. *Alcohols*

When t-butyl alcohol-d_1 is added to HOD/D_2O the ratio R (Section
2.1.2) increases over the range $0 < x_2 < 0.04$ (Fig. 8), the superimposed
spectra showing isobestic points (Fig. 1). This behavior is indicative of an
effective cooling of the water by added alcohol, the equilibrium between
free O—H and bonded O—H favoring the latter as x_2 increases. For this
equilibrium $\Delta H^\ominus$ is not markedly affected by added alcohol, showing that
it is the amount of structured water which is affected rather than the stability
of the structured water in pure water. The intensity of a band at 5814 cm^{-1},
assigned to an overtone of a C—H stretching vibration, obeys Beer's law
in the range $0 < x_2 < 0.03$. The increase in R can be understood in terms
of the clathrate hydrate model (see above). The t-butyl groups occupy

cavities in the water host-lattice, to which the alcohol is hydrogen-bonded. The increase in R can be matched, for example, against the decrease in the relative partial molar volume, $V_2 - V_2^\ominus$ of the alcohol with increase in x_2.[524,751,1015,1016] The latter behavior indicates an increasing economy of volume which can be linked with a mutual reenforcement of water–water interactions with increase in x_2, the increase in R resulting from this mutual effect. At higher temperatures the increase in R with increase in x_2 becomes less marked until near 340°K, where R is almost insensitive to change in x_2.[1432] At each fixed temperature R is essentially a linear function of x_2, e.g., $0 \leq x_2 \leq 0.03$ at 298°K. The slope dR/dx_2 decreases with increase in temperature, although there are clear indications of a hesitation in this fallaway near 310°K. Changes in the properties of water and aqueous solutions are observed at or near this temperature, as shown by the dependence on temperature of the heat capacity $C_P^\bullet$ for water,[444] activation energy for the flocculation of PVA solutions in aqueous suspensions,[714] $\Delta C_P^\ominus$ for acid dissociation in water,[306,475,701] and in NMR spectra.[557] One interpretation[475] considers a "structural melting" in water in this temperature range, but the trends in R (and in E_{max} for iodide,[189] see below) indicate some retention of structure in this range, i.e., a resistance to a structure breaking brought about by an increase in temperature.

An effective cooling of water by added t-butyl alcohol also accounts for the high-energy shift in E_{max} and decrease in half-bandwidth for the ctts absorption band of iodide in water. If these spectroscopic changes were a consequence of differences in solvent polarity, the most marked changes would be expected to occur when small amounts of water are added to iodide dissolved in the pure alcohol, preferential solvation of iodide by water in the alcohol-rich medium producing a marked low-energy shift. The observed changes are quite the reverse. The fact that the dominant change is a band shift (but see below) rather than a band change has been used[187] to counter a suggestion[785] that the absorption stems from a charge transfer from iodide to one solvent molecule S within a charge-transfer complex of the type I⁻S, i.e., $I^-S \xrightarrow{h\nu} I^\bullet S^-$. If such a linear movement of charge occurred, one might expect that E_{max} would depend on the Z value (see above) of the solvent. This is not observed.[602] However, the precise mechanism by which the alcohol produces the high-energy shift in E_{max} is not clear cut. Historically, the complexities of the absorption spectra of iodide in aqueous mixtures were first noted when it was found[1233] that dioxan added to iodide in water produced a high-energy shift in E_{max} even though $E_{max}(I^-$ in dioxan$) < E_{max}(I^-$ in water$)$. It was suggested[1233] that dioxan "dissolved" preferentially in the disordered water (zone B in the

Frank and Wen model for ionic hydration[512]) outside the electrostricted layer of water molecules hydrating iodide. Consequently, the iodide ion is shielded from hydrating water–bulk water interactions, the radius of the potential energy well decreases, and E_{max} increases (see also Ref. 192). However, this model does not account (at least with reference to iodide in alcohol plus water mixtures) for the links in patterns of behavior between E_{max}, R, $V_2 - V_2^\ominus$, and other thermodynamic properties. It appears, therefore, that added alcohol brings about changes in water–water interactions to which the spectra of iodide are sensitive.

The increase in $E_{max}(I^-)$ following addition of t-butyl alcohol to iodide in water could result from (a) a stabilization of the ground state, (b) a destabilization of the excited state, or (c) changes in the energies of both states. The confined model[187,1233,1234] for ctts spectra favors model (b) because, according to this theory, E_{max} is not directly a function of the energy of the ground state but is a function of the radius of the potential energy well which defines the excited state. The diffuse model,[187,1248,1249] probably favors model (a) because here E_{max} is related to the energy of the ground state, iodide in water. Some evidence against this latter approach comes from thermodynamic data. The enthalpies of solvation of salts become endothermic (relative to solution in water) as small amounts of most cosolvents are added to water.[46,47,49,50] If the ionic enthalpies are affected in the same way, then a low-energy shift for E_{max} would be predicted because the energy difference between ground and excited states would decrease. In all probability, both states are affected [model (c)].

More recent studies have lead to the discovery of new complications. Superimposed spectra of iodide in t-butyl alcohol plus water, $0 \le x_2 < 0.04$ show isobestic points[202] (Fig. 10). An equilibrium is postulated involving two similar groups of iodide solvates of which one describes iodide incorporated into a water lattice having some clathrate hydrate structure.[202]

With increase in temperature E_{max} for iodide in water becomes less sensitive to added t-butyl alcohol. At a fixed temperature E_{max} is close to a linear function of x_2. The slope dE_{max}/dx_2 decreases with increase in temperature but, as with the temperature dependence of dR/dx_2 discussed above, there is evidence[189] for a hesitation in this fallaway near 310°K.

For a range of systems $E_{max}(I^-)$ is a linear function of pressure at a fixed temperature, e.g., iodide in acetonitrile[192] and in methyl alcohol.* For iodide in water near 300°K this behavior is only observed at high pressures (Figs. 8–11). Initially, E_{max} is very sensitive to pressure, then less so, and

* T. Burdett, University of Leicester, unpublished data.

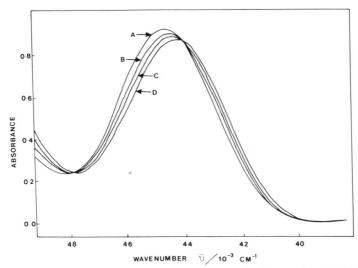

Fig. 10. Charge-transfer-to-solvent spectra (ctts) of potassium iodide ($\sim 5 \times 10^{-5}$ mol dm^{-3}) at 298°K in water containing different mole fractions of t-butyl alcohol; (A) $x_2 = 0.06$, (B) $x_2 = 0.04$, (C) $x_2 = 0.02$, and (D) $x_2 = 0$.

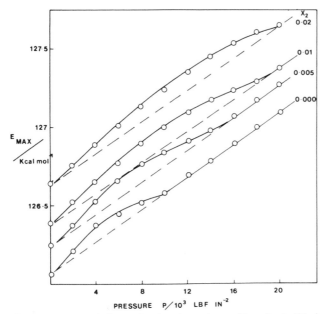

Fig. 11. Energy of absorption band maximum E_{max} for iodide in water at 303°K as a function of applied pressure and mole fraction of added t-butyl alcohol (1 lbf in^{-2} $\equiv$ 6894.757 N m^{-2}).

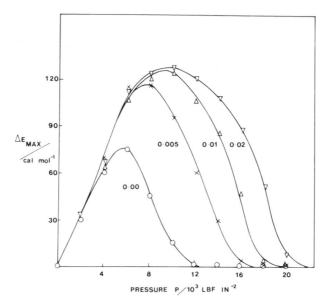

Fig. 12. Dependence of ΔE_{max} on pressure for iodide in water containing different mole fractions of t-butyl alcohol; ΔE_{max} is the difference between observed E_{max} and that given by a linear interdependence of E_{max} on pressure, as indicated in Fig. 11. (Taken from Ref. 1432.)

ultimately a linear interdependence of E_{max} and pressure is observed.[190] With decrease in temperature the extent of deviation from this linear interdependence becomes more marked. Consistent with the observations discussed above small amounts of added t-butyl alcohol increase the deviation when added to iodide in water at a fixed temperature (Fig. 12).

The hypothesis that added alcohol lowers the effective temperature of water is supported by other spectroscopic data. For example, the proton magnetic resonance spectra of O—H protons in water plus t-butyl alcohol, for $0 < x_2 < 0.04$, depends on x_2 in a manner consistent with this model.[29,572] However, in other cases spectroscopic changes are small, close to experimental error, and therefore uninformative. The Raman spectra of 6 mol dm^{-3} D$_2$O in H$_2$O show a broad band in the 2400–2700 cm^{-1} region. Walrafen[1377] has deconvoluted this band and assigned the 2645–2660 cm^{-1} component to free (nonbonded) OD and the 2525 cm^{-1} component to bonded OD (using a mixture model for water, see Volume 1). When t-butyl alcohol is added to this system the band position hardly changes[1365] (Figs. 13 and 14). Similarly, the position of the infrared

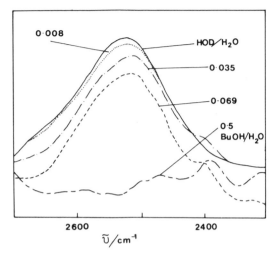

Fig. 13. Raman spectra of 6 mol dm⁻³ D_2O in H_2O (298°K) containing various mole fractions of t-butyl alcohol. Spectra recorded using a Coderg PHo spectrometer with argon-ion laser excitation, 488-nm spectra. (Taken from Ref. 1365.)

absorption band in the 2800–2200 cm⁻¹ region for HOD in H_2O is insensitive to added t-butyl alcohol[1365] (Fig. 15) and trimethylamine.[464] It is probable that where alcohol (ROH) was added to water[1365] conversion of ROH to ROD masks subtle spectroscopic changes which otherwise might have occurred. In the experiments which used trimethylamine[464]

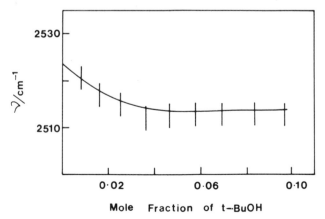

Fig. 14. Dependence of $\bar{\nu}_{max}$ (Raman spectra, Fig. 13) on mole fraction of t-butyl alcohol.

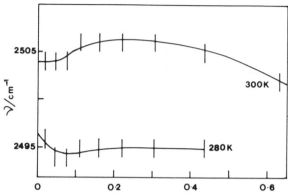

Fig. 15. Dependence of $\bar{\nu}_{max}$ in the infrared absorption
spectra (2500 cm^{-1} region) of 2% HOD in H$_2$O on mole
fraction of t-butyl alcohol. Spectra recorded on Unicam
SP 100. (Taken from Ref. 1365.)

a marked difference was noted between the spectra of solid trimethylamine
hydrate (CH$_3$)$_3$N · 10.25H$_2$O[1070] at 124°K and a liquid of this composition
at 300°K. Indeed, the spectra of the latter system closely resemble spectra
of pure water at 300°K, indicating that, at least so far as infrared spectra
are concerned, the liquid, trimethylamine plus water, has little long-range
structure of the clathrate hydrate form.

In the account given above attention has been concentrated on the
system, water plus t-butyl alcohol. When a range of monohydric alcohols
is examined the effectiveness of alcohols to produce a high-energy shift
for E_{max}(I$^-$ in water) has the order MeOH < EtOH < n-PrOH < t-BuOH.
Such a trend is confirmed by the thermal perturbation spectra (Section
2.3.4) and the sensitivity to added alcohol of the color of solutions containing metal complexes (Section 2.3.3 and Fig. 16).

The effect of added alcohol on water cannot normally be observed by
measuring the (very temperature sensitive) ultraviolet absorption spectra
of water[1213] because alcohols generally have a much higher absorption
intensity in this region. This difficulty is overcome if fluoroalcohols are
used; the spectra of water in aqueous solutions containing trifluoroethanol
and hexafluoro-2-propanol can be measured. The ratio of absorbances at
57,140 cm^{-1} of water in this mixtures to that of pure water decreases with
increase in x_2 to a minimum at 0.4 and 0.10 (298°K) for fluoroethanol
(Fig. 17) and fluoropropanol systems, respectively.* The initial trends are
equivalent to changes brought about by a decrease in the temperature of water.

* M. F. Fox and E. M. Hayon, unpublished data.

Because the far-ultraviolet absorption spectra of alcohols are generally more intense than that for water, the absorption in water may be readily measured in the 54,000 cm^{-1} region. The absorption coefficients of the alcohols at 54,090 cm^{-1} are particularly important in photochemistry, this being the wavenumber corresponding to the 184.9-nm mercury emission line. Quantum yields are calculated from absorbance data and so the latter have been carefully determined at this wavelength.[104a,104b,1381a] At 54,090 cm^{-1} the absorption intensity of an alcohol in water does not obey Beer's law; the molar absorption coefficients of methyl and ethyl alcohols show maxima (Fig. 18) at $x_2 = 0.02$, whereas these coefficients for 2-propyl and t-butyl alcohols decrease over the range $0 < x_2 < 0.1$. The contrast in

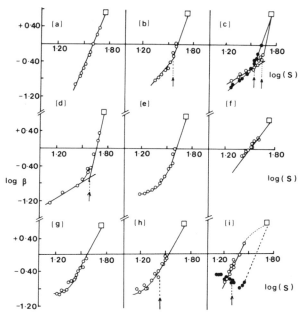

Fig. 16. Effect of added cosolvent on the equilibrium between blue B and yellow Y forms of [Ni(stien)$_2$](AcO)$_2$ in water; $\beta = $ [Y]/[B] and [S] is the total mixed molarity of the solvent (298°K). The cosolvents are (a) methyl alcohol, (b) ethyl alcohol, (c) propyl alcohols (open circle, n; closed circle, iso), (d) t-butyl alcohol, (e) acetone, (f) formamide, (g) N,N'-dimethylformamide, (h) N,N'-dimethylacetamide, and (i) N-methylamides (open circle, N-methylformamide; closed circle, N-methylacetamide). The squares are data points for pure water. (Taken with permission from Ref. 554, p. 2172.)

behavior may be indicative of a tendency for the smaller alcohols to enter, substitutionally, into the water framework but for the larger alcohols to enter as guests in the structure,[516] i.e., to enter interstitially.

The hypothesis that added *t*-butyl alcohol "cools" the water is borne out by the increased lifetimes of asymmetrically solvated *m*-dinitrobenzene radical anions (Section 2.4.3). Where the probe is already present in a cavity in the water structure then addition of a cosolvent which enhances the lifetime of the cavity will have only slight effect on the probe. Such is the case for ClO_2 (Section 2.4.1), where the ESR linewidth hardly changes (pattern B) when alcohol is added to the aqueous solutions. A rather different pattern of behavior is observed in the ratio of *M* and *H* bands for methylene blue (Section 2.3.1). Here the equilibrium between monomer and dimer adjusts to favor the monomer when alcohol is added, the efficiency of the cosolvent in producing this change having the order *t*-BuOH > EtOH > MeOH. The change in the M/H ratio produced by an increase in x_2 at fixed temperature is similar to that brought about by an increase in temperature when $x_2 = 0$. Such trends could be taken to indicate that, say, *t*-butyl alcohol is a water structure breaker, but this conclusion conflicts

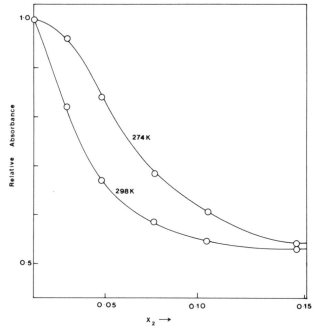

Fig. 17. Effect of added trifluoroethyl alcohol on absorbance of water at 57,140 cm^{-1} at 274 and 298°K.

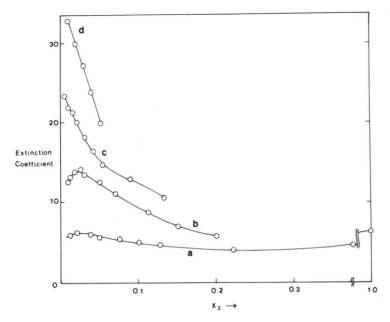

Fig. 18. Effect of added alcohol, mole fraction x_2, on the molar extinction coefficient at 54,090 cm^{-1}; (a) methyl alcohol, (b) ethyl alcohol, (c) isopropyl alcohol, and (d) t-butyl alcohol; $T = 298°$K.

with the weight of evidence given above. It seems more likely that hydrophobic association between dye monomers now competes with hydrophobic association of alcohol and dye monomer, $D + n\text{-ROH} \rightleftharpoons D\text{-}n\text{-ROH}$. The latter equilibrium favors the formation of D-n-ROH over dye dimers because the concentration of alcohol far exceeds that of the dye. The spectra of dyes in these aqueous mixtures provide examples where changes occur by direct involvement of the added cosolvent rather than through a modification of water–water interactions.

As more alcohol is added the patterns of behavior change, e.g., x_2 $\simeq 0.03$ for t-butyl alcohol at 298°K, $E_{\max}$ for iodide, and R for HOD in D$_2$O (Sections 2.3.2 and 2.1.2, respectively) become less sensitive and then almost completely insensitive to further increase in x_2. The mole fractions at which these changes occur are found in nearly all spectroscopic studies to depend on the alcohol, having the order t-BuOH $< n$-PrOH $<$ EtOH $<$ MeOH. It was suggested above that this change can be taken as that mole fraction where the integrity of the simple clathrate structure breaks down. It seems reasonable to argue that the greater the integrity of the original liquid clathrate structure, the more dramatic will be the ensuing cataclysmic collapse when more cosolvent is added. It seems possible, for example, that this collapse and associated composition fluctuations produce the intense ultrasonic absorption (Chapter 9) and X-ray scattering in the intermediate composition range. Further, the *mole fractions* corresponding to this disorder have the order t-BuOH $<$ n-PrOH $<$ EtOH but the *extent* of disorder has the order t-BuOH $>$ n-PrOH $>$ EtOH.

Spectroscopic, particularly ESR, data support this model. Thus the radical ClO_2, which was accommodated in a cavity, $x_2 < 0.04$, for t-butyl alcohol plus water, loses this guest site when $x_2 = 0.1$ such that the line-width becomes sensitive to solvent viscosity. The changes in asymmetric solvation of a radical such as m-dinitrobenzene occur more frequently, i.e., the rate of migration of the asymmetry increases. In the case of 2,6-dimethylsemiquinone radical anion (Fig. 9) these effects maximize where the disorganization and fluctuations are maximized before falling away in the alcohol-rich region [*cf.* sound absorption (Chapter 9) and X-ray scattering[78]]. It is not always obvious why some observables, such as E_{max} for iodide, in pattern A remain insensitive to changes in composition over this region. Some variables do not remain constant. Thus for the Lifschitz salts (Fig. 16) the plot of log β against log S shows two straight lines (a shallow curve?) and the intersection seems to occur at a composition where the disorganization facilitates ligand exchange.

3.1.2. Acetone

When acetone is added to HOD (6 mol dm^{-3}) in D_2O the ratio R (Section 2.1.2), increases[500a,1432]* until $x_2 = 0.07$ (298°K) before leveling off or decreasing slightly. At lower temperatures the plot of R against x_2 (Fig. 19) shows a maximum at $x_2 = 0.05$ (278°K).[†] E_{max} for iodide in these mixtures[1432] reaches an extremum when $x_2 = 0.07$ before shifting in the opposite direction to lower energies.

Absorption spectra of acetone in water can be measured by reference to the low-energy $n \rightarrow \pi^*$ transition. This transition for both acetone and carbonyl groups in general is sensitive to solvent. For pure acetone $\bar{\nu}_{max}$ is at 36,400 cm^{-1} and is at 37,730 cm^{-1} for an infinitely dilute solution of acetone in water.[500a] For intermediate solvent compositions $\bar{\nu}_{max}$ is not a linear function of x_2. If such linearity defines an ideal shift, the differences between observed and ideal shift equal excess shifts, $\Delta\bar{\nu}_{max}$. With increase in x_2, $\Delta\bar{\nu}_{max}$ shows a positive extremum (298°K) near $x_2 = 0.05$ and negative

---

* D. Waddington, unpublished data.

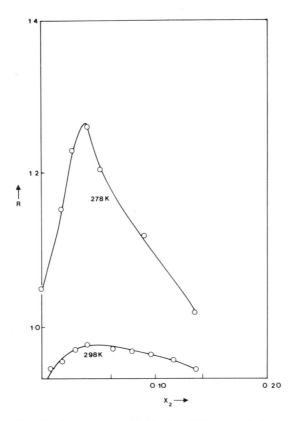

Fig. 19. Dependence of R (Section 2.1.2) on mole fraction of acetone (Ref. 500a).

extrema at $x_2 = 0.20 - 0.25$ and 0.45 (Fig. 20). The intensity of the absorption band is insensitive to x_2 when $x_2 < 0.05$ before decreasing to a
minimum near $x_2 = 0.2$.

When acetone is added to D_2O (6 mol dm^{-3}) in H_2O the band maximum
in the 2600 cm^{-1} region of the Raman spectra is shifted and the bandwidth
increases (Figs. 21 and 22).[1365] The band maximum of the infrared absorption band of D_2O/H_2O at 298°K is shifted to higher energies when acetone
is added. There is some indication of a change in the sensitivity to acetone
mole fraction near $x_2 = 0.4$ (Fig. 23), close to where the ultrasonic absorption (measured as the ratio α_a/ν^2, where α_a is the amplitude absorption coefficient and ν the frequency of the sound wave) is a maximum (Chapter 9).

These trends, especially for the very dilute solutions of acetone in
water, can be understood in terms of the clathrate hydrate model outlined

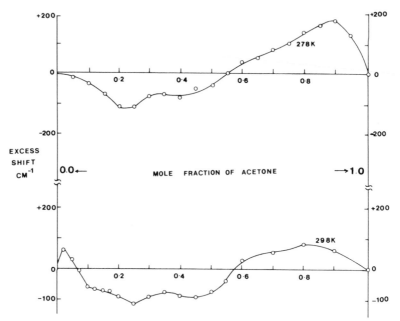

Fig. 20. Dependence of $\bar{\nu}_{max}$ for the $n \to \pi^*$ transition of acetone in water as a function of acetone mole fraction; the quantity ploted is $\Delta\bar{\nu}_{max} = \bar{\nu}_{max}$(observed) $-\bar{\nu}_{max}$(ideal), as described in the text (Ref. 500a).

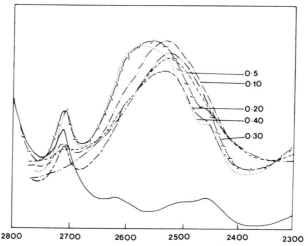

Fig. 21. Raman spectra of 6.0 mol dm^{-3} D$_2$O in H$_2$O as a function of acetone mole fraction (298°K). For further details see caption to Fig. 13.

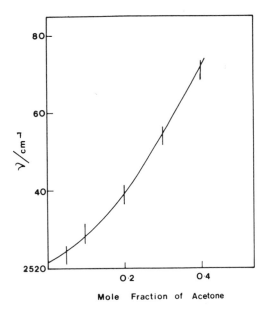

Fig. 22. Dependence of $\bar{\nu}_{max}$ on mole fraction of acetone for Raman spectra of 6.0 mol dm^{-3} D$_2$O in H$_2$O (taken from spectra in Fig. 21).

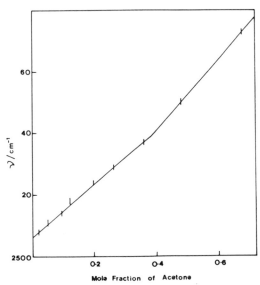

Fig. 23. Dependence of $\bar{\nu}_{max}$ on mole fraction of acetone from infrared absorption spectra of 2% HOD in H$_2$O ($T = 298°$K).

above; in fact, a solid acetone hydrate can be prepared.[1124] The ensuing breakdown of water structure and disorganization produced when more acetone is added seems to maximize in the $x_2 = 0.4$ region. If the intensity of sound absorption is taken as a guide (Chapter 9), the disorganization is not so marked as for water plus n-propyl alcohol mixtures. It is noteworthy that the dependence of $\log \beta$ on $\log S$ (Fig. 16) is much more gradual for aqueous acetone systems than for aqueous n-propyl alcohol systems.

Several other systems have been investigated, although not in such detail. For example, the changes in the near-infrared spectra of HOD in D_2O when hexamethylenetetramine is added[98] support conclusions from other nonspectroscopic data that HMT is a water structure former. Indeed, HMT forms a crystalline clathrate hydrate.[937]

3.2. Typically Nonaqueous Mixtures

As noted above, this rather large group of binary aqueous mixtures can be subdivided into two groups.

The first group ($G^E < 0$) contains those cosolvents where the dominant factor is intermolecular hydrogen bonding between water and cosolvent. While this interaction is very important in the case of, say, t-butyl alcohol plus water (otherwise this system would not be completely miscible at room temperature), it has been shown how many properties of this system reflect the importance of hydrophobic alkyl groups on water–water interactions. However, as the hydrophilic character of the cosolvent increases, the importance of the water–cosolvent interactions rapidly increases. The extreme cases correspond to those systems discussed by Warner[1379] where the solute has hydrophilic groups so arranged that the solute can be accommodated without significant distortion of the water structure.[514] For such systems the spectroscopic properties of water may show little change. For example, the viscosity of water is dramatically changed when polyethylene oxide is added but little change is observed in the R value for the HOD/D_2O system[188] (Section 2.1.2). Similarly, E_{max} for iodide in water (Section 2.3.2.) shows no high-energy shift when ammonia is added and the thermal perturbation spectra of ATA (Section 2.3.3) show no marked extrema when either glycerol or ethylene glycol is added to water.

The second group may be typified by acetonitrile where ΔH_M is positive over most of the composition range[994] (exothermic for $0 < x_2 < 0.4$ at 298K). The effect of acetonitrile on the near-infrared spectra of HOD in D_2O is quite different to that when t-butyl alcohol is added.[44] No isosbestic points are observed (see Fig. 1) and, as noted above, the analysis in terms of changes in R values is not directly relevant. The apparent R value decreases with increase in x_2. Some slight shift in E_{max} to higher energies is observed when acetonitrile is added to iodide in water, indicating a possible enhancement of water–water interactions. Nevertheless, the dominant changes are consistent with a breakdown of water–water interactions by added acetonitrile, a trend agreeing with conclusions drawn[44] from a lack of minimum in the relative partial molar volume, $V_2 - V_2^\ominus$. A similar conclusion[194] is drawn following analysis of the spectroscopic properties of water plus 1,3-dioxalone mixtures.

For many systems, however, no clear pattern may emerge when all the data are examined. This would appear to be the case for dimethylsulfoxide (DMSO) plus water. Franks[514] describes this system as typically non-aqueous, although others[1127] describe DMSO as a water structure former, and so by implication a typically aqueous solute. The latter assignment is based on an analysis of enthalpies of mixing[1127] and to some extent this assignment seems in agreement with NIS data. The latter show that added DMSO enhances and sharpens the peaks in the NIS spectra of pure water.[1171] On the other hand, the relative insensitivity of $\bar{V}_2 - V_2^\ominus$ to x_2 for DMSO plus water is indicative[514] of no cooperative effects in either a structure-breaking or -making context. When DMSO is added to HOD in D_2O the spectra do not show isosbestic points but the apparent R values (Section 2.1.2) increase with increase in x_2 to a maximum at $x_2 = 0.25$ (298°K) and 0.30 (278°K).* These trends are consistent with a dominant intercomponent association with little gross modification of water–water interactions in bulk water. However, the absorption spectra of aqueous DMSO in the 35,000 cm^{-1} region, assigned to a very weak $n \rightarrow \pi^*$ transition on the long-wavelength edge of a much more intense $\pi \rightarrow \pi^*$ transition, is very sensitive to composition* (Fig. 24). The minimum near $x_2 = 0.25$ is indicative of extensive intercomponent association. The low-frequency spectra show that DMSO in aqueous solutions can exist in two local environments, hydrogen-bonded and nonhydrogen-bonded DMSO,[226] the presence of the former species being invoked to account for the effect of added DMSO on the Raman spectra of water[1378] (Section 2.1.1).

Intermolecular association between water and added solute may also account for the large "solvation numbers" obtained for sucrose (21) and dextrose (10) in water.[210] These numbers were obtained from analysis of the effect of added solute on the absorption band in water at 958 μm (assigned to the $2\nu_1 + \nu_3$ combination band of water). By way of contrast,

* M. F. Fox, J. M. Pollock, and K. P. Whittingham, unpublished data.

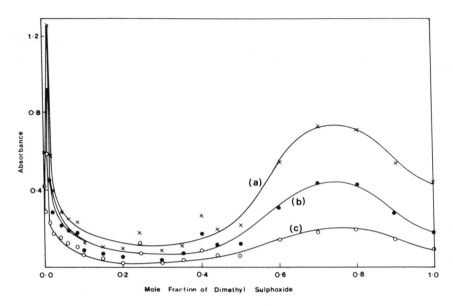

Fig. 24. Variation of absorption at (a) 37,000, (b) 35,000, and (c) 33,000 cm^{-1} for water plus dimethylsulfoxide as a function of mole fraction of DMSO. Path length of cell is 10 mm; absorptions normalized with respect to that for mixture where $x_2 = 0.8$. (From M. F. Fox, unpublished data.)

sym-dimethylurea, propylene carbonate, and ethylene carbonate, appear to act as structure breakers.[209] This may also account for the low solvation number (2.5) calculated[210] for urea in water. Urea plus water is a complicated system, but here the conclusions reached by Walrafen[1376] (Section 2.1.1) following analysis of the Raman spectra are consistent with urea being a structure breaker.[485,509] Indeed a mixture model for water can be used in this context such that urea dissolves in regions of nonbonded water, thereby diluting this "phase," and so some fraction of the water clusters break down to maintain this equilibrium between the two states of water.[509]

ACKNOWLEDGMENTS

We thank Prof. M. C. R. Symons, and Drs. T. A. Claxton, M. J. Wootten, D. Waddington, and D. Jones for many valuable discussions.

CHAPTER 9

Acoustic Properties

M. J. Blandamer

Department of Chemistry
The University
Leicester, England

1. GENERAL INTRODUCTION

When a sound wave passes through a real fluid the intensity of sound decreases with distance traveled. For small pressure amplitudes (and in the absence of cavitation[493]) it can be assumed that the attenuation (damping) is uniform, such that the intensity I is an exponential function of distance, i.e., $I(x = x) = I(x = 0)e^{-2\alpha_a x}$, where α_a is the amplitude attenuation coefficient.[173,205] The absorption per unit wavelength, μ, is given by the product $\alpha_a \lambda$, where λ is the wavelength of sound. For real systems μ is a function of frequency and in a "well-behaved" system a plot of μ against the frequency of sound wave ν has a maximum value μ_{max} at a particular frequency, the relaxation frequency ν_c. The relaxation time τ $[= (2\pi\nu_c)^{-1}]$ describes the dynamics of response of the system to a pressure perturbation under adiabatic conditions, while μ_{max} provides information concerning the extent of response. The frequency of the sound wave can, in principle, be varied over a wide range from a few Hz through the audible ($\nu < 14$ kHz) to the ultrasonic (14 kHz $< \nu \lesssim 800$ MHz) and the hypersonic ($\nu > 800$ MHz) ranges. However, quite formidable experimental difficulties are encountered in making measurements of sound velocity and sound absorption over the complete frequency range. The majority of sound absorption measurements have been made in the ultrasonic range, $1 \lesssim \nu \lesssim 250$ MHz. Nevertheless this is an important frequency range because molecular processes having relaxation times in the range $10^{-6} \lesssim \tau \lesssim 10^{-9}$ sec can be

495

conveniently probed. Of the more widely used techniques, only dielectric relaxation covers this range, although the conditions governing the coupling between perturbation and system are obviously very different. Before examining some of the acoustic properties of aqueous mixtures some of the underlying theory (see Volume 1, Chapter 12) is briefly outlined.[173,205,663]

When a sound wave passes through a system the latter is subjected to a dynamic stress and the response of the system, the strain, can be analyzed in terms of a series of stress–strain relationships.[776] The present discussion will be limited to situations where the pressure wave is a plane dilational wave, the motions of the molecules being along the direction of the propagation.

The elastic properties of a perfectly elastic isotropic medium can be described using two moduli: (a) A bulk modulus K $[= -P/(\partial V/V)]$ describes the change in volume resulting from an applied pressure, P being the acoustic pressure, the difference between the instantaneous and equilibrium confining pressure. (b) A shear or rigidity modulus G describes the resistance of the system to a change in shape. The velocity of a dilational wave equals $[(K + \frac{4}{3}G)/\varrho]^{1/2}$ and the amplitude attenuation coefficient α_a is zero.

Real systems are not perfectly elastic and a sound wave is attenuated. For example, with respect to the bulk modulus, the establishment of an equilibrium volume strain by the compressional wave is not instantaneous, the rate being described by relaxation parameters. In simple systems this relaxation is described by a single discrete relaxation time, characteristic of the system. Another analytical method describes a system using the isentropic compressibilities $\varkappa_s$ when the rate of response of the system to compression is described by the relaxation time τ_{sp}. A relaxation time τ_s describes the dynamic shear characteristics of the system.

The response of a system to a dynamic perturbation, a pressure wave, can be resolved into components which are in phase and those which are $\pi/2$ out of phase with the perturbation. This out-of-phase component results in absorption of energy by the system which is a maximum at the relaxation frequency ν_c. In the following discussion most of our interest concerns those processes involved in the relaxation of the isentropic compressibility. The relaxation times associated with shear moduli are usually very short in the aqueous mixtures discussed in this chapter. A notable exception is the mixture glycerol plus water.[1229] Here the relaxation time τ_s of water can be increased by adding glycerol, e.g., $\tau_s = 0.20 \times 10^{-9}$ sec when the concentration of water is 46%, whereas in pure water τ_s is estimated to be approximately 0.001×10^{-9} sec ($T = 273°K$). In most liquid mixtures the attenuation of sound in the ultrasonic frequency range resulting from the shear contribution can be calculated from the viscosity of the liquid η_s. Thus $(\alpha_a/\nu^2)_s$ can be calculated from η_s using the following relation, where ϱ is the density and c is the velocity of sound:

$$(\alpha_a/\nu^2)_s = 8\pi^2\eta_s/3\varrho c^3 \tag{1}$$

This frequency-independent contribution to the observed α_a/ν^2 is sometimes called the classical absorption. Another contribution to this classical absorption arises from the nonzero thermal conductivity of the liquid, the compression/decompression being not precisely adiabatic. However, for the systems considered here, this contribution to α_a/ν^2 is extremely small and can be ignored. The difference between the measured α_a/ν^2 and the calculated $(\alpha_a/\nu^2)_s$ for a given system is sometimes said to arise from a frequency-dependent volume viscosity. Possible interpretations of this excess absorption are considered in this chapter. In order to show the kind of information which can be derived from this type of experiment, it is useful to outline briefly the absorption properties of a simple system.

It is supposed that the sound wave perturbs a simple chemical equilibrium $X \rightleftharpoons Y$, where X and Y are, say, two conformational isomers which form an ideal liquid mixture. The dependence of α_a/ν^2 on frequency (at fixed temperature and pressure) can be fitted[1131] to the following general equation:

$$\alpha_a/\nu^2 = A[1 + (\nu/\nu_c)^2]^{-1} + B \tag{2}$$

This is the equation of the curve shown in Fig. 1(a). The sum of the rate constants, $k_1 + k_{-1}$, for the forward and reverse conversion processes is calculated from the relaxation time τ ($2\pi\nu_c\tau_{sp} = 1$). The parameter B measures the classical absorption (see above) together with contributions to α_a/ν^2 from processes having relaxation frequencies much higher than ν_c. The parameter A is related to μ_{max}, which in turn is related to the relaxation strength r: $\mu_{max} = \frac{1}{2}Ac\nu_c = (\pi/2)r$. In this example μ_{max} is related to the thermodynamic parameters describing this equilibrium[379,1052]:

$$r = \frac{C_{P(A=0)}}{C_{P(\xi)}} \frac{V}{RT\varkappa_{S(A=0)}} \left\{ \frac{\exp(-\Delta G^{\ominus}/RT)}{[1 + \exp(-\Delta G^{\ominus}/RT)]^2} \right\}$$

$$\times \left[\frac{\Delta V^{\ominus}}{V} - \frac{\alpha_{(A=0)} \Delta H^{\ominus}}{C_{P(A=0)}} \right]^2 \tag{3}$$

where $C_{P(A=0)}$ and $\alpha_{A=0}$ are the equilibrium (low-frequency) heat capacity and expansibility, respectively; $\varkappa_{S(A=0)}$ is the equilibrium isentropic com-

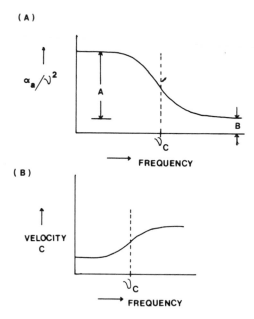

Fig. 1. Dependence of (A) α_a/v^2 and (B) sound velocity c on frequency of sound wave for a system having a single discrete relaxation frequency v_c and relaxation time $\tau_{s,p}$.

pressibility; $C_{P(\xi)}$ is the instantaneous (high-frequency) heat capacity; and $\Delta G^{\ominus} = -RT \ln K$, where K is the equilibrium constant, while $\Delta V^{\ominus}$ and $\Delta H^{\ominus}$ are the differences between volumes and enthalpies, respectively, of reactants and products in their standard states. More complicated equations can be derived for equilibria with different stoichiometries but a sigmoidal dependence of α_a/v^2 on frequency is readily interpreted in these terms even where the system is as complicated as an aqueous mixture (see below). If this approach is used, then we can write the following generalized statements; $\mu_{max} = f(\Delta G^{\ominus}, \Delta H^{\ominus}, \Delta V^{\ominus})$ and $v_c = f(\text{rate parameters})$. Actually even in the simplest systems some simplifying assumptions must be made because eqn. (3) cannot be unambiguously solved for both $\Delta V^{\ominus}$ and $\Delta H^{\ominus}$ quantities. Thus for conformational changes[824,1052,1447] it is often assumed that $\Delta V^{\ominus} = 0$ (i.e., thermally induced changes), although the validity of this assumption now seems doubtful.[351] For equilibria involving ion association/dissociation in water ($T \simeq 277°K$) it is often assumed that the volume change is more important, i.e., the changes are isothermal.[443,1263] Thus even where the interpretation of the absorption and relaxation is straightforward the quantitative analysis is not, especially if account is taken of the nonideality (thermodynamic) of the solution. Further, there may be grounds for questioning the characterization of a system using a single discrete relaxation time, τ_{sp}. Analysis in terms of a distribution of relaxation times may be more attractive.[632] In other instances, a plot of α_a/ν^2 against frequency may show more than one point of inflection, so that the curve can only be fitted using more than one relaxation time. In most instances the derived times describe coupled processes, but if the relaxation frequencies differ by a factor of ten or more, the processes may be treated as independent. In addition to the changes in absorption, the velocity of sound also changes over the same frequency range. However, in comparison to the changes in α_a/ν^2 the change in velocity is often small (Fig. 1b). Generally, velocities are determined at low frequencies with the aim of determining equilibrium isentropic compressibilities $\kappa_{S(A=0)}$.

2. EXPERIMENTAL TECHNIQUES[264,443,584,1052]

A primary objective in this field of research has been the measurement of α_a/ν^2 over a wide range of fixed frequencies for a liquid system at fixed pressure and at a series of fixed temperatures. Indeed, there have been relatively few reports of absorption properties of liquids at pressures other than atmospheric.[122]

A significant proportion of the experimental information discussed later in this chapter has been obtained using the pulse technique,[31,34,60, 438,1085] which can be conveniently used over the range $5 \lesssim \nu \lesssim 250$ MHz. A diagrammatic representation of a liquid sample container is shown in Fig. 2. Different mechanical arrangements and ancillary electronic equipment have been used but the general principles are much the same. An RF signal excites the top transducer, producing an acoustic signal at the desired frequency. This signal passes down the delay rod, through the liquid in the form of a parallel beam (in the Fresnel region), down the lower delay rod, and is reconverted into an RF signal at the lower transducer. This signal, having a smaller amplitude than the in-going signal, is passed through a standard variable attenuator and the amplitude of the final signal is compared with that of the original. The liquid path length is now increased by raising the upper delay rod. The attenuation of the standard attenuator is now diminished such that the final signal has constant amplitude. A plot is produced showing change in path length against change in attenuation, from which α_a is calculated. Because the signals are pulsed,

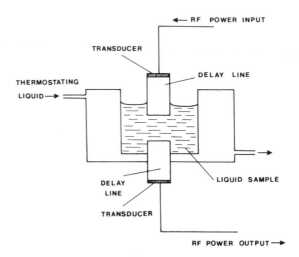

Fig. 2. Typical arrangement of sample container for
measurement of amplitude attenuation coefficient α_a
using pulse technique.

the receiver can discriminate the acoustic signal which has made one pass
through the system from those signals which have made 3, 5, 7, ... passes.
The delay rods increase the time taken for each pulse to travel between
transducers and so the receiver can distinguish between signals received
from the lower transducer and a radiated signal from the oscillator which
would otherwise swamp the detector system. In order to obtain a measurable decrease in signal amplitude at low frequencies, long liquid path
lengths are required and so a lot of sample is required. Further, the sound
wave may no longer propagate as a parallel beam over the complete path
length, resulting in additional loss of signal amplitude at the detector. In
this (Fraunhofer) region corrections must be made for these diffraction
losses.[266,267,438,1102] Under favorable circumstances these corrections can
extend the lower frequency limit of the pulse technique.[108] At the other
end of the range, $v > 250$ MHz, the attenuation of sound by the liquid is
so strong that small liquid path lengths must be used or the signal is lost in
"noise." These small path lengths must be precisely measured and the two
faces of the delay rod (Fig. 2) immersed in the liquid must be kept parallel.
The frequency range can be extended to approximately 800 MHz.[688] A
pulse apparatus can often be used for measurement of sound velocities.[32]

Extensive efforts have been made to develop accurate and convenient
methods for measuring absorption properties below 1 MHz. In the resonance method[989,1054,1417] ($5 \lesssim v \lesssim 50$ kHz) the liquid sample is held in

a spherical glass container which is suspended inside a thermostated container. A transducer fixed to the outside of the vessel is connected to a pulsed RF supply and one of the resonant frequencies of the vessel is intermittently excited. A transducer monitors the decay of the mechanical reverberation; α_a is calculated from the decay time, obtained in some cases by averaging over many resonance decay curves. At higher frequencies, $\nu > 50$ kHz, the resonant frequencies are very close together. Therefore in the reverberation technique[728,729,1006,1008] the pulsed signal excites all resonant frequencies in a band around a selected frequency, e.g., in a band 8 kHz wide centered on 200 kHz. The decay time for each band of frequencies is measured as outlined above. However, despite considerable efforts by experimentalists, these two techniques are not very satisfactory and other techniques have been examined, including acoustic streaming,[1090] comparison,[271] and tuning fork methods.[35] More recently an acoustic resonator technique has been developed[442] for the range 10 kHz $\lesssim \nu \lesssim 50$ MHz. A liquid sample is contained in a cylindrical cell, the ends being sealed with matched transducers. Standing waves (resonant state) are set up in the cell and by measuring the Q of the system, α_a can be calculated. Measurements must be made at frequencies other than the natural frequencies of the transducer. One advantage is that only small quantities of liquid sample are required. However, construction of the cell, particularly the seating of the transducers in the cell, poses several design problems.

   For the other extreme of the frequency range, $\nu \gg 300$ MHz, optical methods have been developed. Two techniques are widely quoted.

   When a sound wave passes through a liquid, regions of compression and dilation are produced at half-wavelength intervals. If a beam of light is passed through the liquid in a direction perpendicular to the progressive sound wave, these regions operate as a diffraction grating.[389] The light intensity in a diffracted order can be measured using, say, a photocell. With increase in signal amplitude the intensity of light in higher orders increases. Consequently, measurement of light intensity in the first-order diffraction as a function of distance along the sound path leads to an estimate of α_a (when $\nu < 80$ MHz).[255]

   The Brillouin scattering technique can be used for measurement of α_a at very high frequencies, e.g., in the GHz range. In a liquid thermally excited acoustic modes produce fluctuations in refractive index. Light (e.g., from a laser source) incident on a liquid is scattered. The Rayleigh-scattered light contains Brillouin components at frequencies both slightly smaller and larger than the centrally scattered frequency. The attenuation coefficient α_a can be calculated from the width of the Brillouin component.[138,221,285,286,544,1129]

3. PRESENTATION OF EXPERIMENTAL RESULTS

With reference to the ultrasonic absorption properties of binary aqueous systems, there are four important experimental variables; (a) composition, (b) frequency, (c) temperature, and (d) pressure. Consequently for a given system, e.g., an alcohol plus water mixture, three types of plots are generally encountered.[203] For convenience, these plots are labeled A, B, and C.

In plot A, α_a/ν^2 is plotted against mixture composition at a fixed frequency and temperature. A typical plot is shown in Fig. 3 for the mixture water plus t-butyl alcohol.[182,184] Families of curves on one figure can summarize the absorption properties of different binary mixtures at fixed temperatures or the same mixture at different temperatures. It is important to state the frequency. If the system has a single relaxation frequency where μ_{max} is dependent on composition but ν_c is not, then a plot of α_a/ν^2 against composition would show no change in α_a/ν^2 when $\nu \gg \nu_c$ but a maximum when $\nu < \nu_c$. In Fig. 3, α_a/ν^2 is plotted against mole fraction of the organic cosolvent, x_2. Quite marked changes in the general appearance of these plots can result if the liquid composition is expressed in another

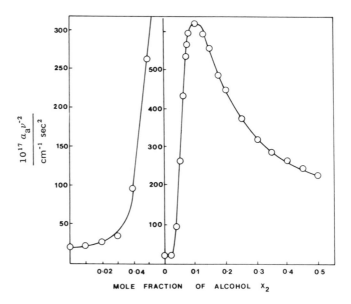

Fig. 3. An example of plot A; dependence of α_a/ν^2 on mole fraction of alcohol x_2 for t-butyl alcohol plus water at 70 MHz and 298°K.

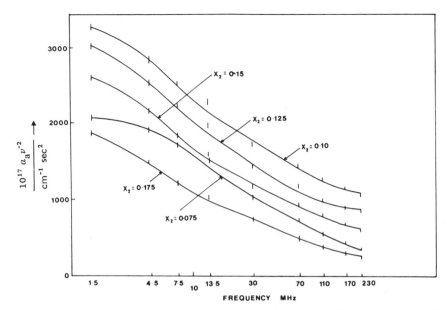

Fig. 4. An example of plot B; dependence of α_a/v^2 on frequency for water plus t-butyl alcohol mixtures at 298°K. The full lines are best-fitted relaxation curves using eqn. (2) modified for two discrete relaxation frequencies. Each successive curve following that for $x_2 = 0.175$ has been displaced by 200×10^{-17} sec² cm⁻¹.

way, e.g., volume fractions. In some reports the dependence of (α_a/v^2) $-(\alpha_a/v^2)_s$ is plotted against liquid composition, where $(\alpha_a/v^2)_s$ is calculated from the shear viscosity [eqn. (1)].

In plot B, α_a/v^2 is plotted against the frequency (or log v) of the sound wave for liquid mixtures at fixed composition and temperature. Typical plots are shown in Fig. 4 for t-butyl alcohol plus water.[184] Analysis of these relaxation curves presents formidable problems (see below). Probably a complete interpretation will only be achieved when data are available for a larger frequency range than are generally available at present. Furthermore, many of the current interpretations may need serious modifications as these data become available, particularly information concerning absorption properties at lower frequencies.

In plot C, α_a/v^2 is plotted against temperature for a system at fixed composition and frequency. These plots are usually constructed for a mixture where α_a/v^2 is at or close to the maximum (plot A, see Fig. 3). An example of this type of plot[199] is shown in Fig. 5, which also emphasizes the change in α_a/v^2 as the temperature approaches a critical solution temperature, an LCST for the system described in Fig. 5.

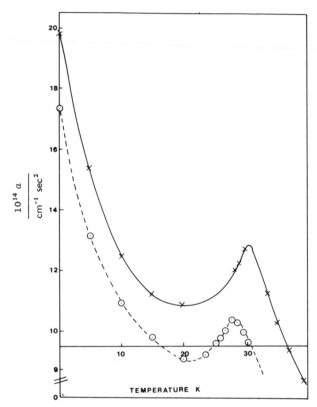

Fig. 5. An example of plot C; dependence of α_a/ν^2 on temperature for water plus [t-amyl alcohol plus t-butyl alcohol, 40 : 60 molar ratio] at fixed frequency, 70 MHz, and mole fraction of water of 0.925 ($\times$) and 0.935 ($\bigcirc$).

4. GENERAL SURVEY OF ACOUSTIC PROPERTIES OF LIQUID MIXTURES

If a liquid shows a strong ultrasonic absorption, then this absorption decreases rapidly when another liquid is added.[663] Indeed, if two strongly absorbing liquids, e.g., benzene and carbon tetrachloride, are used, then α_a/ν^2 for certain mixtures may be less than that for either pure component.[1215,1216]

In 1941 Willard reported[1408] the presence of a maximum in α_a/ν^2 (plot A; $\nu = 10$ MHz) for acetone plus water mixtures at an intermediate composition, $x_2 \simeq 0.5$. Willis confirmed[1413] this and found the same phenomenon in ethyl alcohol plus water mixtures but not in either acetone plus ethyl alcohol or water plus glycerol over the range $3.8 \leq \nu \leq 19.2$

MHz, $T = 299°K$. No velocity dispersion was observed in this range. Burton,[255] in 1948, showed that similar maxima (plot A; 5 MHz and 298°K) were present in binary aqueous mixtures where the organic cosolvent was ethyl alcohol, acetone, t-butyl alcohol, ethyl glycol ether, n- and iso-propyl alcohols,[1217] but not where the cosolvent was methanol, glycol, and dioxan (but see below). Further, no maximum was observed for the mixture t-butyl alcohol plus methyl alcohol. Where there was a maximum in α_a/ν^2 the velocity increased more rapidly than did α_a/ν^2 when the organic cosolvent was first added to water and passed through a maximum at slightly lower mole fractions than the mole fraction at $(\alpha_a/\nu^2)_{max}$. Of the systems examined by Burton,[255] t-butyl alcohol plus water had the most intense maximum in α_a/ν^2 at the lowest mole fraction (see Fig. 3). The composition corresponding to the maximum in α_a/ν^2 is almost characteristic of a given binary mixture. Andreae and co-workers[33] have called this composition the PSAC, peak sound absorption composition, and have summarized PSAC values for a wide range of aqueous systems. In addition, these systems often show a range where α_a/ν^2 of water is insensitive to added cosolvent (see Fig. 3) before α_a/ν^2 rises rapidly to a maximum. Although attention was first drawn to this behavior with reference to t-butyl alcohol plus water mixtures,[182,184] a similar region is indicated in the plots reported by Thamsen[1290] for chloroethanol plus water at 273°K and a series of fixed frequencies over the range $12 \leq \nu \leq 92$ MHz. Nevertheless, most reports of the absorption properties of a particular binary liquid mixture make special reference to the presence of a maximum in plots of α_a/ν^2 against composition. Further, in most cases water is one of the components. Thus as Burton reported, a mixture of two alcohols showed no evidence for a maximum in plot A. Another example[199] is noted in Fig. 6, where α_a/ν^2 is plotted against composition for a series of binary mixtures where diethylamine is the common component. However, it would be wrong to generalize. Several binary mixtures of nonaqueous liquids show maxima in plot A,[663] e.g., ethyl alcohol plus ethyl chloride,[1238] acetic acid plus chlorobenzene,[1091] and t-butyl alcohol plus cyclohexane[302] and various n-alcohol plus organic solvent mixtures.[825]

In the wider context efforts have been made to generalize these observations and to identify some criterion from which to predict the presence of a maximum in plot A. Probably the most successful[1238] notes that maxima occur if the mixture shows a positive deviation from thermodynamic ideality. But it cannot be concluded that if G^E is negative, no maximum in plot A will be observed. Thus for ethylenediamine plus water G^E is negative[1162] but α_a/ν^2 is a maximum[195] at $x_2 \simeq 0.35$ ($T = 273°K$ and $\nu = 70$ MHz).

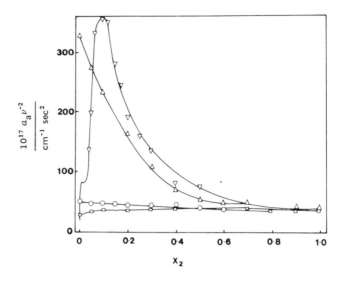

Fig. 6. Dependence of α_a/ν^2 on composition for a series of binary liquid mixtures where the common component is diethylamine: the cosolvents are water ($\bigtriangledown$, $-298°$K and 70 MHz), t-butyl alcohol ($\triangle$, $-296°$K and 30 MHz), ethyl alcohol, ($\bigcirc$, $298°$K and 70 MHz), and methyl alcohol ($\square$, $298°$K and 70 MHz).

   The excess function G^E is negative for many aqueous mixtures and here the sign of G^E is "controlled" by a negative H^E (exothermic mixing), e.g., hydrogen peroxide plus water.[1162] In this group of systems the co-solvent–water interactions dominate the mixing characteristics and the co-solvents can be called "hydrophilic." These are typically nonaqueous systems.[514] For typically aqueous mixtures. $T\,|\,S^E\,| > |\,H^E\,|$, e.g., alcohols plus water, whereas in typically nonaqueous mixtures $|\,H^E\,| > T\,|\,S^E\,|$. Typically aqueous mixtures have a tendency to (and some do) demix at an LCST, while some typically nonaqueous mixtures (with positive G^E) often form a UCST. Indeed, various links have been noted between ultrasonic absorption properties of liquid mixtures and critical phenomena. Several studies[16,301,1218,1451] have shown how the sound absorption of a water-rich mixture of triethylamine plus water (LCST $= 291°$K, at $x_2 \simeq 0.13$) increases rapidly as T_c is approached. McKellar and Andreae[921] used two relaxation frequencies to account for the frequency dependence of α_a/ν^2 for aqueous mixtures containing triethylamine and methyldiethylamine and linked the lower relaxation frequency [e.g., 3.9 MHz when $T = 288°$K and x_2 (triethylamine) $= 0.125$] with some phase separation process. However, the dependence of α_a/ν^2 on frequency for a system close to a critical temperature is probably not given by a sigmoid-shaped curve (see below). The temperature dependence of α_a/ν^2 in these systems is quite interesting. With increase in temperature α_a/ν^2 for water plus methyldiethylamine* decreases initially and passes through a minimum before increasing to large values at and near the LCST (see also Fig. 5). A similar trend is predicted for triethylamine plus water but it is not observed because, it has been suggested,[195] the LCST is low and the freezing point is high. Sette[1219] has drawn attention to the fact that for liquid systems a considerable excess absorption is observed at temperatures some way removed from T_c, in contrast to the behavior near a vapor/liquid critical temperature (see, e.g., Ref. 1194). In the wider context intense absorptions are also observed near the critical temperature (usually UCST) for binary nonaqueous mixtures,[543,1219] e.g., aniline plus cyclohexane[366] nitrobenzene plus n-hexane,[367] and nitrobenzene plus isooctane.[25]

5. WATER

At room temperature and in the MHz frequency range the measured value of α_a/ν^2 for water ($\simeq 25 \times 10^{-17} \text{ cm}^{-1} \text{ sec}^2$ at 293°K) exceeds that calculated from the shear viscosity[663] ($\simeq 8.5 \times 10^{-17}$). The difference is called the structural contribution. As with most associated liquids, α_a/ν^2 decreases with increase in temperature. The associated relaxation has a very short relaxation time and at present this time can only be estimated by extrapolation from the values obtained for glycerol plus water mixtures,[1229] i.e., 2.13×10^{-12} sec at 293°K. A detailed review of the acoustic properties of water[378,663] is given in Volume I. In simple terms the excess absorption can be understood using a simple two-state mixture model for water (see Volume I) which envisages a dynamic equilibrium between open low-density hydrogen-bonded water molecules, $(H_2O)_b$, and dense non-bonded water molecules, $(H_2O)_u$; for $(H_2O)_b \rightarrow (H_2O)_u$, $\Delta V^\bullet < 0$ and $\Delta H^\bullet > 0$. Other models have also been used (see Ref. 381) but the above model suffices. For example, it can be used to account for the effect of added urea on the absorption, α_a/ν^2 of water.

6. AQUEOUS SOLUTIONS CONTAINING UREA

When urea is added to water α_a/ν^2 decreases over the range $0 < C < 8$ mol dm^{-3} at fixed frequency ($18 < \nu < 175$ MHz), the decrease being

---

* J. H. Andreae, unpublished data.

more dramatic at 283 than at 298°K.[39,633] When small amounts of urea are added, $C < 0.5$ mol dm^{-3}, α_a/ν^2 changes slowly but then drops rapidly, the curve flattening out at higher concentrations, e.g., up to 15 mol dm^{-3} at 298°K.[41]

A similar trend is observed for guanidinium chloride.[1035,634] No evidence is found for a relaxation in the aqueous urea solutions over the range $5 \leq \nu \leq 175$ MHz.[39,633] The velocity of sound[41] ($\nu = 5$ MHz) increases with increase in concentration, pressure,[122] and temperature,[1035] but when $C = 8$ mol dm^{-3} the velocity is independent of temperature over the range $293 \leq T < 313$°K.[39]

If the contribution to the observed α_a/ν^2 by the viscosity is subtracted [eqn. (1)], the structural contribution shows a sharp decrease with increase in urea concentration before leveling out. The concentration at which the curve flattens out increases as the temperature is lowered.[122] These trends can be rationalized if urea acts as a water structure breaker. In this respect the effect of added urea is similar to that observed[219] when alkali halides are added to water. The mechanism for this structure breaking by urea is not completely established. One proposal[509] suggests that urea "dissolves" in regions of dense nonhydrogen-bonded water because its molecular shape debars it from entering regions of associated water. Consequently, the non-bonded water is diluted and in order to maintain equilibrium between $(H_2O)_b$ and $(H_2O)_u$, some of the $(H_2O)_b$ species breaks down. The ultrasonic absorption results indicate that urea achieves its maximum structure-breaking effect when $C \simeq 6$ mol dm^{-3} (cf. 3–4 mol dm^{-3} of guanidinium chloride). A similar structure-breaking model accounts for the absorption properties when 1,3-dimethylurea[1183] and acetamide[634] are added to water. However when 1,3-diethylurea is added the structural contribution increases with addition of solute.[1183] This change in behavior has been attributed to an increase in the hydrophobic character of the solute relative to that of urea such that the structure forming action of the ethyl group is dominant. However, the interpretation is probably not so straightforward. Although t-butyl alcohol is a water structure former[514,516] (Chapter 10) in dilute aqueous solution, this in itself does not account (see below) for the increase in α_a/ν^2 when t-butyl alcohol is added to water (see below and Fig. 3).

A similar analysis to that outlined above indicates that the structural contribution to the observed α_a/ν^2 for dimethyl formamide decreases with increase in cosolvent mole fraction even though α_a/ν^2 passes through a small maximum at $x_2 \simeq 0.25$ ($T = 293$°K; $\nu = 70$ MHz).*

---

* T. R. Burdett, unpublished work.

7. GENERAL FEATURES IN THE ABSORPTION PROPERTIES OF TYPICALLY AQUEOUS MIXTURES

7.1. Low Cosolvent Mole Fractions

For the sake of brevity, the mole fraction range over which α_a/v^2 is relatively insensitive to added cosolvent (cf. Fig. 3) is called the plateau. Actually in most systems the measured value of α_a/v^2 depends on x_2, the mole fraction of organic cosolvent. In some cases a slight decrease and in others a slight increase is observed as x_2 increases (see Fig. 3[304]). For t-butyl alcohol plus water there is an appreciable increase in α_a/v^2 on going from $x_2 = 0.01$ to 0.04 but this increase is small when set against the subsequent change. For this mixture the two trends, plateau and sharp rise to a $(\alpha_a/v^2)_{max}$, are readily apparent because $(\alpha_a/v^2)_{max}$ is large and the PSAC small. The presence of a plateau in the case of ethyl alcohol plus water is not so obvious because $(\alpha_a/v^2)_{max}$ is smaller and PSAC is larger at higher alcohol mole fractions. However, because of the possible importance of this feature to the understanding of absorption processes, attempts have been made to quantify the limits of this plateau. For each curve (plot A) two straight lines have been drawn which contain most of the data points on the low-mole-fraction (plateau) and on the rising parts of the curve. The mole fraction corresponding to the intersection of these two lines is taken as the plateau length x_2^* (Fig. 7).[1306] This analysis was not possible for amine plus water mixtures. It seems probable that a "plateau region" is masked by an absorption attributable to an acid–base equilibrium in water.[183,240,443,1447] In these mixtures the increase in α_a/v^2 to $(\alpha_a/v^2)_{max}$ is not so smooth[195] and this indicates a change in the nature of the phenomenon responsible for the absorption over the range $0 < x_2 < $ PSAC. For systems containing triethylamine some part of the absorption may stem from a rotational isomerism[824,1052,1447] (i.e., rotation about C—N bonds) but for aqueous mixtures near the PSAC this contribution is small and can be ignored.[195]

A summary of x_2^* values for various aqueous mixtures is given in Table I. For a given mixture x_2^* increases with decrease in temperature and when H_2O is replaced by D_2O. At a fixed temperature x_2^* increases with decrease in "size" of the alcohol. The velocity of sound in alcohol plus water mixtures at low alcohol mole fractions increases with increase in x_2 at fixed temperature and increases in temperature at fixed x_2. At higher mole fractions the reverse trend is observed (see below). For a given alcohol plus water mixture a family of velocity isotherms cross near mole fractions

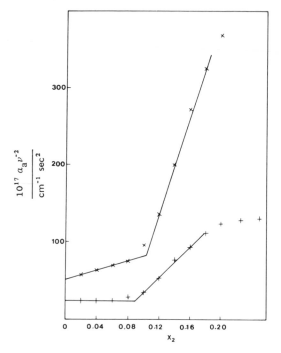

Fig. 7. Dependence of α_a/ν^2 on alcohol mole fraction x_2 for ethanol plus water mixtures at 70 MHz and at 273 ($\times$) and 298°K ($+$). Mole fraction at intersection of straight lines is equated to plateau length x_2^*. (Taken from Ref. 1306.)

close to x_2^*. The crossover appears within a small region[813] rather than at a point as has sometimes been suggested[551,1255] (see also velocities of sound in chloroethanol plus water mixtures.[1289]).

7.2. Region of Excess Absorption: PSAC Region

In this region the velocity passes through a maximum and then decreases with increase in x_2. The velocity decreases (at fixed x_2) with increase in temperature. As the absorption (α_a/ν^2) increases (plot A) appreciable velocity dispersion is observed,[813] the extent of the dispersion being dependent on x_2 and increasing with decrease in temperature.

A summary of PSAC values (i.e., x_2 where α_a/ν^2 is a maximum) for some aqueous mixtures is given in Table II. Actually, the PSAC value for a given mixture may depend slightly on the frequency if the latter is in the range where α_a/ν^2 depends on frequency. For example, [1306] the PSAC

TABLE I. Estimated Values of Plateau Lengths for Various Binary Aqueous Mixtures

			t-Butyl alcohol plus water[304]				
		Water				Deuterium oxide	
T, °K	278	298	318	328	278	298	318
x_2^*	0.049	0.040	0.035	0.034	0.056	0.048	0.043

Cosolvent	T, °K	x_2^*	Ref.
Ethyl alcohol	273	0.106	1306
	298	0.089	1306
n-Propyl alcohol	273	0.056	197, 1306
	298	0.046	197, 1306
Isopropyl alcohol	298	0.057	197, 304
Isobutyl alcohol	298	0.018	304
n-Butyl alcohol	298	0.014	304
t-Amyl alcohol	298	0.016	304
Allyl alcohol	273	0.068	1306
2-Chloroethanol	298	0.046	([a])
2-Bromoethanol	298	0.042	198, 1306
Acetone	298	0.06	([b])

[a] T. R. Burdett, unpublished work.
[b] D. Waddington, unpublished data.

for n-propyl alcohol plus water changes smoothly from $x_2 = 0.10$ to $x_2 = 0.15$ as the frequency is changed from 1.5 to 230 MHz. Nevertheless, Table II highlights some general trends. The PSAC decreases and $(\alpha_a/v^2)_{max}$ increases with increase in hydrophobic character of the organic cosolvent e.g., the PSAC values have the order EtOH < PrOH < t-BuOH. When the methyl group of n-propyl alcohol is replaced by bromine $(\alpha_a/v^2)_{max}$ decreases and the PSAC increases slightly, indicating that the bromine atom in bromoethanol retains appreciable hydrophobic character. If this CH_3 is replaced by either CN to form cyanoethanol or OH to form glycol, no PSAC is observed. In these terms, the shape of plot A can be used as a qualitative indication of the hydrophobic/hydrophilic character of an organic cosolvent.

TABLE II. Some PSAC Values for Binary Aqueous Mixtures[a]

Organic cosolvent	ν, MHz	T, °K	PSAC	Ref.[b]
Alcohols				
Methyl alcohol	23	263	0.3	255, 1218
Ethyl alcohol	22.5	298	0.24	780, 1257
	70	298	0.2	1075 ([c])
n-Propyl alcohol	70	298	0.1	197, 1218
Isopropyl alcohol	70	298	0.18	197, 1218
t-Butyl alcohol	70	298	0.1	184, 1218
Alkyl alcohol	70	273	0.15	198
2-Chloroethanol	70	298	0.15	198
2-Bromoethanol	70	298	0.12	198
1-Chloropropan-2-ol	70	298	0.10	1306
2-Isobutoxyethanol	20.8	298.8	20 wt%	469
Amines				
Dimethylamine	70	273	0.2	([c])
Ethylamine	70	273	0.17	([c])
Diethylamine	70	273	0.09	33 ([c])
Triethylamine	70	273	0.08	([c])
n-Propylamine	30	273	0.10	([c])
i-Propylamine	30	273	0.11	([c])
n-Butylamine	30	273	0.06	([c])
i-Butylamine	30	273	0.06	([c])
s-Butylamine	30	273	0.06	([c])
t-Butylamine	30	273	0.08	([c])
Methyldiethylamine	—	273	0.075	33
1,2-Diaminoethane	70	273	0.35	([c])
1,3-Diaminopropane	70	273	0.25	([c])
n-Amylamine	—	278	0.05	33
Diethylenetriamine	—	283	0.32	33
Triethylenetetramine	—	293	0.23	33
Pyridine	7.5	273	0.05	([c])
2-Methylpyridine	70	273	0.20	([c])

TABLE II. (*Continued*)

Organic cosolvent	ν, MHz	T, °K	PSAC	Ref.[b]
2,6-Dimethylpyridine	70	273	0.40	([c])
Piperidine	70	273	0.10	([c])
N-Ethylpiperidine	70	273	0.10	([c])
Ethers				
1,4-Dioxan	7.5	283	0.5	40, 630
1,3-Dioxalone	70	273	0.35	1306
Tetrahydrofuran	4.5	273	0.3	33
Ketones				
Acetone	7.5	273	0.35	33
	70	273	0.38	([d])
Acetonylacetone	70	273	0.2	([d])
Cyanide				
Methyl cyanide	7.5	298	0.375	44
Others				
Butyric acid	7	298	0.1	1264, 1402
Mono-, di-, and triglycine	—	—	—	631

[a] Taken from compilations[33] made by Andreae *et al.* and by Dr. N. J. Hidden.
[b] Source of data.
[c] N. J. Hidden, unpublished data.
[d] D. Waddington, unpublished data.

Some preliminary measurements have been made of the PSAC values for a mixture of water and two monohydric alcohols.[1306] Figure 8 shows a gradual change in both PSAC and α_a/ν^2 as the ratio n-PrOH/EtOH is changed. The absorption α_a/ν^2 for t-butyl alcohol plus water at 70 MHz, 298°K, and $x_2 = 0.04$ is also sensitive to added salt.[304] No clear-cut pattern emerges in a comparison[304] of the effects of different salts, although it appears that salts with the most marked effects in terms of increasing α_a/ν^2 are also quite effective in diminishing the mutual solubility of glycerol triacetate plus water.[1130]

7.3. Temperature Dependence: PSAC Region

Although the PSAC is relatively independent of temperature, $(\alpha_a/\nu^2)_{max}$ is markedly temperature-dependent. In most cases $(\alpha_a/\nu^2)_{max}$ decreases with increase in temperature (Fig. 9). However, if the temperature is close to an LCST, $(\alpha_a/\nu^2)_{max}$ usually increases with increase in temperature (see above and Fig. 5), although the mixture of water plus 2-isobutoxyethanol seems to be an exception.[469]

7.4. Frequency Dependence: PSAC Region

With increase in frequency α_a/ν^2 decreases for mixtures near the PSAC (cf. Fig. 4). The form adopted in the analysis of the frequency dependence depends in part on the theoretical model employed. However, this dependence is important because it contains dynamic (kinetic) information about the aqueous mixture. For most systems α_a/ν^2 shows a tendency to approach frequency-independent values as the frequency is lowered. However, for

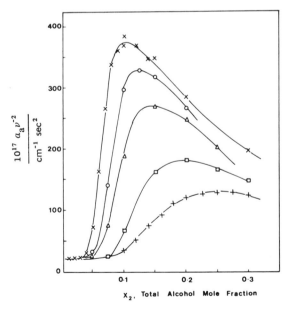

Fig. 8. Dependence of α_a/ν^2 on alcohol mole fraction x_2 at 70 MHz and 298°K. Alcohol comprises n-propyl alcohol plus ethyl alcohol in mole ratios 100 : 0 ($\times$); 70 : 30 ($\bigcirc$); 50 : 50 ($\triangle$); 20 : 80 ($\square$); and 0 : 100 ($+$). (Taken from Ref. 219).

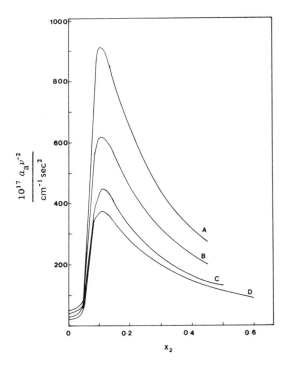

Fig. 9. Dependence of α_a/ν^2 ($\nu = 70$ MHz) on temperature for n-propyl alcohol plus water; $T = 273$ (A), 283
(B), 293 (C), and 298°K (D). (Taken from Ref. 1306.)

near-critical systems α_a/ν^2 generally increases with decrease in frequency,
such that $\alpha_a/\nu^2 \to \infty$ as $\nu \to 0$ (see, however, water plus isobutoxy-
ethanol[469]).

8. TYPICALLY NONAQUEOUS MIXTURES

A system where G^E is negative generally shows no PSAC in plot A.
For example, mixtures of polyhydric alcohols (e.g., 1,2- and 1,3-propane
diols and 1,5-pentane diol) show no PSAC. A similar trend is observed
for aqueous mixtures where the organic cosolvent is ethanolamine, and
cyanoethanol.

The mixture of acetonitrile plus water has a positive G^E because H^E
is large and positive (for further details see Refs. 44, 994, 1162). This
system shows an intense PSAC at $x_2 = 0.38$, α_a/ν^2 increasing sharply as
the temperature approaches the UCST.

9. INTERPRETATION OF ULTRASONIC DATA

In principle, a satisfactory theoretical treatment of the ultrasonic absorption properties of binary aqueous mixtures should account for all the features outlined above. Further, the theory should be able to account for the change in absorption characteristics as the conditions become critical, e.g., as $T \rightarrow T_c$. However, such a detailed theory is not available. Most analyses have concentrated on accounting for the PSAC but these treatments often fail to account for the shape of plot A over the complete composition range $0 \leq x_2 \leq 1$.[1238]

9.1. Low Cosolvent Mole Fractions

9.1.1. *Monohydric Alcohols*

It is informative to contrast the insensitivity of α_a/v^2 to increase in x_2 over the plateau region against the marked sensitivity in other properties in the same region. Here we include both properties of the mixture together with properties of solutes in these mixtures. A few examples serve to emphasize the point (for reviews see Refs. 694, 899, and 1271 and Chapter 8). For example, the velocity of sound increases (see above) while the relative partial molar volume $V_2 - V_2^{\bullet}$ decreases.[516,524] Kinetic studies have pinpointed changes in derived pseudothermodynamic parameters for solvolytic reactions, the change in $\Delta C_P^{\ddagger}$ for t-butyl chloride hydrolysis being a notable example.[1148] Marked changes in spectroscopic parameters[189,1271] are also observed in this plateau range. These and many other observations indicate that alcohol in water $(x_2 \rightarrow 0)$ acts as a water structure former. The actual mechanism of this structure-forming action is uncertain. It may simply involve an effective cooling of the water. Recent nuclear inelastic scattering data have been interpreted in this way,[519] the water being picturesquely described as a "glass-berg." This description is very close the original Bernal and Fowler description[161] of water structure forming. A rather more attractive model (in a conceptual sense) is the liquid clathrate model.[516,564] It is argued that on entering water, alcohol molecules accumulate the void volumes in water and become accommodated in cavities similar to those found for guest molecules in clathrate hydrates[709] (Chapter 3). In so doing, the hydroxyl group hydrogen-bonds into the water lattice forming the cavity wall and the hydrophobic alkyl group is the guest in the cavity, stabilizing the cavity but not specifically interacting with the water. Although no simple solid clathrates containing alcohols are known, molecular models for such systems having t-butyl alcohol as guest show that this hypothesis is certainly a possibility.[513] Dielectric data indicate the existence of an ethanol hydrate.[1114]

If this latter approach is accepted, then a situation is envisaged in which as more alcohol is added, zones of enhanced water structure in the cosphere of each alcohol molecule overlap, mutually reenforcing water–water interactions. Eventually a situation is reached corresponding to maximum enhancement of these interactions and maximum occupancy of cavity sites in the hydrate structure by guest organic cosolvent. This situation is thought to correspond to the end of the plateau. In opposition to these effects is the thermal disordering which militates against the establishment of this clathrate arrangement. Therefore on raising the temperature, the amount of cosolvent which can be accommodated in this way decreases, i.e., the plateau length decreases. At a fixed temperature the available void volume is filled at lower mole fractions by larger organic cosolvents and so $x_2{}^*$ (Table I) decreases on going from ethyl to t-butyl alcohol. Therefore two cosolvents having roughly the same volume for the hydrophobic moiety have roughly the same values for $x_2{}^*$. A comparison[1306] of $x_2{}^*$ values and cosolvent volumes shows that $x_2{}^*$ is smaller for branched chain than for n-alkyl derivatives, indicating that more water is required to build a cage for the latter species. Further, the maximum value for $x_2{}^*$ (possibly for methyl alcohol) is estimated to be 0.17 at 298°K. In this context the mole fraction of vacancies in water which can be occupied by water molecules is thought by some authors to be around 0.17 (e.g., Pauling's model[1080]). There are also indications that when $n \geq 5$ in $C_nH_{2n+1}OH$, $x_2{}^* = 0$. It is noteworthy that Némethy claims that it is impossible to form complete water cages around a hydrocarbon larger than butane.[1024]

9.1.2. Amines

The absence of a real plateau for amine plus water mixtures has been attributed to an absorption stemming from an acid–base reaction (see above). This is unfortunate because there is good evidence for amine clathrate hydrates (Chapter 3).

9.1.3. Acetone

A solid hydrate has been reported[1124] where the composition varies over the range 9–14 wt%. The upper limit of this range agrees closely with $x_2{}^*$ (Table I).

9.1.4. *Acetonitrile*

Although mixtures of water plus acetonitrile show an intense PSAC, there is no dramatic change in α_a/ν^2 with increase in x_2 which defines a value for x_2^* in a clear-cut fashion as found for the water plus alcohol mixtures.[186] The absorption α_a/ν^2 is small at low values of x_2. This lack of definition of a plateau is certainly in agreement with a lack of a minimum in the plot of $V_2 - V_2^{\bullet}$ against x_2,[44] with reports that no clathrate hydrate is formed[1114] and spectroscopic data which shows that methyl cyanide does not enhance water structure.[44]

9.2. The PSAC Region

9.2.1. *Early Models*

Many explanations are to be found in the literature for the presence of an intense absorption at the PSAC. Burton[255] reviewed in 1948 some early explanations. They include explanations based on a hysteresis in adiabatic compressibility and anisotropy of the liquid resulting from molecular groups. Herzfeld noted that the sound wave could produce a local demixing of the liquid mixture, but the estimated absorption is too small.[662] Burton considered the possibility that the absorption was due to liquid crystal formation but rejected it on the grounds that no evidence could be found for such structure in the X-ray diffraction pattern for *t*-butyl alcohol plus water at $x_2 = 0.1$, the PSAC. Burton also considered the possibility that "molecular association" between components was the underlying cause of the absorption but rejected this approach because dielectric data showed no evidence for such association. However, this latter approach has been widely adopted. There is no possibility that the absorption and associated relaxation are simply that of water itself. For example, the required shift in relaxation frequency is too large.

9.2.2. *Chemical Equilibria*

A detailed review of simple equilibrium models has been given by Andreae *et al.*[33] (see also Ref. 824). Prior to this work Storey,[1257] following an examination of the absorption properties of water plus ethyl alcohol, suggested that the absorption could be linked with the formation of a structure or complex in the liquid. Storey emphasized that the term complex indicated a situation having short-range or local order rather than a definite compound. This local order was envisaged as being sensitive to the temperature variation. For ethanol plus water the complex was written as $C_2H_5OH \cdot 4H_2O$. Inflections in the freezing point curves for the mixture were taken as partial confirmation for the formation of the complex. Barfield and Schneider[89] (see also Ref. 1196) accounted for the maximum in α_a/ν^2 for diethylamine plus water mixtures using the following equilibrium:

$$AA + BB \rightleftharpoons 2AB \qquad (4)$$

where A refers to amine and B to water. Thus on compression $A \cdots A$ hydrogen bonds and $B \cdots B$ hydrogen bonds will tend to break and $A \cdots B$ hydrogen bonds are formed which, it was assumed, is accompanied by a decrease in volume. Thus the dominant role was assigned to the change in volume ΔV on the grounds that when A and B are mixed significant changes in volume occur, the latter arising from A–B-type interactions. Barfield and Schneider were able to predict the presence of a PSAC at slightly higher mole fractions than that observed. Following these two proposals Andreae et al. examined the problem in more detail. They accepted that the absorption process was associated with the perturbation of a chemical equilibrium. Actually these authors were concerned with identifying the process responsible for the relaxation where $\nu_c > 20$ MHz. (For aliphatic amine plus water mixtures two relaxation frequencies were used to describe the dependence of α_a/ν^2 on ν and the lower frequency was linked to a phase separation process; see below.) It was argued that when the organic cosolvent was added to water the first effect was to break down the water structure, represented by displacement of the equilibrium $B \rightleftharpoons B^*$, where $B \equiv (H_2O)_u$ and $B^* \equiv (H_2O)_b$. The velocity of sound increases because $(H_2O)_u$ is less compressible than $(H_2O)_b$. With increase in the amount of B (following addition of more A) an equilibrium is established between B and water intermolecularly associated with A, the relaxation frequency for this equilibrium being in the MHz region. In its most general form this latter equilibrium has the following form:

$$A + mB \rightleftharpoons AB_m \qquad (5)$$

In the simplest case $m = 1$. Since the PSAC for acetone is close to $x_2 = 0.5$, it was assumed that for this system $m = 1$. Even with this and other simplifications the analysis is quite complex. The most direct approximation neglects the equilibrium $B \rightleftharpoons B^*$ and simply considers the equilibrium given above where m can be calculated from the experimental data. Thus it can be shown that the PSAC occurs at $(1 + m)^{-1}$, e.g., for acetone $m = 1$ and for diethylamine $m = 9$. A detailed analysis of the absorption properties showed that near the PSAC the absorption was dominated by the enthalpy $\Delta H^{\ominus}$ rather than the volume difference $\Delta V^{\ominus}$, i.e., the sound propagation is not isothermal in these aqueous mixtures.[33]

Hammes and Knoche[630] have used a similar model to account for the ultrasonic absorption properties of dioxan plus water mixtures. Two equilibria, $2H_2O + D \rightleftharpoons D \cdot H_2O$ and $D + D \cdot 2H_2O \rightleftharpoons D_2 \cdot 2H_2O$, where D = dioxan, describe the formation of intermolecular hydrogen bonds. Two equilibria were used because the dependence of α_a/ν^2 on frequency ($10 \leq \nu \leq 185$ MHz at 283 and 298°K) could not be fitted using a single relaxation frequency. Both $\Delta V^{\ominus}$ and $\Delta H^{\ominus}$ terms contribute to the overall absorption. Two relatively large activation energies (20–40 kJ mol^{-1}) were accounted for in terms of an enhanced stability of water–water interactions around the hydrophobic parts of dioxan molecules.

Implicit in the analyses reviewed above is the assumption that in the absence of this equilibrium the mixture would be ideal. If an equilibrium model is accepted, one has to assume that $\Delta G^{\ominus}$, $\Delta H^{\ominus}$, and $\Delta V^{\ominus}$ are independent of composition. An analysis in terms of nonideality of A, B, and AB_m would be too difficult. The precise nature of the species AB_m is not well defined, particularly when m is large, e.g., $m = 19$ for n-amylamine. It is noteworthy that this form of analysis when applied to ethyl alcohol plus ethyl chloride mixtures requires that the complex has the formula Et-OH · 95EtCl. It would seem that this particular form of analysis has reached its limit of usefulness. There is no method of independently confirming derived thermodynamic parameters, e.g., $\Delta V^{\ominus}$ and $\Delta H^{\ominus}$. Further, such equilibrium models do not account adequately for other properties, thermodynamic and spectroscopic, of these binary aqueous mixtures. Despite these criticisms, Andreae et al. drew attention to the importance of water in controlling the acoustic properties of these systems. Indeed the structure breaking of added cosolvent is probably an important feature of mixtures in the region of the PSAC. However, the stress on intercomponent association would suggest that for these systems G^E is negative rather than positive.

A more recent analysis[184,203] has used the equilibrium concept in another way, possibly with even more stress on the role of water. The rise in α_a/ν^2 at the end of the plateau is taken to signify that insufficient water is available to satisfy the demands of the liquid hydrate model. This will occur at lower mole fraction x_2 the larger is the organic cosolvent. Again emphasis is placed on the importance of the hydrophobic nature of the cosolvent. The dependence of α_a/ν^2 on frequency for a wide range of binary aqueous mixtures could not be fitted to a simple relaxation frequency and so two were used for the range $1.5 < \nu \leq 230$ MHz. Nevertheless, the calculated relaxation frequencies and relaxation strengths show similar dependences on mixture composition and temperature.[196] This was taken to indicate that even in the region of the PSAC the absorption process is water-controlled. It is suggested that in these systems water attempts to reestablish a three-dimensional hydrogen-bonded arrangement and achieves this by enforcing the association of cosolvent molecules, this "hydrophobic" association being represented as follows:

$$(ROH)_{aq} + (ROH)_{aq} \rightleftharpoons (HOR \cdot ROH)_{aq}$$

This association may involve one or more steps; alternatively, other equilibria may involve association of three or more organic molecules. With increase in temperature the stability of such associated species may decrease and so the absorption will fall away. Further, the larger the R group, the larger the difference between the two states, and so the absorption will increase. This type of model seems quite attractive for the typically aqueous mixtures but it does not account for the PSAC observed for methyl cyanide plus water. Even for the typically aqueous mixtures there is a notable lack of any quantitative tie-up between the measured absorption properties and other data for these systems. There are nevertheless indications here of a link between the ultrasonic absorption and a tendency for cosolvent association in these mixtures.

9.2.3. Concentration Fluctuations

In a binary aqueous mixture where the mole fraction of cosolvent is x_2 the mole fraction of cosolvent in a small macroscopic volume dV may be slightly smaller or larger than x_2. For a system which is not at or close to a critical point these fluctuations are small because they correspond to an overall increase in the Gibbs function G for the system. The distribution of concentration fluctuations for the whole system will be a characteristic of the system for a given temperature and pressure. When the system is adiabatically compressed a new equilibrium distribution is defined and the rate at which a system achieves this is controlled by diffusion processes in the system. Thus, as for a description of ultrasonic absorption in terms of a perturbed chemical equilibrium, a description in terms of concentration fluctuations involves both a thermodynamic part related to the intensity of absorption and a kinetic part related to the relaxation parameters. Similarly, the power absorption is a consequence of the diffusion control of the response rate to the pressure–temperature variations.

One of the first explanations of ultrasonic absorption in these terms was given by Nomoto.[1034] It was argued that the traveling pressure wave results in breakdown of water clusters, $(H_2O)_b$, to form nonbonded water, $(H_2O)_u$. The cosolvent molecules now diffuse into the region containing $(H_2O)_u$, the relaxation time being controlled by the time required for an equilibrium concentration of cosolvent molecules to be established in this $(H_2O)_u$ region, i.e., τ describes the kinetics of molecular rearrangement.

A more recent theoretical model has been given by Romanov and Solovyev.[1153,1154] Briefly, the analysis assumes that in the absence of concentration fluctuations the molar Gibbs function for a local volume dV of the system equals G_0 where the mole fraction is $x_2°$. In a real mixture the molar Gibbs function at mole fraction x_2 is given by

$$G(x_2) = G°(x_2) + \tfrac{1}{2}g''\langle(x_2 - x_2°)^2\rangle \qquad (6)$$

where $g'' = \partial^2 G/\partial x_2^2$. The quantities h'' and V'' are functions of the molar enthalpy and volume, respectively, i.e.,

$$h'' = \partial^2 H/\partial x_2^2 \qquad \text{and} \qquad V'' = \partial^2 V/\partial x_2^2$$

The relaxation time τ is related to an interaction length l and a diffusion coefficient D by $\tau = l^2/2D$, the overall response of the system being described using a distribution of relaxation times which defines a parameter l_m, the minimum interaction length. The complete expression[186] has the following form:

$$\alpha_a/v^2 = (\alpha_a/v^2)_s + Qf(l_m, D, v) \qquad (7)$$

where

$$Q = (8\pi^2 V^2 kT/cK_s(g'')^2)[(V''/V) - (\alpha h''/C_P)]^2 \qquad (8)$$

where α is the expansibility. Equation (8) shows a close resemblance to eqn. (3) and in the same way it is not possible to calculate from the absorption data alone the values of the volume and enthalpy quantities contained in the square brackets. Romanov and Solovyev showed that the PSAC for ethyl alcohol plus water and n-propyl alcohol plus water could be predicted from the thermodynamic data for these mixtures with fair accuracy using this approach but their calculated values of α_a/v^2 were too small. They noted that for these systems the dominant parameter was V''/V [cf. eqn. (8)]. The major advance provided by this theory is to be found in the definition of Q, eqn. (8). Thus Q can in principle be calculated from the thermodynamic properties of the mixture. However, the analysis is not completely satisfactory because it involves the calculation of second derivatives of G, H, and V parameters. Even with the aid of curve-fitting programs and computer analysis, the derived parameters can only give some idea of the trend rather than an accurate quantitative indication of absorption properties. However, this form of analysis has now been applied to several binary aqueous mixtures, including t-butyl alcohol plus water,[203] acetone plus water,* acetonitrile plus water,[186] and urea plus water.*,†

For the first three systems Q shows a maximum in the region of the PSAC (Fig. 10). It turns out that the plateau region can be linked to extremely large values of g'' [see denominator in eqn. (8)]. However, g'' decreases with increase in x_2 and the Q increases. For the aqueous t-butyl alcohol and acetone mixtures (Fig. 11) the increase in Q and the maximum is produced by more dramatic changes in V''/V than in the enthalpy term, a trend possibly expected for typically aqueous mixtures. The reverse trend (i.e., V''/V is small) is found for acetonitrile plus water mixtures, again possibly anticipated for a typically nonaqueous system.

A complete analysis in terms of eqn. (7) has only been carried out for acetonitrile plus water. Good qualitative agreement is obtained between Q values calculated from thermodynamic data at 298°K and from the absorption at 273°K. As originally formulated and applied to this system, the dependence of α_a/v^2 on frequency is characterized by a distribution of relaxation times, the plot still having a sigmoid shape. The distribution was calculated on the basis of a Debye–type distribution for small wavenumbers. Alternative distributions or discrete relaxation frequencies could be used in the analysis but this would present new analytical and computational difficulties. Nevertheless, the proposed theory was used in an analysis of the data for acetonitrile plus water and as anticipated by the theory, the diffusion coefficient is a minimum at the PSAC, but the precise significance of the calculated interaction length is not clear-cut. On these grounds one can predict that the diffusion coefficients in, for example, t-butyl alcohol plus water pass through a minimum in the region of the PSAC.

With reference to the urea plus water system analysis of published thermodynamic data shows no increase in Q over the range $0.0 < c_2 < 5.0$ mol dm^{-3}.

While the analysis of absorption data using a theory based on a treat-

---

* T. R. Burdett, unpublished work.
† D. Waddington, unpublished data.

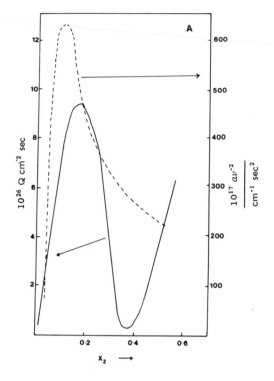

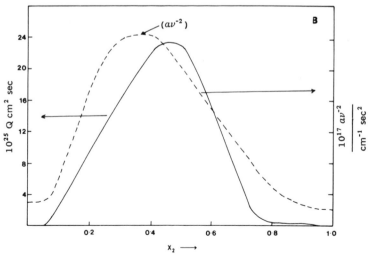

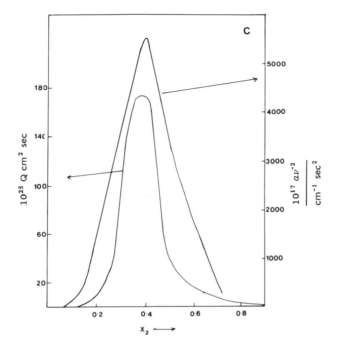

Fig. 10. Comparison of the dependence on mole fraction of the measured α_a/ν^2 ($\nu = 70$ MHz and $T = 298°$K) and the Q parameter [eqn. (8)] calculated from the thermodynamic data for (A) t-butyl alcohol plus water, (B) acetone plus water and (C) acetonitrile plus water ($\nu = 4.5$ MHz and $T = 273°$K for α_a/ν^2).

ment of concentration fluctuations has much to commend it, analysis of these effects using molecular models presents difficulties. However, it is possible that the association processes described in the chemical equilibrium model is in effect describing the same phenomenon. Thus a large value of m calculated from eqn. (5) may mean that each cosolvent molecule effects the arrangement of a large number of water molecules. Such effects may result from the clustering of cosolvent molecules. Actually the evidence in favor of a concentration fluctuation approach to absorption properties is quite strong. A plot of the dependence of the intensity of scattered light on mixture composition is often quite similar in shape to plot A. Recently it has been shown[78] that the small-angle X-ray scattering for t-butyl alcohol plus water has a maximum at $x_2 = 0.12$ ($T = 298°$K). Comparisons were drawn between this phenomenon, and the sound absorption data, and the fact that the X-ray scattering will be large when clustering produces large density fluctuations.

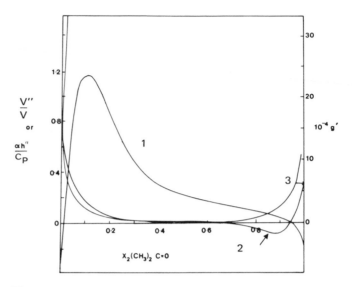

Fig. 11. Dependence on acetone mole fraction ($T = 298°K$) of the functions (1) V''/V, (2) $\alpha h''/C_P$, and (3) g'' which are used in the calculation of Q, eqn. (8).

9.2.4. Critical Phenomena

With increase in temperature the intensity of the X-ray scattering at the maximum also increases, indicating that the extent of clustering also increases.[78] A similar conclusion is also reached with reference to pairwise association of monohydric alcohols in water following analysis of the composition and temperature dependence of the rational solvent activity coefficient.[786] In the limit of extensive clustering, phase separation will occur at an LCST. The question is raised therefore concerning the extent to which the intense absorption at a PSAC reflects a tendency toward phase separation. Several observations back up this view, e.g., the effect of added salt on the magnitude of α_a/v^2 (see above). Further, the PSAC values for many systems, e.g., amine plus water, are close to the compositions at the LCST's even where the LCST is at much higher temperatures than that at which the PSAC is measured.[199]

If this were correct, then one would expect that $(\alpha_a/v^2)_{max}$ at the PSAC should increase with increase in temperature. As noted above, this is not generally so. Such a trend is only observed when the temperature is close to the critical temperature. Then the absorption increases rapidly with increase in temperature. The reasons for the changing temperature dependence of α_a/v^2 are not known.

At the critical temperature the absorption is infinitely large. Indeed, eqn. (7) requires this to be so since a condition of a critical point is that $g'' = 0$. The absorption is too large to be accounted for in terms of an anomalous shear viscosity.[16] For a system close to a critical point the fluctuations in composition are particularly intense; such fluctuations result in only a small change in energy and so the driving force which otherwise would oppose such fluctuations is weak. Two models are generally advanced for the intense absorption near a critical point. The first identifies the absorption as resulting from a "passive" scattering of the sound wave.[233,1230] However, if this were the case then α_a/ν^2 should increase with increase in frequency.[1219] This is not observed. The second model assumes that there is an "active" coupling between system and sound wave. Theoretical analysis of this coupling presents a mammoth problem. Recent experimental work has generally used the model advanced by Fixman.[488,489] This theory[488,489,747] examines the coupling between the oscillating temperature associated with the pressure wave and the fluctuations in heat capacity produced by concentration fluctuations in the neighborhood of a critical point. If these fluctuations are in-phase, no energy is absorbed; absorption of energy results from an out-of-phase component of the fluctuating heat capacity, the rate of response of the system being controlled by diffusion processes. Unfortunately the final equation contains a number of unknown quantities which characterize the system. However, this equation can be approximated to yield the following generalized expression:

$$\alpha_a/\nu^2 = C\nu^{-5/4} + D \tag{9}$$

This equation requires that α_a/ν^2 increases steadily to an infinite value as $\nu \to 0$. Ultrasonic absorption data for several binary systems have been fitted to this equation when the temperature is close to T_c.[199,802] For example, Kruus has reexamined previously published data for triethylamine plus water $(T = 290.2°K, \; T_c = 290.9°K)$ and shown that the data are adequately accounted for using eqn. (9). However, in at least one system α_a/ν^2 does not apparently increase as $T \to T_c$ but decreases steadily with increase in temperature, i.e., the mixture water plus 2-isobutoxyethanol (LCST $\simeq 298°K$). Even at 298°K (homogeneous system) the dependence of α_a/ν^2 on ν $(3.5 \leq \nu \leq 31.8)$ is sigmoid, rather than that required by eqn. (9).[469] For acetonitrile plus water, which has a UCST, the dependence of α_a/ν^2 on ν changes from a sigmoid to a $\nu^{-5/4}$ dependence as $T \to T_c$. (The Romanov–Solovyev treatment predicts a sigmoid dependence and in the limit could yield a single relaxation frequency depending on the form

used for expressing the distribution of relaxation times. As outlined originally a Debye distribution of wavenumbers is employed.) A similar change-
over in frequency dependence for other aqueous mixtures having LCST's
is expected.

ACKNOWLEDGMENTS

I thank Prof. M. C. R. Symons and Dr. N. J. Hidden for many valuable
discussions. I also thank Drs. D. E. Clarke, M. J. Foster, and N. C. Treloar
together with T. Burdett and D. Waddington for their contributions.

CHAPTER 10

NMR Spectroscopic Studies

M. D. Zeidler

Institut für Physikalische Chemie und Elektrochemie
Universität Karlsruhe
Karlsruhe, Germany

1. THEORETICAL BACKGROUND FOR THE INTERPRETATION OF NMR PARAMETERS

In the first volume of this series nuclear magnetic resonance studies on water and ice were reviewed (Chapter 6). A short survey on theoretical and experimental foundations was presented, covering relaxation phenomena and chemical shifts. A large number of books and review articles dealing with these topics exists. For introductory purposes textbooks at different levels can be recommended,[36,269,453,471,648,870,906,1110,1226,1228,1259] while the standard reference book for NMR work is Abragam's text.[1] In the series of "Advances in Magnetic Resonance" basic treatments about relaxation[411,689,1137] and chemical shift[1010] have appeared. Another series, "Progress in NMR Spectroscopy," contains articles which are also of interest for our purpose.[412,656,833]

In the remainder of this section the theory which is necessary for an understanding of the later sections will be summarized briefly.

1.1. Relaxation Rates

An assembly of nuclear magnets, if subjected to a strong external magnetic field, will show a resulting macroscopic magnetization. When the system has gained thermal equilibrium there is only a component parallel to the external field but no perpendicular components exist. Any non-equilibrium state produced by an external perturbation will disappear with time. There will be an approach to the equilibrium and this behavior is termed relaxation. In general one has to observe that this approach to equilibrium can be different for the parallel and perpendicular components of the magnetization, and it must be characterized by different relaxation times T_1 and T_2.

Obviously relaxation has to be brought about by transitions of the nuclear magnets between distinct energy levels which arise from spin quantization along the external field. For such transitions to take place it is necessary for there to be some interaction with the surrounding medium, which is called the lattice (loosely speaking). Only two out of several possible interactions need to be considered for our purposes: dipolar and quadrupolar interaction. For a given relaxation mechanism the basic problem is to relate the relaxation times T_1 and T_2 to molecular dynamic properties.

We will begin by considering the *magnetic dipole–dipole interaction* between two nuclei with identical spin and magnetogyric ratio. The random motion of neighboring nuclei produces a time-dependent magnetic field at the position of a reference nucleus. These motions are described by time-correlation functions, and their Fourier transforms at the resonance frequency are related to the inverse relaxation times, the so-called relaxation rates. The basic formula is

$$1/T_1 = (8\pi/5)\gamma^4\hbar^2 I(I+1)\left[\int_0^\infty \overline{(Y_2^1/r^3)(0)(Y_2^{-1}/r^3)(t)}e^{i\omega_0 t}\,dt\right.$$
$$\left. +4\int_0^\infty \overline{(Y_2^2/r^3)(0)(Y_2^{-2}/r^3)(t)}e^{i2\omega_0 t}\,dt\right] \tag{1}$$

Here γ is the nuclear magnetogyric ratio, I is the nuclear spin, and ω_0 is the Larmor frequency. The functions Y_2^m are the normalized spherical harmonics of second degree and mth order. Their arguments specify the position of the internuclear vector in the magnetic field coordinate system. The above result allows for the variations of the internuclear distance r, i.e., relative translational motion of the spins; therefore we call this rate the *intermolecular relaxation rate*.

In general a correlation function for stationary processes can be rewritten

$$\overline{f(0)f(t)} = \int\int f(y_1)f(y_2) \cdot p(y_1) \cdot P(y_1; y_2, t)\,dy_1\,dy_2 \tag{2}$$

where f is any function of the random variable y and $p(y_1)$ is the probability that at time zero the random variable has the value y_1. The bar denotes the ensemble average. The time dependence of the correlation function is now transferred to the propagator $P(y_1; y_2, t)$, which is the conditional probability (per unit volume) that the random variable has the value y_2 at time t if it was equal to y_1 at time zero.

In most liquids at room temperature the correlation functions decay rapidly compared to the time $2\pi/\omega_0$ and therefore we can put the exponentials in eqn. (1) equal to unity (extreme narrowing case). In this case the relaxation times T_1 and T_2 become equal for the relaxation mechanisms under consideration and therefore we may confine ourselves to just one constant. It is then apparent from eqn. (1) that the relaxation rates measure the integral over the correlation function, and it was shown[685] that the latter does not depend on the order m of the spherical harmonics.

Torrey,[1299] using Chandrasekhar's work,[281] has treated the translational random walk process and thus was able to derive an expression for the propagator P of eqn. (2). Besides the macroscopic translational self-diffusion constant D, the mean time between translational jumps τ appears in the treatment. Inserting for $p(y_1)$ into eqn. (2) the radial distribution function which is correct only if the nuclei occupy central positions in spherical molecules (this way rotational effects on the intermolecular relaxation rate are neglected; for inclusion of this effect see Ref. 684), we obtain, with $P(r_1) = g(r)/V$, where V is the total sample volume,[1051] for the correlation functions in eqn. (1)

$$\overline{(Y_2{}^m/r^3)(0)(Y_2{}^{-m}/r^3)(t)} = (1/V) \int_0^\infty \left[\exp\left(-\frac{2D\varrho^2}{1 + D\tau\varrho^2} t \right) \right]$$
$$\times \left[\int_0^\infty r^{-3/2} g^{1/2}(r) J_{5/2}(\varrho r)\, dr \right]^2 \varrho\, d\varrho \qquad (3)$$

The time integral is easily evaluated and inserted into eqn. (1). The total intermolecular relaxation rate after allowing for the interaction of the $(N' - 1)$ other nuclei with the reference nucleus is then obtained by multiplication with the total number of nuclei in the sample, N':

$$(1/T_1)_{\text{inter}} = 4\pi\gamma^4\hbar^2 I(I + 1)(N/D) \int_0^\infty (1 + D\tau\varrho^2)$$
$$\times \left[\int_0^\infty r^{-3/2} g^{1/2}(r) J_{5/2}(\varrho r)\, dr \right]^2 (d\varrho/\varrho) \qquad (4)$$

where $N = N'/V$ is the spin density. The integral over r involving the Bessel function $J_{5/2}$ can be evaluated only numerically if $g(r)$ is known. However, with a step-function approximation, i.e., $g(r) = 0$ for $r < a$

and $g(r) = 1$ for $r \geq a$, where a is the distance of closest approach, the integrals over r and ϱ can be evaluated in closed form and the result is

$$(1/T_1)_{\text{inter}} = \tfrac{4}{3}\pi\gamma^4\hbar^2 I(I + 1)(N/D)[(D\tau/a^3) + (2/5a)] \qquad (5)$$

Through the relation

$$D = \langle l^2 \rangle / 6\tau \qquad (6)$$

the product $D\tau$ in eqns. (4) and (5) can be identified with the mean jump length $\langle l^2 \rangle$. Further simplification of eqn. (5) is obtained if the jump length is assumed to be very small compared to the distance of closest approach (simple diffusion behavior) so that the first term in the bracket may be neglected.

Looking at the more general formula (4) we see that the experimentally determinable quantity $(1/T_1)_{\text{inter}}D/N$ should be of interest. Changes of this quantity can be attributed to variations in $D\tau$ (jump length) or $g(r)$. The increase of $(1/T_1)_{\text{inter}}D/N$ sometimes observed for nonelectrolytes at low concentration in aqueous solution could be due to preferred association of the solute, which means that the first neighbor peak in $g(r)$ is amplified. A weak point in eqn. (4) is the translational self-diffusion constant which was introduced through the propagator P. Actually the relative diffusion constant between two diffusing molecules is important, but this constant was substituted by $2D$ assuming uncorrelated diffusion, which may be incorrect at very small distances.

If the magnetic dipole–dipole interaction occurs between nuclei within the same molecule, the internuclear distance remains fixed and the *intramolecular relaxation rate* is found from eqn. (1) for the extreme narrowing case:

$$(1/T_1)_{\text{intra}} = 2\gamma^4\hbar^2 I(I + 1)(1/r^6) \int_0^\infty y_2(t)\, dt \qquad (7)$$

Here $y_2(t)$ stand for the normalized [i.e., $y_2(0) = 1$] correlation function of the spherical harmonics the order m of which is immaterial. Equation (7), which was derived for two-spin systems, is valid also for pairwise interaction in multiple-spin systems. There we have for the average rate

$$(1/T_1)_{\text{intra}} = \hbar^2\gamma^4(3/2n) \sum_{\substack{i=1 \\ i \neq j}}^{n} \sum_{j=1}^{n} r_{ij}^{-6}\tau_c \qquad (8)$$

where n is the number of protons per molecule. We have introduced into eqn. (8) the rotational correlation time, which is defined by

$$\tau_c = \int_0^\infty y_2(t)\, dt \qquad (9)$$

i.e., it is the area under the normalized rotational correlation function $y_2(t)$.

The rotational random walk problem was treated by Ivanov[699] and includes the limiting case of simple diffusion studied earlier.[472] In this limit one obtains for the rotational correlation function, using, for example, the propagator valid for anisotropic diffusion (restricted to symmetric top molecules with two components of the diffusion tensor being equal),[1337]

$$y_2(t) = \tfrac{1}{4}(3\cos^2\theta - 1)^2 e^{-6D_\perp t} + 3\sin^2\theta \cos^2\theta\, e^{-[6D_\perp + (D_\parallel - D_\perp)]t}$$
$$+\tfrac{3}{4}\sin^4\theta\, e^{-[6D_\perp + 4(D_\parallel - D_\perp)]t} \qquad (10)$$

Here $D_\perp$ and $D_\parallel$ describe the rotational diffusion perpendicular and parallel to the symmetry axis and θ is the angle between this axis and the internuclear vector. For the special case of isotropic diffusion with $D_\perp = D_\parallel \equiv D_{rot}$ eqn. (10) reduces to

$$y_2(t) = e^{-6D_{rot}t} \qquad (11)$$

and by integration the following relation using eqn. (9) is established:

$$\tau_c = 1/6D_{rot} \qquad (12)$$

In the more general case eqn. (10) is easily integrated and after substitution into eqn. (7) the intramolecular relaxation rate is found.[689] The treatment for internal group rotations is quite similar.[1337,1371,1423]

Since in dielectric relaxation theory the spherical harmonics of first degree are involved, the corresponding correlation function in the diffusion approximation becomes

$$y_1(t) = e^{-2D_{rot}t} \qquad (13)$$

By comparison with eqn. (11), it follows that the dielectric and NMR rotational correlation times are related by a factor of three. The experimentally observed ratio is often discussed in connection with the validity of the diffusion approximation. It may decrease to unity for large-angle rotational random walk.[699]

Although in practically all examples to be discussed later the extreme narrowing condition is fulfilled, we may take a short glance at the frequency-dependent region. If eqn. (11) is fulfilled (single correlation time), it may be substituted into eqn. (1), where r is taken as a constant. Fourier transformation yields a Lorentzian function in the frequency. If plotted against the correlation time, a maximum is obtained for $\omega_0\tau_c = 0.6158$, as proved by differentiation. Thus if the relaxation rate $1/T_1$ is measured for different temperatures and the maximum is observed, the correlation time can be determined directly without a knowledge of the internuclear distance. Furthermore, it can be shown by inserting $\omega_0\tau_c$ at the maximum into eqn. (1) and a corresponding equation for $1/T_2$ that the ratio T_1/T_2 must be 1.6.

So far we have been concerned with the dipolar interaction between nuclei with identical spin and magnetogyric ratio. For interactions between unlike spins the situation is more complex. For our purposes it is merely of interest that, if an unlike spin is substituted in place of a like spin, the relaxation rate is multiplied by a factor

$$\tfrac{2}{3}\gamma_s^2 S(S+1)/\gamma_I^2 I(I+1) \tag{14}$$

where the index s refers to the substituted nucleus and the index I to the observed nucleus.

Electric quadrupolar interaction is usually the predominant relaxation mechanism for nuclei with $I > \tfrac{1}{2}$, since these possess electric quadrupole moments eQ which interact with electric field gradients produced by the surrounding electric charges within the molecule. Random rotational motion of the molecules produces a time-dependent electric field gradient at the nuclear site, and therefore, as in the dipolar case, the relaxation rate is related to Fourier transforms of elements of this field-gradient tensor. The tensor elements expressed in the external field coordinate system are obtained from a transformation of molecule-fixed quantities. If in the molecular system the field gradient has cylindrical symmetry (zero asymmetry parameter), which very often is a good approximation, the problem is considerably simplified. Then only the gradient in the molecule-fixed z axis, q, enters and instead of the total tensor only the position of this vector needs to be specified, which again leads to correlation functions of spherical harmonics. Therefore with the same general definition of the correlation time used in eqn. (9) the quadrupolar relaxation rate in the extreme narrowing case becomes

$$1/T_1 = (3/40)(eQq/\hbar)^2[(2I+3)/I^2(2I-1)]\tau_c \tag{15}$$

Since the correlation times in eqns. (7) and (15) are associated with different vectors within the molecule, they are equal only if the molecule is rigid and moves isotropically. The previous results for rotational diffusion expressed by eqns. (10) and (12) can be introduced into the quadrupolar relaxation rate. More general formulas including the asymmetry parameter but restricted to the diffusion limit are available in the literature.[689,1337]

Thus we notice that the quadrupolar relaxation rate can be used to obtain the area under the correlation function if the quadrupole coupling constant eQq/h is known. Independent measurements of these coupling constants are possible in solids, nematic liquids, and gases.

The quadrupolar relaxation rate is a pure intramolecular rate. Translational motion is not important since the field gradients arise from charge distributions inside the molecule to which the nucleus belongs.

In water and in the organic molecules to be discussed later the following nuclei are contained in natural abundance or isotopic enrichment: ^1H, ^2H, ^{17}O, and ^{14}N. The isotopes ^{12}C and ^{16}O have no spin and cannot be investigated by NMR; ^{13}C resonances have not been reported in the particular systems. Both ^1H and ^{13}C have spin $\frac{1}{2}$ and relax by dipolar interaction (sometimes for ^1H and often for ^{13}C spin–rotational interaction as a competing relaxation mechanism becomes important), whereas the other nuclei relax by quadrupolar interactions.

In proton resonance the intramolecular and intermolecular contributions to the dipolar relaxation rate must be separated. The common technique is the "isotopic dilution method." In this method the neighbors surrounding a given proton-containing molecule are substituted by completely deuterated molecules, i.e., one measures the proton relaxation rate in a mixture of light and heavy compounds and extrapolates to vanishing proton concentration. According to eqn. (14), the intermolecular rate is reduced by a factor of 24. Of course, this method only works if no proton–deuteron exchange is possible. In binary mixtures where one is interested in the intramolecular rate of one of the components, corresponding dilution experiments must be carried out in ternary mixtures containing a fixed ratio of component 1 and 2 but variable ratios of light and heavy molecules of component 1 or 2.

In the binary mixtures to be discussed we can anticipate that in the concentration region where one component dominates the correlation times of molecules of this component vary with the separation from molecules of the other component. That means we have at least two different regions with differing correlation times: the solute-influenced region and the pure solvent region. The observed correlation time will be an average of the regional times and the question is, how must the average be formed? It is clear that the exchange frequency of molecules between the different regions is the decisive factor. If the lifetime τ in each region is very long compared to all correlation times, then they all contribute in proportion to their occurrence, but if the lifetimes are short, the shorter correlation time must dominate, since it corresponds to the fastest decay of the rotational correlation function. Thus we have

$$\tau_c = \sum_i x_i \tau_{c_i} \qquad \tau \gg \tau_{c_i} \quad \text{(long lifetime)}$$

$$\tau_c^{-1} = \sum_i x_i \tau_{c_i}^{-1} \qquad \tau \ll \tau_{c_i} \quad \text{(short lifetime)}$$

(16)

where x_i is the fraction of molecules in the ith region. Since the experimental quantity which provides the correlation times is the relaxation time and since they are inversely proportional to each other the two equations in (16) must be interchanged if the average relaxation time is sought. Of course, this averaging is allowed only for lifetimes much shorter than the relaxation time, otherwise the magnetization no longer approaches its equilibrium value exponentially but is a superposition of several exponentials with different relaxation times. More elaborate theories are necessary for intermediate conditions or if correlation loss during the exchange is to be included.[28,124,655,1457]

In concluding this section on relaxation rates a few remarks should be made about their experimental determination. For direct observation of the approach of the magnetization toward equilibrium pulse methods or adiabatic fast passage experiments are employed. Pulse experiments are divided according to their selectivity or resolution: Nonselective experiments perturb nuclei of the same type irrespective of any nonequivalence which would show up in a high-resolution spectrum. If experiments are performed in an inhomogeneous magnetic field, no separate information for nonequivalent nuclei is obtained,[268,619] but experiments in homogeneous fields do contain this information (Fourier-transform spectroscopy).[1349] On the other hand, selective pulse methods aim at quite distinct equivalent nuclei.[530] Also, in the adiabatic fast passage method separate relaxation times of individual lines in a high-resolution spectrum are obtained.[435,1119] Besides these direct observations, linewidth measurements in a high-resolution spectrum yield the relaxation rate. The relation is

$$1/T_2 = \pi \nu_{1/2} \qquad (17)$$

where $\nu_{1/2}$ is the full width at half-height of the line in frequency units. These data are more open to doubt than direct T_1 results since broadening effects from different sources are possible. Besides instrumental effects, chemical exchange processes[861] resulting in partial collapse of chemical-shifted lines or spin multiplets can be the cause.

1.2. Spin-Echo Self-Diffusion Constants

If the spin system is perturbed by an oscillating magnetic field pulse such that perpendicular components of the magnetization are produced, these decay either with a time constant T_2 or due to magnetic field inhomogeneities (different Larmor frequencies) whichever process is faster. The latter decay is, however, a reversible process and a second appropriate pulse can restore the magnetization in the plane perpendicular to the magnetic field. This phenomenon is called "spin echo." If spin echos are produced at varying times after the first perturbing pulse, their (maximum) amplitude decays with the time constant T_2. A second irreversible process which causes a decrease of the echo amplitude is the self-diffusion of the molecules which contain the nuclei under resonance.

In order to have defined conditions, an artificial inhomogeneity, usually a linear field gradient perpendicular to the axis of the sample tube, is superimposed on the (comparatively) homogeneous static magnetic field. This field gradient may be applied permanently (steady gradient) or only at certain times between the two pulses and between the second pulse and the echo (pulsed gradient). An important condition for the echo maximum to appear at 2τ, where τ is the spacing between the two pulses, is that the area under the gradient versus time curve remain constant for the two time sections before and after the second pulse.

Either by adding to the Bloch equations the macroscopic diffusion term (Fick's second law)[1,1300] or by using the random walk method,[268] the reduction of the echo amplitude by the diffusion process is calculated. The result for the pulsed gradient is[1251]

$$\ln[A(2\tau)/A(0)] = -\gamma^2 D g^2\, \delta^2(\varDelta - \tfrac{1}{3}\delta) \tag{18}$$

where $A(t)$ is the echo amplitude at time t, D is the translational self-diffusion coefficient, g is the amplitude of the linear field gradient, δ is the gradient pulse duration, and $\varDelta$ is the time separation between the gradient pulses at equivalent points. With $\delta = \varDelta = \tau$ the expression for the steady gradient is obtained. Steady gradients can be applied only with amplitudes that are not too large; therefore diffusion constants below $10^{-7}\ \mathrm{cm^2\ sec^{-1}}$ cannot be measured by this technique. Either g or τ may be varied to determine D; in the latter case the echo decay due to relaxation must be eliminated by comparative measurements with $g = 0$. Separate determinations of the gradient amplitude are possible from the echo shape, given by

$$A(t) = A(0)(2/\gamma rgt)J_1(\gamma rgt) \tag{19}$$

where $A(0)$ is the maximum amplitude, r is the radius of the sample tube, which must be cylindrical, and J_1 is the first-order Bessel function. From the first zero point of the Bessel function we find that

$$g = 3.83/\gamma r t_1 \qquad (20)$$

and the time separation between the echo maximum and the first zero point t_1 can be determined experimentally.

The method described for the determination of D, the self-diffusion constant, is a macroscopic one. The diffusion time is given approximately by the gradient pulse separation and a typical value is 10 msec.

One possible relation between the diffusion constant and microscopic properties, for example, the mean jump time, was already presented in eqn. (6). Another equation closely related to eqn. (6) defines D in terms of the velocity correlation function[865]:

$$D = \int_0^\infty \overline{v_x(0) \cdot v_x(t)}\, dt \qquad (21)$$

The correlation function may be normalized by dividing by $\overline{v_x(0)^2} = kT/m$, where m is the mass of the diffusing particle.

If we again have the situation that an observed diffusion constant is an average over different regions in the liquid mixture, the lifetime in a certain region relative to the velocity correlation time [to be defined as the area over the normalized velocity correlation function in analogy to eqn. (9)] is important. The velocity correlation time is of the order 10^{-13} sec, and if the lifetimes are much longer, the slow exchange case is applicable (due to negative contributions in the velocity correlation function, the area is smaller than the actual time for correlation loss, which may be around 10^{-12} sec). According to the first of eqns. (16), the correlation times must be directly averaged, and since D is directly proportional to the correlation time, as evident from eqn. (21), we conclude

$$D = \sum_i x_i D_i \qquad (22)$$

A more detailed derivation may be found in Ref. 655.

1.3. Chemical Shifts

Induced magnetism is responsible for the fact that the magnetic field at the nuclear site is different from the applied magnetic field. Screening as well as amplification effects can occur. The induced magnetic field is of course proportional to the applied field (the same is true for the local field at the nucleus) and the proportionality constant is introduced as "screening constant." The chemical shift then is defined as the difference between screening constants.

Shifts are determined relative to a specified reference resonance. If the screening of nuclei becomes more effective than in the reference state, i.e., if the magnetic field strength at the nuclear site is decreased, a larger magnetic field must be applied to fulfill the magnetic resonance condition. In a high-resolution spectrum obtained by varying the field, the resonance line will lie "upfield" from the reference signal. Correspondingly, a "downfield" shift means decreased screening with respect to the reference nucleus.

In order to measure shifts, the reference signal and the spectrum of the studied sample must be displayed simultaneously. In the aqueous mixtures to be discussed the water proton shift is often measured with respect to some line in the solute spectrum. Since both substances are enclosed in the same sample tube, this constitutes the case of an internal reference. Of course, this method of internal reference is satisfactory only if no specific interactions occur between the two components; otherwise the shift of the reference signal itself obscures the results. Addition of a third component not participating in specific interactions, for example, the frequently employed tetramethylsilane (TMS), might be an alternative. Another possibility is the use of an external standard, which eliminates solvent effects with the reference compound, but the need for bulk susceptibility corrections decreases the reliability. Here sample and reference are enclosed in separate tubes (coaxial-tube arrangement), and because of nonspherical shape of the tube and differing susceptibilities, the magnetic fields induced in the bulk material around the investigated nuclei also differ. For a cylindrical tube whose length is large compared to the radius this effect is equivalent to an additional contribution $(2\pi/3)\chi_V$ (χ_V is the volume magnetic susceptibility, which is negative for diamagnetic substances) to the screening constant, i.e., screening is decreased in diamagnetic and augmented in paramagnetic substances. For unambiguous results magnetic susceptibilities should be determined experimentally. This is especially true for the mixtures described in this chapter. A less satisfactory procedure is to calculate the susceptibilities of the pure components from Pascal's values and then assume ideal behavior, i.e., linear concentration dependence, in the mixtures.

Besides the long-range bulk susceptibility effect, there are other short-range interactions with neighboring molecules which give rise to resonance shifts.[248,833,1133] These include nonspecific effects due to the reaction field or van der Waals forces. Although we are not interested in them, it is not possible to correct for these shifts.

The specific process in the aqueous solutions which is of greatest interest is the formation or breaking of hydrogen bonds either between water molecules or between water and the other component. There is ample evidence from a considerable amount of experimental data, e.g., temperature- or solvent-dependent measurements on hydrogenous systems, that formation of a hydrogen bond is associated with a downfield shift.[856] This is equivalent to a decreased screening around the particular proton. Several effects may be responsible for this variation in the screening constant. The major contribution stems from the action of the electrostatic field associated with the lone-pair electrons of the donor atom,[947] which may be envisaged as attracting the hydrogen nucleus away from its bonding electrons.

In several of the investigated aqueous mixtures protons are exchanged between the solute and the water, for example, hydroxyl protons of alcohols or carboxylic acids. Whether separate resonance signals or only an averaged single line are observed for the proton in its various sites depends on the exchange frequency.[861] In the slow exchange case, i.e., if the exchange frequency is very small compared to the chemical shift (measured in Hz) between the two signals corresponding to the different sites, separate signals are found. At the other extreme of fast exchange the average line is given by

$$\delta = \sum_i x_i \delta_i \tag{23}$$

where δ is the chemical shift.

2. EXPERIMENTAL RESULTS AND INTERPRETATION OF DATA

2.1. Aqueous Solutions of Nonelectrolytes

2.1.1. Alcohols

Aqueous solutions of alcohols have been investigated quite thoroughly. Among these the methanol/water, the ethanol/water and the t-butanol/water systems have received most attention. Next in order are the normal alcohols propanol and butanol. Some results also exist for higher alcohols and polyhydric alcohols.

There exists one severe drawback in alcohol/water mixtures: the exchange possibility of the hydrogen atoms between the alcohol hydroxyl group and the water. For sufficiently rapid exchange one finds in the high-resolution proton (or deuteron) spectrum only a single line which is a weighted average of the alcohol hydroxyl peak and the water peak. Under favorable conditions, such as low temperature or special composition ranges, separate lines may be observed. However, in the fast exchange region the individual effects of the mixture components cannot be determined unambiguously. Moreover, even in the range of line separation all nonselective pulse experiments for relaxation investigations (and they have practically all been performed by nonselective methods) suffer from the same disadvantage since proper isotopic substitutions are possible only for nonexchanging nuclei. Here selective or Fourier-transform pulse experiments or the adiabatic fast passage method should lead to interesting results. But at present there are no separate proton or deuteron relaxation data available for the aqueous component and the alcohol hydroxyl group. Therefore one has to turn to the neighboring and nonexchanging ^{17}O nucleus in order to gain information on the individual behavior of the mixture components.

Turning now to the individual systems, we may first consider *t-butanol/water*, since this has proven to be the most interesting example. Relaxation data are available from Glasel's[558] work on the deuteron resonance in $(CH_3)_3COH/D_2O$ mixtures and from Goldammer et al.,[581,582] who measured the proton resonance in $(CH_3)_3COD/D_2O$ and the deuteron resonance in $(CD_3)_3COH/H_2O$. Goldammer and Hertz[581] also report a value for the limiting slope of the proton relaxation rate in $(CD_3)_3COH/H_2O$. This latter value and Glasel's data are not unambiguous, because of the hydrogen exchange referred to above. Still, we may state safely that the addition of *t*-butanol to water increases the relaxation rate and consequently the correlation times of water molecules, i.e., the molecular motion of the aqueous component in the water-rich region slows down.

No such ambiguity exists for the relaxation data concerning the organic component. By measuring both the proton and deuteron resonances of the appropriate isotopes, Goldammer and Zeidler[582] could separate out the intramolecular relaxation rate over the whole composition range and thus obtain accurate data for the rotational correlation times. They showed that the rotational correlation time of the methyl protons decreases monotonically from 8 psec in pure *t*-butanol to 4 psec in the infinitely dilute mixture at 25°C. An important experimental finding is that the deuteron relaxation rate follows exactly the intramolecular proton rate as seen from a common extrapolation point for the deuteron curve and the total proton rate at vanishing *t*-butanol concentration (neglecting isotope effects) or by independent ternary mixture experiments. The consequence then is that within experimental error no change of the quadrupole coupling constant for the C—D deuterons is observed and that the ratio of proton and deuteron correlation times is constant. This latter conclusion needs to be further discussed. The rotational correlation time of the methyl hydrogens depends on the molecular motion of the whole molecule and on the internal motion of the methyl groups. That such an internal motion exists in pure *t*-butanol was proved by measuring the hydroxyl deuteron rate and comparing it to the methyl deuteron rate.[582] These rates are so very different (43.1 sec^{-1} and 5.75 sec^{-1} for hydroxyl and methyl deuterons at 25°C, respectively) that the difference in quadrupole coupling constants alone cannot account for this effect. If the hydroxyl deuteron rate represents the slow motion (long correlation time) of the whole molecule, then the short correlation time of the methyl deuterons points to a fast internal motion. Of course, the intramolecular proton rate also reflects this internal motion of the methyl groups, but to a different degree, because the correlation times of the corresponding vectors (proton–proton vector for proton relaxation and field gradient vector for deuteron relaxation) depend on the position of these vectors relative to the internal rotation axis. A constant ratio of proton and deuteron correlation times therefore means that the internal motion is not affected in going from pure *t*-butanol to infinitely dilute *t*-butanol in water. Furthermore, one might expect that the concentration dependence of the molecular motion of the whole molecule is reflected by the methyl hydrogen correlation time. Thus it seems that the slow motion in pure *t*-butanol which is due to hydrogen-bonding between alcohol molecules is replaced by increasingly faster rotational motion on dilution with water.

From the data of Goldammer and Zeidler the intermolecular relaxation rate of the methyl protons $(1/T_1)_{inter}$ is also obtained. At the same time they present values for the translational self-diffusion coefficient D of *t*-butanol in water and so the interesting experimental quantity $(1/T_1)_{inter}D/N_H$ where N_H is the proton concentration expressed in protons per cm^3, is known as a function of composition [compare eqn. (4)]. Within the experimental error this quantity seems to be a constant, although an increase at very high water content could be possible. As Goldammer and Hertz[581] discuss in detail, this increase could be evidence for preferred association between *t*-butanol molecules through the nonpolar groups in this water-rich range. They have added experimental data at 0°C to determine $(1/T_1)_{inter}$ $\times D/N_H$ more accurately but still did not obtain conclusive results.

Another question discussed by Goldammer and Hertz is whether in the water-rich range, where an increase of the rotational correlation time of water molecules is certain, stable hydration spheres exist around the alcohol molecules. If the alcohol and its water cage form a rigid (apart from internal methyl group rotation) and long-lived complex (compared to the rotational correlation time) and in addition this complex rotates isotropically, then different vectors in the complex must have equal correlation times. Thus one could compare the correlation time of the hydroxyl deuteron field gradient vector with that of the proton–proton vector of an adjacent water molecule. As should be clear from the above discussion, both quantities are difficult to obtain and are not yet known definitely. Still, it is tempting to make some quantitative estimates. In order to obtain the water correlation time in the hydration sphere, the authors start from a two-region model and assume the slow exchange case represented by the first of eqns. (16). With the reported limiting slope of 0.30 molal^{-1} for the relative proton relaxation rate of water upon t-butanol addition, an estimated coordination number of 25 for the hydration sphere, and the correlation time of pure water of 2.5 psec,[582] a correlation time of 4.1 psec is found. On the other hand, the correlation time of the methyl protons of the t-butanol at infinite dilution was 4.0 psec. The hydroxyl deuteron correlation time is definitely longer (by a factor of six[582] if judged by the pure t-butanol) and therefore the two correlation times to be compared are of different magnitudes.

A similar conclusion was reached by Goldammer and Hertz from their translational self-diffusion data. They studied the diffusion coefficients at 0°C for both components using the proton resonance in the mixtures $(CD_3)_3COH/H_2O$ and $(CH_3)_3COH/D_2O$ between 0 and 14 m t-butanol. Both diffusion coefficients decrease with increasing t-butanol concentration. Neglecting complications due to hydrogen exchange but correcting for the viscosity difference between H_2O and D_2O leads to the result that the self-diffusion coefficient of t-butanol is much lower than that of water over the whole concentration range investigated. From a coordination number of the hydration sphere of 25 a critical concentration of 2.2 m is estimated above which all the available water molecules constitute "hydration water." Even above this concentration the two diffusion coefficients differ. Differing self-diffusion coefficients of course mean that no long-lived (compared to the velocity correlation decay time of 10^{-12} sec) hydration cages exist. However, in the very dilute region ($<0.1\ m$) long-lived hydration cages could exist. Here one can split the observed self-diffusion coefficient of the aqueous component into contributions from the hydration water and from the free water according to eqn. (22). Assuming at various concentrations that the hydration water has the same diffusion coefficient as the solute, then the contribution of the free water at these concentrations can be calculated. Above 0.1 m unrealistically large values, exceeding the pure water self-diffusion coefficient, are found and therefore such a separation must be rejected, i.e., no equal diffusion coefficients of solute and hydration water are acceptable at these concentrations.

It has been an open question for a long time whether the increase of the rotational correlation time or the decrease of the self-diffusion constant of water upon addition of nonpolar solutes in the water-rich region is evidence for an enhancement of water hydrogen bonding. If this is the case, then chemical shift measurements of the water hydrogen should yield downfield shifts, since hydrogen bond formation decreases the electronic shielding around the hydrogen nucleus. It is now known that the *t*-butanol/water and *i*-propanol/water systems show the largest effects of all cases where such downfield shifts have been observed.

Chemical shift measurements for the *t*-butanol/water system were reported by Mavel[957] at 25 MHz and 25°C, by Glew *et al.*[573] at 60 MHz and 0°C, and by Symons *et al.*[29,200] at 60 MHz and different temperatures. More recently Wen and Hertz[1388] have undertaken more systematic studies at 60 MHz and 0°C. Figure 1 is taken from their work and shows the chemical shift of water protons relative to the methyl protons versus the concentration of *t*-butanol at 0°C. Due to hydrogen exchange, only a single peak is observed for water and hydroxyl protons at low *t*-butanol concentration. This coalesced peak at first moves downfield but soon moves

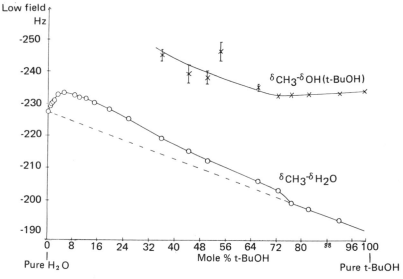

Fig. 1. Chemical shift, in Hz, of the water and hydroxyl protons in a *t*-butanol/water mixture relative to the methyl protons as a function of added *t*-butanol at 0°C and 60 MHz. Below 32 mole % both signals have coalesced due to rapid proton exchange. (From Wen and Hertz.[1388])

upfield with increasing *t*-butanol concentration, exceeding the pure water chemical shift (which of course is an extrapolated value). The maximum downfield shift becomes larger and the concentration at which it occurs (around 5 mole % *t*-butanol at room temperature) moves to slightly higher values as the temperature is lowered. Symons *et al.* found this downfield maximum even at 40°C, but Mavel missed this effect, since he did not look carefully enough at the water-rich region. Beyond a certain concentration of *t*-butanol it is possible to observe the hydroxyl proton separately from the water protons at much lower field. This critical concentration at which the peaks separate can be shifted to lower values by decreasing the temperature, which is equivalent to lowering the hydrogen exchange frequency, or by increasing the resonance frequency so that the shift in Hz becomes large compared to the exchange frequency. At 25°C Symons *et al.* report a critical concentration around 30 mole % *t*-butanol (60 MHz), whereas Mavel's value is 72 mole % (25 MHz). The lowest critical concentration is 20 mole % *t*-butanol at −15°C and 60 MHz as reported by Wen and Hertz. Above the critical concentration both separate signals move further upfield as more *t*-butanol is added.

In most shift measurements the methyl protons of the *t*-butanol were used as internal reference, only Symons *et al.* used chloroform as an external reference. The latter authors also investigated the methyl proton chemical shift relative to external (chloroform) and internal (tetra-N-ethylammonium chloride) standards. They obtained satisfactory agreement for both methods. Their important result is that below 10 mole % of *t*-butanol there is no systematic shift of the methyl proton signal, i.e., methyl groups are not sensitive to the structural change of the surrounding water. At higher *t*-butanol concentrations the methyl protons shift upfield.

It is clear that the observed downfield shift of the coalesced peak around 5 mole % of *t*-butanol cannot immediately be interpreted as an enhancement of water–water hydrogen bonding. The coalesced peak includes the effect of the hydroxyl proton which was termed "solute proton effect" by Wen and Hertz. The solute proton effect can be corrected for by extrapolating the separated hydroxyl proton signal from beyond the critical concentration down to low concentrations of *t*-butanol. At lower temperatures this correction is relatively easy to apply. It is much more difficult to estimate the so-called "polar group effect," which is the water proton shift arising from the fact that one hydrogen atom of a water molecule is now hydrogen-bonded to the oxygen atom of the alcohol molecule rather than to another water molecule. Only after subtracting these two contributions can the proton chemical shift ascribable to the effect of the nonpolar group of t-butanol on the water structure be obtained. Wen and Hertz estimated the polar group effect from hydroxyl proton shifts of alcohols ROH upon addition of t-butanol, where R represents t- or n-butyl, iso- or n-propyl, ethyl, and methyl. In these t-butanol/ROH mixtures the hydroxyl proton signals for both components are separated. As the solvent is changed from butanol to methanol the solvent medium for t-butanol becomes more polar resulting in more pronounced downfield shifts of the ROH signals. Extrapolating this series to a hypothetical alcohol with no carbon atom should, according to Wen and Hertz, yield the polar group effect in water. Since both the solute proton and the polar group effects yield downfield shifts, the influence of the t-butyl group is considerably smaller than the total observed shift. However, a maximum downfield shift of about 3 Hz at 0°C and 60 MHz results but this maximum is now at a lower concentration (around 3 mole %) of t-butanol. It is instructive to recall that the change in proton shift on passing from water vapor to liquid water at 0°C is 4.66 ppm downfield (see Volume 1, Chapter 6), i.e., much larger than the nonpolar group effect on the water structure.

Next we will treat the systems *methanol/water* and *ethanol/water*. Relaxation data were published by Hertz and Zeidler[659] for the deuteron resonance in CH_3OH/D_2O and C_2H_5OH/D_2O, by Fister and Hertz[487] for the ^{17}O resonance of the aqueous component, by Goldammer and Zeidler[582] for the proton resonance in CH_3OD/D_2O and C_2H_5OD/D_2O and the deuteron resonance in CD_3OH/H_2O and C_2D_5OH/H_2O, and by Goldammer and Hertz[581] for the deuteron resonance in CH_3OD/D_2O, C_2H_5OD/D_2O, CD_3CH_2OH/H_2O, and CH_3CD_2OH/H_2O and for the ^{17}O resonance of the alcoholic component. The latter authors performed temperature-dependent measurements and report activation energies for different relaxation rates as a function of composition. In the other publications the temperature was fixed at 25°C. Clifford and Pethica[315,317] reported proton relaxation rates in C_2H_5OH/H_2O mixtures at 32°C. Unambiguous information is obtained for the aqueous component from the ^{17}O relaxation rates and for the alkyl group of the alcoholic component from proton and deuteron rates. In principle, the ^{17}O rates of the alcoholic component should yield definite information on the alcoholic hydroxyl group but Goldammer and Hertz point out that in the pure alcohols CH_3OD and $CH_3{}^{17}OH$ the information from the deuteron and that from the ^{17}O rates are so different that the latter cannot reflect the motion of the hydroxyl group. This discrepancy has been resolved by Versmold and Yoon,[1337a] who detected that the linewidth, and consequently the relaxation rate for ^{17}O, as measured by Goldammer and Hertz, is too large due to an unresolved doublet arising from spin–spin coupling with the hydroxyl proton. They were able to resolve the doublet and fit the measured curve to the theory of spin-multiplet collapse by chemical exchange.[861] Their best parameters for methanol are 86 Hz for the coupling constant, 0.1 sec for the mean lifetime of the hydroxyl proton between exchange events, and 240 sec^{-1} for the relaxation rate $1/T_2$ at 25°C. For the ratio of the relaxation rates of the ^{17}O and deuteron with identical correlation times we therefore expect [see eqn. (15)]

$$\frac{(1/T_2)_{\text{O-17}}}{(1/T_1)_{\text{D}}} = \frac{8}{125} \frac{(eqQ/h)^2_{\text{O-17}}}{(eqQ/h)^2_{\text{D}}}$$

where the spins of ^{17}O ($\frac{5}{2}$) and D (1) were inserted. With the value 3.20 sec^{-1} at 25°C for the deuteron rate[582] we expect a ratio of 34 for the quadrupole coupling constants. Taking the known data for liquid water (no results for methanol are available), i.e., 7.7 MHz for ^{17}O and 230 kHz for D,[1242] the ratio is 33.5, which is in complete agreement with the experimental finding in methanol. We might therefore argue that the quadrupole coupling constants in water and alcohol for ^{17}O and hydroxyl deuterons should be very similar and use the water data for the evaluation of rotational correlation times for the hydroxyl group in alcohols. For pure methanol we then obtain a correlation time of 4 psec. It is also interesting that the ^{17}O proton spin coupling constant in water vapor is 79 ± 2 Hz,[492] which is close to the value found by Versmold and Yoon for methanol. In the following discussion we shall therefore not consider the ^{17}O data by Goldammer and Hertz since the experimental situation is not yet clear. Probably the same error is present in the results for aqueous methanol mixtures and the ethanol/water systems.

Unambigous information on the aqueous component is derived from the ^{17}O resonance in $H_2^{17}O$. Fister and Hertz found an increase in the relaxation rate of water on addition of methanol or ethanol. The increase is stronger for ethanol. The limiting slope of the relative relaxation rate is 0.06 molal^{-1} and 0.15 molal^{-1}, respectively. These data are in principle superior to the corresponding deuteron results of Hertz and Zeidler (0.09 molal^{-1} and 0.21 molal^{-1}) or Goldammer and Hertz (0.11 molal^{-1} and 0.18 molal^{-1}) since the relaxation of the alcohol hydroxy deuterons, known to be rather strong from results on pure alcohols, tends to yield values which are too high even for the limiting slope of the deuteron rate in alcohol/D_2O

mixtures because of extrapolation from higher alcohol concentrations. However, Fister and Hertz covered such a small concentration region that no definite consistency check is possible with the water correlation time derived by Goldammer and Hertz. These authors proceeded in the following way: The correlation time obtained from the deuteron rate of CH_3OD/D_2O and C_2H_5OD/D_2O mixtures, using the deuteron quadrupole coupling constant for water, is given by the average

$$\bar{\tau}_c = [2x_1/(2x_1 + x_2)]\tau_c(D_2O) + [x_2/(2x_1 + x_2)]\tau_c(OD)$$

of the water and hydroxyl group correlations; x_1 and x_2 are the mole fractions of water and alcohol, respectively. With the assumption that $\tau_c(OD)$ shows the same concentration dependence as the alkyl group correlation time in the aqueous mixtures but that its magnitude is different by the same factor as was reported for the pure alcohol, the water correlation time could be calculated. The authors found that for methanol/water mixtures the correlation time increases from the pure water value of 2.5 psec (calculated by Goldammer and Zeidler[582] from the intramolecular proton relaxation rate and corresponding to a deuteron quadrupole coupling constant of 248 kHz) to about 7 psec at 75 mole % methanol and then slightly decreases. In the ethanol/water system the correlation time reaches a maximum of 8 psec around 60 mole % ethanol. These detailed results are still preliminary but qualitatively the increase of the water correlation time is well established.

For the alcoholic component unambiguous information is obtained from Goldammer and Zeidler's measurements. However, this is restricted to the alkyl part of the alcohols. In the pure alcohol the alkyl proton correlation time is found to be much lower (0.9 psec for methanol and 2.2 psec for ethanol, all data at 25°C), than the hydroxyl group correlation time (4 psec and 8 psec, respectively) so that fast internal motion must be present, as may be concluded also from the different activation energies reported by Goldammer and Hertz. This fast motion seems to be preserved in the aqueous mixtures, as concluded from common extrapolation points of the proton and deuteron rates (see the corresponding discussion for t-butanol). The change of the alkyl proton correlation times in the aqueous mixtures is rather small for methanol and somewhat larger for ethanol, the infinite dilution values being 1 psec and 1.6 psec, respectively. Broad and flat maxima appear around 20–40 mole % alcohol. Similarly, the activation energies reported by Goldammer and Hertz change little. The latter authors added more detailed information for the correlation times in the ethyl group by looking separately at the methyl and methylene groups. Whereas the absolute values are slightly different, the methylene correlation time being lower by a factor 0.8, their composition dependence in the aqueous mixture is the same.

Spin-echo self-diffusion measurements for the alcoholic component are available from Goldammer and Zeidler's[582] work. Goldammer and Hertz[581] have investigated the water-rich concentration range in the ethanol/water system more closely and determined the self-diffusion constants of both components. The observed difference indicates, as for the case of t-butanol/water, that no long-lived hydration cages exist.

Several chemical shift measurements of the protons in the ethanol/water system have been performed. Weinberg and Zimmerman[1383] measured the water and hydroxyl proton signal relative to the methyl peak at room temperature and 40 MHz. They obtained separate peaks at concentrations above 57 mole % ethanol; below this critical concentration no dependence of the coalesced line was observed. Mavel,[957] on the other hand, did not obtain separate peaks at 25 MHz and 25°C but noticed that the coalesced peak is shifted downfield by addition of ethanol. Ageno and Indovina[8] measured the same signals at 60 MHz and 32°C and again found separate peaks above 24 mole % ethanol. At 0°C the separation becomes apparent already at 21 mole % ethanol. In the water-rich region the coalesced peak is shifted downfield, more at 0°C than at 32°C, but the separated peaks move in opposite directions: The hydroxyl proton peak which lies downfield from the water peak moves further downfield; the water peak, on the other hand, shifts upfield. Furthermore, the authors could show that the separate water peak and the coalesced peak join without discontinuity. Further shift measurements, confined to the water-rich region with the aim of investigating the water structure around the alkyl groups, were performed by Clifford and Pethica,[316] Rüterjans and Scheraga,[1167] and more recently Wen and Hertz,[1388] referring in their work to unpublished data by J. Oakes. In these investigations the methanol/water system was included because the effect of increasing the alkyl group was of interest. At 0°C and below 10 mole % alcohol the coalesced water-hydroxyl proton signal is shifted downfield in both systems but more so for ethanol. Wen and Hertz corrected in a rough way for the solute proton effect by using shift data of the hydroxyl proton of pure alcohols relative to the water resonance and assuming similar concentration dependence as for t-butanol. The polar group effect then was obtained as previously discussed for t-butanol, by extrapolation from less polar solvents for the solute methanol or ethanol to the polar solvent water. After subtracting the polar effects from the measured shifts an upfield shift for methanol but a slight downfield shift for ethanol below 5 mole % results. Rüterjans and Scheraga performed their measurements at 36°C and here the experimental results are reversed, the shift of the coalesced signal moving stronger downfield more rapidly on addition of methanol than of ethanol. The same trend may be found in the rather inaccurate 25° data of Mavel.[957] Since at these high temperatures no corrections were applied for polar effects, no definite statement on the effect of alkyl group on the water hydrogen bonding is possible.

Apart from the above proton shift measurements there exist ^{17}O chemical shift data at 25°C.[487] A downfield shift in pure water is observed when the temperature is lowered and this is the same direction as observed for the proton shift. Thus hydrogen bonding also decreases the electronic shielding around the oxygen nucleus. Upon addition of methanol or ethanol to water the ^{17}O signal of water moves upfield in the water-rich region. The authors did not find definite differences between methanol and ethanol mixtures.

All other alcohol/water systems, for which much fewer experimental data are available, will be discussed together. Experimental relaxation studies of the homologous series of normal alcohols include the systems *propanol/water*[659] and *butanol/water*[315,317] As for the lower alcohols, the water rotational correlation time increases on addition of alcohol and the effect increases with longer chain length in the water-rich region. At higher concentration reverse effects seem possible. The chemical shift data of Wen and Hertz,[1388] Clifford and Pethica,[316] and Rüterjans and Scheraga[1167] were also extended to higher alcohols. After allowing for the polar effects, increasing downfield shifts are found in the order[1388] *n*-butanol, methanol, *i*-butanol, ethanol, *n*-propanol, *i*-propanol, and *t*-butanol. This is true at 0°C in the water-rich region below 4 mole % alcohol. The shifts in *n*-butanol and methanol solutions are still upfield. It is interesting to compare the results for two isomeric alcohols, for example, *t*-butanol and *n*-butanol. The specific downfield shift is larger for an alcohol with branched alkyl groups than one with unbranched groups, i.e., it is larger when the nonpolar group is more spherical in shape. This effect was verified by extended measurements on branched alcohols with five and six carbon atoms. For example, *t*-butanol gives a downfield shift much larger than *n*-butanol and comparable to *i*-butanol. At 36°C, on the other hand, the upfield shift in the series methanol, ethanol is continued through *n*-propanol and *n*-butanol.[1167] Mavel[957] has studied the chemical shifts of water or coalesced water-hydroxyl signals in several other alcoholic systems at 25°C.

2.1.2. *Ethers*

Cyclic ethers are soluble in water. In the homologous series $(CH_2)_xO$ the members ethylene oxide $(X = 2)$, trimethylene oxide $(X = 3)$, tetrahydrofuran $(X = 4)$, and tetrahydropyran $(X = 5)$ have been investigated in aqueous solution. Two methyl substitutes of these, propylene oxide

$$CH_3—CH—CH_2—O$$

and α-methyl tetrahydrofuran

$$CH_3—CH—(CH_2)_3—O$$

were also included. From the cyclic ethers with two oxygens the compounds p-dioxan, $(CH_2)_4O_2$, and 1,3-dioxolan, $(CH_2)_3O_2$, were selected. Most of the available data refer to the systems tetrahydrofuran/water and dioxan/water.

Detailed relaxation studies on the system *tetrahydrofuran (THF)/water* at 25°C were published by Goldammer and Zeidler[582] and extended to other temperatures by Goldammer and Hertz.[581] Contrary to the case with alcohol/water mixtures, the behavior of both components can be studied without interference by exchange phenomena. In the systems $(CH_2)_4O/D_2O$ and $(CD_2)_4O/H_2O$ both the deuteron and proton relaxation rates were measured; the results are shown in Fig. 2. Since the deuteron rates represent an intramolecular effect, they might also reflect the composition dependence of the intramolecular proton relaxation rate. If this assumption is correct, both curves should extrapolate to a common value at vanishing concentration of the respective component (neglecting isotope effects on the correlation times) since in this limit the (total) proton rate degenerates to an intramolecular rate. On the other hand, in the pure organic component the intramolecular relaxation rate could be found independently by isotopic dilution experiments in the system $(CH_2)_4O/(CD_2)_4O$. In this way both end points on the composition scale were fixed and the measured deuteron rate was found to behave in the same way as the expected intramolecular proton rate. At the same time this result proved that no marked change of the C—D quadrupole coupling constant occurred. In pure water no separate determination of the intramolecular proton rate is possible; therefore both curves for the aqueous component could bejoined only at vanishing water concentration. The end point of the lower deuteron curve at 100% water then gave an estimate of the intra-

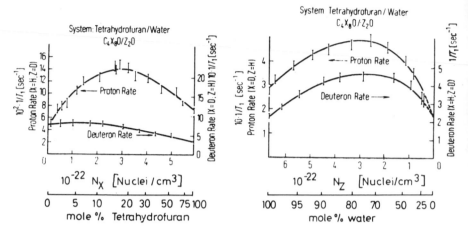

Fig. 2. Proton and deuteron relaxation rates for tetrahydrofuran and water in the mixtures $(CH_2)_4O/D_2O$ and $(CD_2)_4O/H_2O$ as a function of composition at 25°C. The proton rates refer to the left-hand ordinates, the deuteron rates to the right-hand ordinates. The left diagram refers to tetrahydrofuran, the right one to the aqueous component. (From Goldammer and Zeidler.[582])

molecular proton rate from which a rotational correlation time of 2.5 psec and a deuteron quadrupole coupling constant of 248 kHz for water were derived. As is apparent from the figure, the water correlation time increases from this value when THF is added and reaches a maximum of 5 psec at 25 mole % THF. The rotational correlation time of the THF varies between 0.6 psec in the pure liquid and 1.3 psec in the water-rich range where a very slight maximum around 5 mole % THF occurs. Thus the THF molecules rotate much faster than the water molecules at all concentrations, including the more structured water-rich region. One is led to the conclusion that water molecules associate preferably with one another rather than with the organic solute. The same conclusion was reached from the rather different activation energies for the rotational correlation times of both components.

Self-diffusion measurements by the spin-echo technique were also reported by Goldammer and Zeidler.[582] Below 30 mole % THF the diffusion coefficient of the solute is considerably lower than that of the water. This was interpreted by Goldammer and Hertz[581] as discussed above in connection with t-butanol, as indicating that no long-lived hydration cage exists.

The chemical shift of the water protons in THF/water was investigated over the whole composition range by Mavel[956] and Clemett,[312] whereas

Glew et al.[573] and Wen and Hertz[1388] restricted their experiments to the water-rich region and Holmes et al.[677] to the other limit of small water content. Clemett measured the proton chemical shifts of water and also of the α and β methylene protons of THF at about 34°C using internal (TMS) and external (C_6H_6) references. The common result is that the water proton shift is 2.31 ppm upfield going from pure water to the infinitely dilute limit in THF, whereas the ether methylene protons shift upfield by less than 0.2 ppm over the same range. Actually this latter shift is practically negligible below ether concentrations of 10 mole %. Thus, as in the case of t-butanol, the nonpolar groups are not sensitive to changes in the environment in the water-rich region. The large upfield water shift disagrees with Mavel's value (1.40 ppm), measured, however at 22°C, but it compares quite well with Holmes' data at 29°C (2.16 ppm). Also, the extrapolation value of the water shift for vanishing water concentration as reported by Clemett, i.e., 2.42 ppm downfield from TMS, checks satisfactorily with results obtained by Takahashi and Li[1276] in ternary systems of water/THF/cyclohexane at 35°C. The upfield shift of the water must be explained by the fact that water–water hydrogen bonds are broken and replaced by much weaker hydrogen bonds between water and THF. This upfield shift of the water proton signal from the pure water to the limiting concentration in any electron-donating substance is usually smaller as the hydrogen bond strength between water and electron donor increases, except for strongly basic solvents. A usual technique is to correlate the limiting shift at low water content with thermodynamic and infrared data which give a direct measure of the hydrogen bond strength.[312,677] It is an unsolved question how this shift due to hydrogen bonding with the electron donor changes when instead of donor molecules other water molecules form the surrounding medium. That means the limiting shift in the electron donor may not be correlated with the polar group effect in the water-rich range. However, if this were the case, then any deviations from the linear shift dependence between zero water concentration and pure water could be attributed to specific water–water interactions. In the case of the THF/water system this might be approximately true. There are deviations in the direction of downfield shifts which point to a greater degree of water–water hydrogen bonding. It is more conclusive to compare these deviations in the series ethylene oxide, trimethylene oxide, and THF since here the number of methylene groups per oxygen atom increases. Clemett observed a stronger downfield deviation in this order although all shifts are upfield. He also found that the electron donor strength of THF is intermediate between trimethylene oxide and ethylene oxide, the latter being the weakest donor.

The picture is quite different at 0°C. At this temperature Glew *et al.* were the first to observe a downfield shift of the water proton signal upon addition of THF. A maximum downfield shift of about 3 Hz (at 60 MHz) around 4 mole % is reported. Wen and Hertz found a similar but smaller shift (less than 1 Hz) at the same temperature and resonance frequency. In both cases the methylene protons of THF were used as internal reference. In addition the latter authors estimated the polar group effect by measuring the hydroxyl proton shift of methanol in methanol/THF mixtures. This peak moves upfield with a linear concentration dependence in the range 0–17 mole % THF, which would indicate an even larger specific nonpolar downfield contribution to the water shift. Furthermore, this polar group effect is in the same direction as was suggested above for higher temperature from the limiting shift in THF, assuming ideal behavior.

Concerning the system *p-dioxan/water*, relaxation studies were performed by Hertz and Zeidler,[659] Hindman *et al.*,[671] and Clemett.[314] Clemett measured the proton rates in the systems $(CH_2)_4O_2/H_2O$, $(CH_2)_4O_2/D_2O$, and $(CD_2)_4O_2/H_2O$ over a composition range 0–40 mole % of dioxan at 34°C. Since he used the adiabatic fast passage method, both components could be studied even in the nondeuterated mixture. By comparing the relaxation rates at equal compositions in $(CH_2)_4O_2/H_2O$ and $(CH_2)_4O_2/D_2O$ the intermolecular effect of the aqueous component on the dioxan rate was obtained. The intermolecular relaxation formula based on the simple diffusion equation was then used to discuss the experimental data. The relative diffusion coefficient, in this case between dioxan and water molecules, which appears in this formula was taken as the mean of the individual diffusion constants. Thus the distance of closest approach could be calculated and two continuously joined regions with different results were found: 1.5 Å below 20 mole % dioxane and 4.8 Å above this critical concentration. Consistent results were obtained by the corresponding experiment for the aqueous component using the mixtures $(CH_2)_4O_2/H_2O$ and $(CD_2)_4O_2/H_2O$. As must be the case, the same distances of approach were found with a similar variation around the above-mentioned concentration. Alternatively, instead of attributing the change in the intermolecular relaxation rate to a change in the distance of closest approach, a variation of the radial distribution function in the light of eqn. (4) could explain the effect. This would mean that at low dioxan content preferred distribution of water molecules around dioxan occurs. Further results from Clemett's study are derived from ternary mixtures $(CH_2)_4O_2/(CD_2)_4O_2/D_2O$ where at fixed dioxan content (10 mole %) the intramolecular proton rate was determined from the limiting value of $1/T_1$ at vanishing $(CH_2)_4O_2$ concentration. This value is almost the same as the limiting proton rate of dioxan in D_2O. On the other hand, the proton rate in $H_2O/(CD_2)_4O_2$ mixtures changes appreciably, and this was attributed to a large change in the intramolecular relaxation rate of water. This latter conclusion was verified by measurements of the deuteron relaxation rates in $(CH_2)_4O_2/D_2O$.[659,671] Therefore we can conclude that below 10 mole % dioxan the correlation time of the dioxan molecules is constant whereas the water correlation time changes by a factor of 1.5. Apparently the correlation time of dioxan is numerically larger than that of water in this range. Clemett did not determine the intramolecular relaxation rate at higher dioxan content, but Hindman et al. covered the whole composition range. Their deuteron rates indicate that the water correlation times decrease appreciably at high dioxan content even below the pure water value. These authors also report activation energies for the deuteron relaxation over the whole concentration range and a maximum in the water-rich range is apparent.

Self-diffusion coefficients of both components were published by Clemett[313] and that for the nonaqueous component by Fratiello and Douglass.[528] The diffusion coefficients of both components decrease in the water-rich range upon addition of dioxan and a minimum is reached around 20 mole % dioxan. This same behavior was discussed for the THF/water systems, as was the fact that the water diffusion coefficient was always larger than that of the nonaqueous component in this concentration region.

Several authors[312,528,573,677,956] have reported upfield chemical shifts of the water proton signal on addition of dioxan at temperatures between 0°C and 34°C. The magnitude of the total shift between pure water and the limiting water concentration in dioxan is around 2.2 ppm at 30°C, which is comparable to the corresponding value in THF. Thus the electron donor strengths of the two ethers are much alike. There are downfield deviations from the ideal shift dependence but even at 0°C no absolute downfield shifts could be observed. In this connection it is interesting to note that in dioxolan, with one methylene group less than in dioxan, the shift in the water-rich region is even more downfield, although the limiting value at low water content is the same as for dioxan. However, different polar group effects at large water concentration could be responsible for reversing the order. Clemett[312] and Fratiello and Douglass[528] measured the chemical shift of the methylene protons in dioxan over the total composition range. Starting from low dioxan concentration, there is no shift up to 10 mole % dioxan and then the methylene peak moves upfield, the total shift between infinitely dilute and pure dioxan being 0.2 ppm at 34°C. Even at 80°C the total shift amounts to only 0.24 ppm, although this difference could be due to errors in the susceptibility corrections. Still it is interesting that there is some weak interaction between the water and the methylene protons either directly or via the oxygen sites.

To conclude this section on water/ether systems, observations on other systems will be mentioned briefly. As already pointed out, Clemett[312] included in his study chemical shift measurements of the aqueous component in *ethylene oxide/water, trimethylene oxide/water,* and *dioxolan/water.* The ethylene oxide/water system was also studied in the water-rich region at 0°C by Glew *et al.*[570] and the downfield deviation of the water proton shift from the ideal shift dependence is much stronger at this low temperature. In the *propylene oxide/water* system the components are not miscible over the whole composition range. At 0°C[573] and 34°C.[312] downfield deviation at high water concentration are again observed. Wen and Hertz[1388] compared the water shifts in α-methyl–THF, THF, and tetrahydropyran at low ether concentration and 0°C. The methyl substitution produces an enhanced downfield shift, whereas for tetrahydropyran effectively no shift is observed.

2.1.3. *Amines and Nitriles*

In aqueous mixtures of amines or nitriles one can get additional information about the organic component by observing the resonance of the ^{14}N nucleus which is naturally present in large amounts. Except for the systems pyridine/water and acetonitrile/water, very little work on amines or nitriles is available.

Relaxation studies in the *pyridine/water* system at 25°C were published by Goldammer *et al.*[581,582] The same procedure as discussed for tetrahydrofuran/water was used to determine the intramolecular proton relaxation rates of both components from measurements of the proton and deuteron relaxation rates in C_5H_5N/D_2O and C_5D_5N/H_2O. The intramolecular proton rate in pure pyridine was also obtained from isotopic dilution studies in C_5H_5N/C_5D_5N. Evidently the deuteron rates represent the intramolecular proton relaxation rates quite well; thus the deuteron quadrupole coupling constants of both components cannot be affected markedly in the mixture. On the other hand, the nitrogen relaxation rate probably has a concentration dependence different from the deuteron rate: Goldammer and Hertz[581] found larger values for the nitrogen correlation times in the aqueous mixtures; Kintzinger and Lehn[758] obtained results closer to the deuteron times but still a little larger, although they confined their study to only two concentrations. In spite of the unsettled experimental situation, one may attribute this discrepancy either to a change in the nitrogen quadrupole coupling constant or to anisotropic rotational motion of the pyridine molecule. Certainly the nitrogen coupling constant is affected by hydrogen bonding with water molecules but one would expect a decrease. Recently Kintzinger* measured the deuteron rates of differently substituted pyridines in water and for γ-deuterons found larger rates (similar to the concentration dependence of the nitrogen rate) than for α- and β-deuterons, whereas in pure pyridine both are approximately equal.[759] This finding supports the conclusion of anisotropic rotation and indicates faster rotation around the in-plane axis through the γ-deuteron and nitrogen atom than perpendicular to it, consistent with the picture of hydrogen bonding at the nitrogen site. On the other hand, in pure pyridine anisotropic motion with a faster rotation around an axis perpendicular to the plane seems to be present.[759] Anyhow, the pyridine correlation times for deuterons as well as for nitrogen increase strongly when pyridine is diluted with water. This increase also has a line-sharpening effect on the nearby α protons through spin–spin coupling, as already had been noticed earlier.[300] A maximum correlation time is reached around 25 mole % pyridine and the infinite dilution values remain considerably above the correlation times in pure pyridine. Similarly, on addition of pyridine, the water correlation time increases from the pure water value but the maximum is at 50 mole % for this component. The interesting effect, which is quite different from the behavior found in THF/water, is the strong variation of both correlation times and the larger infinite dilution values for pyridine as compared to 2.5 psec of pure water. This is evidence for strong pyridine–water interaction also in the water-rich region: Here the pyridine molecules must associate with water molecules in order for their rotational motion to be retarded. The same conclusion was obtained from the concentration dependence of the activation energies for rotational correlation times which are similar for both components.

Self-diffusion data of both components in the pyridine/water mixture have been published.[528,582] The typical initial decrease of the diffusion coefficients from the value in the pure substances on the addition of the other component is again observed, as is the fact that the diffusion coefficient of the organic component remains below that of water in the water-rich region.

The proton chemical shift of the aqueous component was observed by Mavel[956] and Fratiello and Douglass.[528] Starting from pure water the

* J. P. Kintzinger, private communication.

initial shift is downfield, reaching a maximum of 0.3 ppm (30°C) at about 15 vol% water. At very low water content the shift is upfield even with respect to pure water, the infinite dilution shift being 0.47 ppm at 29°C.[677] On the basis of the hydrogen bond strength between water and pyridine as derived from independent data, this upfield limiting shift is too large, a typical deviation for strongly basic substances. Reasons for this behavior have been discussed by Holmes et al.[677] The observed downfield shift at larger water concentrations is probably due, at least partly, to the polar group effect arising from water hydrogen bonding at the nitrogen site. In this connection it is important to note that in piperidine (hexahydro-pyridine) the proton bonded directly to the nitrogen shows a very large downfield displacement upon dilution with deuterium oxide. Strong interaction of the components is also evident from shifts of the pyridine protons.[300,528] The effect on the α protons is quite different from the other protons: the α-proton signal first moves upfield on dilution of pyridine with water and then downfield at higher water content, whereas the β- and γ-proton signals shift continuously downfield. The total downfield shifts between pure pyridine and the infinitely dilute state are 0.4 and 0.8 ppm for α and β (or γ) protons at 30°C. Fratiello and Douglass attribute this downfield shift in part to a reduction of intermolecular ring-current effects (other molecules, if placed above or below the plane of the aromatic ring, have their protons more shielded).[1110] In order to test the influence of the ring-current effect, the proton shifts in pyridine/cyclohexane and piperidine/D_2O were measured. In the first mixture only the ring-current effect should be operative and in the last mixture no such effect can exist. Thus it was shown that besides a large ring-current effect additional low-field displacements of the pyridine protons seem to be present, indicating direct interactions with the water. Similar shifts of the protons in 2-methyl-pyridine were observed.[283]

   Acetonitrile/water has also been investigated in more detail. Figure 3 is taken from the work of Goldammer and Hertz[581] and it shows the composition dependence of the rotational correlation times of both components at 25°C obtained from deuteron and nitrogen relaxation rates in CD_3CN/H_2O and CH_3CN/D_2O using known quadrupole coupling constants. The large difference of the acetonitrile correlation times, as derived from deuteron and nitrogen rates over the whole composition range, is remarkable. This difference was also observed in pure acetonitrile by Bopp[211] and Woessner et al.[1424] and was explained on the basis of aniso-tropic diffusional rotation [compare eqn. (10)]. Much faster reorientation around the symmetry axis of the acetonitrile molecule is responsible for

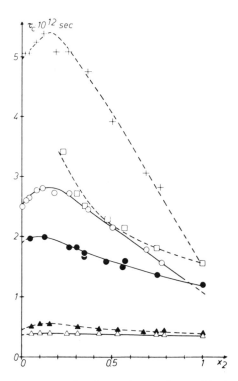

Fig. 3. Rotational correlation times of both components in the acetonitrile/water mixture derived from different relaxation rates. From deuteron relaxation in CD_3CN/H_2O ($\triangle$) 25°C, ($\blacktriangle$) 5°C. From nitrogen relaxation in CH_3-CN^{14}/D_2O, ($\bullet$) 25°C, ($\square$) 5°C. From deuteron relaxation in CH_3CN/D_2O ($\bigcirc$) 25°C, ($+$) 5°C. Here x_2 is the mole fraction of acetonitrile. (From Goldammer and Hertz.[581])

the decrease of the methyl deuteron rate compared to the nitrogen rate since the latter is sensitive only to reorientations perpendicular to the symmetry axis (assuming the z axis of the nitrogen field gradient with cylindrical symmetry to lie along this symmetry axis). Differing temperature dependences of both rates support this conclusion. The concentration dependence of the correlation times then shows that the anisotropic rotation is preserved in the aqueous mixtures and a small influence is noticed only in the perpendicular rotational motion which is slowed down in the water-rich region. The water correlation time still remains above the nitrogen time except at very high acetonitrile content, where both become equal. As usual, the rotational correlation time of the water increases from its pure water value below 20 mole % acetonitrile, indicating an increase of water–water interaction, in agreement with earlier data.[659] Goldammer and Hertz also determined activation energies for the correlation times. These results confirm the anisotropic rotation of acetonitrile in the aqueous mixture. The water activation energy is higher but Hindman[671] reports an activation energy for water at infinite dilution comparable to the nitrogen data.

The chemical shift of the water protons in acetonitrile/water was determined between 20°C and 90°C by Hertz and Spalthoff[658] using the methyl signal as internal reference. Compared to pure water, the shift is always upfield. The water shift at infinite dilution was found to be 3 ppm upfield from water,[677] in agreement with the hydrogen bond strength estimated from other sources.

The system *propionitrile/water* was studied in this connection because the effect of increasing nonpolar chain was of interest. In C_2H_5CN/D_2O an increase of the deuteron relaxation rate above that for acetonitrile shows that the nonpolar groups slow down the rotational motion of water.[659] On the other hand, the substitution of C_2H_5CN for CH_3CN results in higher upfield water shifts.[658] Both investigations were carried out at 25°C.

Proton shifts of water in several *amine/water* mixtures were recently investigated by Wen and Hertz[1388] at 0°C. Secondary amines (diethyl and diisopropyl) yield much stronger downfield shifts than primary amines (*t*-butyl and *n*-butyl). The same is true for cyclic secondary amines; here the five-membered ring (pyrrolidine) gives a greater downfield shift than the six-membered one (piperidine). Because of rapid exchange, no separate signals for the water and amine protons (bonded to nitrogen) can be observed and thus no solute proton effect can be determined. As mentioned before in connection with piperidine, this is most probably responsible for large downfield shifts. In addition, as shown by Wen and Hertz, the polar group effect also gives a downfield contribution.

2.1.4. *Acids*

In the homologous series of normal carboxylic acids the first three members have been investigated: acetic, propionic, and butyric acids. The same trouble as with alcohols exists insofar as the carboxyl proton rapidly exchanges with the water protons. Therefore no separate information exists on the polar group in aqueous solutions. Oxygen-17 data which could yield such information have not been published. Salts of these carboxylic acids will be discussed in the next section.

Relaxation data for the nonaqueous component in *acetic acid/water* were published by Goldammer and Zeidler.[582] By combining proton and deuteron relaxation rates in CH_3COOD/D_2O and CD_3COOH/H_2O the rotational correlation times were obtained. Additional measurements in the pure acid showed that the polar group has a much larger correlation time than the methyl group; thus the latter performs rapid internal rotation. This is probably preserved in the aqueous solution, as concluded from the constant ratio of correlation times derived from deuteron and proton rates. A strong decrease of the methyl correlation time in infinitely dilute aqueous solutions as compared to the pure acid (from 3 psec to 1.5 psec at 25°C) may be explained by a gradual dissociation of acid polymers. There occurs a maximum of the correlation time around 60 mole % acid.

The corresponding decrease of the self-diffusion coefficient is much less marked.[582] From these data and the intermolecular proton relaxation rate, i.e., a plot of the quantity $(1/T_1)_{inter}D/N_H$, preferred acid–acid association through the methyl groups in the water-rich region might be concluded.

Deuteron relaxation rates in CH_3COOH/D_2O at low acid concentrations suggest increasing correlation times[659] for the aqueous component. Proceeding to propionic and butyric acids, the correlation times of water increase systematically.

Detailed information on the systems *propionic acid/water* and *butyric acid/water* became available very recently. Tutsch[583] studied the proton and deuteron relaxation rates of all methyl and methylene groups of the acids separately using the proper isotopically substituted compounds. Following the previously outlined procedure, he derived the intramolecular and intermolecular proton rates of all different groups. The general conclusion is the following: The rotational correlation times decrease continuously in the direction away from the polar carboxyl group. This is true for the pure acids and the aqueous solutions except at very low acid concentrations, where the different values seem to approach each other. The typical concentration dependence is similar to that in acetic acid/water, the correlation times in the pure acids lie above the infinite dilution values, and there is a maximum at relatively high acid concentration. The interpretation leads directly to the picture of dissociation of acid polymers by water and of greater flexibility of alkyl segments away from the polar groups in the associated species. The intermolecular proton rates also show an interesting behavior: If the quantity $(1/T_1)_{inter}D/N_H$ is plotted against concentration, a systematic increase at low acid content is again observed, depending on the position of the investigated group. It is stronger for alkyl segments close to the polar groups, showing the same trend as the rotational correlation time. Again the high intermolecular rate in the water-rich region may be interpreted on the basis of preferred acid–acid interaction via the non-polar groups, supporting the view of hydrophobic interaction postulated for these carboxylic acids.[1023] It is not simple to decide why the inter-molecular rate is larger for segments closer to the polar end: Either indeed these approach each other more closely or the rapid rotational motion of the other groups reduces their effective relaxation rates. As mentioned earlier, the rotation of different segments in the alkyl chain was also studied in the case of ethanol. Here, however, the picture was quite different: The methy-lene group had a correlation time a little shorter than the methyl group. Steric reasons might be responsible for the different behavior in acids and alcohols.

Chemical shift measurements of the coalesced signal for water and carboxyl protons at 36°C in the water-rich region have been published.[1167] The observed behavior is not systematic: All solutes produce strong down-field shifts but for propionic acids it is less than for acetic acid, while butyric acid yields the largest downfield shift.

2.1.5. Ketones

Acetone/water is one of the most studied systems. The few papers dealing with other aqueous mixtures of ketones include the aqueous systems methyl–ethyl ketone, methyl–isopropyl ketone, and cyclobutanone.

Relaxation studies include proton and deuteron rates in CH_3–CO–CH_3/D_2O and CD_3–CO–CD_3/H_2O[558,581,582,659,1139] and ^{17}O rates in CH_3–$C^{17}O$–CH_3/H_2O and CH_3–CO–CH_3/$H_2^{17}O$.[487,581] Additional isotopic dilu-tion experiments in pure acetone for the separation of the intramolecular proton rate have been performed. Unfortunately, no independent measure-ment of the ^{17}O quadrupole coupling constant in acetone is available; there-fore the value for the coupling constant in formaldehyde was substituted instead. With this constant a rotational correlation time of 1.1 psec at 25°C is calculated, in good agreement with the dielectric relaxation time of 2.9 psec if microstep rotational diffusion occurs. On the other hand, the cor-relation time of the methyl protons is 0.6 psec, so that internal rotation of the methyl groups is evident. Furthermore, it is found that the latter cor-relation time hardly changes with composition, whereas the correlation time derived from the ^{17}O rate is always longer and even increases at high water content. The infinite dilution value is very close to the correlation time of pure water, if a constant quadrupole coupling constant is assumed. We therefore conclude that there is some acetone–water interaction via the polar group, but that the methyl groups perform internal rotations which are not affected in the aqueous medium. That the interaction is weak was also shown by the large difference in activation energies for the deuteron rates of both components over the entire composition range. Therefore the observed increase of the water correlation time with increasing acetone concentration below 20 mole %, which is apparent from the deuteron and the ^{17}O relaxation rates of water, must be due to stronger water–water interaction in this water-rich range. Above 30 mole % acetone the water correlation time decreases and an infinite dilution value of 2.3 psec is found.

The corresponding decrease of the water self-diffusion coefficient is also observed; similarly, a shallow minimum of the acetone diffusion constant around 10–20 mole % acetone is found.[581,582,911] The considerably smaller value for acetone precludes there being any long-lived hydration cage around the organic molecule.

The chemical shift of the water protons in acetone/water is upfield from the pure water value around room temperature,[658,677,956,1276] whereas at 0°C Glew et al.[573] observed an initial downfield shift up to about 2 mole % acetone. This downfield shift could not, however, be verified by Wen and Hertz.[1388]

In comparison to acetone/water, the relaxation rates and chemical shifts of the aqueous component were measured in the system *methyl–ethyl ketone/water* in the water-rich region, and the relaxation rate (deuteron resonance) is increased, i.e., the larger alkyl group yields an additional slowing down of the rotational motion of water.[659] This behavior correlates well with a weaker upfield water proton shift observed at 0°C,[1388] but at 25°C this effect is reversed: The water protons suffer a larger upfield shift in methyl–ethyl ketone than in acetone.[658]

Wen and Hertz[1388] also investigated the water proton shift at 0°C in the systems *methyl–isopropyl ketone/water* and *cyclobutanone/water*, where the former is slightly downfield and the latter upfield. These authors also estimated the polar group effects and found them to be upfield in general for ketones, so that all nonpolar groups are indeed producing downfield water proton shifts.

2.1.6. Miscellaneous

One of the more thoroughly studied systems is *dimethylsulfoxide/water*. Packer and Tomlinson[1061] measured the proton relaxation rates in $(CH_3)_2SO/D_2O$ and $(CD_3)_2SO/H_2O$ at different temperatures. Apart from these total rates, they also performed ternary mixture experiments in order to determine the intramolecular proton rate of dimethylsulfoxide. The information for the aqueous component is also complete: Combining the total proton rates with Glasel's[558] deuteron rates, i.e., extrapolating them to a common point at vanishing water concentration, provides an estimate of the intramolecular proton rate for water. The correlation times derived by this procedure show features similar to the pyridine/water system: At high water concentration the dimethylsulfoxide correlation times are higher than the water values, indicating strong interaction between components also in this range. For the aqueous component the values are 2.2 psec in pure water and 5.4 psec in the infinite dilution limit at 31°C, and a maximum of about 9 psec is reached for 40 mole % dimethylsulfoxide. For the organic component the values are 3.2 psec (in the pure liquid) and 2.7 psec, respectively, with a maximum of 4.8 psec at the same composition determined, however, at 26°C. Packer and Tomlinson also investigated the low-temperature range and observed a maximum in the relaxation rate for water and a leveling off for dimethylsulfoxide around 200°K and 70 mole % water. It is probable that for dimethylsulfoxide this effect arises from methyl group rotation, and then also at higher temperatures the internal motion should be excited. The correlation times for the polar group are therefore expected to be much larger than the ones cited for the methyl group and must lie definitely above the water values. So there is thus no doubt about strong interaction between dimethylsulfoxide and water, and the difference in activation energies also supports this conclusion.

Packer and Tomlinson also studied the self-diffusion coefficients of both components and noted a strong decrease of the water value upon addition of small amounts of dimethylsulfoxide. The diffusion coefficient of the organic component always lies below that of water and the same is true for the activation energies of self-diffusion.

Chemical shift measurements at very low water concentration are also available for the water protons.[677] The limiting shift is 1.26 ppm upfield relative to pure water at 29°C, which is much too large as judged from the hydrogen bond strength between water and dimethylsulfoxide as estimated from other sources. This is the same anomaly as observed for pyridine and triethylamine.

Another system studied in some detail is *urea/water*, for which correlation times and proton chemical shifts of both components at different concentrations and temperatures are available. Finer *et al.*[485] recently reported ^{14}N relaxation rates for urea, which partly were measured directly by pulse experiments and partly were derived from proton line shapes affected by the nitrogen relaxation through spin–spin coupling. Assuming a constant quadrupole coupling constant, which was taken 5% above the solid-state value, the rotational correlation time of urea was calculated. It was found to increase with increasing urea concentration in the range 0–8 m at 17°C approximately by a factor of two, the infinite dilution value being 3.3 psec. Similarly, the activation energy for the rotational correlation time increases with rising urea concentration. The same authors measured the ^{17}O linewidth of water in a 4.7 m urea solution in the range 2–65°C. At lower temperatures, where the measurements are probably not complicated by urea decomposition (resulting in pH increase and slower proton exchange), no difference between urea solution and pure water was observed. Glasel,[558] on the other hand, measured an increase of deuteron relaxation rates in urea/D_2O at 31°C, but this could be ascribed to a contribution from the deuterons in the urea molecule instead of enhanced water reorientation. Thus from relaxation measurements a slight increase of the rotational motion of urea but no significant change for the aqueous component are concluded.

The proton chemical shifts of both components could be studied separately at 220 MHz and 60 MHz since the proton exchange is sufficiently slow at pH = 7. The common result obtained at 20°C with internal and external standards is that the water protons shift upfield whereas the urea protons shift slightly downfield with increasing urea concentration. The upfield water shift is stronger at lower temperatures and disappeared above 60°C.

The conclusions to be derived from these experimental facts are the following: The long-range order characteristic for pure water is partly destroyed by introducing urea molecules, as evident from the water upfield shift, indicating breakage of hydrogen bonds. Instead, weak, short-range interactions between water and urea are present which compensate the expected decrease of water correlation times (due to structure-breaking) so that no net change is observed. A corresponding downfield contribution in the observed upfield shift of the water protons could be demonstrated. The existence of stable urea polymers could be ruled out by quantitative considerations of the magnitude of the urea correlation time. Thus urea shows a behavior quite different from the other nonelectrolytes discussed so far, which always lengthened the water correlation time in the water-rich region.

It is of considerable interest to know the rotational correlation time of water molecules present as monomers in a noninteracting surrounding. On the basis of limiting water proton shifts the systems *benzene/water*, *carbon tetrachloride/water*, and *chloroform/water* are appropriate, where the upfield shifts from pure water are comparable to the vapor shift. For water in benzene the published results differ somewhat: 4.3 ppm (external standard, 29°C)[321] and 4.77 ppm (internal standard, 25°C).[1040] The latter value is even larger than the vapor shift (4.37 ppm at 25°C; Volume 1, Chapter 6), and this may be understood in terms of the aromatic ring-current effect (protons above or below the benzene ring are more strongly shielded).[1110] The corresponding shifts in carbon tetrachloride and chloroform are 3.8 and 3.6 ppm, respectively.[321] These data may be compared to the respective rotational correlation times of the water molecules in these solvents at 25°C: 0.42 psec in benzene, 0.44 psec in carbon tetrachloride, and 0.51 psec in chloroform, calculated from deuteron rates with quadrupole coupling constants around 300 kHz (the gas value is 310 kHz).[671] To enhance the solubility of water, *nitromethane* was added as a third component, but this solvent is still rather inert as corresponding water shifts[677] and correlation times[671] show.

The water correlation times in the inert solvents may be used to estimate the monomer concentration in pure water from the experimental correlation time.[581,671] The averaging of the correlation times for different species depends on their lifetimes and is given by eqn. (16). With reasonable estimates for the correlation times of clusters, a monomer concentration of 7 mole % (long lifetimes) or 20 mole % (short lifetimes) is estimated at 25°C.

Chemical shifts were reported for *formamide/water* and the N-methyl substitutes of formamide at 30°C by Fratiello.[527] By use of an external standard, it was found that the water protons shift upfield more strongly in dimethylformamide than in the other solutes; obviously the former is a weaker electron-donor. The methyl protons in dimethylformamide show a considerable downfield shift in the aqueous mixtures with respect to the pure phase, just like the nitrogen-bonded protons in formamide. However, this does not indicate interactions with the water, because the same effect is observed on dilution with carbon tetrachloride, but must arise from a decrease of self-association of amides. Self-association in the pure state is also evident from self-diffusion data: Dimethylformamide has a larger constant than the monosubstitute, but in aqueous solution the order is reversed.

Finally, recent relaxation studies in *monosaccharide/water* systems should be mentioned.[1275] Here the water–^{17}O resonance was investigated in hexose and pentose solutions at several concentrations and temperatures. The relaxation rate of water always increases on addition of the solute, which is explained by hydrogen bonding between water and the hydroxyl groups of the monosaccharides.

2.2. Aqueous Solutions of Electrolytes with Alkyl Groups

2.2.1. *Alkylammonium Salts*

Although this chapter is actually devoted to nonelectrolytes and a separate chapter dealing with nuclear magnetic resonance in electrolyte solutions appears in this volume, the electrolytes with alkyl groups are discussed under this heading. The present arrangement may be justified by the argument that the nonpolar groups are the topic with which the investigations are usually concerned, the charged polar groups being merely necessary for the solubility in water. To exclude the influence of the charge, homologous series of alkyl groups are mostly studied, and of special interest were always tetraalkylammonium salts, where the positive charge is substantially shielded.

In aqueous solutions of alkylammonium salts the following resonances can and have been studied: proton, deuteron, and ^{17}O of water, proton, deuteron, and ^{14}N of the cation, and anion resonances. The nitrogen relaxation rates which are available in the literature are not direct measurements, but their influence on the proton line shape through spin–spin coupling was utilized to obtain the desired data.

Studies related to the aqueous component might be considered first. A relaxation study of the normal tetraalkylammonium salts from the methyl to the butyl component was conducted by Hertz and Zeidler.[659] In most systems only the deuteron rate of heavy water, in which the nondeuterated salts were dissolved, was measured, with the exception of tetramethylammonium, where the deuterated salt was available and the proton resonance of H_2O could be studied. The relaxation rate of water is increased by addition of the salts and the increase is proportional to the chain length of the alkyl group, at least at concentrations below 3 *m*. The results are presented in Fig. 4. Comparison of the relative proton and deuteron rates in the tetramethyl compound shows that the former is somewhat larger, due to the intermolecular contribution. Similar but less well-defined results on two salts were reported by Danyluk and Gore,[364] who investigated the proton relaxation rate in solutions where both components contained protons. Thus we can again conclude that nonpolar groups slow down the rotational motion of water in the water-rich concentration range.

This conclusion was questioned by Eley and Hey,[448] who argued that the quadrupole coupling constant of the water deuterons may be altered considerably from its normal value of pure liquid water in the neighborhood of nonpolar groups. Breaking of hydrogen bonds would

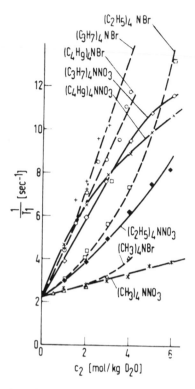

Fig. 4. Relaxation rates of water deuterons in
solutions of tetraalkylammonium salts at 25°C.
(From Hertz and Zeidler.[659])

increase the coupling constant and consequently the deuteron relaxation
rate. They proposed an alternative procedure to obtain the desired information from the water proton relaxation rate without the use of deuterated
solutes. The water proton rate, to be obtained separately from the solute
proton rate by appropriate experimental methods, contains an additional
unwanted intermolecular contribution from the solute protons which must
be subtracted. To account for this effect, the water molecules around a
given H_2O molecule are substituted by D_2O molecules, thus eliminating
the relaxation contribution from other water molecules and retaining only
the wanted solute contribution. In practice the proton rate is measured in
H_2O/D_2O mixtures of varying ratio but with a fixed concentration of the
solute and the limiting rate at vanishing proton content is determined.
Eley and Hey used for their experiments two diquaternary ammonium
bromides, $(CH_3)_3N^+-R-N^+(CH_3)_3$, with R representing three and five

methylene groups, respectively; they could then deduce the influence of two additional methylene groups on the water relaxation rate. The total water proton rate is increased by the additional groups at all investigated temperatures, but the extrapolated values in H_2O/D_2O mixtures differ by an even larger amount (the plots of the water proton rate versus the mole fraction of H_2O for the two salts diverge in the direction from pure H_2O to pure D_2O!), so the effect of the methylene groups is therefore to decrease the proton rate (after subtraction of the solute contribution). The conclusion that methylene groups increase the mobility of water molecules in the water-rich range is contrary to all previous experience, and although the outlined method is certainly an intriguing one, experimental difficulties, especially the correct estimation of the contributions to the rate measurements, are probably encountered.

Confirmation of the water deuteron results is also obtained from ^{17}O relaxation rates of the aqueous component.[487] Only tetramethyl and the tetrabutyl ammonium bromide were investigated, but the agreement with the deuteron results is satisfactory.

Some relaxation studies referring to the alkylammonium cation have also been reported. Hertz et al.[657,659] studied proton and deuteron resonances, and Larsen[828,829] studied the nitrogen rate by an indirect method. The proton rates were measured in D_2O, whereas for the deuteron rates H_2O was used. The latter rates, used in conjunction with the infinite dilution proton data, show the concentration dependence of the intramolecular proton rates, and they were followed for the tetramethyl and tetraethyl compounds up to 6 m. The infinite dilution proton rates and consequently the rotational correlation times increase with the cation size, the values ranging from 2 to 33 psec for the tetraalkylammonium cations from methyl to butyl. However, some caution is indicated regarding these data, since recently more than one relaxation time was discovered in the tetraethyl compound which have to be ascribed to different nonequivalent proton groups.* This effect had not been not noticed before. The intermolecular proton rates obtained from the proton and deuteron data in the tetramethylammonium and tetra-ethylammonium ions show an interesting behavior: Using the appropriate self-diffusion constants, the distance of closest approach turns out to be smaller for the larger tetraethyl compound than for the methyl derivative. This seems to indicate cation–cation association in the former case.

Larsen measured the nitrogen rates of tetraalkylammonium salts $(CH_3)_3NR^+$, with R standing for methyl, ethyl, i-propyl, t-butyl, n-hexyl,

---

* M. Holz, private communication.

and some longer chains, at a fixed concentration (0.05 m). From methyl to ethyl the rate increases by more than fivefold, which may be contrasted to an increase of the intramolecular proton rate from the tetramethyl to the tetraethyl salt by a factor of 3.5.[659] Thus a change of the nitrogen quadrupole coupling constant is very likely and for all asymmetric compounds departures from tetrahedral symmetry is the reason for such a change. The longer-chain derivatives should have approximately equal coupling constants, and here the variation in the relaxation rate must arise from different correlation times. A systematic increase from ethyl to n-decyl is observed, but at longer chain length more complicated effects like micelle formation enter.

Anion resonances of chlorine, bromine, and iodine in different alkylammonium salts were studied quite extensively by Lindman et al.[853,854,1393] The interesting effect common to all these systems and already observed earlier[653] is the enormous increase of the anion relaxation rates with alkylammonium cation concentration as compared to alkali halide solutions. Obviously the alkyl groups are responsible for this effect since longer alkyl chains or tetra as compared to monoderivatives favor this effect. The anions relax by quadrupolar interaction, the field gradients arising from electric dipoles or charges external to the ion.[652,1368] Since exchange of D_2O for H_2O increases the relaxation rate by an amount (20%) to be expected for the change of the rotational correlation time, the interaction must occur with surrounding water. On the other hand, the relaxation rate for ^{79}Br in an aqueous 1 m $(C_2H_5)_4NBr$ solution is ten times higher than the infinite dilution value,[853] whereas the water deuteron rates[659] in the same concentration interval are only increased by a factor of 1.3, both compared at 25°C. This finding leads to the conclusion that the change in correlation time is not responsible for this effect but that a considerable change in the field gradient must occur. Obviously the alkyl groups must modify the water around the anion in such a way that the field gradient at the anion is strongly increased, although no statement concerning the range and the kind of the modification of the water can be made. Probably this behavior is common to all nonpolar groups, similar results having recently been found for t-butanol.* Anyhow, this is a typical anion effect; the rate increases are somewhat stronger for iodine than for bromine or chlorine, but added cations, for example, rubidium,[657,1393] show the normal dependence in agreement with the correlation time. The temperature dependence of the anion relaxation times can satisfactorily be explained by

---

* H. G. Hertz and M. Holz, private communication, to be published.

the variation of correlation times, as is shown by a comparison of the activation energies with those of the deuteron rates.

Spin-echo diffusion measurements were reported at 25°C for the symmetric tetraalkylammonium ions from the methyl to the butyl derivative.[657] Here the self-diffusion coefficients of the tetraalkylammonium ions as well as those of water were measured in a concentration range 0–7 m. As described in connection with t-butanol/water mixture, the aim was to investigate the lifetime of the hydration spheres around the nonpolar groups. The water diffusion coefficient in this case is an average from at least three regions connected with the cation-sphere, anion-sphere, and unaffected water. By use of chloride anions, which change the water diffusion coefficient in their sphere negligibly from the pure water value, the average is reduced to two regions. Both measured diffusion constants decrease with increasing salt concentration, but the diffusion constant of the cation is always much smaller than the water value, decreasing with increasing length of the alkyl groups. From the inequality of the diffusion constants one concludes again that no long-lived hydration cages exist, including the low concentration range. Since the decrease of the water diffusion coefficients becomes stronger in the direction methylammonium to propylammonium, butylammonium being an exception, the effect of alkyl groups is to slow down the translational motion as well as the rotation of the water molecules.

Nearly all chemical shift measurements are restricted to the water protons. Hertz and Spalthoff[658] investigated the tetraalkylammonium (methyl through butyl) bromides and nitrates at 25°C and used the methyl peak of the solutes as an internal standard. All shifts are upfield with increasing concentration, less for nitrates than for bromides, and increasing with the alkyl chain length. The plots show a linear behavior. This upfield shift seems to indicate a breaking of hydrogen bonds by the nonpolar groups, opposite to the expected behavior. Many later studies tried to clarify this unexpected situation, mostly confirming the original measurements.[638,738a] The same conclusion was drawn by Eley and Hey[448] from their chemical shift data in the diquaternary ammonium bromides. Recent reinvestigations by Wen et al.* with external standard and by Hertz and Pfliegel† with internal standard show some new features. The shifts of the chlorides and bromides are still upfield (in contrast to one report[588] where downfield shifts for tetramethylammonium chloride were

---

* W. Y. Wen, P. T. Inglefield, Y. Lin, and G. Minott, private communication.
† H. G. Hertz and H. Pfliegel, private communication.

found) but the linearity and the previous order (at concentrations below 1 m) are changed. The concentration dependence of the tetramethylammonium salts is definitely nonlinear (stronger curvature at low concentration) whereas the butyl derivative still showed the linear behavior. The other salts show intermediate behavior, but here the slope increases at higher concentrations. Thus the different curves partly cross each other and the trends become irregular. But the limiting slope of the concentration curves is less upfield in the order butyl, methyl, propyl, and ethylammonium. It is interesting that at least from methyl to ethylammonium the expected downfield trend is observed. Hertz and Pfliegel also measured the water proton shifts at 0°C and 12°C. The lower the temperature, the better is the expected order methyl, ethyl, propyl, and butyl followed.* Wen *et al.* included in their study the fluorides and these yield absolute downfield shifts, increasing in the expected order. Thus it is evident that the anion is responsible for the gross direction of the shift. The latter authors also noticed that the methyl signal in tetrabutylammonium bromide solution is shifted slightly downfield, and this shift could account for some inconsistencies with the internal reference method.

The water ^{17}O chemical shift was compared in the tetramethylammonium and tetrabutylammonium bromide solutions at 25°C.[487] Although the shift is downfield from pure water with increasing concentration in both cases, it is less so for the butyl derivative. This would indicate a breakage of water hydrogen bonds by the compound with the longer alkyl chain.

2.2.2. Miscellaneous

In addition to the study of aqueous solutions of carboxylic acids, their salts were also investigated. Solutions of the normal potassium salts (potassium ions do not change the water relaxation rate at 25°C), from acetate to butyrate, in heavy water were used to study the deuteron relaxation rates of the water and the proton rates of the carboxylates at room temperature.[659] Water ^{17}O rates are available for the acetate and propionate.[487] In all cases the water rates increase with concentration and increasing chain length of the salts, confirming the previous results for the acids. The limiting values of the proton rates at vanishing salt concentration yield the intramolecular contributions, and corresponding correlation times were presented which increase with the anion size. Chemical shifts of the

---

* Recent measurements by Davies *et al.*[382] yield the order methyl, butyl, propyl, and ethyl at 25°C and the same behavior as Hertz and Pfliegel report at 0°C.

water protons in solutions of the corresponding sodium salts are all down-field with respect to pure water but the trend in the homologous series is irregular.[638]

In a series of papers Clifford et al.[315-317] reported chemical shifts of water protons and proton relaxation rates of water and solute in solutions of sodium alkyl sulfates around 30°C. The water relaxation rates were determined in solutions of nondeuterated sulfates by the adiabatic fast passage method and the water–solute interaction was not taken into account. Better–defined results were presented for the solute rates since in this case deuterium oxide was used as the solvent. Quantitative evaluation of these rates indicates that rapid internal motion must take place in the CH_2 chains. The water rates, as usual, increase with concentration and chain length, and the water proton shift is upfield from pure water in these solutions, more so for the longer alkyl groups, as is mostly found at these high temperatures.

3. GENERAL CONCLUSIONS OBTAINED FROM NMR SPECTROSCOPIC STUDIES

A complete relaxation study of an aqueous mixture leads to a knowledge of the individual rotational correlation times of all components over the entire composition range. Such complete investigations have been carried out for the aqueous systems of acetone, acetonitrile, tetrahydrofuran, pyridine, and dimethylsulfoxide (see Figs. 3 and 5). If different nuclear resonances can be observed, correlation times of various vectors within the component molecules are available, which is the case in acetone (proton–proton vector and deuteron field-gradient vector in the methyl group, oxygen field-gradient vector in the polar group), acetonitrile (methyl and polar groups), and pyridine (average correlation time of the proton–proton vectors, nitrogen field-gradient vector, and field-gradient vectors of different deuterons). The knowledge of rotational correlation times is necessary for a qualitative interpretation of intermolecular interactions between the components in the studied mixtures, and even information about the anisotropies of such interactions which are reflected in anisotropic rotational motions becomes available.

All nonelectrolytes (except urea) and electrolytes with alkyl groups lengthen the water correlation times in the water-rich range. This is evident from relaxation rates for proton, deuteron, and ^{17}O resonance of water.

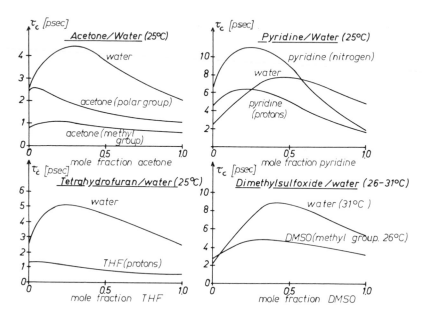

Fig. 5. Rotational correlation times of both components in different mixtures of nonelectrolytes with water. Relaxation rates from various nuclei as reported by different authors[558,581,582,1061] were combined.

In Table I the initial slope at zero concentration of the plot giving the dependence of the relative water relaxation rate $(1/T_1)/(1/T_1)_0$, where $(1/T_1)_0$ is the relaxation rate in pure water, on the molal concentration of the nonaqueous component is listed for several solutes and different resonant nuclei in water. Preferably the relaxation rates of quadrupole relaxing nuclei are summarized, since these are intramolecular rates and directly related to the rotational correlation time. A systematic behavior is apparent: The slopes increase in a homologous series with increasing nonpolar group of the solute. Presentation of these experimental data should suffice, since further evaluation of regional correlation times must be based on certain assumptions. It is necessary to divide the aqueous component into a solute-influenced coordination sphere and the residual water the correlation time of which remains unaltered. Usually the slow exchange case [eqn. (16)] is taken for the calculation of correlation times of water molecules in the solute coordination sphere, and typical results indicate a lengthening by a factor two relative to pure water. The slower rotational motion must reflect stronger intermolecular interaction and whether this is water–water or water–solute interaction can be decided from additional correlation time measurements of the solute.

TABLE I. Relaxation Data of the Aqueous Component at 25°C[a]

Compound	Resonant nucleus in water	$\left[\dfrac{d(1/T_1)/dc_2}{(1/T_1)_0}\right]_{c_2\to0}$	$\left[\dfrac{1/T_1}{(1/T_1)_0}\right]_{x_1\to0}$	Ref.
CH_3OD	D	0.11	—	581
CH_3OH	^{17}O	0.06	—	487
C_2H_5OD	D	0.22	—	581
C_2H_5OH	^{17}O	0.15	—	487
C_3H_7OH	D	0.34	—	659
$(CD_3)_3COH$	H	0.30	—	581
C_4H_8O	D	0.21	1.0	582
$C_4H_8O_2$	D	0.10	0.6	659, 671
CH_3CN	D	0.03	0.4	581
C_2H_5CN	D	0.08	—	659
C_5H_5N	D	0.13	2.0	582
CH_3COOH	D	0.08	—	659
C_2H_5COOH	D	0.13	—	659
C_3H_7COOH	D	0.21	—	659
$(CH_3)_2CO$	D	0.12	0.9	582
$(CH_3)_2CO$	^{17}O	0.07	—	487
$C_2H_5COCH_3$	D	0.17	—	659
$(CD_3)_2SO$	H	0.14	—	659
$(CH_3)_4NBr$	D	0.13	—	659
$(CH_3)_4NBr$	^{17}O	0.17	—	487
$(CD_3)_4NNO_3$	H	0.18	—	659
$(C_2H_5)_4NBr$	D	0.34	—	659
$(C_3H_7)_4NBr$	D	0.88	—	659
$(C_4H_9)_4NBr$	D	0.88	—	659
$(C_4H_9)_4NBr$	^{17}O	1.05	—	487

[a] Column 3 gives the limiting slope of the relative water relaxation rate, where c_2 is the molal concentration of the nonaqueous component and $(1/T_1)_0$ is the relaxation rate of pure water. Column 4 gives the limiting relative water relaxation rate, where x_1 is the mole fraction of water.

TABLE II. Relaxation Data of the Nonaqueous Component[a]

Compound	Solvent	t, °C	Resonant nucleus	$(1/T_1)_{x_2=1}$, sec^{-1}	$(1/T_1)_{x_2 \to 0}$, sec^{-1}	Ref.
CD$_3$OH	H$_2$O	25	D	0.19	0.22	582
CH$_3$OD	D$_2$O	25	H	0.11	0.05	582
CD$_3$CH$_2$OH	H$_2$O	25	D	1.06	0.82	581
CH$_3$CD$_2$OH	H$_2$O	25	D	0.86	0.62	581
C$_2$H$_5$OD	D$_2$O	25	H	0.18	0.07	582
C$_4$H$_9$OD	D$_2$O	32	H	—	0.13	315
(CD$_3$)$_3$COH	H$_2$O	25	D	5.75	2.7	582
(CH$_3$)$_3$COD	D$_2$O	25	H	0.85	0.24	582
C$_4$D$_8$O	H$_2$O	25	D	0.33	0.76	582
C$_4$H$_8$O	D$_2$O	25	H	0.07	0.05	582
C$_4$H$_8$O$_2$	D$_2$O	34	H	—	0.15	314
CD$_3$CN	H$_2$O	25	D	0.15	0.16	581
CH$_3$CN	D$_2$O	25	H	0.06	—	581
CH$_3$CN	D$_2$O	25	^{14}N	246	394	581
C$_5$D$_5$N	H$_2$O	25	D	0.96	2.70	582
C$_5$H$_5$N	D$_2$O	25	H	0.06	0.03	582
C$_5$H$_5$N	D$_2$O	25	^{14}N	675	2030	581
CD$_3$COOH	H$_2$O	25	D	0.76	0.38	582
CH$_3$COOD	D$_2$O	25	H	0.23	0.08	582
CD$_3$CH$_2$COOH	H$_2$O	25	D	1.4	1.1	583[b]
CH$_3$CD$_2$COOD	D$_2$O	25	H	0.20	0.11	583[b]
CH$_3$CD$_2$COOH	H$_2$O	25	D	2.1	1.1	583[b]
CD$_3$CH$_2$COOD	D$_2$O	25	H	0.21	0.06	583[b]
CD$_3$CH$_2$CH$_2$COOH	H$_2$O	25	D	1.5	1.2	583[b]
CH$_3$CD$_2$CD$_2$COOD	D$_2$O	25	H	0.26	0.14	583[b]
CH$_3$CD$_2$CH$_2$COOH	H$_2$O	25	D	2.7	2.0	583[b]
CD$_3$CH$_2$CD$_2$COOD	D$_2$O	25	H	0.21	0.10	583[b]
CH$_3$CH$_2$CD$_2$COOH	H$_2$O	25	D	4.0	1.7	583[b]
CD$_3$CD$_2$CH$_2$COOD	D$_2$O	25	H	0.34	0.09	583[b]

TABLE II. (*Continued*)

Compound	Solvent	t, °C	Resonant nucleus	$(1/T_1)_{x_2=1}$, sec^{-1}	$(1/T_1)_{x_2 \to 0}$, sec^{-1}	Ref.
$(CD_3)_2CO$	H_2O	25	D	0.20	0.26	582
$(CH_3)_2CO$	D_2O	25	H	0.06	0.05	582
$(CH_3)_2CO$	D_2O	25	^{17}O	163	359	581
$(CH_3)_2SO$	D_2O	26	H	0.31	0.15	1061
$(CH_3)_4N^+$	D_2O	25	H	—	0.10	657, 659
$(CD_3)_4N^+$	H_2O	25	D	—	0.71	657
$(C_2H_5)_4N^+$	D_2O	25	H	—	0.35	657, 659
$(C_2D_5)_4N^+$	H_2O	25	D	—	2.35	657
$(C_3H_7)_4N^+$	D_2O	25	H	—	0.95	659
$(C_4H_9)_4N^+$	D_2O	25	H	—	2.10	659
$C_2H_5SO_4^-$	D_2O	32	H	—	0.07	315
$C_4H_9SO_4^-$	D_2O	32	H	—	0.17	315
$C_6H_{13}SO_4^-$	D_2O	32	H	—	0.27	315
$C_8H_{17}SO_4^-$	D_2O	32	H	—	0.32	315

[a] x_2 is the mole fraction of the nonaqueous component.
[b] R. Tutsch, private communication.

Relaxation data of the nonaqueous components are summarized in Table II. The rates in the pure liquid component and at infinite dilution in water for different nuclear resonances are presented. It should be noted that the proton rates in the pure phase include both the intramolecular and intermolecular contributions. Using the intramolecular proton rates, correlation times were calculated, by use of eqn. (8), from known internuclear distances and the results are collected in Table III. Since we know the quadrupole coupling constants for deuterons are hardly affected in aqueous solutions, their corresponding correlation times can also be obtained from the deuteron relaxation rates with a constant coupling constant around 170 kHz. Less certain are the correlation times derived from other quadrupole relaxing nuclei, since the magnitude or concentration dependence of the corresponding quadrupole coupling constants are not so well established. It is seen from Table III that the correlation times at infinite dilution

TABLE III. Correlation Times of the Nonaqueous Component at 25°C in D_2O,
Derived from Proton Relaxation Rates[a]

Compound	$10^{-10}F$, sec^{-2}	$\tau_c{}^0$, psec	$\tau_c{}^\infty$, psec
CH_3OD	5.28	0.9	1.0
C_2H_5OD	4.83	2.2	1.6
$(CH_3)_3COD$	5.88	7.9	3.7
C_4H_8O	3.81	0.6	1.3
C_5H_5N	0.60	1.7	4.5
CH_3COOD	5.28	3.0	1.5
$(CH_3)_2CO$	5.55	0.6	0.8
$(CH_3)_2SO^b$	5.55	3.2	2.7
$(CH_3)_4N^+$	6.51	—	1.5
$(C_2H_5)_4N^+$	6.48	—	5.4
$(C_3H_7)_4N^+$	5.67	—	17

[a] $F = (1/T_1)_{intra}\tau_c^{-1}$ [see eqn. (8)] calculated from known internuclear distances; $\tau_c{}^0$ is the correlation time in the pure nonaqueous liquid; and $\tau_c{}^\infty$ is the correlation time in the infinite dilution limit.
[b] 26°C.

in water are usually smaller or comparable to the pure liquid values, tetrahydrofuran and especially pyridine being exceptions. It is more instructive to observe the total concentration dependence as shown in Figs. 3 and 5. Taking the behavior of the single-component correlation times as a criterion, the mixtures can be classified as weakly or strongly interacting systems. In the weakly interacting systems acetone, acetonitrile, and tetrahydrofuran the correlation times of the components are quite different and no crossing-over effects, as are seen in the strongly interacting systems pyridine and dimethylsulfoxide, occur. In acetone and acetonitrile, where the correlation times of the polar groups could also be derived, some interaction between the polar groups and water is evident, but even here the correlation times of the polar groups remain below those of water over the entire composition range. On the other hand, the proton correlation times are much lower and practically unaltered over the whole range, and this suggests that internal motions of the methyl groups exist which are not restricted by the surrounding water. The question which was raised before as to the origin of the correlation time increase in the aqueous component in the water-rich range must then be answered that interactions between water molecules other than solute–water interactions are the cause. Strong interactions between water and pyridine at the nitrogen site are responsible for anisotropic rotational motion of pyridine molecules in this system, as can be deduced from the difference in nitrogen and proton correlation times. The strong interaction in the dimethylsulfoxide/water system is certainly through the polar group, since evidence exists for internal rotation of the methyl groups, but no polar group correlation times are available yet. In particular, we may therefore conclude that the nonpolar methyl or larger alkyl groups do not interact directly with water but that they cause the neighboring water molecules to interact more strongly with each other, as is also indicated by the increase of water correlation times in homologous series of increasing alkyl chain length.

Just as the nonelectrolytes increase the water relaxation rates, so the self-diffusion constant of the aqueous component is always decreased in the water-rich range. The limiting slopes of the relative water diffusion versus solute concentration curves at zero solute concentration are collected in Table IV. These data correspond roughly to the equivalent relaxation results from Table I. Thus, besides the rotational motion, the translation of water molecules is also restricted due to the stronger water–water interactions. Again it could be shown that solute–water interactions, i.e., common translational motion of the solute together with all neighboring water molecules, is improbable down to rather low concentrations of the solute. This was concluded from a comparison of the self-diffusion constants for both components in the water-rich range, where the solute diffusion constants always remain considerably below the water values (the self-diffusion constants relative to the water diffusion constant of the monaqueous component in the pure liquid and at vanishing concentration in water are also included in Table IV). Thus in contrast to the rotational motion of the solutes in their water cage, the translational motion is always slower than that of water, as is to be expected from the larger masses of the solute molecules.

The increased water–water interaction in the water-rich range caused by nonpolar solutes could be rationalized as due to an increase or strengthening of water hydrogen-bonding. However, hydrogen-bonding produces typical downfield shifts of corresponding signals in a high-resolution spectrum. Downfield shifts of the water proton signals relative to pure water in solutions of nonelectrolytes are in fact often found, mostly at lower temperatures, but a superposition of various other contributions to the

TABLE IV. Spin-Echo Self-Diffusion Data from Proton Resonance[a]

Compound[b]	t, °C	X = D; Z = H		X = H; Z = D		Ref.
		$[(dD/dc_2)/D_0]_{c_2 \to 0}$	$(D/D_0)_{x_1 \to 0}$	$(D_2/D_0)_{x_2=1}$	$(D_2/D_0)_{x_2 \to 0}$	
C_2X_5OZ	25	−0.12	—	0.44	0.5	581, 582
$(CX_3)_3COZ$	0	−0.29	—	—	0.3	581
C_4X_8O	25	−0.13	0.9	1.4	0.4	582
$C_4X_8O_2$	34	−0.08	—	0.55	0.4	313
CX_3CN	25	−0.02	—	2.1	0.8	581
C_5X_5N	25	−0.10	0.8	0.76	0.4	582
$(CX_3)_2CO$	25	−0.07	1.7	2.1	0.6	581, 582
$(CX_3)_2SO$	26	−0.10	0.4	0.33	0.3	1061
$(CX_3)_4NCl$	25	−0.11	—	—	0.44	657
$(C_2X_5)_4NCl$	25	−0.24	—	—	0.29	657
$(C_3H_7)_4NCl$	25	−0.83	—	—	0.21	657
$(C_4H_9)_4NCl$	25	−0.73	—	—	0.16	657

[a] Column 3 gives the limiting slope of the relative water diffusion constant, with c_2 the molal concentration of the nonaqueous component and D_0 the diffusion constant of pure water. Column 4 gives the limiting relative water diffusion constant, with x_1 the mole fraction of water. Column 5 gives the relative diffusion constant of the nonaqueous component, with x_2 the mole fraction of the nonaqueous component. Column 6 gives the limiting relative diffusion constant of the nonaqueous component.

[b] Diffusion data for the aqueous component are measured in mixtures of the deuterated solute in H_2O (X = D and Z = H) with the exception of tetrapopylammonium and tetrabutyl-ammonium chloride, whereas the diffusion data for the nonaqueous component are determined in mixtures of the light solute (X = H, Z = D) in D_2O.

shift often obscures the hydrogen-bonding effect. In Table V the initial slope of the plot representing the dependence of the water proton or ^{17}O shift (relative to pure water) on the molal concentration of the solute at zero concentration is listed for several nonelectrolytes at different temperatures. The behavior is often irregular although in homologous series the expected downfield trend is sometimes detected. Selecting the series of alcohols at 0°C and correcting for the solute proton (alcoholic hydroxyl proton) and polar group effect (interaction of the hydroxyl group with the water), the nonpolar group effect is deduced. If the average shift per water molecule in the nearest-neighbor coordination sphere around the alcohol molecule is calculated using molecular models to estimate coordination

TABLE V. Chemical Shift Data for the Aqueous Component[a]

Compound	t, °C	Resonant nucleus in water	$(d\delta/dc_2)_{c_2 \to 0}$	Ref.
CH_3OH	0	H	−0.019	1388
CH_3OH	36	H	−0.012	1167
CH_3OH	25	^{17}O	+0.2	487
C_2H_5OH	0	H	−0.035	1388
C_2H_5OH	36	H	−0.008	1167
C_2H_5OH	25	^{17}O	+0.15	487
$n\text{-}C_3H_7OH$	0	H	−0.048	1388
$n\text{-}C_3H_7OH$	36	H	−0.008	1167
$i\text{-}C_3H_7OH$	0	H	−0.065	1388
$n\text{-}C_4H_9OH$	0	H	−0.020	1388
$n\text{-}C_4H_9OH$	36	H	+0.004	1167
$i\text{-}C_4H_9OH$	0	H	−0.041	1388
$(CH_3)_3COH$	0	H	−0.060	29, 1388
$(CH_3)_3COH$	25	H	−0.024	29
C_2H_4O	34	H	+0.028	312
$CH_3\text{-}C_2H_3O$	34	H	+0.028	312
C_3H_6O	34	H	+0.023	312
$C_3H_6O_2$	34	H	+0.030	312
C_4H_8O	0	H	−0.009	1388
C_4H_8O	34	H	+0.020	312
$C_4H_8O_2$	34	H	+0.031	312
$C_5H_{10}O$	0	H	−0.001	1388
CH_3CN	25	H	+0.029	658
C_2H_5CN	25	H	+0.047	658
$(C_2H_5)_2NH$	0	H	−0.083	1388
$(C_2H_5)_3N$	0	H	−0.089	1388
$[(CH_3)_2CH]_2NH$	0	H	−0.103	1388
$(CH_3)_3CNH_2$	0	H	−0.015	1388
CH_3COOH	36	H	−0.061	1167

TABLE V. (*Continued*)

Compound	t, °C	Resonant nucleus in water	$(d\delta/dc_2)_{c_2 \to 0}$	Ref.
C_2H_5COOH	36	H	−0.053	1167
C_3H_7COOH	36	H	−0.076	1167
$(CH_3)_2CO$	0	H	0	1388
$(CH_3)_2CO$	25	H	+0.009	658
$(CH_3)_2CO$	25	^{17}O	+0.15	487
$C_2H_5COCH_3$	0	H	−0.003	1388
$C_2H_5COCH_3$	25	H	+0.017	658
$(CH_3)_4NCl$	0	H	+0.072	(^{b})
$(CH_3)_4NCl$	25	H	+0.036	(^{b})
$(CH_3)_4NBr$	0	H	+0.112	(^{b})
$(CH_3)_4NBr$	25	H	+0.081	(^{b})
$(CH_3)_4NBr$	25	^{17}O	−1.2	487
$(C_2H_5)_4NBr$	0	H	+0.078	(^{b})
$(C_2H_5)_4NBr$	25	H	+0.063	(^{b})
$(C_3H_7)_4NCl$	22	H	+0.032	(^{c})
$(C_3H_7)_4NBr$	0	H	+0.067	(^{b})
$(C_3H_7)_4NBr$	25	H	+0.076	(^{b})
$(C_4H_9)NCl$	22	H	+0.074	(^{c})
$(C_4H_9)_4NBr$	0	H	+0.063	(^{b})
$(C_4H_9)_4NBr$	25	H	+0.094	(^{b})
$(C_4H_9)_4NBr$	25	^{17}O	−0.7	487
$C_2H_5SO_4Na$	34	H	+0.09	316
$C_4H_9SO_4Na$	34	H	+0.10	316
$C_6H_{13}SO_4Na$	34	H	+0.13	316
$C_8H_{17}SO_4Na$	34	H	+0.20	316

[a] Column 4 gives the limiting slope of the chemical shift, with c_2 the molal concentration of the nonaqueous component and δ the shift in ppm relative to pure water, upfield shift positive, downfield shift negative.
[b] H. G. Hertz and H. Pfliegel, private communication.
[c] W. Y. Wen, P. T. Inglefield, Y. Lin, and G. Minott, private communication.

numbers, interesting figures, depending on the solute size, are obtained. These results are shown in Fig. 6, where the excess proton shifts per water molecule are plotted against the two principal axes (a and b) of the ellipsoid of rotation representing the alcohol molecule (a is along the C—O bond, b is perpendicular to it) in a three-dimensional picture. This admittedly crude picture represents a trough; i-propanol and t-butanol, with the largest downfield shifts, are at the bottom of the trough. Obviously alcohole molecules which are spherical in shape (with a diameter around 5 Å) favor most effectively the hydrogen bonding of the water molecules making up their cage.

As compared to the water shifts, the shifts of protons in nonpolar solute groups were mostly found to be negligibly small, at least in the water-rich range, which again supports the view of weak solute–water interaction. Pyridine/water, also based on this experimental criterion, deserves its classification as a strongly interacting system.

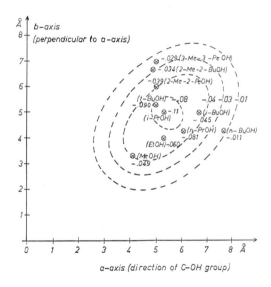

Fig. 6. Dependence of the water proton downfield shifts at 0°C and infinite dilution of alcohols on the geometry of the solute molecule. The numbers indicate the magnitude of the downfield (negative) shift to be attributed to the nonpolar group after subtraction of polar group effects per water molecule in the first coordination sphere around the solute. Definite numbers of water molecules in the coordination sphere were estimated to derive these figures. (From Wen and Hertz.[1388])

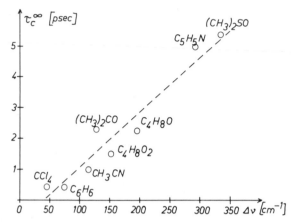

Fig. 7. Relation between the correlation time τ_c^∞ and the OH stretching vibration frequency shift $\Delta\nu$ (relative to water vapor) of water at infinite dilution in various organic solvents. The infrared data are taken from Refs. 599 and 677.

So far attention has been focused primarily on the water-rich range. With decreasing water concentration the stronger water–water interaction is replaced by the weaker solute–water interaction and consequently the rotational correlation times of the water molecules decrease after having passed through a maximum. The limiting correlation times at vanishing water concentration (see the deuteron rates in Table I or the data in Figs. 3 and 5) are then a measure of the solute–water interaction just as are infrared shifts, and a plot of these times against water infrared shifts, similar to the discussed limiting chemical shift correlations, is given in Fig. 7. A satisfactory correlation is indeed observed.

Some indication for hydrophobic interaction in aqueous mixtures of carboxylic acids, i.e., solute–solute interaction through the nonpolar groups in the water-rich region, was obtained from particularly large intermolecular proton relaxation rates at low acid concentrations. On the other hand, the correlation times of the nonpolar groups are much lower in the aqueous environment than in the pure liquid acids. Thus stable acid polymers present in the pure phase, which are held together by strong interactions of the polar groups, must be dissociated in water in favor of associates held together by weaker hydrophobic interaction.

CHAPTER 11

Molecular Theories and Models of Water and of Dilute Aqueous Solutions

A. Ben-Naim

Department of Inorganic and Analytical Chemistry
The Hebrew University of Jerusalem
Israel

1. INTRODUCTION

"Models are, for the most part, caricatures of reality, but if they are good, then, like good caricatures, they portray, though perhaps in distorted manner, some of the features of the real world."[721]

To begin with, models for aqueous solutions are basically models for water. Therefore most of this chapter is concerned with models for water. A few topics, however, will apply specifically to dilute aqueous solutions.

Excellent reviews of the various models which have been applied to water and also to aqueous solutions are available[157,444,507,700,734]* and there seems to be no need for a further review of this sort. Therefore the emphasis in this chapter will not be on surveying the literature and discussing diverse models for water. Instead we shall focus attention on some fundamental concepts in the study of the properties of liquid water and aqueous solutions.

The employment of "models" in the study of aqueous solutions is certainly not an unfamiliar practice. Yet there seems to be a wide range of views of what a model is or should be. We shall not attempt to construct a definition of the term "model"; this is a philosophical question and for

---

* Also see Volume 1, Chapter 14.

the purpose of this chapter we feel we cannot do better than to quote M. Kac on this matter, as given above.

In a superficial fashion we may distinguish between two major classes of models. The first one comprises models which are constructed in such a way as to include the most important features of water and therefore are expected to reveal behavior similar to the real system. The term "models of water" is appropriate in this case and as a matter of fact most common models employed in the study of aqueous solutions fall in this class.

The second class consists of somewhat less ambitious models. These are from the outset not designed to simulate the behavior of the real system. Instead they are aimed at the study of a specific aspect which is "isolated" or "extracted" from the general problem concerning the real system. Models in this class may be referred to as "models of phenomena," examples of which are various one-[126,871] and two-dimensional[127,148] models which deal with some specific and well-defined problems relevant to aqueous fluids.

The ultimate test of the validity of a model in the first class is its ability to predict the experimental behavior of the real system. It is more difficult to state the criterion for judging the usefulness of a model in the second class. In fact there seems to be no general agreement among scientists on the merits of such models. In the author's opinion a model in this class is considered useful if, as a result of its study, we have gained a new insight into at least one feature of the behavior of the real system. This is surely a very subjective matter and different authors may judge differently the utility of such a model and the amount of information it is capable of conveying. We shall return to a specific example in Section 4, where we shall try to indicate in what sense such a model is indeed useful. One should recognize that the whole domain of "models of phenomena" has been created because our systems are too complicated, and hence we are compelled to resort to simplified and artificial models where all we can hope to achieve is partial answers to some of our questions.

The next section is devoted to a detailed discussion of the so-called mixture model approach to water and aqueous solutions. We believe that this topic deserves a profound scrutiny since it has played a central role in developing our understanding of aqueous fluids. We shall begin with a general and rigorous construction of a mixture model approach, and then proceed to examine some specific ad hoc mixture models for water. Some questions of fundamental importance in connection with the current controversy on the question of whether water is or is not a mixture are discussed.

In Section 3 we shall further elaborate on some concepts such as the "structure of water," "structural changes," and "local structure around a solute particle." All of these concepts are ubiquitous in the literature of aqueous solutions, yet it seems that they lack a unified definition and interpretation.

Section 4 treats a single example of a model belonging to the second class mentioned above. The purpose of this section is twofold: In the first place we shall demonstrate the very general purpose of a "model of a phenomenon." Second we shall illustrate in a more specific fashion some of the general concepts which were introduced in previous sections.

2. THE MIXTURE MODEL APPROACH

2.1. Introduction

The idea of viewing liquid water as a mixture of various species is probably very old.* It is also quite safe to state that the so-called mixture model approach has held a central role in the development of our understanding of the properties of water and its solutions. Recently, however, a vigorous debate has been taking place regarding the validity of this approach. Some serious reservations have been put forward regarding the practice of viewing liquid water as a mixture of a number of species or components. The major argument raised against this approach is that no direct experimental evidence exists in its favor, hence, one concludes, its application must be viewed as being basically speculative.

The purpose of this section is to examine the frequently raised question, "Is water a mixture?" The general answer is in the affirmative, if one wants to view it in that way, and in fact this is so for any fluid. Yet one should clearly distinguish between the general ideal of the mixture model approach on the one hand, and the applications of some specific mixture models proposed for water on the other. In the following two subsections we shall provide a firm basis for the general idea of the mixture model approach; in doing so we hope to revive the confidence in this procedure, which we believe has been quite useful for interpreting various properties of aqueous solutions. The central theme of this analysis is that the so-called "mixture model" and "continuous" approaches are equivalent from the formal point of view. Therefore experimental information cannot be used as evidence

* For a review of earlier theories of water and aqueous solutions see Ref. 277, for example.

either to support or to refute any particular approach. The second question that should be examined is the validity of some specific mixture models that have been proposed for water. The formalism carried out in the next section will be sufficiently general so that any conceivable mixture model can be embedded in its framework. In doing so, we shall be able to judge various models in the light of the general and exact formalism. Each specific mixture model is viewed as an approximation derived from the general formalism by neglecting certain regions in the configurational space. The validity of the various approximations cannot as yet be assessed by using any absolute criterion.

2.2. A Discrete Mixture Model

The present discussion will be carried out for a system of N molecules contained in a volume V and at temperature T. We shall employ classical statistical mechanics throughout. The procedure described in this section will be quite general and may be applied to any fluid.

Let X_i be the generalized vector comprising all the coordinates required to specify the configuration of the ith molecule. Assuming for simplicity that the molecules are rigid, it will be sufficient to employ six coordinates for the location and orientation of the molecule; thus

$$X_i = (R_{ix}, R_{iy}, R_{iz}, \phi_i, \theta_i, \psi_i) \tag{1}$$

Let $X^N = X_1, \ldots, X_N$ stand for a specific configuration of the system; then the probability density of observing such a configuration in the TVN ensemble is[669]

$$P(X^N) = \frac{\exp[-\beta U_N(X^N)]}{\int \cdots \int dX^N \exp[-\beta U_N(X^N)]} \tag{2}$$

where $\beta = (kT)^{-1}$, with k the Boltzmann constant and $U_N(X^N)$ the potential energy of interaction among the N molecules being at the specified configuration X^N.

Let $Q(X^N)$ be any function of the configuration X^N; then the average value of this function in the TVN ensemble is given by

$$\bar{Q} = \int \cdots \int dX^N P(X^N) Q(X^N) \tag{3}$$

We shall now apply relation (3) to some specific functions, the outcome of which will enable us to view a one-component system as a mixture of various species.

For a specific configuration of the system $\mathbf{X}^N$ we define the concept of neighbors in the following manner. We pick a specific molecule, say the ith one, and draw a sphere of radius R_c about its center $\mathbf{R}_i$. Each molecule j $(j \neq i)$ whose center falls within the R_c sphere of molecule i will be called a neighbor to molecule i. The number of neighbors of molecule i, as defined above, may be referred to as the coordination number (CN) of that molecule. Clearly, for any specific configuration $\mathbf{X}^N$ one may determine the CN of all the molecules in the system. A two-dimensional demonstration of this procedure is depicted in Fig. 1.

So far we have not specified the value of R_c, which, in principle, may be chosen at will. However, in order to conform with the current meaning of the term CN, it is reasonable to make the choice of R_c somewhere in the range

$$\sigma \lesssim R_c \lesssim 1.5\sigma \tag{4}$$

where σ is the effective diameter of the molecules comprising our system.

Mathematically, the CN of the ith molecule at the specific configuration $\mathbf{X}^N$ is defined by

$$A_i(\mathbf{X}^N, R_c) = \sum_{j=1,j\neq i}^{N} H(|\mathbf{R}_j - \mathbf{R}_i| - R_c) \tag{5}$$

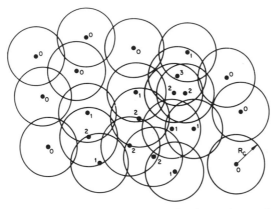

Fig. 1. A two-dimensional example of counting to obtain the coordination number of each molecule for a specific configuration of the whole system. A sphere of radius R_c is drawn about the center of each molecule, indicated by the dark circles. The numbers written near each center are the coordination numbers corresponding to this particular example. There are altogether 23 particles distributed in the following manner: $N_0 = 9$, $N_1 = 7$, $N_2 = 6$, $N_3 = 1$.

where $H(x)$ is a unit step function defined as

$$
\begin{aligned}
H(x) &= 0 \quad \text{if} \quad x > 0 \\
&= 1 \quad \text{if} \quad x \leq 0
\end{aligned}
\tag{6}
$$

Clearly, each molecule j which happens to be in the R_c sphere of molecule i will contribute unity to the sum on the rhs of eqn. (5). Therefore $A_i(\mathbf{X}^N, R_c)$ simply gives the CN of the ith molecule at the configuration $\mathbf{X}^N$. The value of the CN for each molecule is indicated in Fig. 1 next to the corresponding centers.

Next we define the "counting function"

$$
N_K(\mathbf{X}^N) = \sum_{i=1}^{N} \delta[A_i(\mathbf{X}^N, R_c) - K]
\tag{7}
$$

Here $\delta(M - K)$ stands for the Kronecker delta function

$$
\begin{aligned}
\delta(M - K) &= 1 \quad \text{if} \quad M = K \\
&= 0 \quad \text{if} \quad M \neq K
\end{aligned}
\tag{8}
$$

The meaning of $N_K(\mathbf{X}^N)$ (the dependence on R_c, presumed to have been fixed, is suppressed) is the following: For each configuration $\mathbf{X}^N$ we scan the CN of all molecules; those which have CN equal to K will contribute unity to the sum on the rhs of eqn. (7). Therefore the quantity $N_K(\mathbf{X}^N)$ counts the number of molecules the CN of which is equal to K. The values of $N_K(\mathbf{X}^N)$ corresponding to the specific example of Fig. 1 are listed in the caption to this figure.

Finally, the counting function defined in eqn. (7) is employed in the general relation (3) to produce the corresponding average number of molecules possessing CN equal to K:

$$
\begin{aligned}
N_K &= \int \cdots \int d\mathbf{X}^N \, P(\mathbf{X}^N) N_K(\mathbf{X}^N) \\
&= \sum_{i=1}^{N} \int \cdots \int d\mathbf{X}^N \, P(\mathbf{X}^N) \, \delta[A_i(\mathbf{X}^N, R_c) - K] \\
&= N \int \cdots \int d\mathbf{X}^N \, P(\mathbf{X}^N) \, \delta[A_1(\mathbf{X}^N, R_c) - K]
\end{aligned}
\tag{9}
$$

(for simplicity of notation we shall not use a bar over N_K to indicate that this is an average quantity). The third form on the rhs of eqn. (9) arises from the fact that all particles are equivalent; therefore the sum over the index i produces N equal terms. The mole fraction of molecules with CN equal to K is

$$x_K = N_K/N = \int \cdots \int d\mathbf{X}^N \, P(\mathbf{X}^N) \, \delta[A_1(\mathbf{X}^N, R_c) - K] \qquad (10)$$

with the normalization condition

$$\sum_{K=0}^{\infty} x_K = \int \cdots \int d\mathbf{X}^N \, P(\mathbf{X}^N) = 1 \qquad (11)$$

Note that since the particles exert strong repulsive forces when brought very close to each other, the occurrence of many neighbors to a given molecule will be a very improbable event. Therefore the sum over K in eqn. (11) will effectively extend from $K = 0$ to, say, $K = 12$ for the particular choice of R_c made in eqn. (4), i.e., all x_K with larger values of K will be negligibly small.

It is instructive to pause for a reinterpretation of the integral on the rhs of eqn. (10). We recall that $P(\mathbf{X}^N)$ is the probability density for observing a specific configuration $\mathbf{X}^N$. Therefore if we sum over all configurations for which a specific molecule, say 1, has CN equal to K, we get the probability of that event, i.e.,

$$P(K) = \underbrace{\int \cdots \int}_{\begin{Bmatrix} \text{over configurations} \\ \text{for which molecule 1} \\ \text{has CN equal to } K \end{Bmatrix}} d\mathbf{X}^N \, P(\mathbf{X}^N) = \int \cdots \int d\mathbf{X}^N \, P(\mathbf{X}^N) \, \delta[A_1(\mathbf{X}^N, R_c) - K]$$

$$(12)$$

Thus we have the identity

$$P(K) = x_K \qquad (13)$$

which is also quite obvious on intuitive grounds; the mole fraction of molecules having a certain property is equal to the probability that a specific particle, say 1, will have that property. In this example the property is the CN, but the statement made is more general and we shall encounter some other examples in the following.

The normalization condition (11) simply means that the probability that molecule 1 will possess any value of CN, 1, 2, ..., must be equal to unity. On the other hand, the individual quantities x_K defined in eqn. (10) correspond to a split of the entire configurational space of our system into subspaces; each subspace is characterized by the condition that a specific molecule, say 1, has CN equal to K. A schematic description of

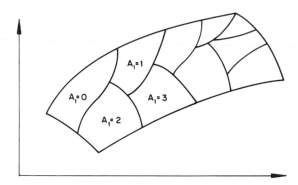

Fig. 2. A schematic description of the split of the entire
configurational space into subspaces; each subspace is
characterized by the condition that particle 1 has a certain coordination number, as indicated in the various
regions.

such a split is depicted in Fig. 2. The latter meaning of the quantities x_K
will be employed in Section 2.5 to examine various specific mixture models
for water. For the present we shall focus our attention on the meaning of
x_K as the mole fractions of certain species. The vector

$$\mathbf{x} = (x_0, x_1, \dots) \tag{14}$$

gives the "composition" of the system when viewed as a *mixture* of various
species. These species may be referred to as quasicomponents; they should
be distinguished from real components in two respects: In the first place
the quasicomponents do not differ in their chemical composition or structure but are characterized by the nature of their local environment, which
in this case has been the CN. The more important difference is that the
composition vector $\mathbf{x}$ cannot be "prepared" in an arbitrary manner. In a
real mixture of the composition N_1 molecules of one species, N_2 molecules
of a second species, etc. we can change the numbers N_i in an arbitrary
manner; this is not the case in our mixture since there exists a strong
dependence among the N_i (or the x_i). One consequence of this restriction
is that the quasicomponents have no existence in a pure state.

There are some analogies and some differences between a mixture of
quasicomponents and a mixture of chemically reacting species. For simplicity, let us observe the following dimerization reaction

$$2A \rightleftharpoons A_2 \tag{15}$$

In general, the numbers of molecules N_A and N_{A_2} of each species are not independent variables. The condition of chemical equilibrium imposes a relation between these two quantities. In the same manner there exists an equilibrium condition on the various x_i in eqn. (14).

In some cases, however, it is possible to find an inhibitor to the reaction (15) (or, equivalently, to remove an indispensable catalyst for the reaction); then the numbers N_A and N_{A_2} become virtually independent, i.e., we can prepare a system with any conceivable composition.

The subtle difference between our system of quasicomponents and a mixture of chemically reacting species is the following. Suppose we could find a hypothetical inhibitor such that the transformation among the various species is "frozen in." Then the equilibrium condition does not impose dependence among the x_i. But we shall still face another kind of dependence which arises from the very definition of our species, i.e., we shall still be unable to prepare a mixture with any arbitrary composition $\mathbf{x}$. The reason is that our species are distinguished by their local environment, which is built of molecules themselves belonging to some species. Therefore if we make a choice of some of the N_i (or the x_i), then there is a strong restriction on the values attained by the remaining N_i. A simple example will make this difficulty clear.

Let N be the total number of molecules and R_c be chosen as in eqn. (4). The composition $(N_0 = N - 1,\ N_{12} = 1,\ N_1 = N_2 = \cdots = N_{11} = N_{13} = \cdots = 0)$ is impossible.* Having one molecule with 12 neighbors means that the neighboring molecules themselves must have at least one neighbor and therefore we cannot require that $N_0 = N - 1$. Similarly there exist many other impossible sets $(N_0, N_1, \ldots)$ which are inconsistent with the definition of our species. Therefore this restriction should be kept in mind whenever this particular mixture model point of view is adopted. We should like to note that other rigorous mixture models can be defined which avoid the kind of restrictions made above. For instance, one can systematically classify molecules in clusters of various sizes (or shapes) and thereby construct different mixture model approaches. We feel, however, that classification into single molecular species is simpler and more appropriate for the purposes of this chapter.

Because of the meaning of the vector $\mathbf{x}$ as a distribution function we have recently[152] referred to it as a "quasicomponent distribution function" (QCDF).

---

* Note that a set of numbers $N_1(\mathbf{X}^N), N_2(\mathbf{X}^N), \ldots$ may be impossible for a specific configuration $\mathbf{X}^N$. However, the same set of numbers $N_1, N_2, \ldots$ may be attainable as average quantities.

2.3. A Continuous Mixture Model

In this section we shall generalize the procedure carried out in the previous section and include a classification procedure according to a continuous parameter. Two examples will be discussed which are believed to be of special relevance to current mixture model theories of water.

We recall the main steps of the procedure carried out in the previous section. First we have defined a local property of the ith molecule for each configuration $\mathbf{X}^N$. Next we have introduced a counting function, eqn. (7), into relation (3) to obtain the average number of molecules having the specified local property.

The same procedure can now be followed but with a different choice of the local property. The two properties discussed here will be the volume of the *Voronoi polyhedron* (VP) and the *binding energy* (BE) of the molecules.

Let $\mathbf{X}^N$ be a specific configuration of a system in the TVN ensemble. The Voronoi polyhedron (VP) or Dirichlet region of the molecule i is defined as follows[160,342,429]: Let l_{ij} be the segment connecting the centers of molecules i and j. Let P_{ij} be the plane perpendicular to l_{ij} and bisecting it. The plane P_{ij} divides the entire space into two halves. Let V_{ij} be the half that includes the center of molecule i. The VP of molecule i is defined as the intersection of all V_{ij}:

$$\bigcap_{j=1, j \neq i}^{N} V_{ij}(\mathbf{R}_i, \mathbf{R}_j) \tag{16}$$

An example of a construction of a VP is depicted in Fig. 3. It is clear from

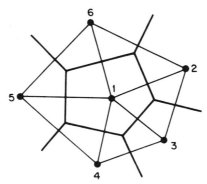

Fig. 3. Construction of the Voronoi polygon of particle 1 in a two-dimensional system of particles.

definition (16) that the VP of molecule i includes all the points in space which are nearer to $\mathbf{R}_i$ than to any other $\mathbf{R}_j$ ($j \neq i$).

Let us denote by $\psi_i(\mathbf{X}^N)$ the volume of the VP of particle i at a given configuration $\mathbf{X}^N$. This property is now employed for the classification into quasicomponents. In analogy with eqn. (7) we define the "counting function" corresponding to this property,

$$N(\phi, \mathbf{X}^N)\, d\phi = \sum_{i=1}^{N} \delta[\psi_i(\mathbf{X}^N) - \phi]\, d\phi \qquad (17)$$

which is the number of molecules whose VP has a volume between the values of ϕ and $\phi + d\phi$. In eqn. (17) the Dirac delta function replaces the Kronecker delta function employed in eqn. (7).

Finally, the average number of molecules in the TVN ensemble having a VP of volume between ϕ and $\phi + d\phi$ is

$$N(\phi)\, d\phi = d\phi \int \cdots \int d\mathbf{X}^N\, P(\mathbf{X}^N) \sum_{i=1}^{N} \delta[\psi_i(\mathbf{X}^N) - \phi]$$

$$= N\, d\phi \int \cdots \int d\mathbf{X}^N\, P(\mathbf{X}^N)\, \delta[\psi_1(\mathbf{X}^N) - \phi] \qquad (18)$$

where, to obtain the second form on the rhs of eqn. (18), we have again used the fact that the molecules are equivalent.

The mole fraction of molecules having a VP of volume between ϕ and $\phi + d\phi$ is

$$x(\phi)\, d\phi = N(\phi)\, d\phi / N \qquad (19)$$

and fulfills the normalization condition

$$\int_0^{\infty} x(\phi)\, d\phi = 1 \qquad (20)$$

The procedure carried out so far enables one to view a one-component system as a mixture of an infinite number of quasicomponents. The ϕ component is the one that has a VP of volume between ϕ and $\phi + d\phi$ and its concentration is given by eqn. (19). Thus $x(\phi)$ may be viewed as a continuous QCDF.[152]

A comment on nomenclature is now in order. A distinction is frequently made in the literature between the so-called "mixture model" and "continuous" approaches to liquid water. We feel that this terminology is inappropriate and somewhat misleading. A better classification distinguishes between a "one-component" and a "multicomponent" point of view;

within the latter a reclassification may be made between "discrete" (finite or infinite) and "continuous" cases according to the nature of the parameter employed in the classification of molecules into quasicomponents.

We shall now turn to a second example of a continuous mixture model which may have some relevance to specific models for water. The property chosen here is the binding energy (BE) of molecule i at configuration $\mathbf{X}^N$, defined by

$$B_i(\mathbf{X}^N) = \sum_{j=1, j \neq i}^{N} U_{ij}(\mathbf{X}_i, \mathbf{X}_j) \tag{21}$$

This is simply the total interaction energy of molecule i with the rest of the system at the specified configuration $\mathbf{X}^N$. The definition in eqn. (21) implies pairwise additivity of the total potential energy, i.e.,

$$U_N(\mathbf{X}^N) = \sum_{1 \leq i < j \leq N} U_{ij}(\mathbf{X}_i, \mathbf{X}_j) \tag{22}$$

It is possible to generalize the definition of the BE in eqn. (21) to include higher-order potentials, but we shall not need that here.

The corresponding "counting function" is [see eqn. (17)]

$$N(v, \mathbf{X}^N)\, dv = dv \sum_{i=1}^{N} \delta[B_i(\mathbf{X}^N) - v] \tag{23}$$

where $N(v, \mathbf{X}^N)\, dv$ is the number of molecules having BE values between v and $v + dv$ at the specific configuration $\mathbf{X}^N$. The average number of such molecules is [see eqn. (18)]

$$N(v)\, dv = N\, dv \int \cdots \int d\mathbf{X}^N\, P(\mathbf{X}^N)\, \delta[B_1(\mathbf{X}^N) - v] \tag{24}$$

and the corresponding mole fraction is defined by

$$x(v)\, dv = N(v)\, dv/N \tag{25}$$

Hence $x(v)$ is another continuous QCDF.

The form of the function $x(v)$ may be of particular interest in the study of liquid water, at least in so far as to contrast it with a simple fluid such as liquid argon. Let us examine some special cases.

2.3.1. A System of Hard Spheres

In this case the BE of any given particle may have one of two values: zero or infinity. It is zero when the particle under observation does not "touch" any other particle and it is infinity whenever it comes closer than σ to any other particle (σ being the diameter of the spheres). Since the latter event has zero probability [because $P(\mathbf{X}^N)$ vanishes in this case], it follows that $x(v)$ is a Dirac delta function

$$x(v) = \delta(v) \tag{26}$$

2.3.2. A System of Simple Spherical Molecules

In this case the pair potential may be written as

$$U(R) = U^{HS}(R) + U^{SI}(R) \tag{27}$$

where $U^{HS}(R)$ is the hard-sphere potential and $U^{SI}(R)$ is the soft part of the potential. The latter may be defined once $U(R)$ and the effective hard-core diameter of the molecule have been determined. It is expected that when $U^{SI}/kT \ll 1$ the major form of $x(v)$ will be as described in Fig. 4(b), i.e., a single sharp peak centered at the value of the average BE of the molecules, namely

$$\bar{v} = 2\bar{U}_N/N \tag{28}$$

where $\bar{U}_N$ is the average potential energy of the system. A recent Monte Carlo computation of $x(v)$ in a two-dimensional system confirms the as-

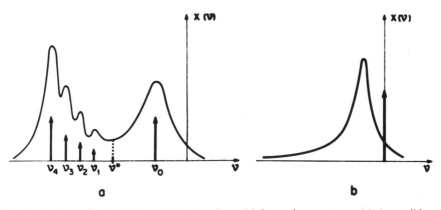

Fig. 4. A schematic description of the function $x(v)$ for various systems. (a) A possible form for water. The various peaks correspond to molecules with different numbers of hydrogen bonds. The point v^* may serve as a cutoff point for constructing a two-structure model. The arrows at the points $v_0, \ldots, v_4$ correspond to the idealized mixture model discussed in Section 2.5. (b) Possible form for a simple fluid like argon. The arrow at $v = 0$ represents a delta function which corresponds to a hard-sphere fluid.

sertion made above on the form of this function for simple spherical particles.

2.3.3. *Water*

In this case we expect that $x(v)$ will exhibit more "structure," i.e., it will convey information on the variety of local environments a given molecule may experience. A schematic description of the possible form of $x(v)$ for liquid water is depicted in Fig. 4(a). The details of the curves in this figure are essentially imaginary, yet they reflect the current belief that most water molecules at room temperature are fully hydrogen-bonded; these are represented by the peak at the left-hand side. Some molecules, which may be considered to be "nonbonded," may be represented by the peak on the right-hand side. Other molecules with intermediate BE values should find their representation somewhere between the two extreme peaks. It should be emphasized that the concept of the hydrogen bond does not feature explicitly in the definition of the function $x(v)$; however, it certainly plays a dominant role in determining the particular form of $x(v)$ in water, as we have indeed done in guessing a possible form of this function as drawn in Fig. 4.

Although we cannot suggest an experimental procedure to determine the various functions defined in eqns. (10), (19), and (25), we should like to indicate that all of these functions may be computed by means of a simulation technique such as the Monte Carlo or the "molecular dynamics" methods. These are in a sense "computer experiments" where we can follow each configuration of the system, and therefore averages of the form (3) may be evaluated.

2.4. An Exact Two-Structure Model (TSM)

The simplest version of the mixture model approach, and the one that has been most extensively applied to aqueous solutions, is the so-called "two structure model" (TSM). A few TSM's will be discussed in the next section. At present we should like to show how an exact version of the TSM may be constructed from any of the QCDF's defined previously. Some examples are the following:

(a) The QCDF defined in eqn. (10) gives the mole fraction x_K of molecules having CN equal to K. Instead we may distinguish between two groups of molecules, those whose CN is smaller than some K^* and those whose CN is larger than or equal to K^*. Thus we define the new mole fractions

$$x_L = \sum_{K=0}^{K^*-1} x_K \tag{29}$$

$$x_H = \sum_{K=K^*}^{\infty} x_K \tag{30}$$

where x_L may be referred to as the mole fraction of molecules with a low (L) local density and x_H refers to molecules with a relatively high (H) local density. Clearly, one can construct many TSM's simply by chosing different values of K^*. An additional possibility of a TSM, with particular relevance to water, may be

$$x_A = x_4, \qquad x_B = \sum_{K \neq 4} x_K \tag{31}$$

i.e., one component A is identified with those molecules with CN equal to four whereas all the other molecules are classified as B molecules.

(b) The QCDF based on BE has been defined in eqn. (25). A possible TSM constructed from this function is

$$x_s = \int_{-\infty}^{v^*} x(v)\, dv \tag{32}$$

$$x_w = \int_{v^*}^{\infty} x(v)\, dv \tag{33}$$

Here a cutoff value v^* has been chosen, say on the negative side of the v axis. Therefore x_s and x_w may be referred to as the mole fractions of molecules with a strong (s) BE and a weak (w) BE, respectively. (Note that molecules with a positive BE are included in the latter group.) An example of a cutoff point v^* is indicated in Fig. 4. Again we should like to note that a more complicated TSM may be constructed from $x(v)$. Let A and B be two nonoverlapping sets such that their union exhausts the entire v axis; then one can define the two mole fractions

$$x_A = \int_A x(v)\, dv, \qquad x_B = \int_B x(v)\, dv \tag{34}$$

where the integration is carried over the corresponding sets A and B.

Similarly, one can use the QCDF defined in eqn. (19) as well as any other QCDF to construct new TSM's for the system. The important point that should be borne in mind is that a TSM may be constructed in an exact fashion, and that this can be done in infinitely many ways. The next section is devoted to an examination of the relation of some existing *ad hoc* mixture models to the general exact formalism.

2.5. Embedding Existing Models into a General Framework

In this section we shall dwell upon some specific *ad hoc* mixture models that have been applied to water. Only representative examples will be discussed. It is hoped that the general argument will become clear, so that its employment in any other example will be a straightforward matter.

The simplest cases of mixture models which are quite well defined are the interstitial models, such as the Pauling[510,1080] or the Samoilov[1176] models. For concreteness we shall discuss the latter. Liquid water is viewed as consisting of "framework molecules" with the ice structure and "interstitial" water molecules filling some of its cavities. This model may be embedded in any of the exact mixture models discussed in previous sections. As an illustration let us employ the distribution defined in eqn. (19). The "idealized" interstitial model corresponds to replacement of the actual distribution $x(\phi)$ by the sum of two delta functions, namely,

$$x(\phi) \rightarrow x_{ice}\,\delta(\phi - \phi_{ice}) + x_{int}\,\delta(\phi - \phi_{int}) \qquad (35)$$

where ϕ_{ice} is the volume of the VP of a water molecule in the ice lattice and ϕ_{int} is the volume of the VP of an interstitial molecule, and x_{ice} and x_{int} are the corresponding mole fractions of the two species. ϕ_{ice} and ϕ_{int} may be functions of P and T,* but otherwise are presumed to be fixed by the geometry of the ordinary ice structure. Note that in (35) the VP is constructed by using the centers of the framework molecules only. If this is not the case, one must include at least one more species, i.e., framework molecules which are adjacent to interstitial molecules.

The replacement of the actual distribution $x(\phi)$ by a sum of two suitable weighted delta functions [eqn. (35)] may be reasonable if it turns out that $x(\phi)$ has indeed two sharp peaks centered at ϕ_{ice} and ϕ_{int}, respectively. A schematic illustration is depicted in Fig. 5. If this is not the case, however, the adoption of such a TSM may not be a good approximation. A somewhat better procedure would be to replace the two delta functions by two Gaussian functions centered at ϕ_{ice} and ϕ_{int}, respectively.

---

* Assuming that $x(\phi)$ is defined in the *TPN* ensemble (see Section 3.2).

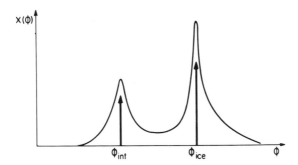

Fig. 5. A schematic illustration of the form of $x(\phi)$ presumed in the interstitial model. The curve represents a "guessed" form of $x(\phi)$ for water, and the two arrows represent the two delta functions used in relation (35).

It is quite clear that an assessment of the approximation made in applying eqn. (35) can only be made if the actual curve $x(\phi)$ is known. However, in some cases the application of such a simplified TSM may be useful as a first step in gaining a qualitative understanding of certain phenomena.

The second example that we should like to give is the Frank and Franks' TSM.[509] Here the classification is made into two components: a "bulky" and a "dense" one. The most natural place to embed this model is within the distribution defined in eqn. (10). As noted earlier, such a TSM may formally be made exact. For a suitable choice of K^* in eqns. (29) and (30) one can identify the low local density component with the "bulky" species and the high local density component with the "dense" species of Frank and Franks. It should be emphasized that some further assumptions are usually introduced in applying the model for a specific problem. In the author's opinion the most serious objection is the assumption that the mixture formed by the two species is an ideal one (in the symmetric sense). This topic has been discussed in detail elsewhere.[152]

As a final example, the Némethy and Scheraga model will be discussed. The classification made here is based on the concept of the hydrogen bond. Clearly, a proper definition of the hydrogen bond must precede the classification procedure. (In the context of this chapter only the *configuration* of the molecules is employed for the classification procedure. Therefore a definition of the hydrogen bond for our purposes can be satisfactory if we prescribe a rule which enables us to recognize whether or not a pair of molecules i and j at configuration $\mathbf{X}_i$, $\mathbf{X}_j$ are hydrogen-bonded. An example of such a rule has been given in Volume 1, Chapter 11. However, other

definitions which take into account the relative kinetic energy of the pair
are possible.) In addition, if we acknowledge the fact that a water molecule
may participate in at most four hydrogen bonds, then the classification
procedure would lead to the distribution function

$$\mathbf{x} = (x_0, x_1, x_2, x_3, x_4) \qquad (36)$$

where x_n is the mole fraction of molecules each of which is connected to its
environment by n hydrogen bonds (according to the definition adopted).
Clearly this model may be made exact and no further experimental evidence
should be supplemented. However, in the actual computations carried out
by Némethy and Scheraga the important parameter has been the "energy
levels" rather than the number of hydrogen bonds (although the former
were estimated from the latter). In a qualitative manner the "energy levels"
can be embedded in the distribution function $x(\nu)$ based on the binding
energy, namely the essential replacement made by Némethy and Scheraga is

$$x(\nu) \to \sum_{i=0}^{4} x_i\, \delta(\nu - \nu_i) \qquad (37)$$

where ν_i is the BE of a water molecule participating in i hydrogen bonds
and x_i is the corresponding mole fraction. This replacement is indicated in
Fig. 4(a). The full curve represents the actual function $x(\nu)$. The five delta
functions are represented by the five arrows at the various values of ν_i.
Clearly, if the actual distribution $x(\nu)$ has five sharp peaks at $\nu_0, \ldots, \nu_4$,
then the replacement made in eqn. (37) may be a good approximation.
Nevertheless, it may be a good starting point for a theory of water if we
recognize that these five species are indeed the most important ones in
determining the properties of liquid water.

   We have so far surveyed some specific mixture models that could have
been embedded in the framework of the distribution functions described
in previous sections. The general procedure is schematically described in
Fig. 6. Suppose for concreteness that we have classified molecules according
to the number of hydrogen bonds. Figure 6(a) shows a split of the entire
configurational space into five subspaces. Each subspace includes all the
configurations for which a specific molecule, say number 1, has a precise
number of hydrogen bonds as indicated in Fig. 6. The five subspaces
exhaust the entire configurational space. Figures 6(b) and 6(c) demonstrate
two possible *exact* TSM's derived from Fig. 6(a). In Fig. 6(b) one component comprises the nonbonded molecules whereas all other molecules
are regrouped in the second component. In Fig. 6(c) the distinction is made

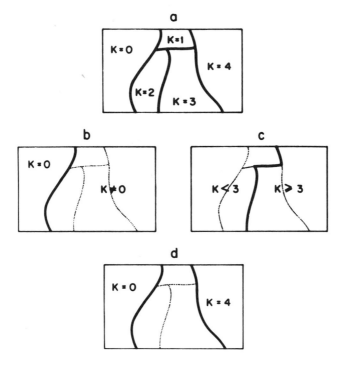

Fig. 6. A schematic description of various two-structure models. (a) An exact split of the entire configurational space into five regions; each region corresponds to all configurations for which a specific molecule, say number 1, participates in K hydrogen bonds. (b, c) Examples of exact two-structure models derived by regrouping species into new classes. (b) One species includes all nonbonded molecules and the second species comprises all the others. (c) The two species are distinguished according to whether the molecule participates in less than three hydrogen bonds or more than three hydrogen bonds (inclusive). (d) An example of an approximate two-structure model. Nonbonded and fully bonded molecules are considered. The entire space is clearly not exhausted by this classification.

between molecules with two hydrogen bonds or less and all the others. Figure 6(d) gives an example of an *approximate* TSM. Here we recognize the nonbonded and the fully bonded species.

Clearly, such a TSM may *not* be a *good approximation*. Nevertheless it may still be a *useful* model if it turns out that the two species thus chosen are indeed the most important ones in determining the unique behavior of water.

We believe that, in conformity with almost all models for water, the most important two species are a "close-packed" one and an "open structure" one. A specific example is the Frank and Franks[509] model. Some of the general consequences of adopting a model of this kind, without any further commitment to the geometry of the two species, have been discussed recently by the author.[152]

3. STRUCTURE OF WATER AND STRUCTURAL CHANGES

3.1. Possible Definitions of the "Structure of Water"

One of the most widely applied concepts in the study of aqueous solutions is the structure of the solvent and its change under various conditions. There is usually a kind of agreement on the qualitative meaning of this concept. In most cases a measure of the structure is inferred from a specific experimental technique, for instance, one measures the viscosity of the liquid and attributes an increase in viscosity to an increase in "structure." This may be viewed as an operational definition of the structure in terms of a specific experimental procedure. In general one cannot expect that definitions derived from different experimental results will coincide. There is therefore room for a more universal definition for this concept.

In all of the following discussions the concept of structure will be defined as an average quantity in the configurational phase space. The time will not feature explicitly in the definition, and in this respect our procedure differs from the discussion of Eisenberg and Kauzmann.[444] We shall also base our definition on the mixture model formalism developed in the previous section.*

Even after restricting ourselves to the mixture model approach, we still have a great deal of freedom in defining the concept of "structure." In the following we shall describe one such possibility; nevertheless we shall attempt to suggest a definition which we believe is closest to the concept in current usage. We shall employ the CN distribution function for this purpose. Let us choose $R_c \simeq 3.5$ Å in eqns. (5) and (10) and define the "structure of water" in the TVN ensemble as

$$x_{ST} = x_4 = \int \cdots \int d\mathbf{X}^N P(\mathbf{X}^N) \, \delta[A_1(\mathbf{X}^N, R_c) - 4] \qquad (38)$$

---

* In Volume 1, Chapter 11, a different definition in terms of the average number of hydrogen bonds in the system has been suggested. The definition is also an average quantity in the configurational space—it is clear that other, more general definitions can be constructed in the full phase space of the system.

The motivation for adopting this particular definition arises from the following considerations. In the first place we are aware of the fact that in most mixture models the "structure" is qualitatively identified with the concentration of the "icelike" species. To define an icelike molecule would require at least the detailed specification of the geometry of packing of molecules in its immediate environment, i.e., one has to require that an icelike molecule be surrounded with four neighbors and in addition that these molecules be situated at the vertices of a regular tetrahedron at a distance of about 2.76 Å from its center. In some specific models one further specifies the geometry of the extended network beyond the first coordination shell.

The definition suggested in eqn. (38) is somewhat more general since it includes a variety of icelike structures. Yet it is too general in the sense that no geometrical restriction is imposed on the location of the surrounding molecules around the central one, i.e., a molecule which has CN equal to four will contribute to our definition of "structure" even if in fact it may have no "structure" at all (in the sense of having a local tetrahedral geometry).

The disposition of the geometrical requirement in eqn. (38) is based on the information available from the experimental radial distribution function of water. The average CN within a distance of about 3.5 Å is 4.4 at room temperature (see Volume I, Chapter 11). Furthermore, the location of the second peak of the radial distribution function is at 4.5 Å. This indicates that in water at room temperature the majority of molecules have CN equal to four and, with high probability, have local tetrahedral geometry. Therefore, although we have required in eqn. (38) that the structure be identified with the average number of molecules having CN equal to four, we shall in practice count molecules with local tetrahedral geometry. (Of course we could construct a more elaborate definition of x_{ST} in which we explicitly introduce our specification of the local geometry. This may be needed when an actual computation of x_{ST} is carried out.) We shall henceforth adopt the quantity x_{ST} as a measure of the "structure of water."

3.2. Structural Changes and Thermodynamics

A simple and very common practice is to choose a TSM in such a way that the mole fraction of one component coincides with the measure of the degree of structure of the system. For concreteness we may use the TSM based on the distribution of CN and the definition of structure in eqn. (38). Since we shall here be working in the TPN ensemble, a minor modification of the definition of x_K in eqn. (10) is required, namely

$$x_K = \int dV \int \cdots \int d\mathbf{X}^N \, P(\mathbf{X}^N, V) \, \delta[A_1(\mathbf{X}^N, R_c) - K] \qquad (39)$$

where $P(\mathbf{X}^N, V)$ is the probability density of finding a system in the *TPN* ensemble with volume V and configuration $\mathbf{X}^N$,

$$P(\mathbf{X}^N, V) = \frac{\exp[-\beta PV - \beta U_N(\mathbf{X}^N)]}{\int dV \int \cdots \int d\mathbf{X}^N \exp[-\beta PV - \beta U_N(\mathbf{X}^N)]} \qquad (40)$$

The mole fractions of the two components will be defined by

$$x_{ST} = x_4 \qquad (41)$$

$$x_{NS} = \sum_{K \neq 4} x_K \qquad (42)$$

Thus the "icelike" component, which is defined as the four-coordinated molecule, serves also to measure the degree of structure of the system. All other components are collected in x_{NS} and may be referred to as the non-structured species. The meaning assigned to the structured and nonstructured components should not be taken too literally. We believe that the proposed meaning is in qualitative agreement with many specific TSM's, yet it may certainly be subjected to modifications and refinements whenever warranted.

Let E be any extensive thermodynamic quantity defined in the *TPN* ensemble. Viewing the system as a two-component one, we may express E as a function of the new set of variables T, P, N_{ST}, N_{NS}, where

$$N_{ST} = N x_{ST}, \qquad N_{NS} = N x_{NS} \qquad (43)$$

Thus

$$E = E(T, P, N) = E(T, P, N_{ST}, N_{NS}) \qquad (44)$$

The new variables N_{ST} and N_{NS} are not independent, as discussed in Section 2.2. The dependence is imposed by the equilibrium condition

$$\mu_{ST}(T, P, N_{ST}, N_{NS}) = \mu_{NS}(P, T, N_{ST}, N_{NS}) \qquad (45)$$

and

$$N_{ST} + N_{NS} = N \qquad (46)$$

These relations may, in principle, be inverted to express N_{ST} and N_{NS} as a function of the independent variables $T, P,$ and N. This, in general, is difficult to achieve explicity. This point will be further demonstrated in the example discussed in Section 4.

Besides the dependence between N_{ST} and N_{NS} imposed by the eqns. (45) and (46), there will, in general, be some further restrictions on the possible values allowable for N_{ST} and N_{NS}. The origin of these restrictions has been discussed in Section 2.2. Here we should like to note that in the TSM defined in eqns. (41) and (42) the restrictions will be less severe than in the case of a multicomponent system considered in Section 2.2.

Applying the Euler theorem to the function $E(T, P, N_{ST}, N_{NS})$, we get

$$E = N_{ST}\bar{E}_{ST} + N_{NS}\bar{E}_{NS} \tag{47}$$

where the partial molar quantities are defined by

$$\bar{E}_{ST} = \left(\frac{\partial E}{\partial N_{ST}}\right)_{T,P,N_{NS}}, \qquad \bar{E}_{NS} = \left(\frac{\partial E}{\partial N_{NS}}\right)_{T,P,N_{ST}} \tag{48}$$

In view of the dependence between N_{ST} and N_{NS} it is clear that the two partial molar quantities cannot be evaluated experimentally, i.e., one cannot in practice add dN_{ST} and hold N_{NS} constant. Nevertheless these quantities are significant as soon as we have expressed the quantity E as a function of the variables T, P, N_{ST}, N_{NS}, which enables us to carry out the mathematical derivatives in eqn. (48).

We shall now turn to aqueous solutions, where the solvent (water) is viewed as a two-component system and in addition we have N_S molecules of a solute S. Thus the variables employed to describe our system are $T, P, N_{ST}, N_{NS}, N_S$. (Note that in the classification procedure of Section 2.2 we count only the water molecules in the R_c sphere of an observed molecule. Any solute molecule which happens to be in the R_c sphere of a water molecule is not counted as one of its neighbors.)

The partial molar quantity $\bar{E}_S$ associated with the solute S may be written as

$$\bar{E}_S = \left(\frac{\partial E}{\partial N_S}\right)_{T,P,N} = \left(\frac{\partial E}{\partial N_S}\right)_{T,P,N_{ST},N_{NS}} + (\bar{E}_{ST} - \bar{E}_{NS})\left(\frac{\partial N_{ST}}{\partial N_S}\right)_{T,P,N} \tag{49}$$

where the first form on the rhs is simply the definition of $\bar{E}_S$ and the second form employs the representation $E = E(T, P, N_{ST}, N_{NS}, N_S)$ and the condition (46). $\bar{E}_{ST}$ and $\bar{E}_{NS}$ are the partial molar quantities associated with the two species of water molecules. The two forms on the rhs of eqn. (49) are of course equivalent, but the latter may be advantageous if we can guess something about $\bar{E}_{ST} - \bar{E}_{NS}$ and about $(\partial N_{ST}/\partial N_S)_{T,P,N}$. We shall not elaborate on this question here since this has recently been reviewed in some detail.[150]

The TSM formalism offers a simple definition of the concept of "structural changes" in the solvent. This concept is probably quite old (see, for example, the review by Chadwell[277] for earlier references) and it was introduced to explain some anomalous properties of aqueous solutions. Bernal and Fowler[162] introduced the idea of "structural temperature," which may be defined as follows. Suppose we measure some physical quantity η for pure water and for an aqueous solution at temperature T, and then define

$$\Delta\eta = \eta(T, N_S) - \eta(T, N_S = 0) \tag{50}$$

On the other hand, we may measure the change of η for pure water due to the change of temperature only, i.e.,

$$\Delta\eta' = \eta(T', N_S = 0) - \eta(T, N_S = 0) \tag{51}$$

The "structural temperature" of the solution (with N_S molecules of S) is defined as the temperature T' for which

$$\Delta\eta' = \Delta\eta \tag{52}$$

The definition of "structural temperature" in fact does not involve the concept of structure; furthermore, it is likely to depend on the property measured, η. The idea behind this definition is that the "structure" of the solvent is expected to be a monotonically decreasing function of the temperature. Therefore "structural changes" may be detected on the corresponding temperature scale. This idea, although useful for a qualitative operational definition of the concept of "structural changes," is not satisfactory unless one establishes the correct dependence of the "structure" of the solvent on temperature; this in turn will depend on the definition we adopt for the structure of water.

The TSM introduced above avoids this complication. If x_{ST} serves as a measure of the structure, then changes of the structure are defined in a natural way by its derivatives, say $\partial x_{ST}/\partial T$, $\partial x_{ST}/\partial P$, and $\partial x_{ST}/\partial N_S$. The concept of structure-breaking and structure-promoting may be identified with the sign of these derivatives. For instance, a solute S will be said to increase the structure, or stabilize it, if

$$D_S \equiv N(\partial x_{ST}/\partial N_S)_{T,P,N} = (\partial N_{ST}/\partial N_S)_{T,P,N} > 0 \tag{53}$$

that is, when the addition of solute increases the number of molecules which are supposed to be more structured.

In general, the quantity D_S defined in (53) will strongly depend on the concentration of the solute. Therefore, in order to characterize the specific effect of a solute on the structure of the solvent, it is useful to define the following limiting value of D_S:

$$D_S{}^\circ = \lim_{N_S \to 0} (\partial N_{ST}/\partial N_S)_{T,P,N} \tag{54}$$

which may serve as an index expressing the way the solute S changes the structure of the solvent.

Recently a general and exact expression for $D_S{}^\circ$ in terms of molecular distribution functions has been obtained.[151] The general form of $D_S{}^\circ$, as well as the other derivatives of N_{ST}, is the following:

$$(\partial N_{ST}/\partial N_S)_{T,P,N} \propto x_{ST}(1 - x_{ST})\, \Delta Q \tag{55}$$

where ΔQ is the difference in some property Q for the "structured" and "nonstructured" species. The property Q may be the partial molar volume or enthalpy of the two species. The general form of the structural changes as expressed in (55) may be used to indicate when a TSM is expected to be a useful one. From (55) it is clear that a TSM must be defined in such a way that the two components will be considerably different (i.e., large ΔQ). At the same time none of the mole fractions of the two components should be negligeably small, or the rhs of (55) will be small and structural changes unimportant.

The situation may be clearly demonstrated with the help of Fig. 4. If the distribution $x(v)$, for example, is concentrated about a single sharp peak, then any choice of a cutoff point v^* will lead to either two very similar components (i.e., ΔQ small) or two very different components but with the product $x_{ST}(1 - x_{ST})$ small. On the other hand, if the distribution function is like the one depicted in Fig. 4(a), then it is possible to find a cutoff point v^* (as indicated in the figure) in such a way that the mole fractions of the two components are comparable in magnitude, yet the two species differ considerably in their properties.

If this is the case for liquid water, then it may be useful to view it as a mixture of two components (similar arguments may apply for three or more components). There exists a strong analogy between the arguments presented above and the reasoning behind the definition of rotational isomers for hydrocarbons. If the rotational potential about a given C—C bond has a very high barrier (compared with kT), then it may be useful to view the (one-component) system as a mixture of two (or more) isomers. It should be stressed, however, that the split into two isomers can be carried out in any case, even if the barrier for rotation is very low, in which case the formal definition holds, but the whole procedure may be useless.

We shall not elaborate on the theoretical or the experimental evidence that exists in favor of the contention that the structure of water is enhanced by some nonelectrolytic solutes. It is important to realize that such an effect is not a trivial matter. We expect that for large concentrations of any solute x_{ST} will tend to zero. This is simply a consequence of the fact that as water becomes dilute in the system the probability of finding four *water* molecules in the R_c sphere of a given water molecule must tend to zero. The question is how x_{ST} reacts to addition of the solute S at very low concentrations of the solute, x_S, $[x_S = N_S/(N_S + N)]$. Two possible forms of such a dependence are depicted in Fig. 7. There are reasons to believe[150] that for some nonpolar solutes the initial slope is positive, corresponding to the idea of stabilization of the structure.

It must be stressed that the whole concept of structural changes in the solvent is purely a matter of definition, i.e., it depends on how we choose to define the structure of the fluid. Once we have adopted a reasonable definition of the structure then a quantity like $D_S{}^\circ$ may be evaluated by means of a computer simulation technique or estimated by other theoretical or experimental approaches.

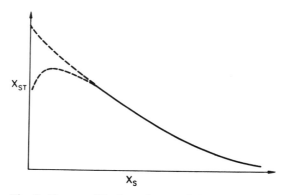

Fig. 7. Two possible dependences of the "structure of water" on the mole fraction of the solute x_S. The limiting behavior must be $x_{ST} \to 0$ as $x_S \to 1$. On the other hand, a positive initial slope at $x_S = 0$ will correspond to a stabilization effect, while a negative slope corresponds to a destabilization effect.

3.3. Distribution of Local Structure around a Solute

So far we have discussed the concepts of structure and structural changes as defined in the bulk of the system. A more penetrating insight could be gained by looking at the change of the structure of the solvent as a function of the distance along a ray originating from a center of a solute molecule. This can be referred to as the distribution of structure around the solute molecule. The qualitative idea underlying this concept has appeared in numerous studies of aqueous solutions, notably in the work of Gurney,[609] Frank and Wen,[512] and Samoilov.[1175,1176] In this section we shall elaborate on the formal definition of this concept. It turns out that for a nonpolar solute the concept of local structure around a solute can easily be constructed by means of an extension of previous concepts. This seems not to be not case, however, for ionic solutes, where further difficulties arise.

In what follows we shall restrict our discussion to very dilute solutions, so that for all our purposes it will be sufficient to follow a single solute particle immersed in the solvent. Also, without loss of generality, we can assume that our simple solute particle is at a fixed position in the liquid. The purpose is to define a function which will measure the change of the structure as a function of the distance from the center of the solute. Let the center of the solute particle $\mathbf{R_S}$ be at the origin of our coordinate system; then we introduce the following probability distribution:

$$x_K(R) = \int dV \int \cdots \int d\mathbf{X}^N \, P(V, \mathbf{X}^N) \, \delta(|\,\mathbf{R}_1\,| - R) \, \delta[A_1(\mathbf{X}^N, R_c) - K] \quad (56)$$

Clearly, the quantity $N x_K(R) \, dR$ is the average number of water molecules having CN equal to K *and* being located at distance between R and $R + dR$ from the center of the solute S. (Note that the integrand includes a Dirac and a Kronecker delta function.)

Adopting again the practical definition of the "structure" in eqn. (38), the function

$$x_{ST}(R) = x_4(R) \quad (57)$$

may serve to describe the distribution of the structure of the solvent around the solute particle S. The function $x_{ST}(R)$, or similar ones, may satisfactorily describe how the structure of the solvent is locally affected by the presence of the solute. The extension of the concept of "structure," x_{ST}, to the function $x_{ST}(R)$ was rendered possible since we believe that the change in the structure produced by a nonpolar solute can be expressed merely as a change in the concentration of the "structured" species. This means that we agree to use the *same* definition of the structure in both pure water and in concentric shells around the solute molecule.

The situation is somewhat more complicated for ionic solutions. Of course we can formally apply the same quantity $x_{ST}(R)$, in this case as well, but this will not be a useful choice. (There is also a minor difficulty arising from the electroneutrality condition, which compels us to introduce at least one pair of oppositely charged ions into the solvent. However, for a sufficiently dilute system we can safely assume that the ions do not see each other, and therefore we can fix their positions as being very well separated from each other.) The reason is that it is difficult to conceive of a single definition of the "structure" which will be valid for the entire range of distances around the ion. This difficulty has been recognized by Frank and Wen[512] and others, who introduced various regions surrounding the center of the ion, in each of which the term "structure" is understood differently. We are here facing an inherent difficulty which stems from the very question of what we really want to mean by "structure." In the case of pure water as well as aqueous solutions of nonelectrolytes we have *adopted* a definition of structure which seems to fit the currently used concept. Given two systems, we can compare the values of x_{ST} and arrive at the relative "structuredness." Such a comparison may in some cases lead to a contradiction with our intuitive meaning of the structure. For instance, in the coordination sphere around the ion we may find $x_{ST} = 0$, yet we shall refuse to accept that it is not "structured." One way to circumvent this difficulty is to redefine the concept of structure so that it will be valid for any region around the solute. It is often agreed that the entropy may serve as a reasonable measure of the structure, hence we can proceed as follows.

Let us fix the solute particle at the origin $\mathbf{R}_S = 0$ and define the following quasicomponent distribution function:

$$x(R) \, dR = dR \int dV \int d\mathbf{X}^N \, P(V, \mathbf{X}^N) \, \delta[|\,\mathbf{R}_1\,| - R] \tag{58}$$

Clearly, $N(R) \, dR = Nx(R) \, dR$ is the average number of water molecules, in the *TPN* ensemble, which are located at distances between R and $R + dR$ from the center of the solute. The total entropy of the system is a function of the variables T, P, N_S, N. However, when the solvent is viewed as a mixture of species with mole fraction $x(R) \, dR$ the entropy is a functional of $N(R)$, $S = S(T, P, N_S, N(R))$. Applying the generalized Euler theorem[152] for this case, we get

$$S = N_S \bar{S}_S + \int_0^\infty \bar{S}(R) N(R) \, dR \tag{59}$$

with

$$N = \int_0^\infty N(R)\, dR \tag{60}$$

$\bar{S}_S$ is the partial molar entropy of the solute S, whereas $\bar{S}(R)$ is the partial molar entropy of the Rth component and is defined by the functional derivative

$$\bar{S}(R) = [\delta S/\delta N(R)]_{P,T,N_S} \tag{61}$$

The function $\bar{S}(R)$ may serve as a measure of the variation of the structure around the solute molecule. This definition does not seem to be useful at present since it is neither experimentally measurable nor calculable through computer simulation techniques. In addition, it is not quite clear whether the partial entropy really measures the degree of structure.

We have endeavored in this section to point out some difficulties rather than suggest answers. We believe that there is an urgent need for a detailed study of the effect of solutes on the "structure" of the solvent. Of course, one is free to define in any reasonable way the concept of structure around the solute. Once this concept has been defined its numerical evaluation for different solutes (or sources of potential function situated at the origin) could be carried out by a computer experiment. Such a study will no doubt help us understand many aspects of the unusual properties of aqueous solutions.

4. EXAMPLE OF A SIMPLE SOLVABLE TWO-STRUCTURE MODEL

4.1. Motivation

The specific model to be treated in this section was originally constructed to test a conjecture about the sign of the partial molar heat capacity (PMHC) of a solute in aqueous solutions.[148] In this respect the model falls into the category of "models of a phenomenon" discussed in Section 1. We shall indeed develop this aspect as the main content of the present section. In doing so we shall have a chance to pause along the way and check a few features of the general approach of the mixture model formalism, for instance: In what sense is the TSM *equivalent* to the one-component point of view? Is the TSM really useful, and if so, in what sense? These and some other questions will now be discussed.

It is important, in order to avoid any misunderstanding, to emphasize from the very outset that the model presented here is *not* a model for water.

It is true that the stimulation for its construction arose from a specific problem related to aqueous solutions. However, the problem itself, as will be stated below, can be reformulated without any relevance to aqueous solutions. It is the general aspect of this problem that will be examined with the aid of the simplified model.

The specific problem which is the main motivation for the model introduced here is the following: It has been experimentally observed that the PMHC of nonpolar solutes in water is large and positive, whereas it is quite small in nonaqueous liquids.[148] Specifically, the value of $\bar{C}_S$ for argon in water (in infinitely dilute solution) at room temperature is about 50 cal deg^{-1} mol^{-1}. The corresponding value in ethanol, methanol, or p-dioxane at the same temperature is nearly zero. The traditional interpretation of these facts rests on the TSM for water. The main arguments are the following:

(a) We assume that equilibrium between two species, say L and H,

$$L \rightleftharpoons H \tag{62}$$

exists, and that the L species has a lower partial molar enthalpy (or energy), i.e.,

$$\bar{H}_L - \bar{H}_H < 0 \tag{63}$$

(b) The high value of the heat capacity of pure water is attributed to the relaxation from one form to the other, i.e., the heat capacity is written[148] as

$$C = (\partial H/\partial T)_{N,P} = x_L \bar{C}_L + x_H \bar{C}_H + (\bar{H}_L - \bar{H}_H)(\partial x_L/\partial T)_{N,P} \tag{64}$$

The last term on the rhs of eqn. (63) has been referred to as the relaxation contribution to the heat capacity in this specific TSM. It can be shown that this term is always positive.[148,152]

(c) It is assumed that a nonpolar solute S shifts the equilibrium (62) in favor of the L form, i.e.,

$$(\partial N_L/\partial N_S)_{T,P,N} > 0$$

(at least for very dilute solutions with respect to S). For the purpose of this section we shall assume that conditions (a)–(c) are fulfilled and proceed to examine the following conventional interpretation of the high value of the PMHC of S:

The large heat capacity of water is attributed to the "melting" of the icelike species (or the L form in our notation). Since S is presumed to stabilize the icelike form, there is more of this form *available* for "melting." Hence, we expect that the heat capacity of the system will be increased by the addition of S.

The above interpretation is quite appealing on qualitative grounds and in fact has been proposed by several authors (for more details see Ref. 148). The weak point is of course the final conclusion, i.e., we do not really know how the total heat capacity of the system depends on the concentration of the icelike form. To assess the validity of such a conclusion for water would not be feasible at present because we do not yet have a reliable model to represent water. Nevertheless, looking through the assumptions (a)–(c), we realize the general character of the problem; therefore we shall try to "extract," so to speak, the problem from its original formulation in terms of concepts relevant to water and reformulate it in more abstract terms.

The new formulation of the problem is as follows: Having a system of two components A_1 and A_2 in chemical equilibrium

$$A_1 \rightleftarrows A_2 \tag{65}$$

for which we know that

$$\bar{E}_1 - \bar{E}_2 < 0 \tag{66}$$

(we shall neglect the difference between enthalpy and energy for the present discussion), what is the *sign* of the PMHC of an added component S if it is given that S shifts the equilibrium (65) in favor of the low-energy form? That is, if we have

$$\partial N_1 / \partial N_S > 0 \tag{67}$$

In this formulation of the problem we have eliminated any specific reference to aqueous solutions, and therefore bypassed the complexities of the original system. We can now search for a simple and solvable model where conditions (a)–(c), or their abstract analogs (65)–(67), are built in. In such a model the sign of the PMHC can be examined exactly and therefore the conclusion we shall reach will be significant for aqueous solutions as well.

4.2. Description of the Model and Its Solution

The system consists of N structureless molecules of type A adsorbed on M localized and independent sites. The sites are of two kinds, L and H, M_L sites of one kind and M_H of the second. With this distinction we may, if we wish, view the one-component system of A molecules as a two-component system. A molecule adsorbed on a L site will be denoted by A_L and similarly A_H denotes a molecule adsorbed on an H site. The equilibrium condition is

$$A_L \rightleftarrows A_H \tag{68}$$

We also require that the binding energy of A to an L site is lower than the binding energy of A to an H site, i.e.,

$$E_L - E_H < 0 \tag{69}$$

For simplicity we have assumed that A has no internal degrees of freedom so that E_L and E_H are the only energies involved in our system.

We now introduce N_S molecules of a second component S to the system; the simplest way to simulate condition (67) is to require that S can be adsorbed solely on sites of type H.

Clearly, since S occupies only sites of type H, its presence reduces the available number of H sites for A and therefore A molecules will be compelled to shift from H to L. This contention will however be formally proved later.

A schematic description of our system is depicted in Fig. 8. The specifications made above are sufficient to simulate all the required conditions of our problem.

The first point that should be noted is that we have started with a one-component system of A molecules, and by using a classification procedure

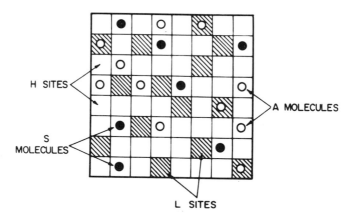

Fig. 8. Schematic description of the model. The dark sites L are those with a lower binding energy than the bright ones H. Open circles are A molecules and full circles are S molecules.

we could have adopted a two-component point of view, A_L and A_H in chemical equilibrium. The classification was based on the type of site at which an A molecule finds itself, which is a simplified example of the more complicated classification procedures described in Section 2.

The canonical partition function of our system is

$$Q = \sum_{N_L + N_H = N} \frac{M_L!}{N_L!(M_L - N_L)!} \; \frac{M_H!}{N_H! \, N_S!(M_H - N_H - N_S)!}$$
$$\times \exp[-\beta(E_L N_L + E_H N_H + E_S N_S)] \tag{70}$$

Here E_S is the binding energy of the S molecule. The first combinatorial coefficient is the number of ways N_L molecules can be arranged on M_L sites. The second combinatorial coefficient is the number of ways N_H molecules of A and N_S molecules of S can be arranged on M_H sites.

The summation is over all possible splits of N into two numbers N_L and N_H such that

$$N_L + N_H = N \tag{71}$$

Let us rewrite eqn. (70) as

$$Q(N, N_S, M_L, M_H, T) = \sum_{N_L + N_H = N} Q(N_L, N_H, N_S, M_L, M_H, T) \tag{72}$$

and

$$Q(N, N_S, M_L, M_H, T) = Q(N_L^*, N_H^*, N_S, M_L, M_H, T) \tag{73}$$

$Q(N_L, N_H, N_S, M_L, M_H, T)$ is the partition function of a system with *fixed* numbers N_L and N_H. In eqn. (73) we have employed a standard procedure to replace the sum on the rhs of eqn. (72) by the largest term. The values of N_L and N_H for which $Q(N_L, N_H, N_S, M_L, M_H, T)$ attains maximal value have been denoted in eqn. (73) by N_L^* and N_H^*.

It is important to realize a second point which is relevant to the idea of the mixture model in general. On the lhs of eqn. (72) we have Q as a function of N and N_S, while N_L and N_H do not appear. This is clearly the result of the fact that we have only two kinds of molecules A and S. On the rhs of eqn. (73) we "suddenly" have N_L^*, N_H^*, and N_S as if we had a three-component system. However, since we have the equilibrium condition

$$\mu_L = \mu_H \tag{74}$$

we have in fact a relation between N_L^* and N_H^* which reads[148]

$$N_H^*(M_L - N_L^*)/(M_H - N_H^* - N_S)N_L^* = \exp[-\beta(E_H - E_L)] = K \tag{75}$$

In spite of this dependence between N_L^* and N_H^* it is sometimes useful to view Q, or any thermodynamic quantity, as a function of N_L^* and N_H^* as if they were virtually independent. Of course, at the end of any computation we should introduce the dependence between N_L^* and N_H^* and ultimately express them in terms of the fundamental variables of our system: N, N_S, M_L, M_H, T. We shall henceforth drop the asterisk from N_L^* and N_H^* for simplicity of notation.

The chemical potentials of the three species are

$$\mu_L = kT \ln[N_L/(M_L - N_L)] + E_L \qquad (76)$$

$$\mu_H = kT \ln[N_H/(M_H - N_H - N_S)] + E_H \qquad (77)$$

$$\mu_S = kT \ln[N_S/(M_H - N_H - N_S)] + E_S \qquad (78)$$

The total energy of the system is

$$E = kT^2 \, \partial(\ln Q)/\partial T = N_L E_L + N_H E_H + N_S E_S \qquad (79)$$

which is simply the sum of all the binding energies of the molecules to their sites.

The partial molar energy of S is defined by

$$\bar{E}_S = (\partial E/\partial N_S)_{T,N} = E_S + (E_L - E_H)(\partial N_L/\partial N_S)_{T,N} \qquad (80)$$

The term E_S may be viewed as the partial molar energy of S in a "frozen-in" system, i.e., when conversion of A_L into A_H is not permitted. The second term on the rhs of (80) may be referred to as the relaxation term; it is the contribution to $\bar{E}_S$ which results from "structural shifts" of A.

By differentiation of the equilibrium condition (75) with respect to N_S at constant N and T, we get

$$\left(\frac{\partial N_L}{\partial N_S}\right)_{T,N} = \frac{1}{M_H - N_H - N_S}$$

$$\times \left(\frac{1}{N_H} + \frac{1}{N_L} + \frac{1}{M_H - N_H - N_S} + \frac{1}{M_L - N_L}\right)^{-1} \qquad (81)$$

Since we always have $N_L \leq M_L$ and $N_H + N_S \leq M_H$, it follows that

$$(\partial N_L/\partial N_S)_{T,N} > 0 \qquad (82)$$

which would be expected to follow from the assumptions of the model. This condition may be referred to as the "stabilization" of the L form by the solute S.

From the equilibrium condition (75) we can also get the explicit dependence of N_L and N_H on the fundamental variables of our system (for more details see Ref. 148), namely,

$$N_H = \frac{-[N - M_L - K(M_H + N - N_S)]}{2(K - 1)}$$

$$- \frac{\{[N - M_L - K(M_H + N - N_S)]^2 - 4(K - 1) KN(M_H - N_S)\}^{1/2}}{2(K - 1)} \qquad (83)$$

We see that in order to get an explicit expression for $\bar{E}_S$ we have to substitute N_H and N_L $(= N - N_H)$ from (83) into (81) and from (81) into (80). This is quite cumbersome, and for the present purposes we shall confine ourselves to one simple case, namely,

$$N = M_L = M_H \qquad (84)$$

$$N_S \ll N \qquad (85)$$

For this case relations (83), (81), and (80) reduce to

$$N_H = N(K - \sqrt{K})/(K - 1), \qquad N_L = N(\sqrt{K} - 1)/(K - 1) \qquad (86)$$

$$\partial N_L/\partial N_S = \tfrac{1}{2}(K - \sqrt{K})/(K - 1) \qquad (87)$$

$$\bar{E}_S = E_S + (E_L - E_H)\tfrac{1}{2}(K - \sqrt{K})/(K - 1)$$

$$= E_S + \tfrac{1}{2}kT(\ln K)(K - \sqrt{K})/(K - 1) \qquad (88)$$

The variations of X_L $(= N_L/N)$ and X_H $(= 1 - X_L)$ as a function of the equilibrium constant K are depicted in Fig. 9. For any value of K in the range $0 \le K \le 1$ [i.e., for $E_H - E_L > 0$, see eqn. (75)] the L sites will be preferred by the A molecules [this is true because of assumption (84)] and hence $X_L > X_H$. As $K \to 1$ the two sites become equivalent and therefore $X_L \to X_H \to \tfrac{1}{2}$.

A more interesting feature of our model is the behavior of the partial molar energy $\bar{E}_S$ as a function of K. Figure 10 shows the dependence of the relaxation term in eqn. (88) on K for $0 \le K \le 1$. Clearly, the relaxation part of $\bar{E}_S$ is always negative (since $E_H - E_L > 0$ and $\partial N_L/\partial N_S > 0$) and it tends to zero at both ends of the range $0 \le K \le 1$. This function has a minimum value of about -0.56 at $K \approx 0.08$. Now suppose that we have chosen a model for which $K \approx 0.08$, so that the partial molar energy is

$$\bar{E}_S \approx E_S - \tfrac{1}{4}kT \qquad (89)$$

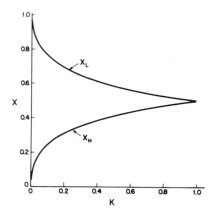

Fig. 9. Variations of the mole fractions X_L and X_H as a function of K for the special case $N = M_L = M_H$ and $N_S \ll N$.

We note that, in principle, the two terms on the rhs of eqn. (89) or eqn. (88) may be of different orders of magnitude. We shall now try to answer the question we posed at the beginning of this section. In what sense is the TSM useful?

Suppose we start with a crystal, on the structure of whose surface we have no information. We first measure the adsorption energy for pure S and find the quantity

$$\Delta \bar{E}(\text{pure S}) = \bar{E}_S - \bar{E}_S{}^g = E_S - \bar{E}_S{}^g \qquad (90)$$

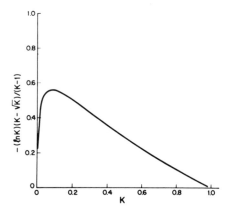

Fig. 10. Variation of the relaxation part of the partial molar energy, $-(\ln K)(K - \sqrt{K})/(K - 1)$, as a function of K for the special case $N = M_L = M_H$ and $N_S \ll N$.

where E_S^g is the average energy per molecule in the gaseous phase. Next we measure the same quantity but in the presence of A, which we know is adsorbed on the same crystal but which does not react with S directly. We may be puzzled to find that now we get

$$\Delta \bar{E}(\text{in presence of A}) = \bar{E}_S - \bar{E}_S^g = E_S - \bar{E}_S^g - \tfrac{1}{4}kT \qquad (91)$$

i.e., we find that at certain concentration of A [see also conditions (84) and (85)] the value of $\Delta \bar{E}$ has changed and, in principle, this change may be very large.

Of course, if we knew the molecular details of our surface we could have solved for $\bar{E}_S$ and found the relations (88) and (89) without ever mentioning a two-structure formalism. We may still wonder about the origin of the second term on the rhs of eqn. (88), as we were puzzled to find the additional contribution to the experimental value of $\Delta \bar{E}$ in eqn. (91). If, on the other hand, we adopt the TSM, we have a somewhat more convenient way of looking at the two terms of eqn. (88). We stress that by adopting this approach we have not changed anything in the physics of the problem. The utility of this point of view is that we can better visualize the two contributions to $\bar{E}_S$. We recognize that E_S is the partial molar energy of pure S on the surface whenever there are no A molecules or when the conversion of A from one site to the other is not permitted. The relaxation term is the contribution to $\bar{E}_S$ arising from "structural changes," i.e., from transfer of A molecules from the H sites to the L sites. Note that the relaxation term does not depend on E_S; it is *induced* by the presence of S but in principle it may have a different order of magnitude from E_S.

We believe that the latter aspect of the model presented here is the most useful in its relevance to the application of the TSM to aqueous solutions. Here, of course, the system is much more complex and we cannot arrive at an explicit expression similar to eqn. (88). Also there is no "natural" way of arriving at a classification into two components. Yet there is a strong analogy between the two cases. For example, one measures the heat of solution of an inert gas in water and finds an unexpectedly large value compared to the heat of solution in other simple liquids. As in the case of the simplified model, it is convenient to adopt a TSM formalism and write the heat of solution in the form[150]

$$\Delta \bar{H}_S = \Delta H_S{}^* + (\bar{H}_L - \bar{H}_H)(\partial N_L / \partial N_S)_{T,P,N} \qquad (92)$$

Such a formalism may satisfy our curiosity if we can show that the dissolution of S induces a rearrangement in the "structure" of the solvent in such a way that further energy is liberated. In a qualitative way, if L is the more hydrogen-bonded species and if we can show that this form is stabilized by S (i.e., $\partial N_L/\partial N_S > 0$) then the relaxation part of eqn. (92) may in principle be of a different order of magnitude from the "static" part ΔH_S^*.

Finally we turn back to our original problem of the sign of the PMHC of S. Again we shall confine ourselves to the case of eqns. (84) and (85) (for the general case see Ref. 148).

The PMHC of S is obtained by differentiation of eqn. (88) with respect to T:

$$\bar{C}_S = \frac{\partial \bar{E}_S}{\partial T} = -\frac{(E_L - E_H)^2}{kT^2}\frac{\sqrt{K}}{2}\frac{(\sqrt{K}-1)^2}{(K-1)} \leq 0 \qquad (93)$$

We see that for this particular case $\bar{C}_S$ is *negative*. This result is sufficient to conclude that in general if conditions (a)–(c) of Section 4.1 are fulfilled, we cannot claim that the PMHC of S should be positive. A more detailed analysis shows that in the general case we cannot predict the sign of the PMHC of S from the information cited in Section 4.1. Therefore the interpretation of the sign of the PMHC of various solutes in water is not valid.

5. CONCLUSION

This chapter has been devoted primarily to a clarification of some fundamental concepts in the studies of aqueous solutions. On going through these concepts, one arrives at the inescapable conclusion that none of the theories of fluids presently available is adequate to describe such complex systems as aqueous solutions. The reason is that most theories of liquids and of solutions are "first-order theories" in certain senses. For instance, the integral equation methods,[1286] notably the Percus–Yevick equation, have been quite successfully applied to simple fluids[1286] and therefore seemed to be promising for more complex liquids as well. The "first-order" character of this theory is due to the fact that only one term in a′Taylor functional expansion is considered. Such an approximation seems to work fairly well for hard-sphere fluids or hard spheres with additional weak attractive forces. Recently[1381] it has become quite clear that for such fluids the repulsive forces play a decisive role in determining the structure of the liquid. This cannot be the case in aqueous fluids, however, where the peculiar structure, i.e., the mode of packing of molecules, is dominated by the strong attractive forces, i.e., the hydrogen bonds. Therefore the prospects of applying such a theory successfully to water are not promising, as has indeed been demonstrated in a recent study.[147]

On the other hand, "first-order" theories of mixtures presuppose that we know everything about each of the pure components and proceed to investigate the properties of the mixture. Here by "first-order" one usually means a small deviation from ideality (in the symmetric sense) which follows whenever there is a slight dissimilarity between the two components. Aqueous solutions of either electrolytes or nonelectrolytes cannot fall into this category of mixtures.

With such gloomy prospects for an analytical treatment of aqueous solutions by any of the available theoretical tools we are left with two possible routes which seem to be helpful in gaining insight into the peculiarities of our system.

The first one is to study simplified models which have some features in common with aqueous solutions. In this way one can isolate various problems and study them thoroughly. We hope we have demonstrated the utility of this approach in Section 4, where we could have reached an exact answer to a specific question.

The second route is to apply "computer experiments" such as the Monte Carlo[1430] or molecular dynamics[11] techniques. These methods have only recently[90,1126] been applied to a sample of "waterlike" particles and can easily be extended to include solute particles as well. It is important to recognize the fact that all of the concepts discussed in this chapter, structure of water, structural changes, and local structure around a solute, are calculable by these methods. Therefore a detailed study of the effect of various solutes on the structure of water can be carried out.

The idea of the mixture model approach has been given prominence throughout this chapter. The reason for doing this is threefold. In the first place the distribution functions introduced may be of importance for the general characterization of complex fluids, in particular, liquid water. Second this approach provides a reasonably qualitative way of interpreting some outstanding phenomena displayed by aqueous solutions.[152] Finally the mixture model approach may serve as a useful starting point for an approximate theory of water. The latter may become feasible once we recognize which are the most important components that are responsible for the peculiar behavior of the macroscopic system. In a sense this attitude had been advocated already by Eyring,[463] whose "significant structure" theory may be embedded in the general framework of the mixture model approach. We feel that this approach is even more justifiable in the case of complex fluids, where intuition can be used to guide us in the choice of the most "significant structures."

The mixture model approach was conceived primarily as a consequence of our inability to apply analytical methods to such complex systems. We nevertheless believe that this approach will survive even after an analytical solution has been achieved for water. At the same time, even when all analytical problems will have been solved and theory will be capable of predicting all experimental quantities with a high degree of accuracy, there will still be a need for a simple and qualitative interpretation for many unusual phenomena of aqueous systems. Interpretation of these will no doubt be based on the mixture model approach.

ACKNOWLEDGMENT

This work was supported by the Central Research Fund of the Hebrew University of Jerusalem, for which the author is very grateful.

References

1. A. Abragam, "The Principles of Nuclear Magnetism," Clarendon Press, Oxford (1961).
2. S. C. Abrahams, *J. Chem. Phys.* **36**, 56 (1962).
3. S. C. Abrahams and J. L. Bernstein, *J. Chem. Phys.* **44**, 2223 (1966).
4. S. C. Abrahams and E. Prince, *J. Chem. Phys.* **36**, 50 (1962).
5. M. Abu-Hamdiyyah, *J. Phys. Chem.* **69**, 2720 (1965).
6. Th. Ackermann, *Z. Phys. Chem.* **27**, 253 (1961).
7. B. L. Afanas'ev, V. I. Kvlividze, and G. G. Malenkov, *Dokl. Akad. Nauk SSSR* **183**, 360 (1968).
8. M. Ageno and P. L. Indovina, *Proc. Nat. Acad. Sci. U.S.* **57**, 1158 (1967).
9. G. Akerlöf, *J. Am. Chem. Soc.* **54**, 4125 (1932).
10. G. Akerlöf and O. A. Short, *J. Am. Chem. Soc.* **58**, 1241 (1936).
11. B. J. Alder and W. G. Hoover, *in* "Physics of Simple Liquids" (H. N. V. Temperley, J. S. Rowlinson, and G. S. Rushbrooke, eds.), North-Holland, Amsterdam (1968).
12. D. M. Alexander, *J. Chem. and Eng. Data* **4**, 252 (1959).
13. D. M. Alexander, *J. Phys. Chem.* **63**, 994 (1959).
14. D. M. Alexander and D. J. T. Hill, *Aust. J. Chem.* **22**, 347 (1969).
15. D. M. Alexander, D. J. T. Hill, and L. R. White, *Aust. J. Chem.* **24**, 1143 (1971).
16. G. F. Alfrey and W. G. Schneider, *Disc. Faraday Soc.* **15**, 218 (1953).
17. K. W. Allen, *J. Chem. Soc. (London)* **1959**, 4131.
18. K. W. Allen, *J. Chem. Phys.* **41**, 840 (1964).
19. K. W. Allen and G. A. Jeffrey, *J. Chem. Phys.* **38**, 2304 (1963).
20. P. S. Allen, *J. Chem. Phys.* **48**, 3031 (1968).
21. S. J. Allen, Jr., *J. Chem. Phys.* **44**, 394 (1966).
22. Z. Alwani, Ph.D. thesis, University of Karlsruhe, Germany (1969).
23. Z. Alwani and G. M. Schneider, *Ber. Bunsenges. Phys. Chem.* **71**, 633 (1967).
24. Z. Alwani and G. M. Schneider, *Ber. Bunsenges. Phys. Chem.* **73**, 294 (1969).
25. A. V. Anantaraman, A. B. Walters, P. D. Edmonds, and C. J. Pings, *J. Chem. Phys.* **44**, 2651 (1966).
26. B. Andersen and F. Gronlund, *Acta Chem. Scand.* **19**, 723 (1965).
27. H. L. Anderson and L. A. Petree, *J. Phys. Chem.* **74**, 1455 (1970).
28. J. E. Anderson, *J. Chem. Phys.* **47**, 4879 (1967).
29. R. G. Anderson and M. C. R. Symons, *Trans. Faraday Soc.* **65**, 2550 (1969).
30. R. J. L. Andon, J. D. Cox, and E. F. G. Herington, *Disc. Faraday Soc.* **15**, 168 (1953).
31. J. H. Andreae, R. Bass, E. L. Heasell, and J. Lamb, *Acustica* **8**, 131 (1958).

32. J. H. Andreae and P. D. Edmonds, *J. Sci. Instr.* **38**, 508 (1961).

33. J. H. Andreae, P. D. Edmonds, and J. F. McKellar, *Acustica* **15**, 74 (1965).

34. J. H. Andreae and P. L. Joyce, *Brit. J. Appl. Phys.* **13**, 462 (1962).

35. J. H. Andreae, C. Jupp, and D. G. Vincent, *J. Acoust. Soc. Am.* **32**, 406 (1960).

36. E. R. Andrew, "Nuclear Magnetic Resonance," Univ. Press, Cambridge (1958).

37. E. R. Andrew and R. Bersohn, *J. Chem. Phys.* **18**, 159 (1950).

38. S. Anwar-Ullah, *J. Chem. Soc. (London)* **1932**, 1176.

39. K. Arakawa and N. Takenaka, *Bull. Chem. Soc. Japan* **40**, 2739 (1967).

40. K. Arakawa and N. Takenaka, *Bull. Chem. Soc. Japan* **42**, 5 (1969).

41. K. Arakawa, N. Takenaka, and K. Sasaki, *Bull. Chem. Soc. Japan* **43**, 636 (1970).

42. R. H. Aranow and L. Witten, *J. Phys. Chem.* **64**, 1643 (1960).

43. H. Aring and U. von Weber, *J. Prakt. Chem.* **30**, 295 (1965).

44. D. A. Armitage, M. J. Blandamer, M. J. Foster, N. J. Hidden, K. W. Morcom, M. C. R. Symons, and M. J. Wootten, *Trans. Faraday Soc.* **64**, 1193 (1968).

45. D. A. Armitage, M. J. Blandamer, K. W. Morcom, and N. C. Treloar, *Nature*, **219**, 719 (1968).

46. E. M. Arnett, *in* "Physico-Chemical Processes in Mixed Aqueous Solvent," (F. Franks, ed.), Heinemann, London (1967).

47. E. M. Arnett, W. G. Bentrude, J. J. Burke, and P. M. Duggleby, *J. Am. Chem. Soc.* **85**, 1350 (1963).

48. E. M. Arnett, W. B. Kover, and J. V. Carter, *J. Am. Chem. Soc.* **91**, 4028 (1969).

49. E. M. Arnett and D. R. McKelvey, *J. Am. Chem. Soc.* **87**, 1393 (1965).

50. E. M. Arnett and D. R. McKelvey, *Rec. Chem. Progr.* **26**, 185 (1965).

51. E. M. Arnett and D. R. McKelvey, *in* "Solute–Solvent Interactions" (J. F. Coetzee and C. D. Ritchie, eds.), Dekker, New York (1969).

52. J. P. Ashmore and H. E. Petch, *Can. J. Chem.* **48**, 1091 (1970).

53. P. Assarsson, Ph.D. Thesis, Polytech. Inst. of Brooklyn, Brooklyn, N.Y. (1966).

54. P. Assarsson and F. R. Eirich, *J. Phys. Chem.* **72**, 2710 (1968).

55. J. G. Aston, S. Isserow, G. J. Szasz, and R. M. Kenedy, *J. Chem. Phys.* **12**, 336 (1944).

56. J. G. Aston, G. J. Szasz, and S. Isserow, *J. Chem. Phys.* **11**, 533 (1943).

57. P. W. Atkins, A. Horsfield, and M. C. R. Symons, *J. Chem. Soc.* **1964**, 5220.

58. P. W. Atkins and D. Kivelson, *J. Chem. Phys.* **44**, 169 (1966).

59. P. W. Atkins and M. C. R. Symons, "The Structure of Inorganic Radicals," Elsevier, London (1967).

60. G. Atkinson, S. Kor, and R. L. Jones, *Rev. Sci. Instr.* **35**, 1270 (1964).

61. M. Atoji and R. E. Rundle, *J. Chem. Phys.* **29**, 1306 (1958).

62. R. P. Auty and R. H. Cole, *J. Chem. Phys.* **20**, 1309 (1952).

63. R. Aveyard and A. S. C. Lawrence, *Trans. Faraday Soc.* **60**, 2265 (1964).

64. P. B. Ayscough, "Electron Spin Resonance in Chemistry," Methuen, London (1967).

65. G. R. Azrak and E. B. Wilson, *J. Chem. Phys.* **52**, 5299 (1970).

66. A. L. Bacarella, A. Finch, and E. Grunwald, *J. Phys. Chem.* **60**, 573 (1956).

67. G. E. Bacon, *Proc. Roy. Soc. (London)* **A246**, 78 (1958).

68. G. E. Bacon, "Neutron Diffraction," Clarendon Press, Oxford (1962).

69. G. E. Bacon and N. A. Curry, *Acta Cryst.* **9**, 82 (1956).

70. G. E. Bacon and N. A. Curry, *Proc. Roy. Soc. (London)* **A266**, 95 (1962).

71. G. E. Bacon and W. E. Gardner, *Proc. Roy. Soc. (London)* **A246**, 78 (1958).

72. R. M. Badger, *J. Chem. Phys.* **2**, 128 (1934).
73. R. M. Badger, *J. Chem. Phys.* **8**, 288 (1940).
74. R. M. Badger and S. H. Bauer, *J. Chem. Phys.* **5**, 839 (1937).
75. A. Baeyer, *Ann. Chim. Phys. Ser.* 2, **27**, 181 (1857).
76. A. Baeyer and V. Villiger, *Ber. Deut. Chem. Ges.* **34**, 736 (1901).
77. S. Baggio, L. M. Amzel, and L. N. Becka, *Acta Cryst.* **B25**, 2650 (1969).
78. H. D. Bale, R. E. Shepler, and D. K. Sorgen, *Phys. Chem. Liquids* **1**, 181 (1968).
79. R. E. Ballard and C. H. Park, *J. Chem. Soc.* A, **1970**, 1340.
80. W. P. Banks, B. O. Heston, and F. F. Blankenship, *J. Phys. Chem.* **58**, 962 (1964).
81. I. M. Barclay and J. A. V. Butler, *Trans. Faraday Soc.* **34**, 1445 (1938).
82. A. J. Barduhn, *Chem. Eng. Progr.* **63**, 98 (1967).
83. A. J. Barduhn, S. L. Colten, R. Fernandez, and J. B. Pangborn, U.S. Office Saline Water Progr. Rep. 230 (1966).
84. A. J. Barduhn, N. Klausutis, R. W. Collette, and J. R. Kass, U.S. Office Saline Water Progr. Rep. 88 (1964).
85. A. J. Barduhn, O. S. Rouher, and D. J. Fontagne, U.S. Office Saline Water Progr. Rep. 366 (1968).
86. A. J. Barduhn, H. E. Towlson, and Y. C. Hu, U.S. Office Saline Water Progr. Rep. 44 (1960).
87. A. J. Barduhn, H. E. Towlson, and Y. C. Hu, *AIChE J.* **8**, 176 (1962).
88. H. P. Barendrecht, *Z. Phys. Chem.* **20**, 234 (1896).
89. R. N. Barfield and W. G. Schneider, *J. Chem. Phys.* **31**, 488 (1959).
90. J. Barker and R. O. Watts, *Chem. Phys. Letters* **3**, 144 (1969).
91. H. M. Barlow and A. L. Cullen, "Microwave Measurements," Constable, London (1950).
92. D. E. Barnaal and I. J. Lowe, *J. Chem. Phys.* **48**, 4614 (1968).
93. A. J. Barnes and H. E. Hallam, *Quart. Rev.* **XXIII**, 392 (1969).
94. A. J. Barnes and H. E. Hallam, *Trans. Faraday Soc.* **66**, 1922, 1932 (1970).
95. A. J. Barnes, H. E. Hallam, and G. C. Scrimshaw, *Trans. Faraday Soc.* **65**, 3150 (1969).
96. A. J. Barnes, H. E. Hallam, and G. F. Scrimshaw, *Trans. Faraday Soc.* **65**, 3172 (1969).
97. A. J. Barnes, H. E. Hallam, and G. F. Scrimshaw, *Trans. Faraday Soc.* **65**, 3159 (1969).
98. G. Barone, V. Cresenzi, and V. Vitagliano, *J. Phys. Chem.* **72**, 2588 (1968).
99. F. Barr-David and B. F. Dodge, *J. Chem. and Eng. Data* **4**, 107 (1959).
100. R. M. Barrer, Inorganic Inclusion Compounds, *in* "Non-Stoichiometric Compounds" (L. Mandelcorn, ed.), p. 310, Academic, New York (1964).
101. R. M. Barrer and A. V. J. Edge, *Proc. Roy. Soc., Ser. A*, **300**, 1 (1967).
102. R. M. Barrer and A. V. J. Edge, *Separation Sci.* **2**, 145 (1967).
103. R. M. Barrer and D. J. Ruzicka, *Trans. Faraday Soc.* **58**, 2239, 2253, 2262 (1962).
104. R. M. Barrer and W. I. Stuart, *Proc. Roy. Soc., Ser. A* **242**, 172 (1957).
104a. J. Barrett, M. F. Fox, and A. L. Mansell, *J. Chem. Soc. A*, **1966**, 487.
104b. J. Barrett, M. F. Fox, and A. L. Mansell, *J. Chem. Soc. B*. **1971**, 173.
105. G. M. Barrow, "Molecular Spectroscopy," McGraw-Hill, London (1962).
106. G. M. Barrow and P. Datta, *J. Phys. Chem.* **72**, 2259 (1968).
107. C. Barthomieu and C. Sandorfy, *J. Mol. Spectry.* **15**, 15 (1965).

108. R. Bass, *J. Acoust. Soc. Am.* **30**, 602 (1958).
109. E. Bauer and M. Magat, *J. Phys. Radium* **9**, 319 (1938).
110. W. H. Baur, *Naturwiss.* **48**, 549 (1961).
111. W. H. Baur, *Acta Cryst.* **17**, 863 (1964).
112. W. H. Baur, *Acta Cryst.* **17**, 1361 (1964).
113. W. H. Baur, *Acta Cryst.* **17**, 1167 (1964).
114. W. H. Baur, *Acta Cryst.* **19**, 909 (1965).
115. W. H. Baur, *Neues Jahrb. Mineral. Monatsh.* **9**, 430 (1969).
116. W. H. Baur, *Trans. Am. Crystal. Assoc.* **6**, 129 (1970).
117. W. H. Baur and A. A. Khan, *Acta Cryst.* **B26**, 1584 (1970).
118. W. H. Baur and B. Rama Rao, *Am. Mineral.* **53**, 1025 (1968).
119. N. S. Bayliss, *J. Chem. Phys.* **18**, 292 (1950).
120. J. Y. Beach, *J. Chem. Phys.* **9**, 54 (1941).
121. I. J. Bear and W. G. Mumme, *Acta Cryst.* **B25**, 1558 (1969).
122. D. V. Beauregard and R. E. Barrett, *J. Chem. Phys.* **49**, 5241 (1968).
123. E. D. Becker, U. Liddel, and J. N. Schoolery, *J. Mol. Spectry.* **2**, 1 (1958).
124. D. Beckert and H. Pfeifer, *Ann. Physik* **16**, 262 (1965).
125. C. A. Beevers and H. Lipson, *Z. Krist.* **82**, 297 (1932).
126. G. M. Bell, *J. Math. Phys.* **10**, 1753 (1969).
127. G. M. Bell and D. A. Lavis, *J. Phys. A: Gen. Phys.* **3**, 427, 568 (1970).
128. R. P. Bell, *Advan. Phys. Org. Chem.* **4**, 1 (1966).
129. R. P. Bell and J. C. Clunie, *Trans. Faraday Soc.* **48**, 439 (1952).
130. L. J. Bellamy, H. E. Hallam, and R. L. Williams, *Trans. Faraday Soc.* **54**, 1120 (1958).
131. L. J. Bellamy and A. J. Owen, *Spectrochim. Acta* **25A**, 319 (1969).
132. L. J. Bellamy and A. J. Owen, *Spectrochim. Acta* **25A**, 329 (1969).
133. L. J. Bellamy and R. C. Pace, *Spectrochim. Acta* **22**, 525 (1966).
134. L. J. Bellamy and R. C. Pace, *Spectrochim. Acta* **25**, 319 (1969).
135. V. P. Belousov, *Vestn. Leningr. Univ.* No. 4, 144 (1961).
136. V. P. Belousov and N. L. Makarova, *Vestn. Leningr. Univ.* No. 22 (1970).
137. W. S. Benedict, N. Gailar, and E. K. Plyler, *J. Chem. Phys.* **24**, 1139 (1956).
138. G. Benedik and T. Greytak, *Proc. IEEE* **53**, 1623 (1965).
139. M. E. Benesh, U.S. Patent 2,270,016 (1942).
140. L. Benjamin and G. C. Benson, *J. Phys. Chem.* **67**, 858 (1963).
141. W. Benjamin, cited in "Physico-Chemical Constants of Binary Systems" (J. Timmermans, ed.), Vol. 4, Interscience, New York (1960).
142. A. Ben-Naim, *J. Chem. Phys.* **42**, 1512 (1965).
143. A. Ben-Naim, *J. Phys. Chem.* **69**, 3240 (1965).
144. A. Ben-Naim, *J. Phys. Chem.* **69**, 3245 (1965).
145. A. Ben-Naim, *J. Phys. Chem.* **71**, 1137 (1967).
146. A. Ben-Naim, *J. Phys. Chem.* **71**, 4002 (1967).
147. A. Ben-Naim, *J. Chem. Phys.* **52**, 5531 (1970).
148. A. Ben-Naim, *Trans. Faraday Soc.* **66**, 2749 (1970).
149. A. Ben-Naim, *J. Chem. Phys.* **54**, 1387, 3696 (1971).
150. A. Ben-Naim, *in* "Water and Aqueous Solutions, Structure, Thermodynamics and Transport Processes" (R. A. Horne, ed.), Wiley, New York (1972).
151. A. Ben-Naim, *Chem. Phys. Letters* **13**, 406 (1972).
152. A. Ben-Naim, *J. Chem. Phys.* **56**, 2864 (1972); **57**, 3605 (1972).
153. A. Ben-Naim and S. Baer, *Trans. Faraday Soc.* **60**, 1736 (1964).

154. A. Ben-Naim and M. Egel. Thal, *J. Phys. Chem.* **69**, 3250 (1965).

155. H. P. Bennetto and D. Feakins, *in* "Hydrogen-Bonded Solvent Systems" (A. K. Covington and P. Jones, ed.), p. 235, Taylor and Francis (1968).

156. H. P. Bennetto, D. Feakins, and K. G. Lawrence, *J. Chem. Soc. A,* **1968**, 1493.

157. H. J. C. Berendsen, *in* "Theoretical and Experimental Biophysics," Vol. 1 (A. Cole, ed.), Arnold, London (1967).

158. G. Bergerhoff, L. Goost, and E. Schultze-Rhonhof, *Acta Cryst.* **B24**, 803 (1968).

159. J. D. Bernal, *J. Chim. Phys. Physicochim. Biol.* **50**, C1 (1953).

160. J. D. Bernal, *in* "Liquids: Structure, Properties and Solid Interactions" (T. J. Hughel, ed.), Elsevier, Amsterdam (1965).

161. J. D. Bernal and R. H. Fowler, *J. Chem. Phys.* **1**, 515 (1933).

162. P. E. M. Berthelot, *Ann. Chim. Phys. Ser. 3* **46**, 490 (1856).

163. J. E. Bertie and J. W. Bell, *J. Chem. Phys.* **54**, 160 (1971).

164. J. E. Bertie, H. J. Labbé, and E. Whalley, *J. Chem. Phys.* **49**, 2141 (1968).

165. J. E. Bertie, H. J. Labbé, and E. Whalley, *J. Chem. Phys.* **50**, 4501 (1969).

166. J. E. Bertie and E. Whalley, *J. Chem. Phys.* **40**, 1637 (1964).

167. J. E. Bertie and E. Whalley, *J. Chem. Phys.* **40**, 1646 (1964).

168. C. Bertinotti and A. Bertinotti, *Acta Cryst.* **B26**, 422 (1970).

169. G. L. Bertrand, J. W. Larson, and L. G. Hepler, *J. Phys. Chem.* **72**, 4194 (1968).

170. P. T. Beurskens and G. A. Jeffrey, *J. Chem. Phys.* **40**, 2800 (1964).

171. G. Beurskens and G. A. Jeffrey, *J. Chem. Phys.* **41**, 924 (1964).

172. G. Beurskens, G. A. Jeffrey, and R. K. McMullan, *J. Chem. Phys.* **39**, 3311 (1963).

173. A. B. Bhatia, "Ultrasonic Absorption," Oxford Univ. Press, Oxford (1967).

174. V. M. Bhatnagar, "Clathrate Compounds," S. Chand, New Delhi (1968).

175. M. Bigorgne, *Compt. Rend.* **236**, 1966 (1953).

176. H. J. Bittrich and G. Kraft, *Z. Phys. Chem. (Leipzig)* **227**, 359 (1964).

177. N. Bjerrum, *Kgl. Dan. Vidensk. Selsk., Mat.-Fys. Medd.* **27**, 3 (1951).

178. N. Bjerrum, *Science* **115**, 385 (1952).

179. M. J. Blandamer, *Quart. Rev.* **24**, 169 (1970).

180. M. J. Blandamer, J. A. Brivati, M. F. Fox, M. C. R. Symons, and G. S. Verma, *Trans. Faraday Soc.* **63**, 1850 (1967).

181. M. J. Blandamer, D. E. Clarke, T. A. Claxton, M. F. Fox, N. J. Hidden, J. Oakes, M. C. R. Symons, G. S. P. Verma, and M. J. Wootten, *Chem. Commun.* **1967**, 273.

182. M. J. Blandamer, D. E. Clarke, N. J. Hidden, and M. C. R. Symons, *Chem. Commun.* **1966**, 342.

183. M. J. Blandamer, D. E. Clarke, N. J. Hidden, and M. C. R. Symons, *Trans. Faraday Soc.* **63**, 66 (1967).

184. M. J. Blandamer, D. E. Clarke, N. J. Hidden, and M. C. R. Symons, *Trans. Faraday Soc.* **64**, 2691 (1968).

185. M. J. Blandamer, M. J. Foster, N. J. Hidden, and M. C. R. Symons, *Chem. Commun.* **1966**, 62.

186. M. J. Blandamer, M. J. Foster, and D. Waddington, *Trans. Faraday Soc.* **66**, 1369 (1970).

187. M. J. Blandamer and M. F. Fox, *Chem. Rev.* **70**, 59 (1970).

188. M. J. Blandamer, M. F. Fox, J. Stafford, and R. E. Powell, *Chem. Commun.* **1968**, 1022.

189. M. J. Blandamer, M. F. Fox, M. C. R. Symons, K. J. Wood, and M. J. Wootten, *Trans. Faraday Soc.* **64**, 3210 (1968).

190. M. J. Blandamer, M. F. Fox, M. C. R. Symons, and M. J. Wootten, *Trans. Faraday Soc.* **66**, 1574 (1970).

191. M. J. Blandamer, T. E. Gough, J. M. Gross, and M. C. R. Symons, *J. Chem. Soc.* **1964**, 536.

192. M. J. Blandamer, T. E. Gough, and M. C. R. Symons, *Trans. Faraday Soc.* **59**, 1748 (1963).

193. M. J. Blandamer, T. E. Gough, and M. C. R. Symons, *Trans. Faraday Soc.* **62**, 296 (1966).

194. M. J. Blandamer, N. J. Hidden, K. W. Morcom, R. W. Smith, N. C. Treloar, and M. J. Wootten, *Trans. Faraday Soc.* **65**, 2633 (1969).

195. M. J. Blandamer, N. J. Hidden, and M. C. R. Symons, *Trans. Faraday Soc.* **66**, 316 (1970).

196. M. J. Blandamer, N. J. Hidden, M. C. R. Symons, and N. C. Treloar, *Chem. Commun.* **1968**, 1325.

197. M. J. Blandamer, N. J. Hidden, M. C. R. Symons, and N. C. Treloar, *Trans. Faraday Soc.* **64**, 3242 (1968).

198. M. J. Blandamer, N. J. Hidden, M. C. R. Symons, and N. C. Treloar, *Trans. Faraday Soc.* **65**, 1805 (1969).

199. M. J. Blandamer, N. J. Hidden, M. C. R. Symons, and N. C. Treloar, *Trans. Faraday Soc.* **65**, 2663 (1969).

200. M. J. Blandamer and M. C. R. Symons, *in* "Hydrogen-Bonded Solvent Systems" (A. K. Covington and P. Jones, eds.), p. 211, Taylor and Francis, London (1968).

201. M. J. Blandamer, M. C. R. Symons, and M. J. Wootten, *Trans. Faraday Soc.* **63**, 2337 (1967).

202. M. J. Blandamer, M. C. R. Symons, and M. J. Wootten, *Chem. Commun.* **1970**, 366.

203. M. J. Blandamer and D. Waddington, *Adv. Mol. Relax. Processes* **2**, 1 (1970).

204. R. Blinc and G Lahajnar, *Phys. Rev. Letters* **19**, 685 (1967).

205. J. Blitz, "Fundamentals of Ultrasonics," Butterworths, London (1967).

206. H. Bode and G. Teufer, *Acta Cryst.* **8**, 611 (1955).

207. R. L. Bohon and W. F. Claussen, *J. Am. Chem. Soc.* **73**, 1571 (1951).

208. M. Bonamico, G. A. Jeffrey, and R. K. McMullan, *J. Chem. Phys.* **37**, 2219 (1962).

209. O. D. Bonner, *J. Phys. Chem.* **72**, 2512 (1968).

210. O. D. Bonner and G. B. Woolsey, *J. Phys. Chem.* **72**, 899 (1968).

211. T. T. Bopp, *J. Chem. Phys.* **47**, 3621 (1967).

212. E. Bose, *Z. Phys. Chem.* **A58**, 505 (1907).

213. C. J. F. Böttcher, "Theory of Electric Polarisation," Elsevier, Amsterdam (1952).

214. J. A. Boyne and A. G. Williamson, *J. Chem. and Eng. Data* **12**, 318 (1967).

215. J. F. Brandts and L. Hunt, *J. Am. Chem. Soc.* **89**, 4826 (1967).

216. E. Braswell, *J. Phys. Chem.* **72**, 2477 (1968).

217. B. Braun, Dissert. University of Bonn (1938).

218. W. Bray, *Z. Phys. Chem.* **54**, 569 (1906).

219. K. G. Breitschwerdt, H. Kistenmacker, and K. Tamm, *Phys. Letters* **24A**, 550 (1967).

220. A. W. Brewer and H. P. Palmer, *Proc. Phys. Soc. London, Sect. B* **64**, 765 (1961).

221. R. G. Brewer and K. E. Rieckhoff, *Phys. Rev. Letters* **13**, 334 (1964).

222. W. S. Brey, Jr. and H. P. Williams, *J. Phys. Chem.* **72**, 49 (1968).
223. F. A. Briggs and A. J. Barduhn, *Advan. Chem. Ser.* **38**, 190 (1963).
224. F. A. Briggs, Y. C. Hu, and A. J. Barduhn, U.S. Office Saline Water Progr. Rep. 59 (1962).
225. G. Brink and M. Falk, *Can. J. Chem.* **48**, 2096 (1970).
226. G. Brink and M. Falk, *J. Mol. Struct.* **5**, 27 (1970).
227. G. Brink and M. Falk, *Can. J. Chem.* **49**, 347 (1971).
228. G. Brink and M. Falk, *Spectrochim. Acta* **27A**, 1811 (1971).
229. K. Bröllos, K. Peter, and G. M. Schneider, *Ber. Bunsenges. Phys. Chem.* **74**, 682 (1970).
230. L. G. S. Brooker, A. C. Craig, D. W. Heseltine, P. W. Jenkins, and L. L. Lincoln, *J. Am. Chem. Soc.* **87**, 2443 (1965).
231. A. C. Brown, Ph.D. Thesis, University of London (1956).
232. A. C. Brown and D. J. G. Ives, *J. Chem. Soc.* **II**, 1608 (1962).
233. A. E. Brown, *Acustica* **18**, 169 (1967).
234. A. S. Brown, P. M. Levin, and E. W. Abrahamson, *J. Chem. Phys.* **19**, 1226 (1951).
235. G. M. Brown and R. Chidambaram, *Acta Cryst.* **B25**, 676 (1969).
236. W. F. Brown, *in* "Handbuch der Physik," Vol. 17, p. 105, Springer, Berlin (1956).
237. S. Brownstein, D. W. Davidson, and D. Fiat, *J. Chem. Phys.* **46**, 1454 (1967).
238. D. A. G. Bruggeman, *Ann. Physik* **24**(5), 636 (1935).
239. G. Brun, *Rev. Chim. Miner.* **5**, 899 (1968).
240. R. S. Brundage and K. Kustin, *J. Phys. Chem.* **74**, 672 (1970).
241. J. Brystead and G. P. Smith, *J. Phys. Chem.* **72**, 296 (1968).
242. D. R. Buchanan and P. M. Harris, *Acta Cryst.* **B24**, 953 (1968).
243. A. D. Buckingham, *Aust. J. Chem.* **6**, 93 (1953).
244. A. D. Buckingham, *Aust. J. Chem.* **6**, 323 (1953).
245. A. D. Buckingham, *Proc. Roy. Soc. A* **248**, 169 (1958).
246. A. D. Buckingham, *Proc. Roy. Soc.* **255**, 32 (1960).
247. A. D. Buckingham, *Trans. Faraday Soc.* **56**, 753 (1960).
248. A. D. Buckingham, T. Schaefer, and W. G. Schneider, *J. Chem. Phys.* **32**, 1227 (1960).
249. K. Buijs and G. R. Choppin, *J. Chem. Phys.* **39**, 2035 (1963).
250. H. B. Bull and K. Breese, *Arch. Biochim. Biophys.* **139**, 93 (1970).
251. F. P. Bundy and J. S. Kasper, *High Temp.–High Pressures* **2**, 429 (1970).
252. K. W. Bunzl, *J. Phys. Chem.* **71**, 1358 (1967).
253. J. Burgess and M. C. R. Symons, *Quart. Rev.* **22**, 276 (1968).
254. J. C. Burgil, H. Meyer, and P. L. Richards, *J. Chem. Phys.* **43**, 4291 (1965).
255. C. J. Burton, *J. Acoust. Soc. Am.* **20**, 186 (1948).
256. A. M. Bushwell, R. L. Maycock, and W. H. Rodebush, *J. Chem. Phys.* **8**, 362 (1940).
257. J. A. V. Butler, *Trans. Faraday Soc.* **33**, 229 (1937).
258. J. A. V. Butler, C. N. Ramchandani, and D. W. Thomson, *J. Chem. Soc.* **1935**, 280.
259. J. A. V. Butler, D. W. Thomson, and W. H. Maclennan, *J. Chem. Soc.* **1933**, 674.
260. S. Cabani, G. Conti, and L. Lepori, *Trans. Faraday Soc.* **67**, 1933 (1971).
261. S. Cabani, G. Conti, and L. Lepori, *Trans. Faraday Soc.* **67**, 1943 (1971).
262. L. Cailletet, *Ann. Chim. Phys., Ser. 5*, **15**, 132 (1878).
263. L. Cailletet and R. Bordet, *Compt. Rend.* **95**, 58 (1882).
264. E. F. Caldin, "Fast Reactions in Solutions," Blackwell, Oxford (1964).

265. L. D. Calvert and P. Srivastava, *Acta Cryst. Sect. A* **25**, S131 (1969).

266. E. F. Carome and J. M. Whiting, *J. Acoust. Soc. Am.* **33**, 187 (1961).

267. E. F. Carome, J. M. Whiting, and P. A. Fleury, *J. Acoust. Soc. Am.* **33**, 1417 (1961).

268. H. Y. Carr and E. M. Purcell, *Phys. Rev.* **94**, 630 (1954).

269. A. Carrington and A. D. McLachlan, "Introduction to Magnetic Resonance," Harper and Row, New York (1967).

270. D. B. Carson and D. L. Katz, *Trans. AIME* **146**, 150 (1942).

271. E. L. Cartensen, *J. Acoust. Soc. Am.* **26**, 858, 862 (1954).

272. J. F. Catchpool, *in* "Structural Chemistry and Molecular Biology" (A. Rich and N. Davidson, eds.), p. 343, Freeman, San Francisco (1968).

273. L. Cavallaro, *Arch. Sci. Biol.* **20**, 567 (1934).

274. L. Cavallaro, *Arch. Sci. Biol.* **20**, 583 (1934).

275. P. J. Ceccotti, *Ind. Eng. Chem. Fundam.* **5**, 106 (1966).

276. F. Cennamo and E. Tartaglione, *Nuovo Cimento* **11**, 401 (1959).

277. H. M. Chadwell, *Chem. Rev.* **4**, 375 (1927).

278. J. E. Chamberlain, G. W. Chantry, H. A. Gebbie, N. W. B. Stone, T. B. Taylor, and G. Wyllie, *Nature* **210**, 790 (1966).

279. S. I. Chan, J. Zinn, and W. D. Gwinn, *J. Chem. Phys.* **34**, 1319 (1961).

280. G. Chancel and F. Parmentier, *Compt. Rend.* **100**, 27 (1885).

281. S. Chandrasekhar, *Revs. Modern Physics* **15**, 1 (1943); *in* "Selected Papers on Noise and Stochastic Processes" (N. Wax, ed.), Dover, New York (1954).

282. D. Chapman, M. D. Barratt, and V. B. Kamat, *Biochim. Biophys. Acta* **173**, 154 (1969).

283. N. Chatterjee and M. Bose, *Mol. Phys.* **12**, 341 (1967).

284. N. V. Cherskii, Yu. F. Makogon, and D. I. Medovskii, *in* "Geol. Str. Neftegazonos. Vost. Chasti Sib. Platformy Prilegayushchikh Raionov, Mater. Vses. Soveshsch. Otsenke Neftegazonos. Territ. Yakutii," 1966, p. 458, Nedra, Moscow (1968).

285. R. Y. Chiao and B. P. Stoicheff, *J. Opt. Soc. Am.* **54**, 1286 (1964).

286. R. Y. Chiao and C. H. Townes, *Phys. Rev. Letters* **12**, 592 (1964).

287. R. Chidambaram, *Acta Crystallogr.* **14**, 467 (1961).

288. R. Chidambaram, *J. Chem. Phys.* **36**, 2361 (1962).

289. R. Chidambaram, *Proc. Nucl. Phys. Solid State Phys. Symp. 13th 1968*, **1969**, 215.

290. R. Chidambaram, Q. O. Navarro, A. Garcia, K. Linggoatmodjo, S. C. Lin, I. H. Suh, A. Sequeira, and S. Srikanta, *Acta Cryst.* **B26**, 827 (1970).

291. R. Chidambaram, A. Sequeira, and S. K. Sikka, *J. Chem. Phys.* **41**, 3616 (1964).

292. R. Chidambaram and S. K. Sikka, *Chem. Phys. Letters* **2**, 162 (1968).

293. W. C. Child, Jr., *J. Phys. Chem.* **68**, 1834 (1964).

294. H. E. Chinworth and D. L. Katz, *Refrig. Eng.* **54**, 359 (1947).

295. G. Chiurdoglu, W. Masschelen, and Y. van Haverbeke Huyskens, *Bull. Soc. Chim. Belg.* **75**, 824 (1966).

296. G. Chottard, J. Fraissard, and B. Imelik, *Bull. Soc. Chim. Fr.* **1967**, 4331.

297. S. D. Christian, H. E. Affsprung, and J. R. Johnson, *J. Chem. Soc.* **1963**, 1896.

298. S. D. Christian, A. A. Taha, and B. W. Gash, *Quart. Rev. Chem. Soc.* **24**, 20 (1970).

299. A. O. Christie and D. J. Crisp, *J. Appl. Chem.* **17**, 11 (1967).

300. R. J. Chuck and E. W. Randall, *J. Chem. Soc. B* **1967**, 261.

301. A. G. Chynoweth and W. G. Schneider, *J. Chem. Phys.* **19**, 1566 (1951).

302. J. Claeys, E. Errera, and H. Sack, *Trans. Faraday Soc.* **33**, 136 (1937).

303. J. R. Clark, *Rev. Pure Appl. Chem.* **13**, 50 (1963).

304. D. E. Clarke, Ph.D. Thesis, Univ. of Leicester (1967).

305. E. C. Clarke, R. W. Ford, and D. N. Glew, *Can. J. Chem.* **42**, 2027 (1964).

306. E. C. Clarke and D. N. Glew, *Trans. Faraday Soc.* **62**, 539 (1966).

307. W. F. Claussen, *J. Chem. Phys.* **19**, 259, 662 (1951).

308. W. F. Claussen, *J. Chem. Phys.* **19**, 1425 (1951).

309. W. F. Claussen and M. F. Polglase, *J. Am. Chem. Soc.* **74**, 4817 (1952).

310. T. A. Claxton, *Trans. Faraday Soc.* **65**, 2289 (1969).

311. T. A. Claxton and D. McWilliams, *Trans. Faraday Soc.* **64**, 2593 (1968).

312. C. J. Clemett, *J. Chem. Soc. A* **1969**, 455.

313. C. J. Clemett, *J. Chem. Soc. A* **1969**, 458.

314. C. J. Clemett, *J. Chem. Soc. A* **1969**, 761.

315. J. Clifford, *Trans. Faraday Soc.* **61**, 1276 (1965).

316. J. Clifford and B. A. Pethica, *Trans. Faraday Soc.* **60**, 1483 (1964).

317. J. Clifford and B. A. Pethica, *Trans. Faraday Soc.* **61**, 182 (1965).

318. W. L. Coburn and E. Grunwald, *J. Am. Chem. Soc.* **80**, 1318 (1958).

319. C. K. Cogan, G. G. Belford, and H. S. Gutowsky, *J. Chem. Phys.* **39**, 3061 (1963).

320. N. D. Coggeshall and E. L. Saier, *J. Am. Chem. Soc.* **73**, 5414 (1951).

321. A. D. Cohen and C. Reid, *J. Chem. Phys.* **25**, 790 (1956).

322. E. Cohen and C. J. F. van der Horst, *Z. Phys. Chem. Abt. B* **40**, 231 (1938).

323. M. D. Cohen and E. Fisher, *J. Chem. Soc.* **1963**, 3044.

324. C. Cohen-Addad, *Acta Cryst.* **B25**, 1644 (1969).

325. C. Cohen-Addad, P. Ducros, and E. F. Bertaut, *Acta Cryst.* **23**, 220 (1967).

326. K. S. Cole and R. H. Cole, *J. Chem. Phys.* **9**, 341 (1941).

327. R. H. Cole, *J. Am. Chem. Soc.* **77**, 2012 (1955).

328. R. H. Cole and D. W. Davidson, *J. Chem. Phys.* **20**, 2389 (1952).

329. R. H. Cole and O. Wörz, *in* "Physics of Ice" (N. Riehl, B. Bullemer, and H. Engelhardt, eds.), Plenum, New York (1969).

330. W. P. Conner, R. P. Clarke, and C. P. Smyth, *J. Am. Chem. Soc.* **64**, 1379 (1942).

331. J. F. Connolly, *J. Chem. Eng. Data* **11**, 13 (1966).

332. B. E. Conway and L. H. Laliberté, *J. Phys. Chem.* **72**, 4317 (1968).

333. B. E. Conway and L. H. Laliberté, *in* "Hydrogen-Bonded Solvent Systems" (A. K. Covington and P. Jones, eds.), p. 139, Taylor and Francis, London (1969).

334. B. E. Conway and R. E. Verrall, *J. Phys. Chem.* **70**, 1473, 3952 (1966).

335. B. E. Conway, R. E. Verrall, and J. E. Desnoyers, *Z. Phys. Chem.* (*Leipzig*) **230**, 157 (1965).

336. B. E. Conway, R. E. Verrall, and J. E. Desnoyers, *Trans. Faraday Soc.* **62**, 2738 (1966).

337. J. L. Copp and D. H. Everett, *Disc. Faraday Soc.* **15**, 174 (1953).

338. P. Coppens and T. M. Sabine, *Acta Cryst.* **B25**, 2442 (1969).

339. C. A. Coulson and D. Eisenberg, *Proc. Roy. Soc. Ser. A* **291**, 445 (1966).

340. J. M. G. Cowie and P. M. Toporowski, *Can. J. Chem.* **39**, 2240 (1961).

341. D. E. Cox, B. C. Frazer, and G. Shirane, *Phys. Letters* **17**, 103 (1965).

342. H. S. M. Coxeter, "Introduction to Geometry," Wiley, New York (1961).

343. F. Cramer, "Einschlussverbindungen," Springer Verlag, Berlin (1954).

344. F. Cramer, "Inclusion Compounds," *Rev. Pure Appl. Chem.* **5**, 143 (1955).

345. B. M. Craven and T. M. Sabine, *Acta Cryst.* **B25**, 1970 (1969).

346. B. M. Craven and W. J. Takei, *Acta Cryst.* **17**, 415 (1964).

347. C. M. Criss, R. P. Held, and E. Luksha, *J. Phys. Chem.* **72**, 2970 (1968).

348. D. T. Cromer, M. I. Kay, and A. C. Larson, *Acta Cryst.* **21**, 383 (1966).

349. D. T. Cromer, M. I. Kay, and A. C. Larson, *Acta Cryst.* **22**, 182 (1967).

350. D. T. Cromer and A. C. Larson, *Acta Cryst.* **15**, 397 (1962).

351. K. R. Crook, E. Wyn-Jones, and W. J. Orville-Thomas, *Trans. Faraday Soc.* **66**, 1597 (1970).

352. C. Cros, M. Pouchard, and P. Hagenmuller, *J. Solid State Chem.* **2**, 570 (1970).

353. P. C. Cross, J. Burnham, and P. A. Leighton, *J. Am. Chem. Soc.* **59**, 1130 (1937).

354. J. F. Crowther, U.S. Patent 2,399,723 (1946).

355. G. L. Cunningham, A. W. Boyd, R. J. Myers, W. G. Gwinn, and W. I. LeVan, *J. Chem. Phys.* **19**, 676 (1951).

356. B. M. Cwilong, *J. Glaciol.* **1**, 53 (1947).

357. A. Czaplinski, *Nafta* (*Katowice*) **12**, 155 (1956).

358. I. G. Dance and H. C. Freeman, *Acta Cryst.* **B25**, 304 (1969).

359. F. Daniels, J. W. Williams, P. Bender, R. A. Alberty, and C. D. Cornwell, "Experimental Physical Chemistry," 6th ed. McGraw-Hill, New York (1962).

360. A. Danneil, K. Tödheide, and E. U. Franck, *Chem. Ingr. Tech.* **39**, 816 (1967).

361. W. Dannhauser and L. W. Bahe, *J. Chem. Phys.* **40**, 3058 (1964).

362. W. Dannhauser and R. H. Cole, *J. Chem. Phys.* **23**, 1762 (1955).

363. W. Dannhauser and A. F. Flueckinger, *J. Chem. Phys.* **38**, 69 (1963).

364. S. S. Danyluk and E. S. Gore, *Nature* **203**, 748 (1964).

365. B. T. Darling and D. M. Dennison, *Phys. Rev.* **57**, 128 (1940).

366. G. D'Arrigo, L. Mistura, and P. Tartaglia, *Phys. Rev. A* **1**, 286 (1970).

367. G. D'Arrigo and D. Sette, *J. Chem. Phys.* **48**, 691 (1968).

368. I. D. Datt, N. V. Rannev, and R. P. Ozerov, *Kristallografiya* **13**, 261 (1968).

369. J. G. David and H. E. Hallam, *Trans. Faraday Soc.* **563**, 2838, 2843 (1969).

370. D. W. Davidson, *Can. J. Chem.* **39**, 571 (1961).

371. D. W. Davidson, *in* "Molecular Relaxation Processes," Chem. Soc. London Spec. Pub. No. 20, p. 33 (1966).

372. D. W. Davidson, *Can. J. Chem.* **46**, 1024 (1968).

373. D. W. Davidson, *Can. J. Chem.* **49**, 1224 (1971).

374. D. W. Davidson and R. H. Cole, *J. Chem. Phys.* **19**, 1484 (1951).

375. D. W. Davidson, M. Davies, and K. Williams, *J. Chem. Phys.* **40**, 3449 (1964).

376. D. W. Davidson and G. J. Wilson, *Can. J. Chem.* **44**, 1424 (1963).

377. P. R. Davies, *Disc. Faraday Soc.* **48**, 181 (1969).

378. C. M. Davies and J. Jarzynski, *Adv. Mol. Relax. Processes* **1**, 155 (1967–68).

379. R. O. Davies and J. Lamb, *Quart. Rev.* **11**, 134 (1957).

380. M. Davies and K. Williams, *Trans. Faraday Soc.* **64**, 529 (1968).

381. C. M. Davis and T. Litovitz, *J. Chem. Phys.* **42**, 2563 (1965).

382. J. Davis, S. Ormondroyd, and M. C. R. Symons, *Chem. Commun.* **1970**, 1426.

383. R. R. Davison, *J. Chem. and Eng. Data* **13**, 348 (1968).

384. H. Davy, *Phil. Trans. Roy. Soc.* (*London*) **101**, 1 (1811).

385. J. C. Deardon, *Nature* **206**, 1147 (1965).

386. W. M. Deaton and E. M. Frost, Jr., U.S. Bur. Mines Monograph No. 8 (1946).

387. P. Debye, "Polar Molecules," Chem. Catalog Co., New York (1929).

388. P. Debye and H. Sack, "Handbuch der Radiologie," Vol. VI, 2nd ed., Leipzig (1934).
389. P. Debye and F. W. Sears, *Proc. Nat. Acad. Sci. U.S.* **18**, 410 (1932).
390. R. de Forcrand, *Compt. Rend.* **90**, 1491 (1880).
391. R. de Forcrand, *Compt. Rend.* **95**, 129 (1882).
392. R. de Forcrand, *Ann. Chim. Phys. Ser. 5,* **28**, 5 (1883).
393. R. de Forcrand, *Compt. Rend.* **134**, 835 (1902).
394. R. de Forcrand, *Compt. Rend.* **134**, 991 (1902).
395. R. de Forcrand, *Compt. Rend.* **135**, 959 (1902).
396. R. de Forcrand, *Ann. Chim. Phys. Ser. 7,* **29**, 5 (1903).
397. R. de Forcrand, *Ann. Chim. Phys. Ser. 7,* **29**, 36 (1903).
398. R. de Forcrand, *Compt. Rend.* **160**, 467 (1915).
399. R. de Forcrand, *Compt. Rend.* **176**, 355 (1923).
400. R. de Forcrand, *Compt. Rend.* **181**, 15 (1925).
401. R. de Forcrand and H. Fonzes-Diacon, *Compt. Rend.* **134**, 229, 281 (1902).
402. J. de Graauw and J. J. Rutten, *in* "Proc. 3rd Int. Symp. on Fresh Water from the Sea" (A. Delyannis and E. Delyannis, eds.), Vol. 3, p. 103, Athens (1970).
403. W. H. de Jeu, *Mol. Phys.* **18**, 31 (1970).
404. A. de la Rive, *Ann. Chim. Phys. Ser. 2* **40**, 405 (1829).
405. M. Deleplanque, A. Kahane, and A. Serra, *Compt. Rend.* **262C**, 52 (1966).
406. J. P. de Loor, Thesis, Univ. of Leydon, 1956.
407. A. H. Delsemme and D. C. Miller, *Planet. Space Sci.* **18**, 717 (1970).
408. A. H. Delsemme and P. Swings, *Ann. Astrophys.* **15**, 1 (1952).
409. B. de Nooijer, D. Spencer, S. G. Whittington, and F. Franks, *Trans. Faraday Soc.* **67**, 1315 (1971).
410. J. E. Desnoyers, M. Arel, G. Perron, and C. Jolicoeur, *J. Phys. Chem.* **73**, 3346 (1969).
411. J. M. Deutch and I. Oppenheim, *in* "Advances in Magnetic Resonance" (J. S. Waugh, ed.), Vol. 3, p. 43, Academic, New York (1968).
412. C. Deverell, *in* "Progress in NMR Spectroscopy" (J. W. Emsley, J. Feeney, and L. H. Sutcliffe, eds.), Vol. 4, p. 235, Pergamon, Oxford (1969).
413. G. Devoto, *Gazz. chim. ital.* **60**, 520 (1930).
414. G. Devoto, *Atti Accad. Lincei* **14**, 432 (1931).
415. G. Devoto, *Gazz. chim. ital.* **61**, 897 (1931).
416. G. Devoto, *Gazz. chim. ital.* **63**, 50 (1933).
417. G. Devoto, *Gazz. chim. ital.* **63**, 119 (1933).
418. G. Devoto, *Gazz. chim. ital.* **63**, 247 (1933).
419. G. Devoto, *Z. Physiol. Chem.* **222**, 227 (1933).
420. G. Devoto, *Gazz. chim. ital.* **64**, 76 (1934).
421. G. Devoto, *Gazz. chim. ital.* **64**, 371 (1934).
422. G. Devoto, *Z. Elektrochem.* **40**, 641 (1934).
423. B. Dickens and W. E. Brown, *Inorg. Chem.* **8**, 2093 (1969).
424. B. Dickens and W. E. Brown, *Inorg. Chem.* **9**, 480 (1970).
425. L. C. Dickenson and M. C. R. Symons, *Trans. Faraday Soc.* **66**, 1334 (1970).
426. G. A. M. Diepen, D. T. A. Huibers, and H. I. Waterman, *Brennst.-Chem.* **38**, 277 (1957).
427. G. A. M. Diepen and F. E. C. Scheffer, *Rec. Trav. Chim. Pays-Bas* **69**, 593 (1950).

428. K. Dimroth, C. Reichardt, T. Siepmann, and F. Bohlmann, *Liebigs Ann. Chem.* **661**, 1 (1963).

429. G. L. Dirichlet, *J. Reine und Angew. Math.* **40**, 216 (1850).

430. W. E. Donath, U.S. Patent 2,904,511 (1959).

431. L. A. D'Orazio and R. H. Wood, *J. Phys. Chem.* **67**, 1435 (1963).

432. D. A. Draegert, N. W. B. Stone, B. Curnutte, and D. Williams, *J. Opt. Soc. Am.* **56**, 64 (1966).

433. R. S. Drago, N. O'Bryan, and G. C. Vogel, *J. Am. Chem. Soc.* **92**, 3924 (1970).

434. R. S. Drago, N. O'Bryan and G. C. Vogel, *J. Am. Chem. Soc.* **92**, 3924 (1970).

435. L. E. Drain, *Phys. Soc. London* **62A**, 301 (1949).

436. K. Drucker and E. Moles, *Z. Phys. Chem.* **75**, 405 (1911).

437. G. Durocher and C. Sandorfy, *J. Mol. Spectry.* **15**, 22 (1965).

438. P. D. Edmonds, V. F. Pearce, and J. H. Andreae, *Brit. J. Appl. Phys.* **13**, 551 (1962).

439. J. T. Edsall, *in* "The Chemistry of Amino Acids and Proteins" (C. L. A. Schmidt, ed.), Springfield, Ill., p. 882 (1938).

440. J. T. Edsall and J. Wyman, *J. Am. Chem. Soc.* **57**, 1964 (1935).

441. Yu. Ya. Efimov and Yu. I. Naberukhin, *Zh. Strukt. Khim.* **12**, 591 (1971).

442. F. Eggers, *Acustica* **19**, 323 (1967–68).

443. M. Eigen and L. de Maeyer, *in* "Techniques of Organic Chemistry" (S. L. Friess, E. S. Lewis, and A. Weissberger, eds.), Vol. 8, Part 2, p. 895, Interscience, New York (1963).

444. D. Eisenberg and W. Kauzmann, "The Structure and Properties of Water," Oxford Univ. Press, London (1969).

445. L. Elbert and J. Lange, *Z. Phys. Chem. (Leipzig)* **130A**, 584 (1928).

446. D. D. Eley, *Trans. Faraday Soc.* **35**, 1281 (1939).

447. D. D. Eley, *Trans. Faraday Soc.* **35**, 1421 (1939).

448. D. D. Eley and M. J. Hey, *Trans. Faraday Soc.* **64**, 1990 (1968).

449. H. D. Ellerton, G. Reinsfelds, D. E. Mulcahy, and P. J. Dunlop, *J. Phys. Chem.* **68**, 398 (1964).

450. S. M. El Sabeh and J. B. Hasted, *Proc. Phys. Soc.* **B66**, 611 (1953).

451. Z. M. El Saffar, *Acta Cryst.* **B24**, 1131 (1968).

452. Z. M. El Saffar, *J. Chem. Phys.* **52**, 4097 (1970).

453. J. W. Emsley, J. Feeney, and H. L. Sutcliffe, "High-Resolution Nuclear Magnetic Resonance Spectroscopy," Pergamon, Oxford (1965).

454. L. Endom, H. G. Hertz, B. Thül, and M. D. Zeidler, *Ber. Bunsenges. Phys. Chem.* **71**, 1008 (1967).

455. A. Engberg, *Acta Chem. Scand.* **24**, 3510 (1970).

456. J. B. F. N. Engberts and G. Zuidema, *Rec. Trav. Chim. Pays-Bas* **89**, 673 (1970).

457. P. Engels, and G. M. Schneider, *Ber Bunsenges. Phys. Chem.* **76**, 1239 (1972).

458. J. Erva, *Suomen. Kem. B* **29**, 183 (1956).

459. A. Eucken, *Z. Elektrochem.* **52**, 264 (1948).

460. D. F. Evans, G. P. Cunningham, and R. L. Kay, *J. Phys. Chem.* **70**, 2974 (1966).

461. D. H. Everett, D. A. Landsman, and B. R. W. Pinsent, *Proc. Roy. Soc.* **215A**, 403 (1952).

462. V. Evrand, *Natuurwetensch. Tijdschr. (Ghent)* **11**, 99 (1929).

463. H. Eyring and M. S. Jhon, "Significant Liquid Structures," Wiley, New York (1969).

464. M. Falk, *Can. J. Chem.* **49**, 1137 (1971).

465. M. Falk and T. A. Ford, *J. Chem. Phys.* **44**, 1699 (1966).
466. M. Falk and E. Whalley, *J. Chem. Phys.* **34**, 1554 (1961).
467. M. Falk and H. R. Wyss, *J. Chem. Phys.* **51**, 5727 (1969).
468. L. Fanfani and P. F. Zanazzi, *Act. Cryst.* **22**, 173 (1967).
469. R. J. Fanning and P. Kruus, *Can. J. Chem.* **48**, 2052 (1970).
470. M. Faraday, *Quart. J. Sci.* **15**, 71 (1823).
471. T. C. Farrar and E. D. Becker, "Pulse and Fourier Transform NMR," Academic, New York (1971).
472. L. D. Favro, *in* "Fluctuation Phenomena in Solids" (R. E. Burgess, ed.), p. 79, Academic, New York (1965).
473. D. Feakins, *in* "Physico-Chemical Processes in Mixed Aqueous Solvents" (F. Franks, ed.), p. 71, Heinemann, London (1967).
474. D. Feakins, B. C. Smith, and L. Thakur, *J. Chem. Soc.* (*A*) **1966**, 714.
475. F. S. Feates and D. J. G. Ives, *J. Chem. Soc.* **1956**, 2798.
476. D. Feil and G. A. Jeffrey, *J. Chem. Phys.* **35**, 1863 (1961).
477. H. Fellner-Feldegg, *J. Phys. Chem.* **73**, 616 (1969).
478. A. Ferrari, A. Braibanti, and A. M. Lanfredi, *Acta Cryst.* **14**, 489 (1961).
479. A. Ferrari, A. Braibanti, A. M. Lanfredi, and A. Tiripicchio, *Acta Cryst.* **22**, 240 (1967).
480. G. Ferraris, *Acta Cryst.* **B25**, 1544 (1969).
481. L. C. Fetterly, *in* "Non-Stoichiometric Compounds" (L. Mandelcorn, ed.), p. 491, Academic, New York (1964).
482. R. A. Fifer and J. Schiffer, *J. Chem. Phys.* **50**, 21 (1969).
483. R. A. Fifer and J. Schiffer, *J. Chem. Phys.* **52**, 2664 (1970).
484. R. A. Fifer and J. Schiffer, *J. Chem. Phys.* **54**, 5097 (1971).
485. E. G. Finer, F. Franks and M. J. Tait, *J. Am. Chem. Soc.*, **94**, 4424, (1972).
486. J. J. Finney, *Acta Cryst.* **21**, 437 (1966).
487. F. Fister and H. G. Hertz, *Ber. Bunsenges. Phys. Chem.* **71**, 1032 (1967).
488. M. Fixman, *J. Chem. Phys.* **36**, 1961 (1962).
489. M. Fixman, *in* "Advances in Chemical Physics" (I. Prigogine, ed.), Vol. VI, p. 175, Interscience, New York (1964).
490. A. N. Fletcher and C. A. Heller, *J. Phys. Chem.* **71**, 3742 (1967).
491. A. N. Fletcher and C. A. Heller, *J. Phys. Chem.* **72**, 1839 (1968).
492. A. E. Florin and M. Alei, *J. Chem. Phys.* **47**, 4268 (1967).
493. H. G. Flynn, *in* "Physical Acoustics" (W. P. Mason, ed.), Vol. Ib, p. 203, Academic, London (1965).
494. A. Foldes and C. Sandorfy, *J. Mol. Spectry.* **20**, 262 (1966).
495. H. Follner, *Acta Cryst.* **B26**, 1544 (1970).
496. V. I. Fomina and S. Sh. Byk, *Zh. Prikl. Khim.* (*Leningrad*) **42**, 2855 (1969).
497. T. A. Ford and M. Falk, *Can. J. Chem.* **46**, 3579 (1968).
498. T. A. Ford and M. Falk, *J. Mol. Struct.* **3**, 445 (1969).
499. D. L. Fowler, W. V. Loebenstein, D. B. Pall, and C. A. Kraus, *J. Am. Chem. Soc.* **62**, 1140 (1940).
500. R. Fowler and E. A. Guggenheim, "Statistical Thermodynamics," Cambridge Univ. Press (1949).
500a. M. F. Fox, *J. Chem. Soc. Faraday Trans.*, I, **68**, 1294 (1972).
501. E. U. Franck, *Angew. Chem.* **73**, 309 (1961).
502. E. U. Franck, *Ber. Bunsenges. Phys. Chem.* **73**, 135 (1969).

503. E. U. Franck and F. Meyer, *Z. Elektrochem.* **63**, 571 (1959).
504. H. S. Frank, *J. Chem. Phys.* **13**, 493 (1945).
505. H. S. Frank, *Proc. Roy. Soc. A* **247**, 481 (1958).
506. H. S. Frank, *in* "Chemical Physics of Ionic Solutions" (B. E. Conway and R. G. Barradas, eds.), p. 53, Wiley, New York (1966).
507. H. S. Frank, *Science* **169**, 635 (1970).
508. H. S. Frank and M. W. Evans, *J. Chem. Phys.* **13**, 507 (1945).
509. H. S. Frank .and F. Franks, *J. Chem. Phys.* **48**, 4746 (1968).
510. H. S. Frank and A. S. Quist, *J. Chem. Phys.* **34**, 604 (1961).
511. H. S. Frank and P. T. Thompson, "The Structure of Electrolyte Solutions" (W. J. Hamer, ed.), Chapter 8, Wiley, New York (1959).
512. H. S. Frank and W. Y. Wen, *Disc. Faraday Soc.* **24**, 133 (1957).
513. F. Franks, *Ann. N. Y. Acad. Sci.* **125**, 277 (1965).
514. F. Franks, *in* "Hydrogen-Bonded Solvent Systems" (A. K. Covington and P. Jones, eds.), p. 31, Taylor and Francis, London (1968).
515. F. Franks, M. Gent, and H. H. Johnson, *J. Chem. Soc.* **1963**, 2716.
516. F. Franks and D. J. G. Ives, *Quart. Rev.* **20**, 1 (1966).
517. F. Franks and H. H. Johnson, *Trans. Faraday Soc.* **58**, 656 (1962).
518. F. Franks, M. A. J. Quickenden, D. S. Reid, and B. Watson, *Trans. Faraday Soc.* **66**, 582 (1970).
519. F. Franks, J. Ravenhill, P. Egelstaff, and D. I. Page, *Proc. Roy. Soc. (London) A* **319**, 189 (1970).
520. F. Franks, J. R. Ravenhill, and D. S. Reid, *J. Solution Chem.* **1**, 3 (1972).
521. F. Franks and D. S. Reid, *J. Phys. Chem.* **73**, 3152 (1969).
522. F. Franks and D. S. Reid, *Proc. 1st Int. Conf. Thermodynamics and Thermochemistry, Warsaw 1969*, p. 891.
522a. F. Franks, D. S. Reid, and A. Suggett, *J. Solution Chem.* (in press).
523. F. Franks and H. T. Smith, *Trans. Faraday Soc.* **63**, 2586 (1967).
524. F. Franks and H. T. Smith, *Trans. Faraday Soc.* **64**, 2962 (1968).
525. F. Franks and B. Watson, *Trans. Faraday Soc.* **63**, 329 (1967).
526. F. Franks and B. Watson, *Trans. Faraday Soc.* **65**, 2339 (1969).
527. A. Fratiello, *Mol. Phys.* **7**, 565 (1964).
528. A. Fratiello and D. C. Douglass, *J. Mol. Spectry.* **11**, 465 (1963).
529. A. Fratiello and D. C. Douglass, *J. Chem. Phys.* **41**, 974 (1964).
530. R. Freeman and S. Wittekoek, *J. Magnetic Resonance* **1**, 238 (1969).
531. M. R. Freymann, *Compt. Rend.* **193**, 656, 928 (1931).
532. H. L. Friedman, P. S. Ramanathan, and C. V. Krishnan, *J. Solution Chem.* **1**, 237 (1972).
533. H. L. Friedman, *J. Phys. Chem.* **71**, 1723 (1967).
534. M. E. Friedman and H. A. Scheraga, *J. Phys. Chem.* **69**, 3795 (1965).
535. H. Fröhlich, "Theory of Dielectrics," Oxford Univ. Press, London (1949).
536. H. Frühbuss, Dissert., Bonn (1942).
537. K. Fukushima and H. Kataiwa, *Bull. Chem. Soc. Japan* **43**, 690 (1970).
538. R. M. Fuoss, *in* "Chemical Physics of Ionic Solutions" (B. E. Conway and R. G. Barradas, eds.), p. 463, Wiley, New York (1966).
539. N. Fuson, M.-L. Josien, R. Powell, and E. Utterback, *J. Chem. Phys.* **20**, 145 (1952).
540. K. J. Gallagher, *in* "Hydrogen Bonding" (D. Hadži, ed.), p. 45, Pergamon, London (1959).

541. T. J. Galloway, W. Ruska, P. S. Chappelear, and R. Kobayashi, *Ind. Eng. Chem. Fundam.* **9**, 237 (1970).
542. S. K. Garg, B. Morris, and D. W. Davidson, *J. Chem. Soc. Faraday Trans. II*, **68**, 481 (1972).
543. C. W. Garland, *in* "Physical Acoustics" (W. P. Mason and R. N. Thurston, eds.), Chapter 2, Academic, London (1970).
544. E. Garmine and C. H. Townes, *Appl. Phys. Letters* **5**, 84 (1964).
545. K. Garrison, R. J. Slape, L. L. Snedden, J. A. Hunter, W. W. Rinne, and S. C. Verikios, U. S. Office Saline Water Progr. Rep. 368 (1968).
546. H. J. Gebhart, Jr., E. C. Makin, and E. D. Pierron, *Chem. Eng. Progr. Symp. Ser.* **66**(103), 105 (1970).
547. G. Geiseler and S. Fruwert, *Z. Phys. Chem.* **26**, 111 (1960).
548. E. R. Gerling and T. V. Koltsova, *Radiokhimiya* **5**, 277 (1963).
549. S. Y. Gerlsma, *J. Biol. Chem.* **243**, 957 (1968).
550. S. Y. Gerlsma, *Eur. J. Biochem.* **14**, 150 (1970).
551. A. Giacomini, *J. Acoust. Soc. Am.* **19**, 701 (1947).
552. P. A. Giguère, B. G. Morissette, A. W. Olmos, and O. Knopp, *Can. J. Chem.* **33**, 804 (1955).
553. A. S. Gilbert and N. Sheppard, *Chem. Commun.* **1971**, 337.
554. R. D. Gillard and H. M. Sutton, *J. Chem. Soc. A* **1970**, 1309, 2172, 2175.
555. E. L. Ginzton, "Microwave Measurements," McGraw-Hill, New York (1957).
556. R. M. Glaeser and C. A. Coulson, *Trans. Faraday Soc.* **61**, 389 (1965).
557. J. A. Glasel, *Proc. Nat. Acad. Sci. U.S.* **55**, 499 (1966).
558. J. A. Glasel, *J. Am. Chem. Soc.* **92**, 372 (1970).
559. S. Glasstone, "Thermodynamics for Chemists," Van Nostrand, New York (1947).
560. O. Glemser and E. Hartert, *Naturwiss.* **42**, 534 (1955).
561. O. Glemser and E. Hartert, *Z. Anorg. Chem.* **283**, 111 (1956).
562. D. N. Glew, *Can. J. Chem.* **38**, 208 (1960).
563. D. N. Glew, U.S. Patent 3,058,832 (1962).
564. D. N. Glew, *J. Phys. Chem.* **66**, 605 (1962).
565. D. N. Glew, *Nature* **201**, 292 (1964).
566. D. N. Glew, *Trans. Faraday Soc.* **61**, 30 (1965).
567. D. N. Glew, U.S. Patent 3,415,747 (1968).
568. D. N. Glew and M. L. Haggett, *Can. J. Chem.* **46**, 3857, 3867 (1968).
569. D. N. Glew and D. A. Hames, *Can. J. Chem.* **47**, 4651 (1969).
570. D. N. Glew, H. D. Mak, and N. S. Rath, *Can. J. Chem.* **45**, 3059 (1967).
571. D. N. Glew, H. D. Mak, and N. S. Rath, *Chem. Commun.* **1968**, 264.
572. D. N. Glew, H. D. Mak, and N. S. Rath, *Chem. Commun.* **1968**, 265.
573. D. N. Glew, H. D. Mak, and N. S. Rath, *in* "Hydrogen-Bonded Solvent Systems" (A. K. Covington and P. Jones, eds.), p. 195, Taylor and Francis, London (1968).
574. D. N. Glew and E. A. Moelwyn-Hughes, *Disc. Faraday Soc.* **15**, 150 (1953).
575. D. N. Glew and N. S. Rath, *J. Chem. Phys.* **44**, 1710 (1966).
576. D. N. Glew and N. S. Rath, *Can. J. Chem.* **49**, 837 (1971).
577. J. Glowacki, *Acta Phys. Polonica* **26**, 905 (1965).
578. J. Glowacki and F. Warkusz, *Acta Phys. Polonica* **29**, 595 (1966).
579. J. R. Goates and R. J. Sullivan, *J. Phys. Chem.* **62**, 188 (1958).
580. M. Godchot, C. Canquil, and R. Calos, *Compt. Rend.* **202**, 759 (1936).

581. E. v. Goldammer and H. G. Hertz, *J. Phys. Chem.* **74**, 3734 (1970).

582. E. v. Goldammer and M. D. Zeidler, *Ber. Bunsenges. Phys. Chem.* **73**, 4 (1969).

583. R. Göller, H. G. Hertz, and R. Tutsch, *J. Pure Appl. Chem.* **32**, 149 (1972)

584. G. L. Gooberman, "Ultrasonics," English Univ. Press, London (1968).

585. R. Gopal and M. A. Siddiqi, *J. Phys. Chem.* **72**, 1814 (1968).

586. J. E. Gordon, *J. Phys. Chem.* **70**, 2413 (1966).

587. W. Gordy and Ph. C. Martin, *J. Chem. Phys.* **7**, 99 (1939).

588. S. Goto and T. Isemura, *Bull. Chem. Soc. Japan* **37**, 1693 (1964).

589. S. R. Gough and D. W. Davidson, *J. Chem. Phys.* **52**, 5442 (1970).

590. S. R. Gough and D. W. Davidson, *Can. J. Chem.* **49**, 2691 (1971).

591. S. R. Gough, E. Whalley, and D. W. Davidson, *Can. J. Chem.* **46**, 1673 (1968).

592. A. J. Gow, H. T. Ueda, and D. E. Garfield, *Science* **161**, 1011 (1968).

593. A. Grandchamp-Chaudin, *Compt. Rend.* **243**, 321 (1956).

594. J. P. Greenstein and J. Wyman, *J. Am. Chem. Soc.* **58**, 463 (1936).

595. J. P. Greenstein, J. Wyman, and E. J. Cohn, *J. Am. Chem. Soc.* **47**, 687 (1935).

596. J. Greyson, *J. Phys. Chem.* **71**, 2210 (1967).

597. H. Gränicher, *Phys. Kondens. Mater.* **1**, 1 (1963).

598. H. Gränicher, *Z. Kristogr., Kristallgeometrie, Kristallphys., Kristallchem.,* **110**, 432 (1958).

599. E. Greinacher, W. Lüttke, and R. Mecke, *Ber. Bunsenges. Phys. Chem.* **59**, 23 (1955).

600. H. Greisenfelder and H. Zimmermann, *Ber. Bunsenges.* **67**, 482 (1963).

601. O. H. Griffith and A. S. Waggoner, *Acc. Chem. Res.* **2**, 17 (1969).

602. T. R. Griffiths and M. C. R. Symons, *Trans. Faraday Soc.* **56**, 1125 (1960).

603. W. E. Griffiths, C. J. W. Gutch, G. F. Longster, J. Myatt, and P. F. Todd, *J. Chem. Soc. B* **1968**, 785.

604. J. Griswold and S. Y. Wong, *Chem. Eng. Prog. Symp. Ser.* **48**(3), 18 (1952).

605. F. T. Gucker and F. D. Ayres, *J. Am. Chem. Soc.* **59**, 2152 (1937).

606. L. J. Guggenberger and A. W. Sleight, *Inorg. Chem.* **8**, 2041 (1969).

607. J. Guillermet and A. Novak, *J. Chim. Phys. Physiochim. Biol.* **66**, 68 (1969).

608. Yu. V. Gurikov, *J. Struct. Chem.* **10**, 494 (1969).

609. R. W. Gurney, "Ionic Processes in Solution," McGraw-Hill, New York (1953).

610. C. J. W. Gutch, W. A. Waters, and M. C. R. Symons, *J. Chem. Soc. B* **1970**, 1261.

611. C. M. Guzy, J. B. Raynor, and M. C. R. Symons, *J. Chem. Soc.* **1969**, 2791.

612. C. Haas, *Phys. Letters* **3**, 126 (1962).

613. D. J. Haas, *Acta Cryst.* **17**, 1511 (1964).

614. R. Haase, "Thermodynamik der Mischphasen," Springer, Berlin–Göttingen–Heidelberg (1956).

615. A. Hadni, G. Morlot, and F. Brehat, *Spectrochim. Acta* **24A**, 1167 (1968).

616. D. R. Hafemann and S. L. Miller, *J. Phys. Chem.* **73**, 1392, 1398 (1969).

617. M. Hagan, "Clathrate Inclusion Compounds," Reinhold, New York (1962).

618. G. H. Haggis, J. B. Hasted, and T. J. Buchanan, *J. Chem. Phys.* **20**, 1425 (1952).

619. E. L. Hahn, *Phys. Rev.* **80**, 580 (1950).

620. F. N. Hainsworth and H. E. Petch, *Can. J. Phys.* **44**, 3083 (1966).

621. B. J. Hales, G. L. Bertrand, and L. G. Hepler, *J. Phys. Chem.* **70**, 3970 (1966).

622. D. Hall, *Acta Cryst.* **18**, 955 (1965).

623. S. D. Hamann and F. Smith, *Aust. J. Chem.* **24**, 2431 (1971).

624. W. C. Hamilton, *Acta Cryst.* **15**, 353 (1962).

625. W. C. Hamilton, *Ann. Rev. Phys. Chem.* **13**, 19 (1962).
626. W. C. Hamilton, *in* "Structural Chemistry and Molecular Biology" (A. Rich and N. Davidson, eds.), p. 466, Freeman, San Francisco (1968).
627. W. C. Hamilton and J. A. Ibers, "Hydrogen Bonding in Solids," p. 188, Benjamin, New York (1968).
628. W. C. Hamilton and S. J. LaPlaca, *Acta Cryst.* **B24**, 1147 (1968).
629. E. G. Hammerschmidt, *Ind. Eng. Chem.* **26**, 851 (1934).
630. G. G. Hammes and W. Knoche, *J. Chem. Phys.* **45**, 4041 (1966).
631. G. G. Hammes and C. N. Pace, *J. Phys. Chem.* **72**, 2227 (1968).
632. G. G. Hammes and P. R. Schimmel, *J. Phys. Chem.* **70**, 2319 (1966).
633. G. G. Hammes and P. R. Schimmel, *J. Am. Chem. Soc.* **89**, 442 (1967).
634. G. G. Hammes and J. C. Swann, *Biochemistry* **6**, 1591 (1967).
635. "Handbook of Chemistry and Physics," 48th ed. (R. C. Weast and S. M. Selby, eds.), Chemical Rubber Co., Cleveland, Ohio (1967–68).
636. H. S. Harned and B. B. Owen, "The Physical Chemistry of Electrolyte Solutions," 3rd ed., Am. Chem. Soc. Monograph No. 137 (1958).
637. W. H. Harris, *J. Chem. Soc. (London)* **1933**, 582.
638. K. A. Hartmann, Jr., *J. Phys. Chem.* **70**, 270 (1966).
639. K. O. Hartman and F. A. Miller, *Spectrochim. Acta* **24A**, 669 (1968).
640. K. B. Harvey, F. R. McCourt, and H. F. Shurvell, *Can. J. Chem.* **42**, 960 (1964).
641. K. B. Harvey and H. F. Shurvell, *J. Nuclear Spect.* **25**, 120 (1967).
642. F. X. Hassion and R. H. Cole, *J. Chem. Phys.* **23**, 1756 (1955).
643. J. B. Hasted and S. H. M. El-Sabeh, *Trans. Faraday Soc.* **49**, 1003 (1953).
644. J. B. Hasted, G. H. Haggis, and P. Hutton, *Trans. Faraday Soc.* **47**, 577 (1951).
645. G. Haugen and R. Hardwick, *J. Phys. Chem.* **69**, 2988 (1965).
646. I. Hausser, *Sitzber. Heidelberg. Akad. Wiss., Mat. Naturwiss. Kl.* No. 6 (1935).
647. R. E. Hawkins and D. W. Davidson, *J. Phys. Chem.* **70**, 1889 (1966).
648. H. G. Hecht, "Magnetic Resonance Spectroscopy," Wiley, New York (1967).
649. G. Hederstrand, *Z. Physik. Chem.* **135**, 36 (1928).
650. L. G. Hepler, *Can. J. Chem.* **47**, 4613 (1969).
651. H. W. Herreilers, Dissert., Amsterdam (1936).
652. H. G. Hertz, *Ber. Bunsenges. Phys. Chem.* **65**, 20 (1961).
653. H. G. Hertz, *Ber. Bunsenges. Phys. Chem.* **67**, 311 (1963).
654. H. G. Hertz, *Ber. Bunsenges. Phys. Chem.* **68**, 907 (1964).
655. H. G. Hertz, *Ber. Bunsenges. Phys. Chem.* **71**, 979 (1967).
656. H. G. Hertz, *in* "Progress in NMR Spectroscopy" (J. W. Emsley, J. Feeney, L. H. Sutcliffe, eds.), Vol. 3, p. 159, Pergamon, Oxford (1967).
657. H. G. Hertz, B. Lindman, and V. Siepe, *Ber. Bunsenges. Phys. Chem.* **73**, 542 (1969).
658. H. G. Hertz and W. Spalthoff, *Z. Elektrochemie* **63**, 1096 (1959).
659. H. G. Hertz and M. D. Zeidler, *Ber. Bunsenges. Phys. Chem.* **68**, 821 (1964).
660. G. Herzberg, "Infrared and Raman Spectra Polyatomic Molecules," van Nostrand, New York (1945).
661. G. Herzberg, "Molecular Spectra and Molecular Structure," II, van Nostrand, New York (1951).
662. K. F. Herzfeld, *J. Chem. Phys.* **9**, 513 (1941).
663. K. Herzfeld and T. A. Litovitz, "Absorption and Dispersion of Ultrasonic Waves," Academic, London (1959).

664. J. C. Heson, U.S. Patent 2,704,703 (1955).
665. M. Hess and G. E. Jones, Jr., U.S. Patent 3,119,772 (1964).
666. G. L. Hiebert and D. F. Hornig, *J. Chem. Phys.* **20**, 918 (1952).
667. D. J. T. Hill, Ph.D. Thesis, University of Queensland (1965).
668. N. E. Hill, W. E. Vaughan, A. H. Price, and M. Davies, "Dielectric Properties and Molecular Behavior," Van Nostrand, New York (1970).
669. T. L. Hill, "Statistical Mechanics," McGraw-Hill, New York (1956).
670. D. M. Himmelblau, *J. Phys. Chem.* **63**, 1803 (1959).
671. J. C. Hindman, A. Svirmickas, and M. Wood, *J. Phys. Chem.* **72**, 4188 (1968).
672. J. C. Hindman, A. J. Zielen, A. Svirmickas, and M. Wood, *J. Chem. Phys.* **54**, 621 (1971).
673. Y. Hirose, P. Engels, and G. M. Schneider, *Chem. Ingr. Techn.* **44**, 857 (1972).
674. J. O. Hirschfelder, C. F. Curtiss, and R. B. Bird, "Molecular Theory of Gases and Liquids," Wiley, New York (1954).
675. J. L. Hoard and L. Goldstein, *J. Chem. Phys.* **3**, 645 (1935).
675a. J. D. Hoffman and H. G. Pfeiffer, *J. Chem. Phys.* **22**, 132 (1954).
676. E. G. Hoffmann, *Z. Phys. Chem.* **53**, 179 (1943).
677. J. R. Holmes, D. Kivelson, and W. C. Drinkard, *J. Am. Chem. Soc.* **84**, 4677 (1962).
678. A. Holtzer and M. F. Emerson, *J. Phys. Chem.* **73**, 26 (1969).
679. M. Holzbecher, O. Knop, and M. Falk, *Can. J. Chem.* **49**, 1413 (1971).
680. G. Honjo and K. Shimaoka, *Acta Cryst.* **10**, 710 (1957).
681. J. Horiuti, *Sci. Pap. Inst. Phys. Chem. Res. Tokyo* **17**, 125 (1931).
682. H. J. Hrostowski and G. C. Pimentel, *J. Chem. Phys.* **19**, 661 (1951).
683. C. P. Huang, O. Fennema, and W. D. Powrie, *Cryobiology* **2**, 240 (1966).
684. P. S. Hubbard, *Phys. Rev.* **131**, 275 (1963).
685. P. S. Hubbard, *Phys. Rev.* **180**, 319 (1969).
686. A. Hudson and G. R. Luckhurst, *Chem. Revs.* **69**, 191 (1969).
687. R. Huisgen and H. Walz, *Chem. Ber.* **89**, 2616 (1956).
688. J. L. Hunter and H. D. Dardy, *J. Acoust. Soc. Am.* **39**, 1914 (1964).
689. W. T. Huntress, Jr., *in* "Advances in Magnetic Resonance" (J. S. Waugh, ed.), Vol. 4, p. 1, Academic, New York (1970).
690. J. O. Hutchens, *in* "Handbook of Biochemistry" (H. A. Sober, ed.), Chemical Rubber Co., Cleveland (1968).
691. J. O. Hutchens, K. M. Figkio, and S. M. Cranito, *J. Biol. Chem.* **238**, 1419 (1963).
692. A. J. L. Hutchinson, U.S. Patent 2,410,583 (1946).
693. P. J. Hyde, *Proc. IEE* **117**, 1891 (1970).
694. J. Hyne, *in* "Hydrogen-Bonded Solvent Systems" (A. K. Covington and P. Jones, eds.), p. 99, Taylor and Francis, London (1968).
695. D. A. Ibbitson and L. F. Moore, *J. Chem. Soc.* (*B*) **76**, 80 (1967).
696. "International Tables for X-Ray Crystallography" (N. F. M. Henry and K. Lonsdale, eds.), Vol. 1, p. 44, The Kynoch Press, Birmingham (1952).
697. F. Isambert, *Compt. Rend.* **86**, 481 (1878).
698. J. Itoh, R. Kusaka, Y. Yamagata, R. Kiriyama, and H. Ibamoto, *J. Phys. Soc. Japan* **8**, 293 (1953).
699. E. N. Ivanov, *Soviet Phys.—JETP* **18**, 1041 (1964).
700. D. J. G. Ives and T. H. Lemon, *RIC Rev.* **1**, 62 (1968).
701. D. J. G. Ives and P. D. Marsden, *J. Chem. Soc.* **1965**, 649.

702. C. Jaccard, *Helv. Phys. Acta* **32**, 89 (1959).

703. H. H. Jaffé and M. Orchin, "Theory and Applications of Ultraviolet Spectroscopy," Wiley, London (1962).

704. R. J. Jakobsen, Y. Mikawa, and J. W. Brasch, *Appl. Spectry.* **24**, 333 (1970).

705. G. A. Jeffrey, *Accounts Chem. Res.* **2**, 344 (1969).

706. G. A. Jeffrey, T. H. Jordan, and R. K. McMullan, *Science* **155**, 689 (1967).

707. G. A. Jeffrey and R. K. McMullan, *J. Chem. Phys.* **37**, 2231 (1962).

708. G. A. Jeffrey and R. K. McMullan, *Progr. Inorg. Chem.* **8**, 43 (1967).

709. G. A. Jeffrey and R. K. McMullan, *Prog. Inorg. Chem.* **8**, 43 (1967).

710. G. A. Jeffrey and T. C. W. Mak, *Science* **149**, 178 (1965).

711. S. J. Jensen, *Acta Chem. Scand.* **22**, 647 (1968).

712. S. J. Jensen and M. S. Lehmann, *Acta Chem. Scand.* **24**, 3422 (1970).

713. M. D. Joesten and R. S. Drago, *J. Am. Chem. Soc.* **84**, 3817 (1962).

714. G. A. Johnson, S. M. Lecchini, E. G. Smith, J. Clifford, and B. A. Pethica, *Disc. Faraday Soc.* **42**, 120 (1966).

715. J. R. Johnson, S. D. Christian, and H. E. Affsprung, *J. Chem. Soc. A* **1966**, 77.

716. C. Jolicoeur and H. L. Friedman, *Ber. Bunsenges. Phys. Chem.* **75**, 248 (1971).

717. D. Jones and M. C. R. Symons, *Trans. Faraday Soc.* **67**, 961 (1971).

718. T. H. Jordan and T. C. W. Mak, *J. Chem. Phys.* **47**, 1222 (1967).

719. M. A. Kabayama and D. Patterson, *Can. J. Chem.* **36**, 563 (1958).

720. M. A. Kabayama, D. Patterson, and L. Piche, *Can. J. Chem.* **36**, 557 (1958).

721. M. Kac, *Science* **166**, 695 (1969).

722. B. Kamb, *Science* **148**, 232 (1965).

723. B. Kamb, *in* "Structural Chemistry and Molecular Biology" (A. Rich and N. Davidson, eds.), p. 507, Freeman, San Francisco (1968).

724. B. Kamb and A. Prakash, *Acta Cryst. Sect. B* **24**, 1317 (1968).

725. I. Kampschulte-Scheuing and G. Zundel, *J. Phys. Chem.* **74**, 2363 (1970).

726. J. Karle and L. O. Brockway, *J. Am. Chem. Soc.* **66**, 574 (1944).

727. G. Karnofsky and P. F. Steinhoff, U.S. Office Saline Water Progr. Rep. 40 (1960).

728. J. Karpovitch, *J. Chem. Phys.* **23**, 1767 (1954).

729. J. Karpovitch, *J. Acoust. Soc. Am.* **26**, 819 (1955).

730. J. S. Kasper, P. Hagenmuller, M. Pouchard, and C. Cros, *Science* **150**, 1713 (1965).

731. J. S. Kasper, P. Hagenmuller, M. Pouchard, and C. Cros, *Bull. Soc. Chim. Fr.* **7**, 2737 (1968).

732. J. Kass, Dissert. Syracuse Univ. (1967).

733. J. R. Katz, *Proc. Roy. Acad. Sci. Amsterdam* **13**, 958 (1911).

734. L. Katz, *J. Mol. Biol.* **44**, 279 (1969).

735. W. Kauzmann, *Advan. Protein Chem.* **14**, 1 (1959).

736. J. L. Kavanau, "Water and Solute–Water Interaction," Holden-Day, San Francisco (1964).

737. M. I. Kay, I. Almodovar, and S. F. Kaplan, *Acta Cryst.* **B24**, 1312 (1968).

738. R. L. Kay, *in* "Trace Inorganics in Water," Adv. Chem. Series No. 73, Amer. Chem. Soc. (1968).

738a. R. L. Kay, *Adv. Chem.* **73**, 1 (1968).

739. R. L. Kay, and D. F. Evans, *J. Phys. Chem.* **69**, 4216 (1965).

740. R. L. Kay and D. F. Evans, *J. Phys. Chem.* **70**, 2325 (1966).

741. R. L. Kay, T. Vituccio, C. Zawoyski, and D. F. Evans, *J. Phys. Chem.* **70**, 2336 (1966).

742. W. B. Kay, *Accounts Chem. Res.* **1**, 344 (1968).
743. M. Kehren and M. Rösch, *Fette, Seifen, Anstrichmittel* **59**, 1 (1957).
744. M. Kehren and M. Rösch, *Fette, Seifen, Anstrichmittel* **65**, 223 (1963).
745. H. Kempter and R. Mecke, *Naturwiss.* **27**, 583 (1939).
746. H. Kempter and R. Mecke, *Z. Phys. Chem. Abt. B* **46**, 229 (1940).
747. A. P. Keding, R. H. Bigelow, P. D. Edmonds, and C. J. Pings, *J. Chem. Phys.* **40**, 1451 (1964).
748. G. C. Kennedy, *Econ. Geol.* **45**, 629 (1950).
749. G. C. Kennedy, G. J. Wasserburg, H. C. Heard, and R. L. Newton, Publ. No. 150, Institute of Geophysics, UCLA Cal. (1960).
750. J. Kenttämaa and J. J. Lindberg, *Suomen Kem.* **33B**, 98 (1960).
751. J. Kenttämaa, E. Tommila, and E. Martti, *Ann. Acad. Sci. Fennicae, Ser. A* **11**(93), 3 (1959).
752. I. S. Kerr and D. J. Williams, *Acta Cryst. B* **25**, 1183 (1969).
753. G. Kessler and H. A. Lehmann, *Z. Anorg. Allg. Chem.* **339**, 87 (1965).
754. R. K. Khanna, C. W. Brown, and L. H. Jones, *Inorg. Chem.* **8**, 2195 (1969).
755. A. W. K. Khanzada and C. A. McDowell, *J. Mol. Struct.* **7**, 241 (1971).
756. N. E. Khazanova, *Zh. Fiz. Khim.* **42**, 3000 (1968).
757. V. A. Khoroshilov, B. V. Deggyarev, and E. B. Bukhgalter, *Gazov. Prom.* **15**(11), 18 (1970).
758. J. P. Kintzinger and J. M. Lehn, *Mol. Phys.* **14**, 133 (1968).
759. J. P. Kintzinger and J. M. Lehn, *Mol. Phys.* **22**, 273 (1971).
760. H. H. Kirchner, *Z. Phys. Chem. (Frankfurt am Main)* **73**, 169 (1970).
761. H. Kiriyama and R. Kiriyama, *J. Phys. Soc. Japan* **28** (suppl.), 114 (1970).
762. J. G. Kirkwood, *J. Chem. Phys.* **7**, 911 (1939).
763. P. Klason, *Ber. Deut. Chem. Reg.* **20**, 3409 (1887).
764. T. Kleinert, *Angew. Chem.* **46**, 18 (1933).
765. R. Kleinberg, *J. Chem. Phys.* **50**, 4690 (1969).
766. R. Kling and J. Schiffer, *Chem. Phys. Letters* **3**, 64 (1969).
767. R. Kling and J. Schiffer, *J. Chem. Phys.* **54**, 5331 (1971).
768. V. A. Klyachko, G. D. Pavlov, and I. N. Medvedev, in "Proc. First Int. Symp. Water Desalination," October 1965, Vol. 1, p. 331, and Vol. 3, p. 123 Office of Saline Water, Washington (1967).
769. W. S. Knight, Ph.D. Thesis, Princeton Univ. (1962).
770. W. G. Knox, M. Hess, G. E. Jones, and H. B. Smith, *Chem. Eng. Prog.* **57**, 66 (1961).
771. R. Kobayashi, in "Handbook of Natural Gas Engineering," Chapter 16, McGraw-Hill, New York (1959).
772. F. Kohler, *Monatsh. Chem.* **82**, 913 (1951).
773. F. Kohler, H. Arnold, and R. J. Munn, *Monatsh. Chem.* **92**, 876 (1961).
774. F. Kohler and O. K. Rice, *J. Chem. Phys.* **26**, 1614 (1957).
775. G. T. Koide, Dissert., Univ. of Rochester (1969).
776. H. Kolsky, "Stress Waves in Solids," Clarendon, Oxford (1953).
777. T. V. Koltsova, *Zh. Neorg. Khim.* **3**, 1505 (1958).
778. J. Konicek and I. Wadso, *Acta Chem. Scand.* **25**, 1571 (1971).
779. Koppers Co., Inc., U.S. Office Saline Water, Progr. Rep. 90 and 125 (1964).
780. S. K. Kor and G. S. Verma, *Physica* **27**, 875 (1961).
781. A. E. Korvezee and F. E. C. Scheffer, *Rec. Trav. Chim. Pays-Bas* **50**, 256 (1931).
782. V. S. Koshelev and S. Sh. Byk, *Neftepererab. Neftekhim.* (*Moscow*) **170**(3), 22.

783. V. S. Koshelev, L. P. Kopikova, Y. N. Prokof'ev, V. I. Fomina, S. Sh. Byk, R. S. Bunegina, and A. P. Orlova, *Prom. Sin. Kauch., Nauch.-Tekh. Sb.* **1969**(5), 8.

784. E. Kosower, *J. Am. Chem. Soc.* **80**, 3253 (1958).

785. E. M. Kosower, R. L. Martin, and V. M. Meloche, *J. Chem. Phys.* **26**, 1353 (1957).

786. J. J. Kozak, W. S. Knight, and W. Kauzmann, *J. Chem. Phys.* **48**, 675 (1968).

787. B. Krebs, *Chem. Commun.* **1970**, 50.

788. G. C. Kresheck, *J. Chem. Phys.* **52**, 5966 (1970).

789. G. C. Kresheck and L. Benjamin, *J. Phys. Chem.* **68**, 2476 (1964).

790. G. C. Kresheck and H. A. Scheraga, *J. Phys. Chem.* **69**, 1704 (1965).

791. G. C. Kresheck, H. Schneider, and H. A. Scheraga, *J. Phys. Chem.* **69**, 3132 (1965).

792. S. Kreuzer, *Z. Phys. Chem.* **53**, 213 (1943).

793. J. Kreuzer and R. Mecke, *Z. Phys. Chem. Abt.* **B49**, 309 (1941).

794. I. R. Krichevskii, *Acta Phys. Chim. URSS* **12**, 480 (1940).

795. I. R. Krichevskii and P. E. Bol'shakov, *Zh. Fiz. Khim.* **14**, 353 (1941).

796. B. Krishna, M. L. Bata, and S. C. Srivastava, *J. Sci. Ind. Res.* **24**, 626 (1965).

797. C. V. Krishnan and H. L. Friedman, *J. Phys. Chem.* **73**, 1572 (1969).

798. C. V. Krishnan and H. L. Friedman, *J. Phys. Chem.* **73**, 3934 (1969).

799. C. V. Krishnan and H. L. Friedman, *J. Phys. Chem.* **74**, 2356 (1970).

800. C. V. Krishnan, P. S. Ramanathan, and H. L. Friedman, *J. Solution Chem.* **1**, 237 (1972).

801. D. J. Kroon, *Philips Res. Rep.* **15**, 501 (1960).

802. P. Kruus, *Can. J. Chem.* **42**, 1712 (1964).

803. E. Küchler and J. Derkosch, *Z. Naturforsch.* **21b**, 209 (1966).

804. J. P. Kuenen and W. G. Robson, *Phil. Mag.* **3**, 149, 622 (1902).

805. J. P. Kuenen and W. G. Robson, *Phil. Mag.* **4**, 116 (1902).

806. L. P. Kuhn, *J. Am. Chem. Soc.* **74**, 2492 (1951).

807. L. P. Kuhn, *J. Am. Chem. Soc.* **74**, 2492 (1952).

808. L. P. Kuhn, *J. Am. Chem. Soc.* **76**, 9323 (1954).

809. L. P. Kuhn, *J. Am. Chem. Soc.* **80**, 5950 (1958).

810. L. P. Kuhn, *J. Am. Chem. Soc.* **86**, 650 (1954).

811. L. P. Kuhn and R. A. Wires, *J. Am. Chem. Soc.* **86**, 2161 (1954).

812. M. Kuhn, W. Lüttke and R. Mecke, *Z. Analyt. Chem.* **170**, 106 (1959).

813. R. Kuhnkies and W. Schaffs, *Acustica* **12**, 254 (1962).

814. K. Kume, *J. Phys. Soc. Japan* **15**, 1493 (1960).

815. K. Kume and R. Hoshino, *J. Phys. Soc. Japan* **16**, 290 (1961).

816. T. Kurucsev and U. P. Strauss, *J. Phys. Chem.* **74**, 3081 (1970).

817. Y. Kyogoku, R. C. Lord, and A. Rich, *J. Am. Chem. Soc.* **89**, 496 (1967).

818. M. F. C. Ladd, *Z. Krist.* **126**, 147 (1968).

819. M. F. C. Ladd and W. H. Lee, *Progr. Solid State Chem.* **1**, 37 (1964).

820. M. F. C. Ladd and W. H. Lee, *J. Phys. Chem.* **69**, 1840 (1965).

821. M. F. C. Ladd and W. H. Lee, *J. Phys. Chem.* **73**, 2033 (1969).

822. N. Laiken and G. Nemethy, *J. Phys. Chem.* **74**, 3501 (1970).

823. R. F. Lama and B. C. Y. Lu, *J. Chem. and Eng. Data* **10**, 216 (1965).

824. J. Lamb, *in* "Physical Acoustics" (W. P. Mason, ed.), Vol. 2A, p. 203, Academic, London (1965).

825. J. Lang and R. Zana, *Trans. Faraday Soc.* **66**, 597 (1970).

826. J. Lange, *Z. Phys. Chem.* (*Leipzig*) **168A**, 147 (1934).

827. A. Lannung, *J. Am. Chem. Soc.* **52**, 68 (1930).

828. D. W. Larsen, *J. Phys. Chem.* **74**, 3380 (1970).

829. D. W. Larsen, *J. Phys. Chem.* **75**, 509 (1971).

830. F. K. Larsen, R. G. Hazell, and S. E. Rasmussen, *Acta Chem. Scand.* **23**, 61 (1969).

831. S. D. Larson, Dissert., Univ. of Illinois (1955).

832. K. E. Larsson and U. Dahlborg, *Reactor Sci. Tech.* **16**, 81 (1962).

833. P. Laszlo, *in* "Progress in NMR Spectroscopy" (J. W. Emsley, J. Feeney, L. H. Sutcliffe, eds.), Vol. 3, p. 231, Pergamon, Oxford (1967).

834. W. M. Latimer and W. H. Rodebush, *J. Am. Chem. Soc.* **42**, 1419 (1920).

835. F. S. Lee and G. B. Carpenter, *J. Phys. Chem.* **63**, 279 (1959).

836. I. Lee and J. B. Hyne, *Can. J. Chem.* **46**, 2333 (1968).

837. R. U. Lemieux, A. A. Pavia, J. C. Martin, and K. A. Watanabe, *Can. J. Chem.* **47**, 4427 (1969).

838. J. Lennard-Jones and J. A. Pople, *Proc. Roy. Soc. A* **205**, 155 (1951).

839. H. Lentz, Ph.D. Thesis, Univ. of Karlsruhe, Germany (1969).

840. H. Lentz and E. U. Franck, *Ber. Bunsenges. Phys. Chem.* **73**, 28 (1969).

841. L. Lewin, *J. IEE* **94**, 65 (1947).

842. G. N. Lewis and M. Randall, "Thermodynamics," 2nd ed. (revised by K. S. Pitzer and L. Brewer), McGraw-Hill, New York (1961).

843. U. Liddel and E. D. Becker, *Spectrochim. Acta* **10**, 70 (1957).

844. R. Y. Lin and W. Dannhauser, *J. Phys. Chem. (Ithaca)* **67**, 1805 (1963).

845. Sheng Hsien Lin, *in* "Physical Chemistry" (H. Eyring, D. Henderson, and W. Jost, eds.), Vol. V, Academic, New York, London (1970).

846. J. E. Lind, Jr. and R. M. Fuoss, *J. Phys. Chem.* **65**, 999 (1961).

847. S. Lindenbaum, *J. Phys. Chem.* **70**, 814 (1966).

848. S. Lindenbaum, *J. Phys. Chem.* **72**, 212 (1968).

849. S. Lindenbaum, *J. Phys. Chem.* **73**, 4334 (1969).

850. S. Lindenbaum, *J. Phys. Chem.* **74**, 3027 (1970).

851. S. Lindenbaum and G. E. Boyd, *J. Phys. Chem.* **68**, 911 (1964).

852. S. Lindenbaum, L. Leifer, G. E. Boyd, and J. W. Chase, *J. Phys. Chem.* **74**, 761 (1970).

853. B. Lindman, S. Forsén, and E. Forslind, *J. Phys. Chem.* **72**, 2805 (1968).

854. B. Lindman, H. Wennerström, and S. Forsén, *J. Phys. Chem.* **74**, 754 (1970).

855. O. Lindqvist, *Acta Chem. Scand.* **23**, 3062 (1969).

856. E. Lippert, *Ber. Bunsenges. Phys. Chem.* **67**, 267 (1963).

857. E. L. Lippert, Jr., H. A. Palmer, and F. F. Blankenship, *Proc. Okla. Acad. Sci.* **31**, 115 (1950).

858. E. R. Lippincott and R. Schroeder, *J. Chem. Phys.* **23**, 1099 (1955).

859. H. Lipson, *Proc. Roy. Soc. (London) A* **156**, 462 (1936).

860. H. W. Loeb, P. A. Quickenden, A. Suggett, and G. M. Young, *Ber. Bunsenges. Phys. Chem.* **75**, 1155 (1971).

861. A. Loewenstein and T. M. Connor, *Ber. Bunsenges. Phys. Chem.* **67**, 280 (1963).

862. U. Lohmann and A. Tilly, *Chem. Ingr. Tech.* **37**, 913 (1965).

863. A. Loir, *Compt. Rend.* **34**, 547 (1852).

864. G. G. Longinescu, *Chem. Rev.* **6**, 381 (1929).

865. H. C. Longuet-Higgins and J. A. Pople, *J. Chem. Phys.* **25**, 884 (1956).

866. K. Lonsdale, *Proc. Roy. Soc. Ser. A* **247**, 424 (1958).

867. H. Looyenga, *Molecular Physics* **9**, 501 (1965).

868. H. Looyenga, *Physica* **31**, 401 (1965).

References 647

869. R. C. Lord and R. E. Merrifield, *J. Chem. Phys.* **21**, 166 (1953).

870. A. Lösche, "Kerninduktion," Deutscher Verlag der Wissenschaften, Berlin (1957).

871. R. A. Lovett and A. Ben-Naim, *J. Chem. Phys.* **51**, 3108 (1969).

872. C. Löwig, *Mag. Pharm.* **23**, 12 (1828).

873. C. Löwig, *Ann. Chim. Phys. Ser. 2* **42**, 113 (1829).

874. W. A. P. Luck, *Z. Naturforsch.* **6a**, 191 (1951).

875. W. A. P. Luck, *Z. Naturforsch.* **6a**, 313 (1951).

876. W. A. P. Luck, *Z. Elektrochem.* **65**, 355 (1961).

877. W. A. P. Luck, *Ber. Bunsenges. Phys. Chem.* **67**, 186 (1963).

878. W. A. P. Luck, *Fortschr. Chem. Forsch.* **4**, 653 (1964).

879. W. A. P. Luck, *Ber. Bunsenges.* **69**, 69 (1965).

880. W. A. P. Luck, *Ber. Bunsenges.* **69**, 626 (1965).

881. W. A. P. Luck, *Naturwiss.* **52**, 25, 49 (1965).

882. W. A. P. Luck, *Naturwiss.* **54**, 601 (1967).

883. W. A. P. Luck, *Disc. Faraday Soc.* **43**, 115 (1967).

884. W. A. P. Luck, Habilitationsarbeit, "Spektroskopische Bestimmungen der Wasserstoffbrückenbindungen im Nahen IR," Univ. Heidelberg (1968).

885. W. A. P. Luck, *Z. Naturforsch.* **23b**, 152 (1968).

886. W. A. P. Luck, *Z. Naturforsch.* **24b**, 482 (1969).

887. W. A. P. Luck, *Ber. Bunsenges.* **73**, 526 (1969).

888. W. A. P. Luck, G. Böttger, and H. Harders, *J. Phys. Chem.* **71**, 459 (1967).

889. W. A. P. Luck and W. Ditter, *Ber. Bunsenges.* **70**, 113 (1966).

890. W. A. P. Luck and W. Ditter, *J. Mol. Struct.* **1**, 261 (1967/68).

891. W. A. P. Luck and W. Ditter, *J. Mol. Struct.* **1**, 339 (1967/68).

892. W. A. P. Luck and W. Ditter, *Ber. Bunsenges.* **72**, 365 (1968).

893. W. A. P. Luck and W. Ditter, *Z. Naturforsch.* **24b**, 482 (1969).

894. W. A. P. Luck and W. Ditter, *J. Phys. Chem.* **74**, 3687 (1970).

895. W. A. P. Luck and W. Ditter, *Ber. Bunsenges.* **75**, 163 (1971).

896. W. A. P. Luck and W. Ditter, *Tetrahedron* **27**, 201 (1971).

897. W. A. P. Luck and W. Ditter, *Spectrochim. Acta* (in press).

898. W. A. P. Luck and G. Kortüm, *Z. Naturforsch.* **6a**, 305 (1951).

899. R. Lumry and S. Rajender, *Biopolymers* **9**, 1125 (1970).

900. J.-O. Lundgren, *Acta Cryst.* **B26**, 1893 (1970).

901. J.-O. Lundgren and I. Olovsson, *Acta Cryst.* **23**, 966 (1967).

902. J.-O. Lundgren and I. Olovsson, *Acta Cryst.* **23**, 971 (1967).

903. J.-O. Lundgren and I. Olovsson, *J. Chem. Phys.* **49**, 1068 (1968).

904. V. Luzzati, *Acta Cryst.* **4**, 239 (1951).

905. V. Luzzati, *Acta Cryst.* **6**, 152 (1953).

906. R. M. Lynden-Bell and R. K. Harris, "Nuclear Magnetic Resonance Spectroscopy," Nelson, London (1969).

907. O. Maass and E. H. Boomer, *J. Am. Chem. Soc.* **44**, 1709 (1922).

908. C. McAuliffe, *Nature* **200**, 1092 (1963).

909. C. McAuliffe, *J. Phys. Chem.* **70**, 1267 (1966).

910. W. C. McCabe and H. F. Fisher, *J. Phys. Chem.* **74**, 2990 (1970).

911. D. W. McCall and D. C. Douglass, *J. Phys. Chem.* **71**, 987 (1967).

912. A. C. Macdonald and S. K. Sikka, *Acta Cryst.* **B25**, 1804 (1969).

913. D. D. Macdonald, J. B. Hyne, and F. L. Swinton, *J. Am. Chem. Soc.* **92**, 6355 (1970).

914. C. A. McDowell and P. Raghunathan, *Mol. Phys.* **13**, 331 (1967).

915. C. A. McDowell and P. Raghunathan, *Mol. Phys.* **15**, 259 (1968).

916. C. A. McDowell and P. Raghunathan, *J. Mol. Struct.* **5**, 433 (1970).

917. J. A. McIntyre and D. R. Peterson, *J. Chem. Phys.* **47**, 3850 (1967).

918. R. B. McKay, *Trans. Faraday Soc.* **61**, 1787 (1965).

919. R. B. McKay and P. J. Hillson, *Trans. Faraday Soc.* **61**, 1800 (1965).

920. R. B. McKay and P. J. Hillson, *Trans. Faraday Soc.* **63**, 777 (1967).

921. J. F. McKellar and J. H. Andreae, *Nature* **195**, 778 (1962).

922. V. McKoy and O. Sinanoglu, *J. Chem. Phys.* **38**, 2946 (1963).

923. J. M. McLeod, Jr. and J. M. Campbell, *J. Petrol. Technol.* **13**, 590 (1961).

924. W. McMillan and J. Mayer, *J. Chem. Phys.* **13**, 176 (1945).

925. R. K. McMullan, M. Bonamico, and G. A. Jeffrey, *J. Chem. Phys.* **39**, 3295 (1963).

926. R. K. McMullan and G. A. Jeffrey, *J. Chem. Phys.* **31**, 1231 (1959).

927. R. K. McMullan and G. A. Jeffrey, *J. Chem. Phys.* **42**, 2725 (1965).

928. R. K. McMullan, G. A. Jeffrey, and T. H. Jordan, *J. Chem. Phys.* **47**, 1229 (1967).

929. R. K. McMullan, G. A. Jeffrey, and D. Panke, *J. Chem. Phys.* **53**, 3568 (1970).

930. R. K. McMullan, T. H. Jordan, and G. A. Jeffrey, *J. Chem. Phys.* **47**, 1218 (1967).

931. R. K. McMullan, T. C. W. Mak, and G. A. Jeffrey, *J. Chem. Phys.* **44**, 2338 (1966).

932. E. G. MacRae, *Spectrochim. Acta* **12**, 192 (1958).

933. G. McTurk and J. G. Waller, *Nature* **202**, 1107 (1964).

934. M. M. Maguire and R. West, *Spectrochim. Acta* **17**, 369 (1961).

935. Y. A. Majid, S. K. Garg, and D. W. Davidson, *Can. J. Chem.* **46**, 1683 (1968).

936. Y. A. Majid, S. K. Garg, and D. W. Davidson, *Can. J. Chem.* **47**, 4697 (1969).

937. T. C. W. Mak, *J. Chem. Phys.* **43**, 2799 (1965).

938. T. C. W. Mak and R. K. McMullan, *J. Chem. Phys.* **42**, 2732 (1965).

939. Yu. F. Makogon and G. A. Sarkis'yants, "Preduprezhdenie Obrazovaniya Gidratov pri Dobyche i Transporte Gaza," Nedra, Moscow (1966).

940. Yu. F. Makogon, F. A. Trebin, A. A. Trofimuk, V. P. Tsarev, and N. V. Cherskii, *Dokl. Akad. Nauk SSSR* **196**, 203 (1971).

941. G. N. Malcolm and J. S. Rowlinson, *Trans. Faraday Soc.* **53**, 921 (1957).

942. G. G. Malenkov, *Zh. Strukt. Khim.* **3**, 220 (1962).

943. L. Mandelcorn, *Chem. Rev.* **59**, 827 (1959).

944. P. T. Manoharan and W. C. Hamilton, *Inorg. Chem.* **2**, 1043 (1963).

945. D. W. Marquardt, *J. Soc. Ind. Appl. Math.* **11**, 431 (1963).

946. D. R. Marshall, S. Saito, and R. Kobayashi, *AIChE J.* **10**, 202, 723 (1964).

947. T. W. Marshall and J. A. Pople, *Mol. Phys.* **1**, 199 (1958).

948. J. G. Martin and R. E. Robertson, *J. Am. Chem. Soc.* **88**, 5353 (1966).

949. W. L. Masterton, *J. Chem. Phys.* **22**, 1830 (1954).

950. W. L. Masterton, and M. C. Gendrano, *J. Phys. Chem.* **70**, 2895 (1966).

951. W. L. Masterton and H. K. Seiler, *J. Phys. Chem.* **72**, 4257 (1968).

952. A. McL. Mathieson, D. P. Mellor, and N. C. Stephenson, *Acta Cryst.* **5**, 185 (1952).

953. J. P. Mathieu, *J. Chim. Phys. Physiochim. Biol.* **50**, C79 (1953).

954. J. Matouš, J. Hnčiřik, J. P. Novák, and J. Šobr, *Coll. Czech. Chem. Commun.* **35**, 1904 (1970).

955. T. Matsuo, H. Suka, and S. Seki, *J. Phys. Soc. Japan* **30**,* 785 (1971).

956. G. Mavel, *Compt. Rend.* **248**, 1505 (1959).

957. G. Mavel, *Compt. Rend.* **250**, 1477 (1960).

958. R. M. Mazo, *Mol. Phys.* **8**, 515 (1964).

959. A. Mazzucchelli and R. Armenante, *Gazz. Chim. Ital.* **52**, 338 (1922).

960. R. Mecke, *Phys. Z.* **30**, 907 (1929).

961. R. Mecke, *Naturwiss.* **20**, 657 (1932).

962. R. Mecke, *Z. Physik.* **81**, 313 (1933).

963. R. Mecke, *Z. Elektrochem.* **50**, 57 (1944).

964. R. Mecke, *Z. Elektrochem.* **52**, 274 (1948).

965. R. Mecke, *Z. Elektrochem.* **52**, 280 (1948).

966. R. Mecke and W. Baumann, *Phys. Z.* **33**, 833 (1932).

967. R. Mecke and W. Baumann, *Z. Physik.* **81**, 445 (1933).

968. R. Mecke and A. Reuter, *Z. Naturforsch.* **4a**, 371 (1949).

969. R. Mecke, A. Reuter, and R. L. Schupp, *Z. Naturforsch.* **4a**, 182 (1949).

970. R. Mecke and O. Vierling, *Z. Physik.* **96**, 567 (1935).

971. W. Meinhold, Dissert., Bonn (1948).

972. V. A. Mikhailov, *J. Struct. Chem.* **9**, 332 (1968).

973. V. A. Mikhailov and L. I. Ponomarova, *J. Struct. Chem.* **9**, 8 (1968).

974. B. Miller, *Gas Age* **97**(9), 37 (1946).

975. B. Miller and E. R. Strong, Jr., *Proc. Am. Gas Assoc.* **27**, 80 (1945).

976. F. A. Miller, G. L. Carlson, F. F. Bentley, and W. H. Jones, *Spectrochim. Acta* **16**, 135 (1960).

⨯ 977. K. W. Miller and J. H. Hildebrand, *J. Am. Chem. Soc.* **90**, 3001 (1968).

978. S. L. Miller, *Proc. Nat. Acad. Sci. U.S.* **47**, 1515 (1961).

979. S. L. Miller, *Proc. Nat. Acad. Sci. U.S.* **47**, 1798 (1961).

980. S. L. Miller, *Science* **165**, 489 (1969).

981. S. L. Miller, E. I. Eger, and C. Lundgreen, *Nature* **221**, 468 (1969).

982. S. L. Miller and W. D. Smythe, *Science*, **170**, 531 (1970).

983. F. J. Millero and W. Drost Hansen, *J. Phys. Chem.* **72**, 1758 (1968).

984. N. A. E. Millon, *Ann. Chem. Pharm.* **46**, 281 (1843).

985. N. A. E. Millon, *Ann. Chim. Phys. Ser. 3* **7**, 308 (1843).

986. A. G. Mitchell and W. F. K. Wynne-Jones, *Disc. Faraday Soc.* **15**, 161 (1953).

987. T. Miyazawa, *J. Chem. Soc. Japan* **34**, 202 (1961).

988. M. Mizutani, *Osaku Daigaku Igaka Zasshu* **8**, 1334 (1956).

989. C. J. Moen, *J. Acoust. Soc. Am.* **23**, 62 (1951).

990. R. K. Mohanty, T. S. Sarma, S. Subramanian, and J. C. Ahluwalia, *Trans. Faraday Soc.* **67**, 305 (1971).

991. S. C. Mohr, W. D. Wilk and G. M. Barrow, *J. Am. Chem. Soc.* **87**, 3048 (1965).

992. "Molecular Luminescence" (E. C. Lim, ed.), Benjamin, New York (1969).

993. D. J. Mootz and H. Altenburg, *Acta Cryst.* **B25**, 1077 (1969).

994. K. W. Morcom and R. W. Smith, *J. Chem. Thermodynamics* **1**, 503 (1969).

995. K. W. Morcom and R. W. Smith, *Trans. Faraday Soc.* **66**, 1073 (1970).

996. B. Morosin, *J. Chem. Phys.* **44**, 252 (1966).

997. B. Morosin, *Acta Cryst.* **23**, 630 (1967).

998. B. Morosin, *Acta Cryst. B* **26**, 1203 (1970).

999. J. R. Morrey, *J. Phys. Chem.* **66**, 2169 (1962).

1000. B. Morris and D. W. Davidson, *Can. J. Chem.* **49**, 1243 (1971).

1001. T. J. Morrison, *J. Chem. Soc.* **1952**, 3814.

1002. T. J. Morrison and F. Billet, *J. Chem. Soc.* **1952**, 3819.

1003. J. C. Morrow, *Acta Cryst.* **15**, 851 (1962).

1004. P. Mukerjee and A. K. Ghosh, *J. Phys. Chem.* **67**, 193 (1963).

1005. P. Mukerjee and A. K. Ghosh, *J. Amer. Chem. Soc.* **92**, 6403, 6409, 6413, 6419 (1970).

1006. C. E. Mulders, *Appl. Sci. Res.* **131**, 341 (1949).

1007. E. M. J. Mulders, Dissert. Tech. Hogeschool, Delft (1936).

1008. H. Muller, *Z. Phys. Chem. Leipzig* **215**, 238 (1960).

1009. H. R. Müller and M. von Stackelberg, *Naturwiss.* **39**, 20 (1952).

1010. J. I. Musher, *in* "Advances in Magnetic Resonance" (J. S. Waugh, ed.), Vol. 2, p. 177, Academic, New York (1966).

1011. D. H. Myers, R. A. Smith, J. Katz, and R. L. Scott, *J. Phys. Chem.* **70**, 3341 (1966).

1012. I. Nagata and R. Kobayashi, *Ind. Eng. Chem. Fundam.* **5**, 344, 466 (1966).

1013. T. Nakagawa and T. Shimanouchi, *Spectrochim. Acta* **20**, 429 (1964).

1014. K. Nakamoto, M. Margoshes, and R. E. Rundle, *J. Am. Chem. Soc.* **77**, 6480 (1955).

1015. N. Nakanishi, *Bull. Chem. Soc. Japan* **33**, 793 (1960).

1016. K. Nakanishi, N. Kato, and M. Maruyama, *J. Phys. Chem.* **71**, 814 (1967).

1017. H. Nakayama and K. Shinoda, *J. Chem. Thermodynamics* **3**, 401 (1971).

1018. A. Yu. Namiot, *Zh. Strukt. Khim.* **2**, 408 (1961).

1019. J. L. Neal and D. A. I. Goring, *J. Phys. Chem.* **74**, 658 (1970).

1020. D. Neerink and L. Lamberts, *Bull. Soc. Chim. Belg.* **75**, 473 (1966).

1021. H. D. Nelson and C. L. de Ligny, *Rec. Trav. Chim.* **87**, 528 (1968).

1022. R. Nelson and L. Pierce, *J. Mol. Spectry.* **18**, 344 (1965).

1023. G. Nemethy, *Angew. Chem.* **79**, 260 (1967).

1024. G. Nemethy, *Angew. Chem. Internat. Ed.* **6**, 195 (1967).

1025. G. Nemethy and H. A. Scheraga, *J. Chem. Phys.* **36**, 3382 (1962).

1026. G. Nemethy and H. A. Scheraga, *J. Chem. Phys.* **36**, 3401 (1962).

1027. G. Nemethy and H. A. Scheraga, *J. Phys. Chem.* **66**, 1773 (1962).

1028. D. E. Nicholson, *J. Chem. and Eng. Data* **5**, 309 (1960).

1029. B. A. Nikitin, *Z. Anorg. Allg. Chem.* **227**, 81 (1936).

1030. B. A. Nikitin, *Nature* **140**, 643 (1937).

1031. B. A. Nikitin, *Zh. Obshch. Khim.* **9**, 1167, 1176 (1939).

1032. B. A. Nikitin, *Izv. Akad. Nauk SSSR, Otd. Khim. Nauk* **1**, 39 (1940).

1033. L. J. Noaker and D. L. Katz, *J. Petrol. Technol.* **6**, 135 (1954).

1034. O. Nomoto, *J. Phys. Soc. Japan* **11**, 827 (1956).

1035. O. Nomoto and H. Endo, *Bull. Chem. Soc. Japan* **43**, 2718 (1970).

1036. J. P. Novak and G. M. Schneider, *Ber. Bunsenges. Phys. Chem.* **72**, 791 (1968).

1037. K. Nukasawa, "Some Researches on Biophysics," Monograph Series of the Research Institute of Applied Electricity, No. 2, p. 57 (1951).

1038. J. Oakes and M. C. R. Symons, *Trans. Faraday Soc.* **64**, 2579 (1968).

1039. B. H. O'Connor and D. H. Dale, *Acta Cryst.* **21**, 705 (1966).

1040. L. Ödberg and E. Högfeldt, *Acta Chem. Scand.* **23**, 1330 (1969).

1041. F. Oehme and M. Feinauer, *Chem. Z. Chem. Apparat.* **86**, 71 (1962).

1042. J. D. Offmann and H. G. Pfeiffer, *J. Chem. Phys.* **22**, 132 (1954).

1043. T. M. O'Grady, *J. Chem. Eng. Data* **12**, 9 (1967).

1044. K. Okada, M. I. Kay, D. T. Cromer, and I. Almodovar, *J. Chem. Phys.* **44**, 1648 (1966).

1045. I. Olovsson, *J. Chem. Phys.* **49**, 1063 (1968).
1046. S. Ono and K. Takahashi, *in* "Biochemical Microcalorimetry" (H. D. Brown, ed.), Academic, New York (1969), Chapter 4 and references therein.
1047. L. Onsager, *J. Am. Chem. Soc.* **58**, 1486 (1936).
1048. L. Onsager and M. Dupuis, *Rendi. Scu. Int. Fis. "Enrico Fermi"* **10**, 1 (1960).
1049. L. Onsager and M. Dupuis, *in* "Electrolytes" (B. Pesce, ed.), Pergamon, New York (1962),
1050. L. Onsager and L. K. Runnels, *J. Chem. Phys.* **50**, 1089 (1969).
1051. I. Oppenheim and M. Bloom, *Can. J. Phys.* **39**, 845 (1961).
1052. W. J. Orville-Thomas and E. Wyn-Jones, *in* "Transfer and Storage of Energy by Molecules" (G. M. Burnett and A. M. North, eds.), Vol. 2, Chapter 5, Wiley–Interscience, London (1969).
1053. J. A. Osborn, *Phys. Rev.* **62**, 351 (1945).
1054. T. Oshawa and Y. Wada, *Japanese J. Appl. Phys.* **6**, 1351 (1967).
1055. G. Oster, *J. Amer. Chem. Soc.* **68**, 2036 (1946).
1056. G. Oster and J. G. Kirkwood, *J. Chem. Phys.* **11**, 175 (1943).
1057. H. R. Oswald, *Helv. Chim. Acta* **48**, 600 (1965).
1058. D. F. Othmer and S. L. Levy, *Chem. Eng. Prog. Symp. Ser. 49*, No. 6, 64 (1953).
1059. F. D. Otto and D. B. Robinson, *AIChE J.* **6**, 602 (1960).
1060. H. Otsuki and F. C. Williams, *Chem. Eng. Prog. Symp. Ser. 49*, No. 6, 55 (1953).
1061. K. J. Packer and D. J. Tomlinson, *Trans. Faraday Soc.* **67**, 1302 (1971).
1062. J. F. Padday, *J. Phys. Chem.* **71**, 3488 (1967).
1063. V. M. Padmanabhan, W. R. Busing, and H. A. Levy, *Acta Cryst.* **16**, A26 (1963).
1064. V. M. Padmanabhan, S. Srikanta, and S. M. Ali, *Acta Cryst.* **18**, 567 (1965).
1065. G. E. Pake, *J. Chem. Phys.* **16**, 327 (1948).
1066. G. Paliani, A. Poletti, and A. Santucci, *J. Mol. Struct.* **8**, 63 (1971).
1067. D. E. Palin and H. M. Powell, *J. Chem. Soc. (London)* **1947**, 208.
1068. D. E. Palin and H. M. Powell, *J. Chem. Soc. (London)* **1948**, 815.
1069. H. A. Palmer, Dissertation, University of Oklahoma (1950).
1070. D. Panke, *J. Chem. Phys.* **48**, 2990 (1968).
1071. M. Th. Pâques-Ledent and P. Tarte, *Spectrochim. Acta* **25A**, 1115 (1969).
1072. J. D. Parent, *Inst. Gas Technol. Res. Bull.* **1** (1948).
1073. A. Parker, *Nature* **149**, 184 (1942).
1074. F. S. Parker, *Progr. in Infrared Spectroscopy* **3**, 75 (1967).
1075. S. Parthasarathy and M. Pancholy, *Z. Physik.* **10**, 453 (1958).
1076. F. Paschen, *Ann. Phys.* **51**, 22 (1894).
1077. F. Paschen, *Ann. Phys.* **53**, 334 (1894).
1078. L. Pauling, *J. Am. Chem. Soc.* **57**, 96 (1935).
1079. L. Pauling, *in* "Hydrogen Bonding" (D. Hadzi and H. W. Thompson, eds.), p. 1, Pergamon, New York (1959).
1080. L. Pauling, "The Nature of the Chemical Bond," 3rd ed., Cornell Univ. Press, Ithaca, New York (1960).
1081. L. Pauling, *Science* **134**, 15 (1961).
1082. L. Pauling, *Anesth. Analg. (Cleveland)* **43**, 1 (1964).
1083. L. Pauling and R. E. Marsh, *Proc. Nat. Acad. Sci. U.S.* **38**, 112 (1952).
1084. G. D. Pavlov and I. N. Medvedev, *in* "Proc. First Int. Symp. Water Desalination," Vol. 3, p. 123, October 1965, Office of Saline Water, Washington (1967).
1085. M. E. Pedinoff, *J. Chem. Phys.* **36**, 777 (1962).

1086. J. M. Peterson and W. H. Rodebush, *J. Phys. Chem.* **32**, 709 (1928).

1087. S. W. Peterson and H. A. Levy, *Acta Cryst.* **10**, 78 (1957).

1088. S. W. Peterson and H. A. Levy, *J. Chem. Phys.* **26**, 220 (1957).

1089. S. U. Pickering, *J. Chem. Soc. London, Trans.* **63**, 141 (1893).

1090. J. E. Piercy and J. Lamb, *Proc. Phys. Soc. London* **266A**, 43 (1954).

1091. J. E. Piercy and J. Lamb, *Trans. Faraday Soc.* **52**, 930 (1956).

1092. A. P. Pieron, *Rec. Trav. Chim. Pays-Bas* **74**, 995 (1955).

1093. J. Pieutuchowsky, *Diplomarbeit, Bonn* (1941).

1094. G. C. Pimentel, *Spectrochim. Acta* **12**, 94 (1958).

1095. G. C. Pimentel, *Pure Appl. Chem.* **4**, 61 (1962).

1096. G. C. Pimentel, *in* "Hydrogen Bonding" (R. Blinc and D. Hadzi, eds.), p. 107, Pergamon, London (1959).

1097. G. C. Pimentel, *in* "Formation and Trapping of Free Radicals" (A. M. Bass and H. P. Broida, eds.), Academic, New York (1960).

1098. G. C. Pimentel and S. W. Charles, *Pure Appl. Chem.* **7**, 111 (1963).

1099. G. C. Pimentel and A. L. McClellan, "The Hydrogen Bond," Freeman, San Francisco and London (1960).

1100. G. C. Pimentel and C. H. Sederholm, *J. Chem. Phys.* **24**, 639 (1956).

1101. K. L. Pinder, *Can. J. Chem. Eng.* **43**, 271 (1965).

1102. J. M. M. Pinkerton, *Proc. Phys. Soc. (London)* **B62**, 286 (1949).

1103. E. Pittz and J. Bello, *Trans. Faraday Soc.* **66**, 537 (1970).

1104. J. C. Platteeuw, *Rec. Trav. Chim. Pays-Bas* **77**, 403 (1958).

1105. J. C. Platteeuw and J. H. van der Waals, *Mol. Phys.* **1**, 91 (1958).

1106. J. C. Platteeuw and J. H. van der Waals, *Rec. Trav. Chim. Pays-Bas* **78**, 126 (1959).

1107. F. M. Pohl, *Eur. J. Biochem.* **7**, 146 (1968).

1108. D. Polder and J. N. van Santen, *Physica* **12**, 257 (1946).

1109. J. A. Pople, *Proc. Roy. Soc.* **205**, 163 (1951).

1110. J. A. Pople, W. G. Schneider, and H. J. Bernstein, "High Resolution Nuclear Magnetic Resonance," McGraw-Hill, New York (1959).

1111. G. Poppe, *Bull. Soc. Chim. Belges* **44**, 640 (1935).

1112. R. Pottel and U. Kaatze, *Ber. Bunsenges. Phys. Chem.* **73**, 437 (1969).

1113. R. Pottel and D. Lossen, *Ber. Bunsenges. Phys. Chem.* **71**, 135 (1967).

1114. A. D. Potts and D. W. Davidson, *J. Phys. Chem.* **69**, 996 (1965).

1115. H. M. Powell, *J. Chem. Soc. (London)* **1948**, 61.

1116. H. M. Powell, "Clathrate Compounds," *Rec. Trav. Chim. Pays-Bas* **75**, 885 (1956).

1117. H. M. Powell, "Clathrates" *in* "Non-Stoichiometric Compounds" (L. Mandelcorn, ed.), p. 438, Academic, New York (1964).

1118. J. G. Powles, *J. Chem. Phys.* **20**, 1302 (1952).

1119. J. G. Powles, *Proc. Phys. Soc.* **71**, 497 (1958).

1120. H. J. Prask and H. Boutin, *J. Chem. Phys.* **45**, 3284 (1966).

1121. I. P. Prigogine and R. D. Defay, "Chemical Thermodynamics" (transl. by D. H. Everett), p. 431, Longmans Green, London (1954).

1122. J. E. Prue, A. J. Read, and G. Romeo, *Trans. Faraday Soc.* **67**, 420 (1971).

1123. K. F. Purcell and R. S. Drago, *J. Amer. Chem. Soc.* **89**, 2874 (1967).

1124. A. S. Quist and H. S. Frank, *J. Phys. Chem.* **65**, 560 (1961).

1125. S. W. Rabideau, E. D. Finch, and A. B. Denison, *J. Chem. Phys.* **49**, 4660 (1968).

1126. A. Rahman and F. H. Stillinger, Jr. *J. Chem. Phys.* **55**, 3336 (1971).

1127. F. Rallo, F. Rodante, and P. Silvestroni, *Thermochim. Acta* **1**, 311 (1970).

1128. M. B. Ramiah and D. A. I. Goring, *J. Polymer Sci.* II (Part C) **1965**, 27.

1129. D. Rank, E. Kiero, U. Fink, and T. Wiggins, *J. Opt. Soc. Am.* **55**, 925 (1965).

1130. R. J. Raridon and K. A. Kraus, *J. Colloid Sci.* **20**, 1000 (1965).

1131. J. Rassing and H. Lassen, *Acta Chem. Scand.* **23**, 1007 (1969).

1132. W. Rau, *Schr. Deut. Akad. Luftfahrforsch.* **8**, 65 (1944).

1133. W. T. Raynes, A. D. Buckingham, and H. J. Bernstein, *J. Chem. Phys.* **36**, 3481 (1962).

1134. H. H. Reamer, T. F. Selleck, and B. H. Sage, *J. Petrol. Technol.* **4**(8), 197 (1952).

1135. C. J. Rebert and K. E. Hayworth, *AIChE J.* **13**, 118 (1967).

1136. C. J. Rebert and W. B. Kay, *AIChEJ* **5**, 285 (1959).

1137. A. G. Redfield, *in* "Advances in Magnetic Resonance" (J. S. Waugh, ed.), Vol. 1, p. 1, Academic, New York (1965).

1138. L. W. Reeves, *in* "Progress in Nuclear Magnetic Resonance Spectroscopy" (J. W. Emsley, J. Feeney, and L. H. Sutcliffe, eds.), Vol. 4, p. 193, Pergamon, Oxford (1969).

1139. L. W. Reeves and C. P. Yue, *Can. J. Chem.* **48**, 3307 (1970).

1140. C. Reichardt, *Angew. Chem., Internat. Ed.* **4**, 29 (1965).

1141. D. S. Reid, M. A. J. Quickenden, and F. Franks, *Nature* **224**, 1293 (1969).

1142. D. S. Reid and D. J. Tibbs, to be published.

1143. R. Reynaud, *Bull. Soc. Chim. France* **1967**, 2686.

1144. R. Reynaud, *Bull. Soc. Chim. France* **1968**, 3945.

1145. J. A. Reynolds, Thesis, Univ. of Hull (Quoted in Ref. 1318).

1146. J. E. Ricci, "The Phase Rule and Heterogeneous Equilibrium," Van Nostrand, New York (1951).

1147. O. L. Roberts, E. R. Brownscombe, and L. S. Howe, *Oil Gas J.* **39** (30), 37 (1940).

1148. R. E. Robertson and S. E. Sugamori, *J. Am. Chem. Soc.* **91**, 7254 (1969).

1149. A. L. Robinson, *J. Chem. Phys.* **14**, 588 (1946).

1150. R. A. Robinson, *J. Phys. Chem.* **73**, 3165 (1969).

1151. R. A. Robinson and R. H. Stokes, "Electrolyte Solutions," 2nd ed., Butterworths, London (1959).

1152. R. A. Robinson and R. H. Stokes, *J. Phys. Chem.* **65**, 1954 (1961).

1153. V. P. Romanov and V. A. Solovyev, *Soviet Phys.–Acoust.* **11**, 68 (1965).

1154. V. P. Ramanov and V. A. Solovyev, *Soviet Phys.–Acoust.* **11**, 219 (1965).

1155. H. W. B. Roozeboom, *Rec. Trav. Chim. Pays-Bas* **3**, 26 (1884).

1156. H. W. B. Roozeboom, *Rec. Trav. Chim. Pays-Bas* **4**, 65 (1885).

1157. H. W. B. Roozeboom, "Die heterogenen Gleichgewichte vom Standpunkte der Phasenlehre," Vol. II, Part 2, p. 191, Vieweg und Sohn, Braunschweig (1918).

1158. M. Rösch, *Fette, Seifen, Anstrichmittel* **59**, 745 (1957).

1159. M. Ross, H. T. Evans, Jr., and D. E. Appleman, *Am. Mineral.* **49**, 1603 (1964).

1160. K. Roth, G. M. Schneider, and E. U. Franck, *Ber. Bunsenges. Phys. Chem.* **70**, 5 (1966).

1161. J. S. Rowlinson, "Liquids and Liquid Mixtures," 1st ed., Butterworths, London (1959).

1162. J. G. Rowlinson, "Liquids and Liquid Mixtures," 2nd ed., p. 170, Butterworths, London (1969).

1163. R. Ruepp and M. Käss, *in* "Physics of Ice" (N. Riehl, B. Bullemer, and H. Engelhardt, eds.), p. 555, Plenum, New York (1969).

1164. R. E. Rundle and M. Parasol, *J. Chem. Phys.* **20**, 1487 (1952).

1165. L. K. Runnels, Dissert., Yale Univ. (1964).

1166. G. S. Rushbrooke, "Introduction to Statistical Thermodynamics," p. 243, Clarendon, Oxford (1949).

1167. H. H. Rüterjans and H. A. Scheraga, *J. Chem. Phys.* **45**, 3296 (1966).

1168. H. H. Rüterjans, F. Schreiner, U. Sage, and Th. Ackermann, *J. Phys. Chem.* **73**, 986 (1969).

1169. T. M. Sabine, G. W. Cox, and B. M. Craven, *Acta Cryst.* **B25**, 2437 (1969).

1170. L. Sacconi, P. Paoletti, and M. Ciampolini, *J. Am. Chem. Soc.* **82**, 3828 (1960).

1171. G. J. Safford, P. C. Scheffer, P. S. Leung, G. F. Doebbler, G. W. Brady, and E. F. X. Lyden, *J. Chem. Phys.* **50**, 2140 (1969).

1172. T. A. Saifeev, *Gazov. Delo* **1969**(12), 27.

1173. S. Saito and R. Kobayashi, *AIChE J.* **11**, 96 (1965).

1174. O. Ya. Samoilov, *Zh. Strukt. Khim.* **20**, 1411 (1956).

1175. O. Ya. Samoilov, *Disc. Faraday Soc.* **24**, 141 (1957).

1176. O. Ya. Samoilov, "Structure of Aqueous Electrolyte Solutions and the Hydration of Ions" (transl. by D. J. G. Ives), Plenum, New York (1965).

1177. H. A. Samulon, *Proc. IRE* **39**, 175 (1951).

1178. D. F. Sargent and L. D. Calvert, *J. Phys. Chem.* **70**, 2689 (1966).

1179. T. S. Sarma and J. C. Ahluwalia, *Trans. Faraday Soc.* **67**, 2528 (1971).

1180. T. S. Sarma, R. K. Mohanty, and J. C. Ahluwalia, *Trans. Faraday Soc.* **65**, 2333 (1969).

1181. V. Sarojini, *Trans. Faraday Soc.* **57**, 1534 (1961).

1182. G. Sartori, C. Furlani, and A. Damiani, *J. Inorg. Nucl. Chem.* **8**, 119 (1958).

1183. K. Sasaki and K. Arakawa, *Bull. Chem. Soc. Japan* **42**, 2485 (1969).

1184. K. Sasvari and G. A. Jeffrey, *Acta Cryst.* **20**, 875 (1966).

1185. G. Scatchard, *Chem. Rev.* **19**, 309 (1936).

1186. G. Scatchard, W. J. Hamer, and S. E. Wood, *J. Am. Chem. Soc.* **60**, 3061 (1938).

1187. L. H. Scharpen and V. W. Laurie, *J. Chem. Phys.* **49**, 221 (1968).

1188. F. E. C. Scheffer and G. Meyer, *Proc. Kon. Ned. Akad. Wetensch.* **27**, 1104 (1919) [English version: *Proc. Roy. Acad. Sci. Amsterdam* **21**, 1204, 1338 (1919)].

1189. J. A. Schellman, *Compt. Rend. Lab. Carlsberg, Ser. Chim.* **29**, 223 (1955).

1190. Z. A. Schelly, D. J. Harward, P. Hemmes, and E. M. Eyring, *J. Phys. Chem.* **74**, 3040 (1970).

1191. J. Schiffer, Ph.D. Thesis, Princeton Univ. (1963); *Dissert. Abstr.* **25**, 2786 (1964).

1192. J. Schiffer and D. F. Hornig, *in* "Molecular Dynamics and Structure of Solids" (R. S. Carter and J. J. Rush, eds.), p. 257, NBS Special Publication No. 301, U.S. Gov. Printing Office, Washington, D.C. (1969).

1193. W. Schlenk, *Fortschr. Chem. Forsch.* **2**, 1425 (1951).

1194. W. G. Schneider, *Can. J. Chem.* **29**, 243 (1951).

1195. W. G. Schneider, *J. Chem. Phys.* **23**, 26 (1955).

1196. W. G. Schneider, *Coll. Intern. Centre Nat. Rech. Sci. Paris* **77**, 259 (1959).

1197. G. M. Schneider, *Z. Phys. Chem. (Frankfurt)* **39**, 187 (1963).

1198. G. M. Schneider, *Z. Phys. Chem. (Frankfurt)* **41**, 327 (1964).

1199. G. M. Schneider, *Ber. Bunsenges. Phys. Chem.* **70**, 497 (1966).

1200. G. M. Schneider, *Adv. Chem. Phys.* **17**, 1 (1970).

1201. G. M. Schneider, *Fortschr. Chem. Forsch.* **13**, 559 (1970).

1202. G. M. Schneider and C. Russo, *Ber. Bunsenges. Phys. Chem.* **70**, 1008 (1966).

1203. G. R. Schneider and J. Farrar, U.S. Off. Saline Water Progr. Rep. 292 (1968).

1204. H. Schönert, *Z. Phys. Chem. (Frankfurt)* **61**, 262 (1968).

1205. E. E. Schrier, R. T. Ingwall, and H. A. Scheraga, *J. Phys. Chem.* **69**, 298 (1965).

1206. W. Schroeder, *Sammlung Chem. Chem.-Tech. Vortrage* **29**, 1 (1926).

1207. R. Schroeder and E. R. Lippincott, *J. Phys. Chem.* **61**, 921 (1957).

1208. R. L. Schupp, *Z. Elektrochem.* **53**, 12 (1949).

1209. E. Schwarzmann, *Z. Naturforsch.* **24b**, 1104 (1969).

1210. V. Seidl, O. Knop, and M. Falk, *Can. J. Chem.* **47**, 1361 (1969).

1211. F. T. Selleck, L. T. Carmichael, and B. H. Sage, *Ind. Eng. Chem.* **44**, 2219 (1952).

1212. W. A. Senior and R. E. Verrall, *J. Phys. Chem.* **73**, 4242 (1969).

1213. W. A. Senior and R. E. Verrall, *J. Chem. Phys.* **50**, 2746 (1969).

1214. A. Sequeira, S. Srikanta, and R. Chidambaram, *Acta Cryst.* **B26**, 77 (1970).

1215. D. Sette, *J. Chem. Phys.* **18**, 1592 (1950).

1216. D. Sette, *Nature* **166**, 114 (1950).

1217. D. Sette, *J. Chem. Phys.* **21**, 558 (1953).

1218. D. Sette, *Nuovo Cimento* **1**, 800 (1955).

1219. D. Sette, *in* "Phenomena in the Neighborhood of Critical Points," p. 183, National Bureau of Standards (U.S.), Misc. Publications, 273, Washington, D. C. (1965).

1220. D. J. Shaw, "An Introduction to Colloid and Surface Chemistry," p. 147, Butterworths, London (1966).

1221. C. A. Shelton and D. Panke, *Acta Cryst. Sect. A* **25**, S147 (1969).

1222. S. E. Sheppard and A. L. Geddes, *J. Am. Chem. Soc.* **66**, 1995 (1944).

1223. S. E. Sheppard and A. L. Geddes, *J. Am. Chem. Soc.* **66**, 2001 (1944).

1224. S. K. Sikka and R. Chidambaram, *Acta Cryst.* **B25**, 310 (1969).

1225. S. K. Sikka, S. N. Momin, H. Rajagopal, and R. Chidambaram, *J. Chem. Phys.* **48**, 1883 (1968).

1226. H. Sillescu, "Kernmagnetische Resonanz," Springer, Berlin (1966).

1227. C. M. Slansky, *J. Am. Chem. Soc.* **62**, 2430 (1940).

1228. C. P. Slichter, "Principles of Magnetic Resonance," Harper and Row, New York (1963).

1229. W. M. Slie, A. R. Donfor, and T. A. Litovitz, *J. Chem. Phys.* **44**, 3712 (1966).

1230. A. S. Sliwinski and A. E. Brown, *Acustica* **14**, 280 (1964).

1231. F. A. Smith, E. C. Creitz, *J. Res. Nat. Bur. Std. U.S.* **46**, 145 (1951).

1232. H. G. Smith, S. W. Peterson, and H. A. Levy, *J. Chem. Phys.* **48**, 5561 (1968).

1233. M. Smith and M. C. R. Symons, *Trans. Faraday Soc.* **54**, 338 (1958).

1234. M. Smith and M. C. R. Symons, *Trans. Faraday Soc.* **54**, 346 (1958).

1235. P. K. Smith and E. R. B. Smith, *J. Biol. Chem.* **121**, 607 (1937).

1236. H. Snell and J. Greyson, *J. Phys. Chem.* **74**, 2148 (1970).

1237. L. E. Snell, F. D. Otto, and D. B. Robinson, *AIChE J.* **7**, 482 (1961).

1238. V. A. Solovyev, C. J. Montrose, M. H. Watkins, and T. A. Litovitz, *J. Chem. Phys.* **48**, 2155 (1968).

1239. W. C. Sommerville, *J. Phys. Chem.* **35**, 2412 (1931).

1240. L. D. Sortland and D. B. Robinson, *Can. J. Chem. Eng.* **42**, 38 (1964).

1241. S. Sourirajan and G. C. Kennedy, Nucl. Explos. Peaceful Applic. UC 35, TID-4500, Report UCRL-6175 (1960).

1242. H. W. Spiess, B. B. Garett, R. K. Sheline, and S. W. Rabideau, *J. Chem. Phys.* **51**, 1201 (1969).

1243. S. Srikanta, A. Sequeira, and R. Chidambaram, *Acta Cryst.* **B24**, 1176 (1968).

1244. M. v. Stackelberg, O. Gotzen, J. Pietuchowsky, O. Witscher, H. Fruhbuss, and M. Meinhold, *Fortschr. Mineral.* **26**, 122 (1947).

1245. M. v. Stackelberg and H. R. Müller, *J. Chem. Phys.* **19**, 1319 (1951).

1246. L. A. K. Staveley, "Physics and Chemistry of Inclusion Compounds," *in* "Non-Stoichiometric Compounds" (L. Mandelcorn, ed.), p. 606, Academic, New York (1964).

1247. J. Steigman and J. Dobrow, *J. Phys. Chem.* **72**, 3424 (1968).

1248. G. Stein and A. Treinin, *Trans. Faraday Soc.* **55**, 1086 (1959).

1249. G. Stein and A. Treinin, *Trans. Faraday Soc.* **55**, 1091 (1959).

1250. R. Steiner and E. Schadow, *Z. Phys. Chem. (Frankfurt)* **63**, 297 (1969).

1251. E. O. Stejskal and J. E. Tanner, *J. Chem. Phys.* **42**, 288 (1965).

1252. R. H. Stokes, *Aust. J. Chem.* **20**, 2087 (1967).

1253. R. H. Stokes and R. A. Robinson, *J. Phys. Chem.* **70**, 2126 (1966).

1254. H. Stone, *J. Opt. Soc. Am.* **52**, 998 (1962).

1255. J. Stone and R. E. Pontinen, *J. Chem. Phys.* **47**, 2407 (1967).

1256. T. J. Stone, T. Buckman, P. L. Nordio, and H. M. McConnell, *Proc. Natl. Acad. Sci.* **54**, 1010 (1965).

1257. L. R. O. Storey, *Proc. Phys. Soc.* **B65**, 943 (1952).

1258. M. B. Stout, "Basic Electrical Measurements," 2nd ed., Prentice-Hall, Englewood Cliffs, N.J. (1960).

1259. H. Strehlow, "Magnetische Kernresonanz und chemische Struktur," Steinkopff, Darmstadt (1968).

1260. E. V. Stroganov, I. I. Kozhina, S. N. Andreev, and A. B. Kolyadin, *Vestn. Leningrad. Univ. 15, No. 4, Ser. Fiz. i. Kim.* **1960**(1), 130.

1261. F. Strohbusch and H. Zimmerman, *Ber. Bunsenges. Ges.* **71**, 567 (1967).

1262. J. E. Stuckey and C. H. Secoy, *J. Chem. and Eng. Data* **8**, 386 (1963).

1263. J. Stuehr and E. Yeager, *in* "Physical Acoustics," (W. P. Mason, ed.), Vol. 2A, Chapter 6, Academic, London (1965).

1264. F. B. Stumpf and L. A. Crum, *J. Acoust. Soc. Am.* **39**, 170 (1966).

1265. J. M. Sturtevant, *J. Phys. Chem.* **45**, 127 (1941).

1266. S. Subramanian and J. C. Ahluwalia, *J. Phys. Chem.* **72**, 2525 (1968).

1267. A. Suggett, P. A. Mackness, M. J. Tait, H. W. Loeb, and G. M. Young, *Nature* **227**, 457 (1970).

1268. H. Susi, *Spectrochim. Acta* **17**, 1257 (1961).

1269. P. Süsse, *Z. Krist.* **127**, 261 (1968).

1270. M. C. R. Symons, *J. Phys. Chem.* **71**, 172 (1967).

1271. M. C. R. Symons and M. J. Blandamer, *in* "Hydrogen-Bonded Solvent Systems" p. 211, (A. K. Covington and P. Jones, eds.), Taylor and Francis, London (1968).

1272. Syracuse Univ. Res. Inst., U.S. Office Saline Water Progr. Rep. 70 (1963).

1273. I. Taesler and I. Olovsson, *Acta Cryst.* **B24**, 299 (1968).

1274. I. Taesler and I. Olovsson, *J. Chem. Phys.* **51**, 4213 (1969).

1275. M. J. Tait, A. Suggett, F. Franks, S. Ablett, and P. A. Quickenden, *J. Solution Chem.* **1**, 131 (1972).

1276. F. Takahashi, and N. C. Li, *J. Am. Chem. Soc.* **88**, 1117 (1966).

1277. K. Takahashi and S. Ono, *Abstr. 2nd Japan. Calorimetry Conf. Tokyo* (1966).

1278. S. Takenouchi and G. C. Kennedy, *Am. J. Sci.* **262**, 1055 (1964).

1279. G. Tammann and G. J. R. Krige, *Z. Anorg. Allg. Chem.* **146**, 179 (1925).

1280. J. B. Taylor and J. S. Rowlinson, *Trans. Faraday Soc.* **51**, 1183 (1955).

1281. J. C. Taylor and M. H. Mueller, *Acta Cryst.* **19**, 536 (1965).
1282. J. C. Taylor, M. H. Mueller, and R. L. Hitterman, *Acta Cryst.* **20**, 842 (1966).
1283. W. H. Taylor, *Proc. Roy. Soc. (London)* **A145**, 80 (1934).
1284. "Technique of Microwave Measurements," Vol. 11 of the MIT Radiation Series, reprinted by Boston Technical Publishers, Lexington, Mass. (1964).
1285. G. H. Teletzke, *Chem. Eng. Prog.* **60**, 33 (1964).
1286. H. N. V. Temperley, J. S. Rowlinson, and G. S. Rushbrooke, "Physics of Simple Liquids," North-Holland, Amsterdam (1968).
1287. L. K. Templeton, D. H. Templeton, and A. Zalkin, *Acta Cryst.* **17**, 933 (1964).
1288. Texaco Inc., New York, N.Y., U.S. Patent 3,318,805 (May 9, 1967).
1289. J. Thamsen, *Acta Chem. Scand.* **19**, 1939 (1965).
1290. J. Thamsen, *Acustica* **16**, 14 (1965–66).
1291. G. H. Thomas, O. Knop, and M. Falk, to be published.
1292. E. Tillmanns and W. H. Baur, *Acta Cryst.* **B27**, 2124 (1971).
1293. S. N. Timasheff and H. Inoue, *Biochem.* **7**, 2501 (1968).
1294. J. Timmermans, *J. Chim. Phys.* **20**, 491 (1923).
1295. J. Timmermans, "Les Solutions Concentrées," Masson, Paris (1936).
1296. K. Tödheide, *Ber. Bunsenges. Phys. Chem.* **70**, 1022 (1966).
1297. K. Tödheide, *Naturwiss.* **57**, 72 (1970).
1298. K. Tödheide and E. U. Franck, *Z. Phys. Chem. (Frankfurt)* **37**, 387 (1963).
1299. H. C. Torrey, *Phys. Rev.* **92**, 962 (1953).
1300. H. C. Torrey, *Phys. Rev.* **104**, 563 (1956).
1301. B. H. Torrie, I. D. Brown, and H. E. Petch, *Can. J. Phys.* **42**, 229 (1964).
1302. A. Tovborg-Jensen, "Krystallinske Salthydrater," Doctoral Thesis, Univ. of Copenhagen (1948).
1303. F. A. Trebin, V. A. Khoroshilov, and A. V. Demchenko, *Gazov. Prom.* **11**(6), 10 (1966).
1304. F. A. Trebin and Yu. F. Makogon, *in* "Materialy Respub. Konf. po Gazifik. Uzbekistana," p. 155, Akad. Nauk UzSSR, Tashkent (1961).
1305. F. A. Trebin and Yu. F. Makogon, *Tr. Mosk. Inst. Neftekhim. Gazov. Prom.* **1963**(42), 196.
1306. N. C. Treloar, Ph.D. Thesis, Univ. of Leicester (1970).
1307. D. S. Tsiklis, *Dokl. Akad. Nauk SSSR* **86**, 1159 (1952).
1308. D. S. Tsiklis, "Immiscibility of Gases," Chemistry Publ., Moscow (1969).
1309. D. S. Tsiklis, L. R. Linshitz, and N. P. Gorjunova, *Zh. Fiz. Khim.* **39**, 2978 (1965).
1310. D. S. Tsiklis and V. la Maslennikova, *Dokl. Akad. Nauk SSSR* **157**, 426 (1964).
1311. D. S. Tsiklis and V. la Maslennikova, *Dokl. Akad. Nauk SSSR* **161**, 645 (1965).
1312. D. S. Tsiklis and L. A. Rott, *Russ. Chem. Rev.* **1967**, 351.
1313. A. J. Tursi and E. R. Nixon, *J. Chem. Phys.* **52**, 1521 (1970).
1314. T. Uchida and I. Hayana, *Tokyo Kogyo Shikensho Hokoku* **59**, 382 (1964).
1315. C. H. Unruh and D. L. Katz, *Trans. AIME* **186**, 83 (1949).
1316. S. Valentiner, *Z. Phys.* **42**, 253 (1927).
1317. A. E. van Arkel and J. L. Snoek, *Trans. Faraday Soc.* **30**, 707 (1934).
1318. L. K. H. van Beek, *Prog. Dielectrics* **7**, 69 (1967).
1319. A. van Cleeff, Dissert., Tech.-Hogeschool, Delft (1962).
1320. A. van Cleeff and G. A. M. Diepen, *Rec. Trav. Chim. Pays-Bas* **79**, 582 (1960).
1321. A. van Cleeff and G. A. M. Diepen, *Rec. Trav. Chim. Pays-Bas* **81**, 425 (1962).

1322. A. van Cleeff and G. A. M. Diepen, *Rec. Trav. Chim. Pays-Bas* **84**, 1085 (1965).
1323. A. van Cleeff, G. A. M. Diepen, D. T. A. Huibers, and H. I. Waterman, *Brennst.-Chem.* **41**, 55 (1960).
1324. J. V. Vand and W. A. Senior, *J. Chem. Phys.* **43**, 1873, 1878 (1965).
1325. J. van der Elsken and D. W. Robinson, *Spectrochim. Acta* **17**, 1249 (1961).
1326. J. H. van der Waals, *Disc. Faraday Soc.* **15**, 261 (1953).
1327. J. H. van der Waals, *Trans. Faraday Soc.* **52**, 184 (1956).
1328. J. H. van der Waals and J. C. Platteeuw, "Clathrate Solutions," Adv. Chem. Phys., Vol. 2, p. 1 (1959).
1329. R. van Loon, J. P. Dauchot, and A. Bellemans, *Bull. Soc. Chim. Belg.* **77**, 397 (1968).
1330. N.-G. Vannerberg, *Ark. Kemi* **14**, 107 (1959).
1331. J. N. van Niekerk and F. R. L. Schoening, *Acta Cryst.* **4**, 35 (1951).
1332. M. van Thiel, E. D. Becker, and G. C. Pimentel, *J. Chem. Phys.* **27**, 95 (1957).
1333. M. van Thiel, E. D. Becker, and G. C. Pimentel, *J. Chem. Phys.* **27**, 486 (1957).
1334. W. Vedder and D. F. Hornig, *Adv. Spectroscopy* **2**, 189 (1961).
1335. A. Venkateswaran, J. Easterfield, and D. W. Davidson, *Can. J. Chem.* **45**, 884 (1967).
1336. R. E. Verrall and B. E. Conway, *J. Phys. Chem.* **70**, 3961 (1966).
1337. H. Versmold, *Z. Naturforsch.* **25a**, 367 (1970).
1337a. H. Versmold and C. Yoon, *Ber. Bunsenges. Phys. Chem.* **76**, 1164 (1972).
1338. G. A. Vidulich and R. L. Kay, *Rev. Sci. Instr.* **37**, 1662 (1966).
1339. A. L. Vierk, *Z. Anorg. Chem.* **261**, 283 (1950).
1340. P. Villard, *Compt. Rend.* **106**, 1602 (1888).
1341. P. Villard, *Compt. Rend.* **107**, 395 (1888).
1342. P. Villard, *Compt. Rend.* **111**, 183 (1890).
1343. P. Villard, *Compt. Rend.* **111**, 302 (1890).
1344. P. Villard, *Compt. Rend.* **123**, 377 (1896).
1345. P. Villard, *Ann. Chim. Phys. Ser.* 7, **11**, 289 (1897).
1346. P. Villard, *Compt. Rend.* **176**, 1516 (1923).
1347. P. Villard and R. de Forcrand, *Compt. Rend.* **106**, 849, 1357 (1888).
1348. B. Vincent, M.Sc. Thesis, Univ. of Bristol, 1965.
1349. R. L. Vold, J. S. Waugh, M. P. Klein, and D. E. Phelps, *J. Chem. Phys.* **48**, 3831 (1968).
1350. M. von Stackelberg, *Naturwiss.* **36**, 327, 359 (1949).
1351. M. von Stackelberg, *Z. Elektrochem.* **58**, 162 (1954).
1352. M. von Stackelberg, *Rec. Trav. Chim. Pays-Bas* **75**, 902 (1956).
1353. M. von Stackelberg and H. Frühbuss, *Z. Elektrochem.* **58**, 99 (1954).
1354. M. von Stackelberg and W. Jahns, *Z. Elektrochem.* **58**, 162 (1954).
1355. M. von Stackelberg and W. Meinhold, *Z. Elektrochem.* **58**, 40 (1954).
1356. M. von Stackelberg and B. Meuthen, *Z. Elektrochem.* **62**, 130 (1958).
1357. M. von Stackelberg and H. R. Müller, *Naturwiss.* **38**, 456 (1951).
1358. M. von Stackelberg and H. R. Müller, *J. Chem. Phys.* **19**, 1319 (1951).
1359. M. von Stackelberg and H. R. Müller, *Z. Elektrochem.* **58**, 25 (1954).
1360. M. von Stackelberg and F. Neumann, *Z. Phys. Chem. Abt. B* **19**, 314 (1932).
1361. S. von Wroblewski, *Compt. Rend.* **94**, 212 (1882).
1362. G. Vuillard and N. Satragno, *Compt. Rend.* **250**, 3841 (1960).
1363. G. Wada and S. Umeda, *Bull. Chem. Soc. Japan* **35**, 646 (1962).

1364. G. Wada and S. Umeda, *Bull. Chem. Soc. Japan* **35**, 1797 (1962).

1365. D. Waddington, Ph.D. Thesis, Univ. of Leicester (1972).

1366. A. S. Waggoner, O. H. Griffith, and C. P. Christensen, *Proc. Natl. Acad. Sci.* **55**, 1198 (1967).

1367. A. L. Wahrhaftig, *J. Chem. Phys.* **8**, 349 (1940).

1368. K. A. Walijew and B. M. Chabibullin, *Russ. J. Phys. Chem.* **35**, 1118 (1961).

1369. M. Walker, T. Bednar, and R. Lumry, *J. Chem. Phys.* **47**, 1020 (1967).

1370. T. T. Wall and D. F. Hornig, *J. Chem. Phys.* **43**, 2079 (1965).

1371. D. Wallach, *J. Chem. Phys.* **47**, 5258 (1967).

1372. J. G. Waller, *Nature* **186**, 429 (1960).

1373. J. G. Waller and G. McTurk, *Brit. Patent* 961,115 (1964).

1374. G. E. Walrafen, *J. Chem. Phys.* **40**, 3249 (1964).

1375. G. E. Walrafen, *J. Chem. Phys.* **44**, 1546 (1966).

1376. G. E. Walrafen, *J. Chem. Phys.* **44**, 3726 (1966).

1377. G. E. Walrafen, *J. Chem. Phys.* **48**, 244 (1968).

1378. G. E. Walrafen, *J. Chem. Phys.* **52**, 4176 (1970).

1379. D. T. Warner, *Ann. N. Y. Acad. Sci.* **125**, 605 (1965).

1380. J. S. Waugh and E. J. Fedin, *Fiz. Tverd. Tela* **4**, 2233 (1962).

1381. J. D. Weeks, D. Chandler, and H. C. Andersen, *J. Chem. Phys.* **54**, 5237 (1971).

1381a. J. L. Weeks, C. Meaburn, and S. Gordon, *Rad. Res.* **19**, 559 (1963).

1382. E. G. Weidemann and G. Zundel, *Z. Physik* **198**, 288 (1967).

1383. I. Weinberg and J. R. Zimmerman, *J. Chem. Phys.* **23**, 748 (1955).

1384. H. Weintraub, E. E. Baulieu, and A. Alfsen, *Biochem. Biophys. Acta* **258**, 655 (1972).

1385. A. F. Wells, "The Third Dimension in Chemistry," Oxford Univ. Press, London (1956).

1386. A. F. Wells, "Structural Inorganic Chemistry," 3rd ed., p. 572, Oxford Univ. Press, London (1962).

1387. W. Y. Wen and C. L. Chen, *J. Phys. Chem.* **73**, 2895 (1969).

1388. W. Y. Wen and H. G. Hertz, *J. Solution Chem.* **1**, 17 (1972).

1389. W. Y. Wen and J. H. Hung, *J. Phys. Chem.* **74**, 170 (1970).

1390. W. Y. Wen and K. Nara, *J. Phys. Chem.* **71**, 3907 (1967).

1391. W. Y. Wen and K. Nara, *J. Phys. Chem.* **72**, 1137 (1968).

1392. W. Y. Wen and S. Saito, *J. Phys. Chem.* **68**, 2639 (1964).

1393. H. Wennerström, B. Lindman, and S. Forsen, *J. Phys. Chem.* **75**, 2936 (1971).

1394. G. N. Werezak, *Chem. Eng. Progr. Symp. Ser.* **65**(91), 6 (1969).

1395. W. West and R. T. Edwards, *J. Chem. Phys.* **5**, 14 (1937).

1396. W. West and A. L. Geddes, *J. Phys. Chem.* **68**, 837 (1964).

1397. W. West and S. Pearce, *J. Phys. Chem.* **69**, 1894 (1965).

1398. D. B. Wetlaufer, S. K. Malik, L. Stoller, and R. I. Coffin, *J. Am. Chem. Soc.* **86**, 509 (1964).

1399. E. Whalley, *in* "Physics of Ice" (N. Riehl, B. Bullemer, and H. Engelhardt, eds.), Plenum, New York (1969).

1400. E. Whalley, D. W. Davidson, and J. B. R. Heath, *J. Chem. Phys.* **45**, 3976 (1966).

1401. E. Whalley, J. B. R. Heath, and D. W. Davidson, *J. Chem. Phys.* **48**, 2362 (1968).

1402. P. White, D. Moule, and G. C. Benson, *Trans. Faraday Soc.* **54**, 1638 (1958).

1403. T. A. Whittingham, *J. Phys. Chem.* **74**, 1824 (1970).

1404. E. Wicke, *Angew. Chemie* **78**, 1 (1966).

1405. J. N. Wickert, W. S. Tamplin, and R. L. Shank, *Chem. Eng. Progr. Symp. Ser.* **48**(2), 92 (1952).

1406. W. I. Wilcox, D. B. Carson, and D. L. Katz, *Ind. Eng. Chem.* **33**, 662 (1941).

1407. R. C. Wilhoit, *in* "Biochemical Microcalorimetry" (H. D. Brown, ed.), Academic, New York (1969).

1408. G. W. Willard, *J. Acoust. Soc. Am.* **12**, 438 (1941).

1409. J. M. Williams and S. W. Peterson, *J. Am. Chem. Soc.* **91**, 776 (1969).

1410. V. C. Williams, U.S. Patent 2,914,102 (1961).

1411. V. C. Williams, C. L. Roy, H. Smith, Jr., O. B. Battle, J. A. Hunter, W. W. Rinne, and S. C. Verkios, U.S. Office Saline Water Progr. Rep. 373 (1968).

1412. H. V. Williamson and C. A. Hempel, U.S. Patent 2,683,651 (1954).

1413. F. H. Willis, *J. Acoust. Soc. Am.* **19**, 242 (1947).

1414. D. A. Wilms and A. A. van Haute, *in* "Proc. Third Int. Symp. Fresh Water from the Sea" (A. Delyannis and E. Delyannis, eds.), Vol. 3, p. 117, Athens (1970).

1415. G. J. Wilson, R. K. Chan, D. W. Davidson, and E. Whalley, *J. Chem. Phys.* **43**, 2384 (1965).

1416. G. J. Wilson and D. W. Davidson, *Can. J. Chem.* **41**, 264 (1963).

1417. O. B. Wilson and R. W. Leonard, *J. Acoust. Soc. Am.* **26**, 223 (1954).

1418. R. Wilson and D. Kivelson, *J. Chem. Phys.* **44**, 164 (1970).

1419. H. E. Wirth, *J. Phys. Chem.* **71**, 2922 (1967).

1420. O. Witscher, Dissert., Bonn (1944).

1421. T. A. Wittsruck, W. S. Brey, Jr., A. M. Buswell, and W. H. Rodebush, *J. Chem. Eng. Data* **6**, 343 (1961).

1422. F. Woehler, *Ann. Chem. Pharm.* **33**, 125 (1840).

1423. D. E. Woessner, B. S. Snowden, and G. H. Meyer, *J. Chem. Phys.* **50**, 719 (1969).

1424. D. E. Woessner, B. S. Snowden, and E. T. Strom, *Mol. Phys.* **14**, 265 (1968).

1425. K. L. Wolf, H. Dunken, and K. Merkel, *Z. Phys. Chem.* **46**, 297 (1940).

1426. K. L. Wolf and H. Harms, *Z. Phys. Chem.* **44**, 363 (1939).

1427. K. L. Wolf and G. Metzger, *Ann. Chem. Liebigs* **563**, 157 (1949).

1428. E. O. Wollan, W. L. Davidson, and C. G. Shull, *Phys. Rev. Series 2*, **75**, 1348 (1949).

1429. R. H. Wood and H. L. Anderson, *J. Phys. Chem.* **71**, 1871 (1967).

1430. W. W. Wood, *in* H. N. V. Temperley, J. S. Rowlinson, and G. S. Rushbrooke, "Physics of Simple Liquids," Chapter 5, North-Holland, Amsterdam (1968).

1431. W. A. Wooster, *J. Chim. Phys. Physicochim. Biol.* **50**, C19 (1953).

1432. M. J. Wootten, Ph.D. Thesis, Univ. of Leicester, 1969.

1433. J. D. Worley and I. M. Klotz, *J. Chem. Phys.* **45**, 2868 (1966).

1434. H. Worzala, *Acta Cryst.* **B24**, 987 (1968).

1435. W. Woycicki, *Bull. Acad. Polon. Sci. Ser. Sci. Chim.* **15**, 447 (1967).

1436. W. Woycicki, *Bull. Acad. Polon. Sci. Ser. Sci. Chim.* **15**, 613 (1967).

1437. Y.-C. Wu and H. L. Friedman, *J. Phys. Chem.* **70**, 166 (1966).

1438. Y.-C. Wu and H. Friedman, *J. Phys. Chem.* **70**, 2020 (1966).

1439. O. R. Wulf and U. Liddel, *J. Am. Chem. Soc.* **57**, 1464 (1935).

1440. J. A. Wunderlich, *Bull. Soc. Franc. Miner. Crist.* **81**, 287 (1958).

1441. C. A. Wurtz, *Ann. Chim. Phys. Ser. 3*, **69**, 317 (1863).

1442. J. Wyman, *Phys. Rev.* **35**, 623 (1930).

1443. J. Wyman, *J. Am. Chem. Soc.* **55**, 4116 (1933).

1444. J. Wyman, *J. Am. Chem. Soc.* **56**, 536 (1934).

1445. J. Wyman, *Chem. Rev.* **19**, 213 (1936).
1446. J. Wyman and T. L. McMeekin, *J. Am. Chem. Soc.* **55**, 915 (1933).
1447. E. Wyn-Jones, *Roy. Inst. Chem. Rev.* **2**, 59 (1969).
1448. W. Yellin and W. A. Cilley, *Spectrochim. Acta* **25A**, 879 (1969).
1449. Y. K. Yoon and G. B. Carpenter, *Acta Cryst.* **12**, 17 (1959).
1450. G. V. Yukhnevich, *Opt. i Spektroskiya* **2**, 223 (1963).
1451. S. S. Yun, *J. Chem. Phys.* **52**, 5200 (1970).
1452. R. F. Zahrobsky and W. H. Baur, *Acta Cryst.* **B24**, 508 (1968).
1453. A. Zalkin, H. Ruben, and D. H. Templeton, *Acta Cryst.* **17**, 235 (1964).
1454. R. Zana and E. Yeager, *J. Phys. Chem.* **70**, 954 (1966).
1455. J. Zernicke, *Rec. Trav. Chim. Pays-Bas* **70**, 784 (1951).
1456. J. Zernicke, "Chemical Phase Theory," Kluwer, Deventer, Antwerp, Djakarta (1955).
1457. J. R. Zimmerman and W. E. Brittin, *J. Phys. Chem.* **61**, 1328 (1957).
1458. G. Zundel, "The Hydration of Ions. An IR Investigation of Ion Exchanger Membranes," Academic, New York (1970).

Subject Index

Acoustic absorption
  amplitude attenuation coefficient, 495
  and concentration fluctuations, 488,
    526, 527
  by Brillouin scattering technique, 501
  by pulse technique, 499, 500
  by resonance method, 500, 501
  by reverberation method, 501
  chemical equilibrium interpretation of,
    518–521
  effect of solution composition on, 24,
    518–528
  frequency dependence of, 514, 515
  peak sound absorption composition
    (PSAC), 23, 24, 505–507, 510–514
  relationship with thermodynamic
    behavior, 498, 499
  relaxation time, 496, 497
  temperature dependence of, 514
Activation enthalpy
  in amine hydrates, 192, 193
  in clathrate lattice, 181, 182
  of dielectric relaxation, 35, 438, 450,
    457
  of guest molecule rotation in clathrate,
    206, 207
  of water molecule rotation in clathrate,
    178, 179
Activity coefficient, 368
  from isopiestic measurements, 366, 367
  virial expansion, 31, 32, 360
Angular frequency, 436
Anharmonicity
  due to hydrogen bonding, 289, 304
Anisotropy
  from ESR spectrum, 471
  in rotational diffusion, 558, 559

Apparent compressibility
  adiabatic, 355
  and steric position of functional groups,
    350
Apparent molal quantity, definition of,
  325
Apparent molal volume, an ion–solvent
  interaction, 370
Apparent molar heat capacity, 355
Apparent volume, 355
  and steric position of functional groups,
    350
  of tetraalkylammonium ion, 354
Aquo complex, 58
Association
  constant
    calculation of, 260, 261
    due to hydrogen bonding, 258, 259
    in hydrogen bonded systems, calcula-
      tion of, 250, 251
  equilibria, in liquid alcohols, 409, 410
  of cations, in tetraalkylammonium halide
    solutes, 569, 570
Asymmetry
  in solvation, 472–474, 486
  of infrared bands, due to hydrogen
    bonding, 311, 312
Axial functional groups, their effects on
  hydration properties, 367
Azeotrope, 381
  negative, 383
  positive, 382, 383

Badger–Bauer rule, 264–269, 271, 277,
  285
Barclay–Butler plot, 342, 343

Compound Index

Deuterated analogs are not listed separately in the Compound Index, i.e., CH_3CH_2OH, CH_3CD_2OH, CD_3CH_2OH, CH_3CH_2OD, etc. are all listed under "ethanol." Water of hydration is indicated only where more than one hydrate of a compound is referred to in the text.

675

Formula Index

The entries in the Formula Index are restricted to complex substances, mainly hydrates.

683